Construction Specifications Writing

Principles and Procedures

Fifth Edition

Harold J. Rosen

PE, Hon. CSI

John R. Regener, Jr.

AIA, CCS, CCCA, CSI SCIP

M. D. Morris

PE, Advisory Editor

WILEY

JOHN WILEY & SONS, INC.

This book is printed on acid-free paper.∞

Copyright © 2005 by John Wiley & Sons, Inc. All rights reserved

Published by John Wiley & Sons, Inc., Hoboken, New Jersey
Published simultaneously in Canada

For general information on our other products and services or for technical support, please contact our Customer Care Department within the United States at 800-762-2974, outside the United States at (317) 572-3993 or fax (317) 572-4002.

Wiley also publishes its books in a variety of electronic formats. Some content that appears in print may not be available in electronic books.

Library of Congress Cataloging-in-Publication Data:

Rosen, Harold J.
    Construction specifications writing : principles and procedures / Harold J. Rosen, John R. Regener, Jr. ; M.D. Morris, advisory editor.—5th ed.
        p. cm.
Includes bibliographical references and index.
ISBN 0-471-43204-0 (cloth)
    1. Specification writing.   2. Buildings—Specifications.   I. Regener, John R.   II. Morris, M. D. (Morton Dan)   III. Title.
TH425.R59 2005
692'.3—dc22
2004015428

Printed in the United States of America

10 9 8 7 6 5 4 3 2 1

CONTENTS

PREFACE

The history of construction specifications can be traced back to Noah's Ark. Instructions for construction of the ark were expressed in specifications; there were no drawings.

> So make yourself an ark of cypress wood; make rooms in it and coat it with pitch inside and out. This is how you are to build it: The ark is to be 450 feet long, 75 feet wide and 45 feet high. Make a roof for it and finish the ark to within 18 inches of the top. Put a door in the side of the ark and make lower, middle and upper decks. Genesis, Chapter 6, Verses 14–16 (*The Holy Bible, New International Version,* 1978 by New York International Bible Society)

Specifications sufficed to design and build the Ark. Today, life in the construction industry is more complicated and specifications are more detailed. There are also drawings to follow. Both drawings and specifications have evolved as construction has become more complex. In the early 1900s, architectural drawings became virtually an art form, with ink drafting on cloth. Reproduction of drawings was by "blueprints": white lines on blue backgrounds. Specifications were uncommon except on large projects, but there are examples of "book specs" dating back to the early 1900s. With computerization of graphic and text production and distribution, new media are used to communicate the design criteria and instructions for constructions. But the basic concept of the designer instructing the builder through specifications has remained timeless.

AUTHORS' OVERVIEW AND PREVIEW

Harold J. Rosen, PE, Hon. CSI
Professional Engineer / Honorary Member, The
Construction Specifications Institute
Chevy Chase, Maryland
February 2004

I began writing construction specifications in 1942 while working for the U.S. Army Corps of Engineers, during a period of immense and extremely rapid construction due to World War II. Since then, I have witnessed firsthand for over six decades the countless changes that have occurred in the construction industry and the practice of specifications writing. My experience has included eight years as chief specifications writer for the New York City office of Skidmore, Owings and Merrill and eight years as an independent specifications consultant.

Initially, only governmental agencies had what are called today "master specifications." These specifications were written around standard designs and building products that were available then. Master specifications were typically used on projects undertaken by agencies such as hospitals for the Veterans Administration, post offices, federal courthouses for the Public Buildings Administration, and temporary cantonments, airfields, and similar projects for the Army Corps of Engineers.

Construction in the private sector came to a standstill during World War II. After the war, in the private sector, specifications writing resumed as it had been practiced since the beginning of the twentieth century. There were no standards for how the specification sections were placed in each book of specifications, except that there was an attempt to arrange them on the basis of the chronology of the work—first earthwork, then concrete, masonry, structural steel, carpentry, and so forth. The "and so forth" varied from office to office and even within the same office. In addition, there were no groupings of similar specifications into what we now call "Divisions." In the public sector, interest in standardization carried over from war production. Experienced specifications writers from large public agencies, such as the Army Corps of Engineers and the Naval Facilities Engineering Command, became involved in The Construction Specifications Institute (CSI).

As a result, in 1963, CSI introduced a 16-division format for organizing specifications. In 1965, the American Institute of Architects (AIA) conceived a new title for the book of specifications, calling it the Project Manual because it contained more than the specifications. The Project Manual also contained bidding, construction contract, insurance, administration, and technical documents and specifications. CSI's 16-division format was applied to AIA's Project Manual concept. After four decades, this practice has proven to be effective and the ordering of documents according to national standards continues to evolve to this day, accommodating changes in construction contracting and construction technology.

The production of specifications has undergone major changes over the past five decades. For making multiple copies of specifications, the hectograph (a machine used for making copies of text transferred to a gelatin surface) gave way to the Mimeograph machine (which used a waxed fabric stencil that had typed characters cut by a typewriter, through which ink passed onto paper). IBM "Selectric" typewriters with their golf ball–like, interchangeable type fonts, and other machine-assisted typewriter designs advanced the productivity of typewriters. Nevertheless, specifications production still required a typist pounding text onto a page or through a reproducible stencil.

In the middle to late 1960s, computers began to be used to produce documents. Keypunch cards and tape reels were used with mainframe computers to record text for replay through what was essentially an automated typewriter onto plain paper, for reproduction using the new xerographic electrostatic photo-reproduction process. Later, punch cards and tape reels were replaced by "mag cards" (index card–size polyester sheets with magnetic recording media) and 8-inch "floppy disks" used with early generation mini-computers. In the mid-1980s the desktop personal computer (PC) superseded the minicomputer, and software was developed so that true computer-assisted text production was realized and made available to ordinary architectural and engineering offices. With the introduction of laser-controlled electrostatic printers ("laser printers") in the late 1980s, the typewriter was completely supplanted by the computer for specifications production. Now PCs have capabilities that were unthinkable for even the most powerful mainframe computers four decades ago.

With the acquisition of in-house mainframe computer capabilities by large architectural firms in the mid-1960s, the concept of "master specifications" was developed to make production of similar documents easier. These masters reflected the specific materials and designs that these firms had become familiar with and represented their continued and consistent experience project after project.

A national trend toward the development of master specifications started in the late 1960s and early 1970s, when large public agencies and private organizations, in addition to the AIA and CSI, began to produce standardized master texts for use by individual firms. However, a disturbing trend arose as a result of the availability of these master specifications. Some firms decided that it was unnecessary to have a qualified construction specifier on staff to prepare project specifications. It was assumed that, with the imprimatur of the AIA and CSI on standardized master specifications, project-specific specifications could be produced less expensively by having a project designer, a job captain, or, in some cases, a draftsperson edit the office masters to produce project-specific specifications. The specialized training, aptitude, and skills of a construction specifier were not recognized.

There are countless anecdotal accounts of master specifications being utilized by individuals unschooled and ill prepared to edit master specifications, producing unintended and notably negative consequences. Their inability to edit, fill in blanks, and select options resulted in project specifications subject to construction claims and lawsuits, because the final specification document was replete with errors and was not in sync with the drawings.

While the development of master specifications, the development of more efficient reproduction techniques, and the ease of editing by means of a word processor evolved apace over the past four decades, the education of specifiers lagged. Inquiries regarding the education of architects and engineers in basic specification writing at traditional 4- and 5-year colleges and universities revealed that little if any time is devoted to the subject. Most colleges and universities today offer as little education in this field as they did a half-century ago.

CSI has developed education seminars and certification programs for specifiers, and books on specifications writing principles are available. However, despite these professional development efforts, there are virtually no formal curricula at the college and university level for teaching or training architects and engineers in materials and methods of construction. Specifications writers develop an aura of expertise in construction materials because they interact with them in the specifications writing process. However, few specifications writers have substantial education in technical knowledge and training in materials selection and evaluation.

Perhaps the term "specifications writer" obscures and inadequately describes the technical skills that this individual should possess. This term is reserved for the person in a design office whose function is to write specifications. However, the end product represents all the technical knowledge and experience that the writer has accumulated and that is embodied in the completed specifications.

The specifier of yesteryear prepared specifications for designs that utilized natural materials such as wood and quarried stone, and the designs included such basic materials as concrete, brick, steel, and glass. The specifier required little background in the physical and chemical sciences to understand and specify these traditional materials properly.

Since World War II, a radical change has taken place in the materials of construction. The chemical industry has taken its raw materials and fashioned products from them that are utilized in myriad building components. In the field of metallurgy, tremendous changes occurred in steel alloys for structural steel and electrolytic processes for finishes on aluminum. In the field of chemistry, new coatings were brought to market whose longevity, chemical resistance, appearance, and environmental responsibility are marvelous. Yet, when we look at the curricula being offered at schools to prepare students as specifiers, we find a virtual absence of courses of study that would prepare an individual to understand and select appropriate materials, to compare competitive products, and to utilize materials responsibly. In those colleges and universities that recognize the need to prepare specifiers, teaching of the elements of specifications writing (if such a course is given) is typically limited to a cursory one-semester, two-unit course that also includes professional office practices and working drawings production.

What is the function of a specifications writer? He or she is a resource person who in today's building science must, above all else, have a comprehensive knowledge of the physical and chemical properties of materials. This person must be thoroughly grounded in modern, advanced construction methods and practices. He or she must be conversant with the legal requirements inherent in the General Conditions, the contract forms, and the warranty provisions of the construction contract. Finally, this individual must be a grammarian with the ability to use clear, correct, and concise English. Today's specifications are no longer simply written or verbal instructions to carpenters and masons. Precisely written, explicit specifications on materials and methods must be written for suppliers, fabricators, erectors, applicators, and installers employed by construction contractors, all of whom are working in a cost- and time-driven context. The specifier must know the cost of materials, and must select materials and installation methods commensurate with the needs of the project.

The former concept of the specifications writer who occupied a space in a back room, surrounded by glue pots and scissors, and who put together a set of instructions for master craftsmen to follow in a document called "Specifications," is completely outmoded. What is needed today are comprehensive and explicit specifications embodying the most advanced knowledge of materials, engineering, construction methods, and laboratory testing procedures. Such specifications can be produced only by a highly qualified individual who has had specific preparation either at a technical college or in a learning process over a period of years in a design office and in the field as well. This individual, with his or her specialized knowledge, must be an integral member of the design team who is consulted about design concepts, construction methods, and selection of materials.

The duties of the specifier are varied. This is the person called upon to check details to determine whether the arrangement of materials is compatible and whether the project can be built. He or she is asked to recommend building products for certain applications or building locations and must interpret drawings and specifications whenever a question arises as to their meaning or when discrepancies appear. The specifier is consulted whenever samples are submitted to determine whether they meet specified requirements and is a fount of information about construction methods, materials, and equipment.

Specifiers must also keep abreast of the advances in building technology by maintaining lines of communication with authorities in all the different lines of work so as to become, in effect, a source of information on all phases of building construction. Without these contacts, specifiers will find themselves stranded and alone, as the scope of modern specifications is beyond the experience and knowledge of any one person.

Writing specifications requires analyzing the requirements of a project and determining those ingredients and essentials that are necessary in a specification. Only an individual grounded in the aforementioned disciplines can take a master text, edit it and tailor it to the drawings, and present a coherent set of bidding and construction documents in the Project Manual. This should be the goal of the design profession: the education and development of a resource individual, a "specifier," who can master specifications and edit master specifications.

John Regener, AIA, CCS, CCCA, CSI, SCIP
Architect / Certified Construction Specifier / Certified Construction Contract Administrator Member, The Construction Specifications Institute / Member, Specifications Consultant in Independent Practice
Irvine, California
February 2004

As Harold Rosen described above, specifications production has evolved from the relatively crude production techniques used by architectural and engineering firms up to the early 1980s, when computerization began to take hold in design offices. Specifications production moved from cut-and-paste, retyped documents on mimeo and ditto masters to minicomputer-based, dedicated word processing systems. This meant that revisions, additions, and deletions were more easily accommodated. At the same time, the size of firms doing substantial projects generally increased, and the firms' principals became less involved in production of contract documents and focused on design, marketing, and business management. Computer-assisted drafting (CAD) was introduced and the production of contract drawings supposedly improved, although CAD seems to me to be more of a clumsy drafting machine that produces precise linework and perfect (although sometimes misspelled) notations rather than a mystical instrument for automating drawings production. It has been GIGO: garbage in, garbage out. Until recently, CAD systems in most firms have been operated not by a computer-literate design professional but by (at best) a paraprofessional drafter who knows far more about the tool (CAD program) than the output (architecture and engineering). An acronym has been developed to describe this person: TUCO (totally uninformed CAD operator).

In recent years, CAD has become ubiquitous, and CAD proficiency is now required for taking architecture and engineering license exams. Design professionals are "back on the boards" or, rather, "on the screen." Perhaps the quality of detailing and drafting will improve as more design professionals produce drawings rather than delegating drafting tasks to paraprofessionals.

In the arena of specifications production, as firm principles have become less involved in document production, text editing and specifications usually falls to one of the project architects or engineers, who is also responsible for production of the drawings. Urgency and expedience join together and, unfortunately, it has become common practice to simply modify previous project specifications rather than think through the needs of the project and make the most appropriate selections of materials and products.

Conscientious architects and engineers often organize commonly used construction specifications, as Harold Rosen describes above. Specifications masters are developed that include optional text set off in brackets, and sometimes simple editing notes and instructions are included. These master specifications become the working draft to be edited by a design professional for project-specific specifications.

The semiautomated specification production process, using computer-based word processing programs, swept through architectural and engineering firms in the mid-1980s. Generally, it has worked. Specification writing has become less burdensome since retyping whole documents has been replaced by inserting changes in an electronic document. As word processing programs have been learned by typists and professional and technical staff, the power of functions such as search-and-replace has been used to make repetitive changes to the text. Cut-and-paste within and between electronic documents has drastically reduced straight typing of text. The appearance of text has been freed from the limited typefaces, character sizes, and appearance of the typewriter. Now, boldfacing, underlining, and italicizing of text have become simple matters. With the introduction of laser printers, the time needed for printing has been greatly reduced and the quality of the output has become almost like that of typeset text.

Through the mid-1990s, specification writers began to learn to edit their documents directly on the computer, and even die-hards gave up their typewriters and dedicated word processing machines in order to personally control specifications production. Like architects and engineers who learned CAD, specifiers have learned word processing programs to produce the written documents that accompany the bidding and construction contract drawings. Few full-time specifications writers rely upon handwritten text that a typist transforms into a completed document.

Architects and engineers experienced the tempering effects of several severe economic recessions in the middle and late 1980s and early 1990s. These recessions made design and construction even more competitive. Firms that weathered the economic storms of recession did so by being "lean and mean." The need for greater productivity in construction specifications production continued. At the same time, construction technology continued to be more complex, and the legal environment of construction continued to be increasingly adversarial. Design errors and omissions resulted in and continue to have substantial financial consequences for architects and engineers.

A few medium-sized and almost all large-sized architectural firms have recognized the value of having a full-time specifier on staff despite the overhead expense. Other small to medium-sized architectural firms and a few engineering firms have recognized the value of having a specifier involved in the project and have elected to outsource specifications writing to independent consultants, such as those who are members of the national nonprofit organization Specifications Consultants in Independent Practice (SCIP). The remaining firms, typically, have a project team member, with limited education and training in specifications writing and production, who prepares the construction specifications.

Computer-assisted text production (word processing) programs have made specifications writing easier, but the pressure for greater productivity continues to increase. The desire grows for a computer-assisted specifications program that will not only produce project-specific specifications based on parameters fed into the program but will also automatically identify the products shown on the drawings and make suitable product selections based on superficial information fed in by technically unsophisticated designers and detailers. Needless to say, I do not share the optimism of the publishers of these programs. There is no complete substitute for professional knowledge and technical skills in producing competent construction specifications. Computerized programs can make the process more productive, but they cannot substantially replace the professional knowledge of architects and engineers—at least not for the foreseeable future—in my humble opinion.

Construction technology continues to develop at an ever-increasing rate, leading to more technically complex materials and building systems. Regulatory requirements (building codes) continue to become more complex, with greater requirements for fire and life safety, for resistance to windstorm and seismic forces, and for thermal and acoustical performance. Competition in the construction industry is increasingly feverish, especially on publicly funded projects such as schools. Quality in construction continues to decrease, along with the graduation of would-be architects and engineers whose education continues to be geared to future developments in architecture and engineering rather than preparation to fit into the production-driven environment of architectural and engineering firms. Architecture and engineering school curricula may need to be loaded with true architecture and engineering classes that are theoretical and reflect the expanding complexity of construction technology so that graduates have a sufficient foundation for careers that include lifelong learning that will encompass the skills of drawings and specifications production.

It has been left to on-the-job training to develop production skills in "working drawings" and construction specifications. College and university curriculums offer some remnants of instruction in working drawings and specifications. Some colleges and universities have used previous editions of this book as the text for specifications writing classes. Regrettably, graduates of most architecture and engineering curriculums are essentially ignorant of construction specifications and other construction contract documents. For that reason, this book is addressed to practicing architects and engineers while at the same time remaining suitable for undergraduate classes.

In my opinion, changes in construction technology, regulatory requirements, and construction contracts have overwhelmed the design and production processes of most architectural and engineering firms. Based on my observations and discussions with other specifications writers, with relatively rare exceptions in large firms designing projects with substantial budgets, ordinary medium-sized and small architectural and engineering firms struggle with construction specifications production and are dissatisfied with the quality and expense of their specifications. Project owners are also dissatisfied. Proof of this can be found by asking facility owners and managers what they think of the quality of construction specifications and other bidding and construction documents. The comments I have heard are scathing.

The problems are not new. Problems with construction specifications became acute as early as the late 1960s and early 1970s. Estimators, bidders, contractors, construction administrators, and inspectors complained that it was difficult to find information in the specifications. Each design firm seemed to have unique formats and writing styles. Standardization was needed to make it easier for everyone involved with the specifications to produce and use the documents. Industry standards were sought to improve construction specifications. Fortunately, The Construction Specifications Institute (CSI) was an organization whose pioneering founders and leaders, among whom is Harold Rosen, had the wisdom and vision to create necessary principles and procedures for organizing and producing construction specifications.

From the late 1960s through the mid-1990s, a relatively small group of CSI's leading members gathered and developed formats and instructional materials in response to the needs of the design professions and the construction industry. Many of the instruction materials were based on Harold Rosen's articles published in *Progressive Architecture* magazine and his first (1974) edition of *Construction Specifications Writing: Principles and Procedures.* Harold participated in the development of the now widely recognized 16 divisions of *MasterFormat,* and some of his writings were incorporated into CSI's *Manual of Practice.* Harold Rosen was one of the first to develop the content of Division 1 - General Requirements, which governs much of what is specified in other Divisions of the specifications. I am amazed at and grateful for the wisdom displayed by Harold Rosen and others who developed CSI's seminal formats and practice documents.

Now these classic formats, principles, and practices are being challenged by forces within and outside of CSI that do not fully understand or appreciate the substance and legacy that has developed in the design professions and construction industry for over three decades. At the time of this writing, both of CSI's premier publications, the *CSI Manual of Practice* and *MasterFormat* (1995 edition) are undergoing major revisions that are supposed to accommodate future developments in the construction industry and improve production of construction documents, including construction specifications. The final documents will not be published until shortly before this fifth edition of *Construction Specifications Writing: Principles and Procedures* is published. Therefore, the fifth edition will be generally based on the legacy format of *MasterFormat* (1995 edition).

Changes in CSI's premier documents are controversial, with many specifications writers vehemently voicing opposition to change. One who has supported the changes to *MasterFormat* is Harold Rosen, who was involved in developing the first edition of CSI's *MasterFormat* and the documents that preceded it. From the beginning, he recognized the need to better accommodate engineering profes-

sions and related construction fields, such as process engineering, materials handling, and new electronic technologies, into the formats for construction documents.

I have been very vocal in public forums regarding changes to *MasterFormat,* not to hinder accommodation of new technologies and engineering disciplines, but to preserve the fundamental principles of specifications organization and production. My criticisms have focused on problems with using the new version for *MasterFormat* for project specifications. I have been assured that the final publication will be more functional.

When Harold Rosen invited me to join him in preparing the fifth edition of *Construction Specifications Writing: Principles and Procedures,* I was deeply honored. Then I was more than mildly intimidated. I realized that it would be primarily my responsibility to do the actual writing of the update of this book, which has been the foremost text for learning to write construction specifications, predating the CSI's *Manual of Practice.* After extensive discussions with Harold Rosen and other senior specifiers, I realized that I had deeply absorbed both the theoretical principles and the practical procedures for producing and using construction specifications over my three-decade career, and I have (sometimes zealously) regularly applied these principles and practices.

It has been a great deal of work but very satisfying to produce this fifth edition of *Construction Specifications Writing: Principles and Procedures.*

SPECIFIERS' COMMENTS

To broaden the prefatory comments to this book, the authors have sought insights from several leading specification authorities concerning construction specifications writing and being a construction specifier. Here are the comments of some of them on the present state of construction specifications and their vision of the future for construction specifications and specifications writers.

"My crystal ball is cloudy when it comes to seeing what specs and spec writing will be like in 20 years. But it is clear as a bell that spec writing will not be done in 20 years the way we are doing it today. The current system, which involves the specification writer being given a half-finished, unchecked, uncoordinated set of drawings, and poring over them with a checklist, eyeballing for particular things, is ridiculous in an electronic age. Maybe it's okay today, though, since we are still at an early stage of use of electronic documents. But 20 years from now it's surely not going to be like this. Somehow or other, the drawings and specs are going to trigger each other to assemble the needed information through linkages, through associative key words or key numbers, or something similar.

"While I'm sure that this trend points in the right direction, it concerns me to see what could occur when it happens. CAD has not made the content of drawings better; it has allowed them to become worse. The lines and lettering are precise and they look good, but frequently the thinking process to make them accurate and job-specific is missing. Unless spec writers are a lot smarter than drafters, mechanized specs probably will be no better than CAD drawings are. You can see some of the parallels when you look at specs put out using *Master-Spec, SpecText,* and the other "canned" systems—lots of information, but unless they are edited by skilled, knowledgeable persons, there is little relevance to the project or the project locale. Maybe, with books like this one and 20 years' time to think about it, the 'computerized, automated, mechanized, or whatever' system that develops will work."

Jo Drummond, FCSI, CCS

"In the previous edition of *Construction Specifications Writing: Principles and Procedures,* I said: 'There has been a revolution in how to properly prepare construction specifications. Many architects and engineers don't hear about that revolution until confronted by a lawsuit or a demand for arbitration because of bad specs. Our best protection is to stay up-to-date on the new standards of spec writing.' This is even more true today."

Hans W. Meier, AIA, FCSI
Independent Specifications Consultant, retired

"The future of specifications writing is clear to any who want to open their eyes, but it seems to me that those who could see the future are running as fast as they can away from it.

"Perhaps in order to move forward, what we really need first is to take a step back and understand that words are the essence of specifications and clarity is the goal of every project. We need to work together to make sure our words match our intent so everyone can read the same words and recognize the same concepts."

John A. Raeber, FAIA, FCSI, CCS
Specifications Writer

"For design professionals to serve as the stewards of quality in construction and in contractual rela-

tions, which is our proper role, we must be adept at producing and interpreting the contract documents that govern construction quality and contract issues. Easily 75 percent of the fees paid to us by our clients are for the development and administration of quality technical design, but less than 5 percent of our formal university education is devoted to preparing us for this role. This places a great burden on our profession and on our firms to provide adequate professional development to fill our practitioners' great technical gap. It is also a promising area in which practitioners and firms, through focused professional development, can achieve improved performance and profitability. Specifications are the key tool in this effort."

Phillip W. Kabza, RA FCSI, CCS
SpecGuy.com

"As Director of Specifications for a large multioffice architectural firm for nearly 20 years, I have gradually migrated my role from authoring Project Manuals for individual projects to equipping our staff to take on more of the technical quality management of their projects. My technique for accomplishing this has comprised a blend of educational offerings (delivering some personally, as well as coordinating presentations by experts) and creating and deploying an array of resources upon which staff can draw.

"Recently, as I've pursued these dual missions, I have gravitated toward our company's intranet as my communications medium for reaching the widest audience, given our broad diversity of experience levels, geographic locations, and scheduling constraints. In the interest of efficiency, I've had to become as proficient using Microsoft FrontPage as I had been in Microsoft Word. The goal was to eliminate IT 'clerical work,' just as we had done years ago with the advent of word processing in taking typists out of spec production.

"While I can't predict the next road we will travel with the benefit of yet-to-be-developed technologies, what will not be altered is the need for optimal communication in both content and expediency of information transfer. And isn't that what specs have always been about? The more things change, the more they remain the same."

Martin M. Bloomenthal, FAIA, CSI, CCS
Principal, Director of Specifications
Hillier ARCHITECTURE

"The future of specifications is the nexus of advancing information technology, the increased complexity of product selection options, and the relative conservatism of the specification-writing profession, all in an environment of increased litigation and emphasis on value. Where will it lead? I think it leads in two directions.

"First, there will be practitioners who will choose to consider function, who will develop ways to communicate and manage product application during the design process, and who will ensure that decisions are made by properly qualified design professionals. Their specifications will migrate to a form that is consistent with the way business will be transacted in the future: distributing design models over communications networks, with product information bound tightly in packages that are linked together in logical ways. The manufacturer will become more in control of and responsible for the product information, but the design professional will gain greater control over the product application and selection. These kinds of practitioners will tend to be involved in projects that are innovative, large, or complex.

"Secondly, there will also be practitioners who focus more on the form of specifications, who will spend time arguing that the three-part format of the automated specification is not as satisfactory as their own version. They will continue to automate slowly, and will publish specifications traditionally even as they use electronic communications to make the process more efficient. Their concern with form will move their automation in the direction of using paraprofessionals to write specifications rather than using professionals to select products. They will be more likely to be involved on smaller, less complex, and repetitive design projects.

"The result: Twenty years from now we will still have bound, hard-copy specifications being distributed, much as they are now. Twenty years from now we will also have specifications derived directly from the design model, transmitted electronically to stakeholders who require the information at the time they require it."

Charles A. Shrive, PE
Professional Engineer

"Contrary to popular belief, the future of construction specifications does not lie in their production method or in the technology to link drawing details to a manufacturer's website using a long string of computerese understandable only to the programmers. The future of all construction documents, not just the written specifications, lies in the understanding of the principles, practices, and applications of construction documentation by the preparers as well as the users of these documents.

"Technology is advancing at an unprecedented pace, and continuing education has become the mantra of every professional organization since the late 1980s. There are two and three generations of design professionals that have never drafted manually or been exposed to construction document principles and practices. There are construction managers, contractors, and subcontractors who do not understand the purpose of the Project Manual or its content. There are manufacturers and representatives offering guide specifications who do not understand specification conventions, required technical language, or the legal implications of the specifications. Owners and facility managers are seldom educated in their responsibilities required by the construction documents.

"Specifiers have been the unofficial keepers of the written documents, often serving as educators and mentors, and typically, the individuals tasked with understanding the principles, practices, and applications of the written specifications and their relationship and coordination with the other construction documents.

"As our society and our documents have become more complex and open to litigation, more users of the documents are discovering a growing need to understand the construction documents. We are inundated with education opportunities offering sought-after continuing education credits where we can learn about almost every construction product or code change. We can learn about new computer software programs, how to improve our practice or business, and how to become environmentally sensitive. But where are the education programs and school curriculums teaching the construction documents?

"Regardless of format, the future for not just the written specifications but all of the construction documents lies in quality education programs teaching the principles, practices, and applications of construction documents not only to specifiers but to the users of the documents."

Linda Brown, FCSI, CCS, CCCA, SCIP
Independent Construction Specifications Consultant

"Commercial master specifications and their associated software are written by experienced design professionals having particular expertise in specifications writing. The content of masters is researched and developed in good faith and believed to be correct. They are published as tools for use by professionals.

"Masters contain a wide range of options and alternatives from which to choose to suit various project requirements and conditions. Software tools provide the utility to enable efficient production of project specifications from masters. However, master specifications are just tools, and like all complex and sophisticated tools, master specifications must be used by people who have the necessary knowledge to make selections from among the options and alternatives presented, and the skills to insert additional requirements in the proper location and style for those not presented. Like a carpenter's power tool, master specifications can be dangerous in the hands of the untrained and unskilled.

"Preparation of construction documents, including specifications, requires a professional level of care, no matter what the nature and source of the tools used."

Michael J. King, FCSI, CCS
Director of Engineering Specifications
ARCOM Master Systems

"In the past, there have been references to the 'master architect'—one who has learned from enough experience to be a master of all the aspects of architecture. The tremendous growth in the complexity of design and construction and the exponential growth in the amount of accessible information have made it essentially impossible for an individual to be a master architect; it has become a necessity to operate with teams to master and apply all aspects of the profession. We are far from having mastered how to consistently complete successful projects within a team approach.

"Individuals who were experts in the technical aspects of the profession have disappeared from many offices and have not been replaced. The people with the specification or construction administration responsibilities are filling this void and becoming the prominent 'holders of technical knowledge' and thus the major technical resource for many firms. This broader 'technical resource' responsibility for specifiers will continue to grow in the coming years and will pressure specifiers to broaden their perspective of who they are and what they are responsible for.

"The reality of the current situation has resulted in more and more specialization by individuals and by firms and fewer people who have sufficient understanding and empathy for what other team members are responsible for. Those preparing specifications often do not properly think about the repercussions of how specifications preparation affects the preparation of the drawings. Those preparing drawings do not understand the role of specifications and often include too much specification-type information in

the form of extensive notes that properly belongs in the specifications. They do this with good intentions, but with lack of understanding.

"The complaints about the lack of coordination of the contract documents continue to grow. Lack of coordination among disciplines is obviously a major part of this problem, but it is also the result of the continued separation of the preparation of drawings and specifications. In today's marketplace, there are details and specifications produced by separate organizations that are not coordinated with each other, nor are they within a total coordinated, comprehensive system. The problem will not be solved in the future by just tagging manufacturer and model information in object-oriented CAD files, as we see in examples today. It is a much more complex problem than that. The solution lies in the creation of systems that include both 'graphics' and 'word'-type information in an integrated approach. In terms of current practice, where most firms still produce separate drawings and specifications, this means producing a sufficient set of master reference details that is truly coordinated with a set of master specifications in regard to terminology and division of information. Keynoting is one of the techniques that can be used to facilitate and discipline this coordination. Unfortunately, there is currently very little industrywide effort at creating total integrated systems in our world of specialization."

Robert W. Johnson, AIA, FCSI, CCS, CCCA
Johnson and Johnson Consultants, LLC

"Within the next 10 years, a majority of the specifiers who helped build CSI will have retired. With little guidance, future specifiers will reinvent the wheel, numbering systems, and standard practice. I predict that the keepers of specification systems such as *Masterspec* and ARCAT will become even more essential to design firms.

"Specifications must evolve into targeted documents that allow the designer to integrate the knowledge of the product manufacturer, the project requirements, and the construction process.

"'Fake specs,' which confuse the construction estimator and take no account of the owner's needs, will give way to direct-order entry of precisely specified materials within rapid flash-track project delivery methods.

"Where we are going is where we have been: The concept of the master builder able to find the precise needle in the ever-growing haystack of construction materials and methods."

Mark Kalin, FAIA, FCSI, CCS
Kalin Associates

INTRODUCTION

A knowledge of specification writing principles and procedures is essential to the specifier in an architectural or engineering firm in order to prepare sound, enforceable construction specifications. Unless skills are properly developed to understand and apply these principles, and unless expert knowledge of materials, contracts, and construction procedures is also applied, the architect or engineer cannot communicate successfully with the ultimate users of the specifications: facility owners, general contractors, subcontractors, materials suppliers, code authorities, and quality assurance inspectors.

What, then, constitute the principles of specification writing? Basically, the principles should encompass those factors that permit architects or engineers to understand clearly the relationship between drawings and specifications—between the graphic and the verbal—and should enable them to communicate effectively by setting forth in logical, orderly sequence the information to be incorporated the specifications.

This book presents the principles and procedures for organizing and producing construction specifications. It is intended for students in architecture and engineering curriculums and for practicing design professionals in professional development and continuing education programs. It is also appropriate for others involved in the production and use of construction specifications, including facility managers, construction managers, and building product representatives.

In summary, this book presents principles and procedures for construction specifications writing as follows:

1. The Role of the Specifications. Specifications are one component of the documents used for bidding and construction of a project. Another component is the drawings. The specifica-

tions and drawings are intended to work in harmony to describe what shall be built. Other components are bidding requirements and other contract requirements, which prescribe the duties and responsibilities of the primary parties of the construction contract. Bidding requirements are applicable during the procurement or bidding phase prior to actual construction. Contract requirements apply during fulfillment or execution of the Contract for construction. Each component has distinct purposes. Specifications, as written instructions, are frequently judged by courts as having greater importance than drawings when these documents are in conflict, with judgments based usually on what is contained in the specifications. This means that specifications should be carefully prepared by knowledgeable people. Chapter 1 discusses the role of specifications in detail.

2. The Relationship between Drawings and Specifications. Specifications address qualitative requirements for products, materials, and workmanship, while the drawings indicate relationships between elements and show the location, identification, dimension and size, details and diagrams of connections, and shape and form. There should not be duplication or conflict between these two documents. Instead, they should be complementary. To improve coordination between drawings and specifications, there should be standardization of the information appearing in them. Chapter 2 discusses the purposes of drawings and specifications and their relationship.

3. Organization of Specifications. For many years, specifications were arranged in a series of Sections based on the order or chronology in which various trades appeared on the con-

struction scene. However, it was found that our increasingly complex building structures did not necessarily follow these simple rules, nor was there a uniform nationwide system of specifications. In 1963, the Construction Specifications Institute (CSI) established a uniform arrangement of the various Sections in a division-section organization titled the *CSI Format,* which in subsequent revisions has evolved into the CSI *MasterFormat.*™ The lists of Section numbers and titles in *Master-Format*™ enable construction information to be consistently identified and retrieved. Chapter 3 discusses industry standards, including CSI formats, for organizing specifications.

4. The Project Manual and Specifications Sections. Specifications are included in a book published for the project titled the Project Manual. The Project Manual contains bidding and contract requirements and the construction contract Specifications. The Project Manual is divided into chapter-like Sections organized according to *MasterFormat.*™ Chapter 4 discusses how to determine the level of detail for Specifications and the appropriate Section number and title according to *Master-Format*™ (1995).

5. Format for Specification Sections. Until CSI promulgated the 3-Part *SectionFormat,* there was no universal arrangement of information in an orderly, coherent series of paragraphs dealing with the content of the Specification Section. With the addition of CSI *PageFormat,* there are industry standards for internal organization of the Section and standardized page presentation. Chapter 5 discusses how to organize and present a Section of the Specifications.

6. Types of Specifications. There are four methods of specifying the Work of a construction Contract to be performed by the Contractor. These methods, used individually or in combination, are descriptive specifying, reference standard specifying, proprietary specifying, and performance specifying. Additionally, there are considerations of whether the Specifications are "restrictive" (sole source or limited sources) or "nonrestrictive" (commonly known as "or equal") Specifications. Chapter 6 discusses how to choose and use the various methods of specifying.

7. Specifications Writing Principles. After the formats for specifications and the methods of specifying are understood, the technical and procedural content of the Specifications needs to be determined. The content is described using techniques involving appropriate specifications language, workmanship requirements, and coordination among various Specifications Sections to avoid redundance and conflicting requirements. Chapter 7 discusses these principles.

8. Bidding Requirements. Bidding requirements consist of documents that are used in the solicitation of bids and typically include the Advertisement or Invitation to Bid, the Instructions to Bidders, and the Bid Form. The specifier often prepares these documents based on instructions from the owner. Chapter 8 discusses the content, purposes, and formats for the bidding requirements.

9. General Conditions of the Contract. The Conditions of the Contract define basic rights, responsibilities, and relationships of the entities involved in the performance of the Contract. The Conditions of the Contract are an inherent part of the Owner-Contractor Agreement and are considered to be the "general clauses" of the Agreement. There are generally two types of Conditions of the Contract: the General Conditions, which are found in a standardized, preprinted document, and the Supplementary Conditions, which are project-specific modifications to the standard document. Chapter 9 discusses the General Conditions of the Contract, and Chapter 10 discusses the Supplementary Conditions of the Contract.

10. Supplementary Conditions of the Contract. Each project has unique requirements. In terms of the general clauses or Conditions of the Contract, the unique requirements are presented in the form of Supplementary Conditions of the Contract, which modify the standard preprinted General Conditions. Chapter 10 discusses the typical content of Supplementary Conditions of the Contract.

11. Bonds, Guaranties, and Warranties. To ensure performance by the Contractor and to protect the owner from premature failure of products and workmanship, the Contract Documents include provisions related to bonds, guaranties, and warranties. These are presented in general terms as part of the Contract requirements, preceding the Specifications in the Project Manual, and in Specifications Sections to describe specific provisions. Chapter 11 discusses bonds, guaranties, and warranties.

12. Division 1 - General Requirements. These are Sections of the Specifications that apply generally to all Sections. The use of Division 1 follows one of the prime principles of Specifications writing: "Say it once." Chapter 12 discusses the use and content of Division 1 Specifications.

13. Modifications. It is inevitable that the bidding and Contract requirements, the Specifications, and the Drawings will require revision after being issued. Chapter 13 discusses the procedures and formats for preparing the various types of modification documents.

14. Specification Language. It is imperative to use clear technical language that can be understood by those who use the Specifications. In order to communicate with proper language, the specifier must sufficiently master the tools of specifications language, including grammar, vocabulary, spelling, use of abbreviations and symbols, punctuation, capitalization, sentence structure, and the unique considerations of streamlined writing and specifications detail. The specifier must not only follow hard rules of language but must understand the subtleties of language. Chapter 14 discusses the unique language requirements of construction specifications.

15. Specification Resources. Construction technology, project delivery methods, and sources of construction information change constantly and rapidly. Chapter 15 presents some common resources useful for specifiers.

16. Products Evaluation. Other books address construction technology in much more detail than can be accommodated in this book. Chapter 16, however, discusses fundamental procedures for evaluating products, identifying necessary attributes, and selecting appropriate products to be included in the Specifications.

17. Specification Writing Procedures. Applying the principles of specifications writing is facilitated if there are established procedures for producing Specifications Sections. Chapter 17 discusses those procedures and the use of Specifications checklists when gathering information, researching, and writing.

18. Master Guide Specifications. Master guide specifications are published and nationally marketed to assist specifiers. The publishing organizations have resources to continually research, create, and maintain construction specifications. Many architectural and engineering offices and independent specifications consultants use these master guide specifications to create office-specific master specifications that serve as the basis for project-specific specifications. Chapter 18 discusses the use of master guide specifications published by commercial organizations and public agencies, as well as the development of office-specific masters.

19. Computer-Assisted Specifications. Today, several true computer-assisted specifications programs are in the marketplace. These programs offer automation features beyond word processing programs that enable the specifier to more expediently and accurately create project-specific specifications—or so the marketing materials promise. Chapter 19 discusses the history and current offerings of three of the computer-assisted specifications programs.

20. Preliminary Project Description. During the preliminary design phase of a project, an alternative format, based on building Elements, is sometimes used to provide information for scope descriptions and cost estimating. Based on *Uniformat* rather than *MasterFormat,* a Preliminary Project Description (PPD) is produced from which Specifications are derived at later phases of the project. Chapter 20 discusses the PPD.

21. Outline and Shortform Specification. Used during the preliminary design of a project, Outline Specifications are produced using either the typical 3-Part Section format or an abbreviated format with sequentially identified articles. Outline Specifications describe preliminary product selections and other project-specific requirements. Another abbreviated specifications format, shortform specifications, is used for less complicated projects or those of limited scope where highly detailed information is either unnecessary or inappropriate. Chapter 21 discusses outline and shortform specifications.

Appendices follow the text and provide examples of Specifications and other Bidding and Construction Contract documents.

Chapter 1

The Role of Specifications

DOCUMENTS FOR CONSTRUCTION

When an Owner decides to build, renovate, or reconstruct a facility, the Owner usually engages others to prepare documents describing the Work to be performed and the contractual requirements under which construction and related administrative activities are accomplished. Contemporary construction practices in North America are varied, and construction contracts likewise express varied contractual relationships, procurement (bidding and negotiation) methods, and regional construction practices.

This complexity can be very confusing, so, for the purposes of this book, the discussion will be generally limited to the context of the traditional design-bid-build method for construction procurement, with comments occasionally describing alternative procurement methods, such as design-build, multiple prime contracts, and phased ("fast track") construction. Also, the discussion will be in the context of three primary parties in the construction contract: the Owner, the Architect/Engineer, and the Contractor.

The Architect or Engineer, and his or her various consultants, prepare documents for construction of the facility. These develop over time, from conception of the design, through the gestation of design development, through the birth pangs of bidding/pricing and construction, until delivery of the completed facility at closeout of the construction Contract. Many types of documents are used during design and construction, but for actual construction three basic types of documents are used:

1. *Bidding and Contract Requirements:* Text documents
2. *Drawings:* Graphic documents
3. *Specifications:* Text documents

Combined, these three types of documents are called the "Bidding Documents" (before signing of the Agreement or "contract") and the *Contract Documents* (after signing the Agreement or contract). The difference concerns when the documents are used. Prior to execution (signing) of the Agreement, the combined documents are known as Bidding Documents under traditional design-bid-build projects. Under design-build and certain types of construction management-type projects, the documents prior to execution of the Agreement may be known as "Procurement documents." This is a fine distinction reflecting the process of negotiation for selection of product vendors and subcontractors, but in most cases, even under design-build and construction management-type projects, a competitive bidding process is used. So, "Bidding Documents" will be the term used here for the documents prior to execution of the Agreement. After execution of the agreement, the documents are known as the "contract documents."

BIDDING DOCUMENTS

"Bidding documents" is a term generally used to describe the documents furnished to bidders. For traditional design-bid-build projects, the Architect/Engineer and Owner prepare the set of Bidding Documents, consisting of bidding requirements, Drawings, and Specifications. These are issued to prospective general contractor bidders for competitive bidding and for the Owner to select the Contractor named in the Agreement and referenced in other Contract Documents.

However, there are Projects where the Contractor is selected by a method other than competitive bidding, such as direct selection by the Owner based on qualifications of the Contractor. In such cases, competitive bidding still occurs but it is managed by the Contractor, who issues bidding documents and manages the bidding process.

The primary difference in documents between Bidding Documents and Contract Documents is the inclusion in the Bidding Documents of Bidding Requirements. These generally consist of the Advertisement or Invitation to Bid, the Instructions to Bidders, the Bid Form, and other documents to be submitted to the Owner for the Owner's selection of the Contractor and modifications to the documents issued during bidding (addenda). The

Bidding Requirements are removed after bidding and selection of the Contractor and are replaced by the Contract Requirements. The Contract Requirements typically consist of the executed (signed) Agreement with its related documents, such as insurance forms, bonds, and certifications. Note: the Bidding Documents typically include copies of the Agreement form and the Conditions of the Contract (General Conditions and Supplementary Conditions when industry-standard documents are used). The bidding process is described further in Chapter 9.

As noted above, two other types of documents are included in the Bidding Documents and the Contract Documents: the Drawings and the Specifications. This book will not describe principles and practices for production of the Drawings but will describe the types of information best presented on the Drawings (graphic presentation) and in the Specifications (text presentation), and the discussion will include coordination issues between the Drawings and the Specifications. Chapter 2 discusses the relationship between the Drawings and the Specifications in greater detail.

CONTRACT DOCUMENTS

"Contract Documents" is the term used for documents identified in the Agreement (construction Contract). It appears simple, but, of course, these documents can become complicated as the project delivery process becomes more complex and the relationships between the parties involved in the project become more varied and obscure. Considering the typical design-bid-build project, the Contract Documents consist of the following:

- *Agreement*—A written agreement between the Owner and the Contractor summarizing the work to be performed, the Time in which the Work shall be completed, and the Contract Sum to be paid. Also identified in the Agreement are the Contract Drawings, the Contract Specifications, and other referenced documents such as bond forms, insurance certificates, other certifications, Contractor's qualifications statement, documentation of the Contractor's financial status, subcontractors and suppliers lists, special warranty documents, and just about any other type of written document that the Owner requires. Note: the Agreement is typically prepared by the Owner's legal and insurance counsels or by the Construction Manager if one is involved in the project. The Architect/Engineer typically

does not prepare the Agreement and its attachments. If the Architect/Engineer is involved in producing the Agreement and its attachments, it should be under the direction of the Owner. Architects and engineers are not trained to produce legal instruments (documents) and insurance documents, nor are they licensed to practice law and insurance underwriting.

- *Conditions of the Contract*—Typically, these consist of the General Conditions and the Supplementary Conditions. General Conditions are typically preprinted standard documents prepared by professional societies such as (for architectural projects) the American Institute of Architects (AIA) and (for engineering projects) the National Society of Professional Engineers (NSPE), American Consulting Engineers Council (ACEC), and American Society of Civil Engineers (ASCE). Together, the listed engineering societies jointly publish documents as the Engineers Joint Contract Documents Committee (EJCDC). The Conditions of the Contract are discussed in greater detail in Chapter 9.

- *Drawings*—Graphic descriptions of the Work to be performed by the Contractor. The content of Drawings and the relationship between the Drawings and Specifications are discussed in greater detail in Chapter 2.

- *Specifications*—Written descriptions of the Work to be performed by the Contractor. The types of Specifications, their content, and specification writing principles and practices are discussed in greater detail below and in Chapters 5, 6, and 7.

- *Modifications*—Architect's Supplemental Instructions (for contracts based on AIA *A201, General Conditions of the Contract*), Field Orders (for contracts based on *EJCDC C-700, General Conditions of the Contract*), Construction Change Directives (for contracts based on AIA *A201, General Conditions of the Contract*), *Work Change Directives* (for contracts based on *EJCDC C-700, General Conditions of the Contract*), and *Change Orders* (for contracts based on both AIA and EJCDC *General Conditions of the Contract*). "Modifications" are changes to the documents after execution (signing) of the Agreement. Prior to execution of the Agreement, changes are generally made using *Addenda*. Addenda and Contract Modifications are discussed in greater detail in Chapter 13.

Often the term "construction documents" is used as a synonym for "Contract Documents." This is incorrect. Simply stated, the Contract Documents are the documents identified in the Agreement. The Contract Documents, together with other documents used during construction, may be called construction documents. This is a fine but important distinction contractually. The Contractor, the Architect/Engineer, and the Owner are only obligated to perform according to the Contract Documents. Other documents may be required by the Contract Documents to be produced and used during performance of the Work under the Contract, including shop drawings, construction schedules, construction reports, meeting notes, submittals, installation instructions, test reports, permits, and certificates from authorities having jurisdiction, and operating and maintenance data. However, these are not Contract Documents, although most should become part of the "contract record documents," which describe the completed Work of the Contract and which may be used by the Owner for operation and maintenance purposes.

SPECIFICATIONS

Imagine a movie or video presentation of the construction of a building, park, water or sewage treatment plant, refinery, highway, or bridge. Imagine that all the activities of construction are shown in great detail, from procurement of materials and manufactured products, through fabrication, delivery to the job site, storage and staging on the job site, surface preparation, mixing, application, installation, fitting, and finishing. Also imagine the Owner, the Architect/Engineer, the Contractor's managers and supervisors, the subcontractors, the testing and inspection agency personnel, the manufacturers' representatives, and code authorities meeting and discussing matters related to the construction. Imagine the movie or video presentation without a sound track. There is not only no background music, there is no dialog. It would be very difficult to construct the facility based only on a silent movie. To properly understand the requirements and construct the facility, dialog is essential.

The relationship between the Contract Drawings and the Contract Specifications is similar. The Contract Specifications are essential for complete understanding of the Work to be performed by the Contractor.

Most Conditions of the Contract recognize the significance of construction specifications and refer to the Specifications as part of the Contract Documents, with importance equivalent to that of the Drawings. Because of this, it is imperative that all parties identified in the Agreement (the Owner, the Architect/Engineer, and the Contractor) understand the role of the Specifications and understand how the Specifications are used during bidding and performance of the Work under the Contract. Moreover, the Architect/Engineer should be just as skilled in preparing the Specifications as in preparing the Drawings. The documents are complementary and carry equal weight for interpretation of Contract requirements.

Unfortunately, the education of architects, landscape architects, engineers, specialty designers, construction managers, constructors, inspection personnel, code authorities, manufacturers, fabricators, installers, and applicators rarely includes substantial instruction in written documents for construction, including construction specifications. Perhaps this is because these parties are more familiar and comfortable with graphic documents (drawings) and computations (spreadsheets and calculations). Nevertheless, proper performance of the Work requires clear, correct, and adequate descriptions of the requirements of the project, including written documents called the "Specifications."

While the future appears to hold major changes in the way construction information is managed and presented, including object-oriented computer-assisted drafting (CAD) that blurs the lines between drawings and specifications, the current separation of information into Contract Drawings and Contract Specifications for bidding and construction will continue for many years. It is essential that those who prepare and use these documents understand their purposes and properly integrate them.

To drive this point home, realize that attorneys and some construction managers understand information written on an 8½ by 11-inch page much better than they understand what is shown on a drawing. Although the General Conditions of the Contract may state otherwise, there is a tendency in a dispute to give greater significance to the Specifications than to the Drawings. This is a particularly good reason to apply as much care in preparing the Specifications as the Drawings.

Chapter 2

Relationship Between Drawings and Specifications

WHAT GOES WHERE

The information necessary for construction of a facility is developed by the Architect/Engineer and is presented in two basic types of documents: the Contract Drawings and the Contract Specifications. These two types of documents are a means of communicating information between the Architect/Engineer and the Contractor, but each type uses special forms of communication. One is pictorial or graphic, and the other is verbal or textual. Despite these distinctions, each type of document should complement while not contradicting or duplicating the other. In this way, each type of document fulfills its unique function.

According to *AIA Document A201-1997 - General Conditions of the Contract for Construction,* Paragraph 1.1.5, "The Drawings are the graphic and pictorial portions of the Contract Documents showing the design, location and dimensions of the Work, generally including plans, elevations, sections, details, schedules and diagrams."

According to *AIA Document A201-1997,* Paragraph 1.1.6, "The Specifications are that portion of the Contract Documents consisting of the written requirements for materials, equipment, systems, standards and workmanship for the Work, and performance of related services."

In broad terms, the Contract Drawings are graphical depictions, and the Contract Specifications are written descriptions of the end result of the Work to be performed. Each type of Contract Document, whether Drawings or Specifications, contributes to the overall "story" of construction of a new, remodeled, renovated, or reconstructed facility. To repeat the metaphor used in the preceding chapter, the construction Contract is like a movie or video presentation of a story. The Contract Drawings are like the video portion of the presentation, and the Contract Specifications are the audio portion. Both are necessary for understanding the story. Without the audio portion, it would be a silent movie.

The video and audio tracks need to be synchronized. Imagine the graphic depiction of the jamb condition of an interior hollow metal door frame in a metal stud wall, coupled with the text description of a wood door frame in a wood stud wall. Which is the correct depiction of the Work?

In both definitions for *AIA Document A201-1997,* above, the term "Work" is used. This is a very important term to understand. According to *AIA Document A201-1997,* Paragraph 1.1.3, "The term Work means the construction and services required by the Contract Documents, whether completed or partially completed, and includes all other labor, materials, equipment and services provided or to be provided by the Contractor to fulfill the Contractor's obligations. The Work may constitute the whole or a part of the Project."

Consider that the Work may be simple or complex. Consider the broad range of activities embodied in the Work, from procurement of materials and manufactured products, through fabrication, delivery to the project site, storage and staging at the project site, surface preparation, mixing, application, installation, fitting, and finishing. Consider that the activities include administrative procedures, such as preparation and review of shop drawings, product data and samples. Consider that the Work includes tests and inspections, as well as demonstrations, adjustments and validation of performance, also known as "commissioning." Consider that the Work includes activities and construction that are temporary in nature, such as temporary utilities, barriers, field offices, security and cleaning. Descriptions of the Work need to be detailed to suit the nature of the Work—its simplicity or complexity, its need for careful craftsmanship, its need for monitoring to ensure quality, and its need for compliance with codes, standards, and administrative requirements. Most of these do not lend themselves to graphic depictions. Most are best described in written requirements presented in the Specifications.

To maintain the separate yet complementary nature of these two types of documents, to ensure that they will be interconnected without describing overlapping or contradictory requirements, and to avoid omissions in necessary information, it is essential to understand the nature of the Drawings and Specifications.

THE DRAWINGS

Drawings present a picture or a series of pictures of a project or parts of a project to be constructed. Drawings present the size, form, location, and arrangement of various elements of the project. This information should not be described in the Specifications because it is best described graphically on the Drawings. In fact, a Drawing can be considered a special language or means of communication to convey ideas of construction from one person to another. These ideas cannot be effectively conveyed by words alone.

Drawings should indicate the relationship between elements of the facility and may designate the following for each material, assembly, component and accessory:

- *Location* of each material, assembly, component and accessory.
- *Identification* of components and pieces of equipment. Use only generic names and locations, and coordinate terminology used on Drawings and in Specifications with short keynotes.
- Give *dimensions* of components and sizes of field-assembled components.
- Indicate *interfaces and connections* between materials, detail assemblies and diagram systems. Indicate boundaries between materials of different capacities.
- Show *forms and relationships* of building elements.
- Indicate *limits of Work* and, as applicable, indicate areas of construction phases.
- Indicate *extent of alternates* and indicate "base bid" and "alternate bid" construction so that the scope of each condition is clear.
- Indicate *work to be performed by or for the Owner* under separate contracts.
- On multiple-prime contract projects, indicate *locations, limits and extent of the Work included in separate contracts* and detail interfaces between scopes of Work.
- Identify *applicable Drawing symbols* in a schedule of symbols.
- Indicate the *graphic scale* of Drawings.

Well-prepared Drawings:

- Should not use *comprehensive or too many notes.* Redundancy should be avoided; concise notes enhance the clarity of the Drawings. The Specifications should present information in text form.
- Should not use *notes that define Work* to be performed *by a specific subcontractor or trade* unless required by authorities having jurisdiction. The Contract Documents are addressed to the Contractor, who has overall responsibility for all Work under the Contract. The General Conditions of the Contract typically note that the Contract Documents do not establish trade or subcontract jurisdiction for portions of the Work.
- Should not use *proprietary names and slang terms.* Instead, use proper, generic terms that are coordinated with the terminology used in the Specifications.
- Should not *cross-reference* with specifications *by indicating "SEE SPECS."* The Specifications should always be "seen." Use of this phrase could be interpreted to mean that there is information presented that does not require "seeing" or reading of the Specifications.

The purpose of the Drawings is to convey information regarding the intent of the design and depictions of Work to be accomplished. The Drawings may be in the form of plan views (looking down on the floor), in small- and large-scale sections (looking at a cutaway view), in details (large-scale, limited portion of the work), in diagrams and schedules (such as single-line power diagrams and finish materials schedule) and in notes (tied to specific elements by arrows or as symbols referencing keynotes on a table or listed in the margin of the Drawing.

Throughout the nineteenth century and into the early decades of the twentieth century, Drawings were organized in a simple numerical sequence, without distinctions between design discipline and construction trades. The Drawings were depictions of the end result of construction and included information regarding the structure, the finishes, the fenestrations, the portals, the weather barriers, the heating and ventilating appliances, and the decorative and accessory elements of the facility. Notations were relatively sparse and concise. Yet, the information was sufficient for performance of Work necessary for the project.

As construction technology became more complex and as construction contracting became more competitive, information needed to become more extensive and precise. One response was to segregate information into series of Drawings in the overall set prepared by specialist design professionals. No longer was the Architect a "master builder" who produced all-encompassing drawings. Structural engineers, mechanical engineers, and electrical engineers were engaged by the Architect to design portions of the project and to present graphic depictions of the Work on "S" (structural), "P" (plumbing), "M" (mechanical or heating/ventilating), and "E" (electrical) drawings. Over time, these expanded with inclusion of "C" (civil or site development) and "L" (landscape irrigation and planting) drawings. Specialty products and systems were included for "K" (kitchen or food service equipment), "F" (fire protection), "T" (vertical transport or elevator), and "Q" (equipment such as laboratory, medical, and process equipment).

With advances in reprographics, Drawings sometimes include aerial photography of the site, photographic presentations of existing conditions, and photographs of components to be replicated. Color as well as the more common black-and-white xerographic printing processes are becoming as common as diazo ("blueline," "blackline," and "sepia") reproduction processes. True "blueprints" (white linework on a blue background) are now considered archaic and, if available, are usually prohibitively expensive. With drawings being archived on restricted-access (intranet) project websites for ready access by project team members, the traditional concept of printed drawings is being challenged. The impact of these changes is unclear at this time.

For greater discussion of organization and production of construction contract drawings, see the *National CAD Standard (NCS)*™, available from The Construction Specifications Institute (CSI). The *National CAD Standard (NCS)* has been developed by the National CAD Standard (NCS) Project Committee, a group of representatives from many professional and industry associations, including The Construction Specifications Institute (CSI), The American Institute of Architects (AIA) and the U.S. Department of Defense Tri-Service Computer-Aided Design and Drafting and Geographic Information Systems (CADD/GIS) Center, convened by the National Institute of Building Sciences (NIBS). NCS is an industrywide effort to improve the efficiency of building design, construction, and management throughout the life cycle of building facilities. It includes:

- NCS Project Committee Report
- Introduction and Amendments to Industry Publications
- *Uniform Drawing System (UDS)*™, published by CSI
- Plotting guidelines, developed by the National CAD Standards Project Committee
- *CAD Layering Guidelines*™, published by AIA

THE SPECIFICATIONS

As defined above, from *AIA A201-1997,* Specifications are merely "that portion of the Contract Documents consisting of the written requirements for materials, equipment, systems, standards and workmanship for the Work, and performance of related services." *Webster's Encyclopedic Unabridged Dictionary of the English Language* (Gramercy Books, New York, 1989) defines a specification as "a detailed description of requirements, dimensions, materials, etc., as of a proposed building, machine, bridge, etc." Neither of these definitions suffices.

The definitions of Specifications and Drawings by the AIA, above, and the dictionary definition give no indication of the relationship between the Specifications and the Drawings other than that they are both part of the Contract Documents. Yet, they are closely interrelated, as the metaphor of a movie or video uses both images and sound to tell a story. Both are needed to understand the requirements of the Work under the Contract and the intent of the design for the project. The Drawings and the Specifications each serve distinct purposes in telling the story. Specifications should generally describe the following:

- *Type and quality* of every product in the work, from the simplest material through the functioning system
- Quality of workmanship, including quality during manufacture, fabrication, application, installation, finishing, and adjusting
- Requirements for fabrication, erection, application, installation, and finishing
- Applicable regulatory requirements, including codes and standards applicable to performance of the Work

- Overall and component dimensional requirements for specified materials, manufactured products, and equipment
- Specific descriptions and procedures for allowances and unit prices in the contract
- Specific descriptions and procedures for product alternates and options
- Specific requirements for administration of the contract for construction

Specifications should not overlap or duplicate information contained on the drawings. Duplication, unless it is repeated word for word, is harmful because it can lead to contradiction, confusion, misunderstanding, and difference of opinion.

In broad terms, lines of demarcation should be established between the Drawings and the Specifications for specific elements in the project so that one does not attempt to do what is more suitable to the other. For example, the Drawings should indicate a material such as gypsum board in general terms, using graphic indications or simple notations. It should be left to the Specifications to describe specific attributes of the gypsum board, such as thickness and resistance to fire, impact and moisture.

What if there is more than one type of material, such as gypsum board? The Specifications should assign a "type" indicator to each type of gypsum board and specify the attributes of each type. See Exhibit 2-1 for an example of what the specifications should state.

This is the authors' recommendation for one method of identifying multiple types of similar products. Other specifiers may disagree. How does one decide which process to follow? It is a matter of professional judgment. Each design office should establish design criteria (principles) and formal policies (procedures) for production of the Drawings and Specifications. These policies should be clear and cover all likely conditions. They should be readily available to those who actually produce the documents and should be included in staff training programs.

Principles and procedures for producing Specifications will be addressed in subsequent chapters. First, however, we must discuss the design process and the resolution of conflicts and disputes during construction, which often involve issues touched upon above.

COORDINATING THE DESIGN PROCESS

To achieve proper separation of information between the Drawings and the Specifications, it is essential that development of the Specifications go hand in hand with preparation of the Drawings. At the outset, a member of the design team for each discipline should be made responsible for establishing and keeping an all-important project checklist. This person can be the project architect or project engineer.

The project checklist should establish a schedule of what is to appear on the Drawings, what is to be described in the Specifications, and what is to be itemized and listed in schedules on the Drawings. The checklist should indicate milestones for publishing various versions of the Specifications, including preliminary or outline specifications, design coordination draft Specifications, Specifications for plancheck submissions to code authorities having jurisdiction, Specifications for use in preparing estimates of probable construction costs, Specifications for issuance to bidders, and Specifications included as a portion of the Contract Documents used for actual construction.

Accompanying the Checklist should be listings of decisions made by designers and detailers, with action items noted for matters to be developed or resolved. Changes in design and detailing should be recorded and described, with notations on why the changes were made.

In Masterspec®, published by ARCOM for the AIA, two types of supplementary documents are included with the guide Specifications, titled "Drawing Coordination Checklist" and "Specification Coordination Checklist." These closely follow the principles and practices stated here. They are examples of fundamental coordination items that can serve as the basis for developing more comprehensive, office-specific procedures.

Specifications checklists should be used to ensure that:

1. Necessary items are identified appropriately in the Drawings and Specifications. Specified items need to be consistent with the indications on the Drawings. For example, if acid-resistant sinks are specified for a laboratory, the Drawings should not indicate stainless steel sinks for the laboratory, and the Drawings should indicate and the Specifications should include appropriate drain piping for acid waste.

2. Specified product names and series, models, and catalog numbers are correct. Availability of specified products also needs to be verified, especially for the project location and regulatory requirements such as air quality (VOC emissions).

EXAMPLE OF GYPSUM BOARD TYPES AND KEYNOTE CODES

A. Gypsum Board, Type GB1 [09250.gb1]:

1. Use: Typical walls, unless otherwise indicated.
2. Fire resistance: Fire Resistant, Type X.
3. Thickness: ⅝-inch (15.9 mm).
4. Long edges: Tapered.
5. Description: ASTM C 36, Type X.

B. Gypsum Board, Type GB2 [09250.gb2]:

1. Use: Typical ceilings, unless otherwise directed.
2. Fire resistance: Fire Resistant, Type X.
3. Thickness: ½-inch (12.7 mm).
4. Long edges: Tapered.
5. Description: ASTM C 36, proprietary product having improved fire resistance over standard Type X. Acceptable manufacturers and products:
   a. G-P Gypsum Corp.; Firestop Type C.
   b. National Gypsum Company; Gold Bond Fire-Shield G.
   c. United States Gypsum Co.; SHEETROCK Brand Gypsum

C. Gypsum Board, Type GB3 [09250.gb3]:

1. Use: Exterior gypsum soffit board, for walls and ceilings in damp interior locations such as showers and laundries.
2. Fire resistance: Fire Resistant, Type X.
3. Thickness: ⅝-inch (15.9 mm).
4. Long edges: Manufacturer's standard.
5. Description: ASTM C 931/C 931M, Type X.

D. Gypsum Board, Type GB4 [09250.gb4]:

1. Use: Walls of corridors.
2. Fire resistance: Fire Resistant, Type X.
3. Thickness: ⅝-inch (15.9 mm).
4. Long edges: Tapered.
5. Description: ASTM C 36, proprietary product having impact-resistant facing and Type-X fire-resistant core. Acceptable manufacturers and products:
   a. National Gypsum Company; Gold Bond Hi-Abuse Wallboard.
   b. United States Gypsum Co.; SHEETROCK Brand Abuse-Resistant Gypsum Panels.

Exhibit 2-1. Example of product types and keynote codes.

3. Drawings and Specifications do not contain duplicate and conflicting information. Typically, Specifications supplement and amplify information shown on the Drawings, but they should not repeat the information. For example, the manufacturer and model number of a boiler should be identified in the Specifications and not on the Drawings. The ceramic tile manufacturer, pattern, and colors could be identified in a legend associated with the Finish Materials Schedule included in the Drawings, and the Specifications would then refer to this specific information "as indicated on Legend of Finish Materials Schedule in the Drawings."

4. Cross references in the Specifications are correct, especially the use of Section numbers and titles for related Work specified in other Sections. Sometimes, references are made to Specifications Sections that do not exist.

5. Referenced standards are correct and applicable. Industry associations change names. Standards are superseded or withdrawn.

6. Manufacturers' names are correct and contact information is current. The business world is in constant change, and corporate names, addresses, and telephone numbers change often.

TERMINOLOGY

To ensure correct understanding on the part of users of the Drawings and Specifications, it is essential that standard terminology be employed and used consistently. Often terms are used on the Drawings that do not appear in the Specifications, and vice versa. For example, "service sink" in the Specifications should not be "janitor's sink" on the Drawings, and "G.I. flashing" on the Drawings should not be "galvanized sheet metal" in the Specifications. Worse still, "G.I. coping" on the Drawings should not be "prefabricated aluminum coping system" in the Specifications.

There have been pitched battles in architectural and engineering offices over proper terminology. Old habits die hard. Change is uncomfortable. That is why there needs to be a commitment by upper management to adopt and implement the use of proper industry-recognized terminology and to insist on coordination of terminology between the Drawings and the Specifications. Where there is an in-house specifications writer, this person can be the authority for terminology. Where a consulting specifications writer is used, either this person needs to be given authority or a senior member of the design team needs to implement instructions for the specifications consultant.

These are simple principles, and procedures for implementing them can be simple. However, they require a great deal of care and expenditure of considerable time and energy in order to be a regular part of Drawings and Specifications production.

CONSIDERATIONS FOR USERS OF DRAWINGS AND SPECIFICATIONS

It is a common maxim that authors should know their readers. This also holds for users of Drawings and Specifications. However, catering to all possible users of the documents is not only impossible but may prove to be counterproductive and may increase the professional liability risk of architects and engineers.

First, Drawings and Specifications are prepared as from the Owner to the Contractor. The Owner and the Contractor are the two parties who sign the Agreement (contract), and the Drawings and Specifications are portions of the Contract Documents identified in the Agreement. Thus, the primary purpose is to describe the construction and services required by the Contract Documents and the responsibilities of these two parties in fulfilling their obligations under the Contract. One other party is identified in the Agreement: the Architect/Engineer. Although this person is not a signer of the Agreement, the General Conditions of the Contract and the Specifications prescribe his or her duties and responsibilities. (It is important that the Architect/Engineer ensure that the provisions of the Contract for construction are consistent with the Agreement for professional services between the Owner and the Architect/Engineer.)

Because the Contract Documents are addressed as between the Owner and the Contractor (typically a general contractor), addressing other parties who are not bound to the Owner by the signed Agreement is inappropriate and perhaps creates risk for the one who prepares the Contract Drawings and Specifications. For example, prescribing subcontract responsibilities ("concrete contractor shall prepare floor slab to receive ceramic tile") may be a noble attempt to ensure quality, but it can backfire when subcontractors claim that certain portions of the Work are not included in the scope of their contracts with the (general) Contractor. Or it can be troublesome when the general contractor intends to have an equipment supplier be responsible for making final utility connections, including engaging licensed plumbers and electricians, when the equipment specifications state, "plumbing connections by plumbing subcontractor" or "electrical connections by electrical subcontractor." Work may be omitted from the Contract sum or the Contract sum may include double payment for the same portion of the Work.

In typical AIA and EJCDC General Conditions of the Contract, the means, methods, techniques, and sequences of construction are stated to be solely the responsibility of the Contractor. Managing the Work is the Contractor's responsibility. There is typically a line item in applications for payment from the Contractor for this management task. Usually it is called "Division 1" or it is part of the Contractor's "overhead and profit." When a design professional specifies how the project is managed (what the means, methods, techniques, and sequences of construction shall be, and which trade or subcontractor shall perform which portion of the Work), the design professional is assuming management responsibility over the Work.

Perhaps under some project delivery methods, such as multiple prime contracts and design-build contracts, or when a Construction Manager is involved as one of the parties in the Contract, assignment of subcontracts may be appropriate. But for the purpose of this discussion of preparation of construction specifications, the prohibition against identifying trades and subcontracts will be held.

Assignment of portions of the Drawings and Specifications can be made in the Agreement or another referenced Contract document, leaving the Specifications with the traditional organization and descriptions of the Work.

Despite this, there are users of the Drawings and Specifications whose interests should be appropriately considered by the Architect/Engineer and the specifications writer without violating the principles stated above. It is a matter of organizing—rather than assigning—the information on the Drawings and in the Specifications in such a fashion that these users can find the information they need in order to fulfill their responsibilities. These users are plancheckers, cost estimators, and construction inspectors.

The interests of plancheckers, cost estimators, and construction inspectors can be represented by clear, consistent, and uniform presentation of information. Industry standards for organizing Drawings are found in the *National CAD Standard* (*NCS*). For construction specifications, industry standards are presented in publications of CSI in what are presently known as the *CSI Manual of Practice, SectionFormat*™, *PageFormat*™, and *MasterFormat - Master List of Section Numbers and Titles for the Construction Industry*. These publications are under revision as this book is being written, and current versions may be obtained from CSI in Alexandria, Virginia.

Plancheckers need to readily see compliance with applicable codes. Architects and engineers should use the Drawings and Specifications to explicitly demonstrate compliance. Sometimes, in their quest to ensure compliance with code requirements, plancheckers require specifications to give the code or reference certain sections and paragraphs in the code. This can be troublesome when the same Specifications and Drawings are used for the construction Contract Documents. If the Specifications cite the code, does the Contractor then assume design responsibility for code-mandated provisions? For example, if the Drawings recite code requirements for clearances for accessibility by persons with disabilities, but the dimensions indicated on the Drawings prevent such clearances from being achieved, is the Contractor responsible for reconstruction to comply with the code? This has happened, and the Contractor has had to bear the cost of reconstruction.

Cost estimators need to have a complete understanding of the Work under the Contract. They need coordinated Drawings and Specifications in order to prepare a complete and accurate estimate. For example, when Drawings have notes regarding louvers but no louvers are shown on building elevations or in the Specifications, should louvers be included in or omitted from the estimate or bid? If they are included, for example, and the Drawings note steel louvers but the Specifications include aluminum louvers, which shall it be? If the Specifications indicate factory paint color on the louvers "to be selected by Architect from manufacturer's standard selection" but the Drawings indicate a custom color to match other building components, which instruction does the cost estimator follow? These matters need to be resolved for pre-bid estimates of probable construction cost and for bids to be submitted. When it comes time to buy out the Work (sign actual subcontracts) and when it comes time to make submittals for review by the Architect/Engineer, which requirements apply?

Finally, Drawings and Specifications need to be understandable by construction inspectors and the supervisors for the Contractor and subcontractors. Field representatives of the Owner, Architect, and Engineer also need to have readily understandable documents.

For example, in the Drawings, building exterior elevations and sectional views need to correctly represent the grades around the building. There should be graphic indications of below-grade walls to guide material take-offs. If a basement wall requires waterproofing, how is the area properly determined and waterproofing applied in proper locations if the below-grade conditions are not indicated?

As another example, the Specifications should generally describe preparation, mixing, application, and curing activities for products, although the Specification may reference much more detailed instructions by the product manufacturer that shall be followed.

The principle is to provide indications of activities that inspectors, supervisors, and observers should see for quality assurance. If the activity is not performed ("treat and prime galvanized steel," for example), failure of a painted finish may result that, although solely the Contractor's responsibility to remedy, will cause time-consuming disputes ("you shouldn't have let me do it wrong").

Chapter 3

Organization of Specifications

HISTORY OF SPECIFICATION ORGANIZATION

Specifications in the eighteenth and nineteenth centuries consisted of a single document containing a description of all the work and materials to be included in a building. This was especially true of small, simple structures that were constructed by a general contractor who engaged all the craftsmen and did not sublet or subcontract any parts of the work. An early textbook by T. L. Donaldson titled *Handbook of Specifications* (London, 1860) provided for the arrangement of specifications on a craft basis. The specifications were divided into two general divisions, with subdivisions as follows:

Carcase	Finishing
Excavator	Joiner
Bricklayer	Plasterer
Mason	Plumber
Slater	Painter
Founder and smith	Glazier
Carpenter	Paperhanger
Ironmonger	Smith and bellhanger
Gasfitter	

Toward the end of the nineteenth century, specifications written for substantial buildings often broke the work into several categories or sections, such as masonry, carpentry, and mechanical work, with various allied or related subjects under each section. The masonry section included excavation, concrete, brickwork, stonework, steel columns and lintels, and waterproofing. The carpentry section included roofing, glazing, and painting, as well as carpentry. The mechanical or pipe trades consisted of plumbing, gas, and heating work. When electricity came into use, it was included in the mechanical work.

With the introduction of new materials, products, and construction techniques and equipment in the twentieth century, more than one craft or trade was needed to perform the work described in some sections. Rather than hire workers in more crafts and trades under the contractor's direct supervision, the contractor began to sublet portions of the work to other businesses or "subcontractors," who, in turn, hired workers in several crafts and trades to perform their subcontracted portions of the work. Sections of the specifications were written to describe work responsibilities for each subcontractor and craft or trade.

With growing sophistication of construction materials and methods of construction in the first half of the twentieth century, construction specifications similarly evolved in complexity. Specifications for larger, more complex buildings became book size, with many "chapters" or sections. Mild chaos developed as each architect/engineer used different methods to organize and identify these specifications. Trade jurisdictions and construction practices varied in different locales and regions. With the explosion of new technologies and with highly competitive markets for subcontracting and materials procurement following World War II, the process of organizing construction specifications became problematic for the construction industry. Standardization was needed for construction specifications.

One of the reasons The Construction Specifications Institute (CSI) was formed in 1948 was to address this issue. In the early 1960s the CSI T-1 CSI Format Task Force was formed to develop a common format for organizing construction specifications. In 1961, an initial draft of a guideline for specifications organization, "Format and Arrangement of Specifications and Related Documents," was developed, followed after a conference in 1962 by a second draft. This second draft organized specifications into 22 divisions. Work continued, and after a conference in September 1963, the Task Force's efforts were published in *The CSI Format for Building Specifications.*

The 1963 *The CSI Format for Building Specifications* included 16 divisions to organize the various subjects under each division. The subjects were considered "items of work," and each subject was termed a "section." This endeavor at organizing construction subjects had 314 sections spread among the 16 divisions and arranged alphabetically rather than numerically under each division. In Division 15 (mechanical) there were 20 headings, and in Division 16 there were 17. Also included were headings such as Bidding Requirements, Contract Forms, General Conditions, and a new category, Supplementary General Conditions, which had been worked out with the AIA to address the customizing of construction contracts using AIA's preprinted General Conditions.

In 1964, the original *The CSI Format for Building Specifications* was updated and formally published by CSI. It had 16 divisions and 1010 headings under the various divisions. The headings (sections) were still arranged alphabetically under each division. The published document was 28 total pages, with the divisions and section listings taking up 5 of those pages.

The *CSI Format* continued to grow and began to be widely accepted. In 1972, the *CSI Format*™ was incorporated into the *Uniform Construction Index,* "A System of Formats for Specifications, Data Filing, Cost Analysis and Project Filing." The *Uniform Construction Index* was developed by the Conference on Uniform Indexing Systems, which had as its purpose to develop a "simple, logical and flexible system for rapid classification and retrieval of technical data in the construction industry" (*Uniform Construction Index,* The Construction Specifications Institute, 1972, page 0.3, Introduction, Background). It was preceded by publication in 1966 of *Uniform System for Construction Specifications, Data Filing and Cost Accounting; Title 1 - Buildings.* The *Uniform Construction Index* of 1972 was jointly published by:

- The American Institute of Architects
- Associated General Contractors of America, Inc.
- The Construction Specifications Institute
- Consulting Engineers Council of the United States
- Council of Mechanical Specialty Contracting Industries, Inc.
- Professional Engineers in Private Practice/ National Society of Professional Engineers
- The Producers Council, Incorporated
- Specifications Writers Association of Canada/ Association des Rédacteurs de Devis du Canada

The *Uniform Construction Index* was also endorsed by:

- American Society of Landscape Architects
- Association of Consulting Engineers of Canada

- Canadian Construction Association
- Canadian Institute of Quantity Surveyors
- Royal Architecture Institute of Canada
- Sweet's Division, McGraw-Hill Information Systems Co.—U.S.A./Canada

Contributors to the *Uniform Construction Index* included:

- National Association of Plumbing-Heating-Cooling Contractors
- National Electrical Contractors Association, Inc.
- Sheet Metal and Air Conditioning Contractors' National Association, Inc.

There was broad representation of the construction industry in the *Uniform Construction Index* and they incorporated the *CSI Format.* This attempt to organize construction Specifications, building product information, and construction cost data into an all-encompassing system did not continue with updated versions of the *Uniform Construction Index.* However, the *CSI Format* continued to develop.

In the cost data portion of the *Uniform Construction Index,* five-digit numbers were assigned to various subjects or Sections under each of the 16 Division headings. For example, "08100" was assigned to "Metal Doors & Frames" under Division 8 — Doors & Windows. "09900" was assigned under Division 9 — Finishes to "Painting." Otherwise, for construction specifications, the *Uniform Construction Index* went along with the instruction in the *CSI Format* for the specifier to list applicable Sections in alphabetical order under each Division.

Also in 1972, CSI published an update to the *CSI Format* titled *CSI Format - Master List of Specifications Section Titles.* It had 1220 listings (sections) and introduced a five-digit numbering scheme for sections comparable to the numbering for cost data in the *Uniform Construction Index.* In 1975, *CSI Format - Master List of Specifications Section Titles* was updated and had 1290 section listings.

In 1978, CSI published a major revision to its format, titled *MasterFormat*™ — *Master List of Numbers and Titles.* This document had 2120 section listings and introduced the name *MasterFormat* to replace *CSI Format.* It also introduced another division, "Division 0," which caused controversy since the documents identified for Division 0 were not Specifications. They included Bidding Requirements and Conditions of the Contract. Some professional associations objected to the implication that specifiers should prepare these legal documents.

MasterFormat was revised and published in 1983 with Division 0 numbers titled "Document Number." Under the major (broad-scope or Level Two) section numbers and titles, brief explanations were added. *MasterFormat* was again revised and published in 1988 using the format of the 1983 edition. Headings preceding Division 0 were identified as "documents" rather than sections in order to demonstrate that these were not specifications.

This began what was supposed to be a 5-year cycle for revisions to *MasterFormat.* There was supposed to be a 1993 edition, but it was delayed due to objections by specifiers to major changes in section numbers and titles and reassignment of some Sections to different Divisions.

In early 1996, the 1995 edition of *MasterFormat - Master List of Numbers and Titles for the Construction Industry* (note the subtle name change) was published jointly by CSI and Construction Specifications Canada (CSC). It has been widely adopted and used in the construction industry of North America. With the increasing significance of construction specifications in the construction contract documents, the section numbers and titles of *MasterFormat* have become deeply ingrained in construction specifications, cost data, building product data, and drawing keynoting.

As this is being written, *MasterFormat* is undergoing major revisions to accommodate changes in construction technology and to make it suitable for more than building construction. The final form and content of the next edition of *MasterFormat* are unclear. *MasterFormat* will be expanded to 48 Divisions plus a 00 Division for procurement and contracting requirements. Some Divisions will be unassigned and reserved for future expansion. Section numbers and titles will reflect an approach to specifications that accommodates not just initial construction but also "life-cycle" activities of facilities, including maintenance, renovation, remodeling, and demolition. Because the 2004 edition of MasterFormat is not available, and it is not expected to be implemented until 2005 and 2006, discussions in this book regarding *MasterFormat* will refer to the 1995 edition and its 16 Division format.

ORGANIZING SPECIFICATIONS DIVISIONS AND SECTIONS

The organization of construction specifications as represented in the *CSI Format* was adequate for the 1960s and 1970s, when construction technology was relatively simple compared to the proliferation of new materials and more complex conveying systems, mechanical and electrical systems, and communication systems. With the advent of more complex construction, the need for standardized organization of specifications became more apparent and more critical. It became necessary to increasingly subdivide and organize the large mass of specifications information into more manageable and understandable portions.

For convenience in writing, speed in estimating and ease of reference, it has been found that the most suitable organization of the specifications is in a series of chapters or sections dealing with different portions of the construction work. With increasing complexity, it became impossible to assign or organize sections according to subcontract or trade.

The original concept behind the descriptions of headings (sections) in the *CSI Format* was to describe the likely subcontractor or trade that would perform the identified work. As the *CSI Format* evolved into *MasterFormat,* titles of the sections changed. Where the original titles were subjects such as the verb "carpeting," the new titles were nouns such as "carpet." This represented a shift from emphasis on trades or work activities to materials and products.

In the April 1964 edition of *The Construction Specifier,* CSI's official magazine, the original task force chair, Bernard B. Rothschild, FAIA, wrote an article in which he stated that specifications should not attempt to separate work by union jurisdiction. Since then, there has been confusion and controversy over whether specifications sections should specify only materials, manufactured products, assemblies, and systems or should also specify the trades and subcontractors and their construction activities. Should the titles reflect both products and application, such as "paint" rather than "painting" or "woodwork" rather than "woodworking?" Looking at the familiar *MasterFormat* - 1995 and listening to the controversies about the proposed *MasterFormat* - 2004, with its attempt to use titles based on "work results" (products + fabrication/application/installation), reveals that the matter is far from resolved.

Until April 1963, when the *CSI Format for Construction Specifications* was published, each specifier organized the specifications in a series of divisions or sections that more or less followed a time relationship or chronological order related to the order of performance of various portions of the construction work.

For example, specifications for excavating and backfilling were logically placed at the beginning, followed by specifications for construction of the foundations and building framework, enclosure of the building for thermal and moisture protection, installation of doors and windows, application of building finishes, and installation of specialty products, equipment, and furnishings. Along the way, usually from beginning to end of the construction, plumbing, heating, ventilating, fire protection, electrical power, lighting, and communications systems were constructed.

From office to office, and even within the same office, specifications were written with the order of Sections not uniform. Sometimes the order was the one in which the specifier wrote the section. Sometimes, it was in alphabetical order rather than following the usual sequence of construction. This made finding information difficult for bidders and contractors who had to work with documents for various projects produced by various specifiers. Sometimes the order presented in the specifications was inconsistent with the actual construction sequence. Complex structures often required certain mechanical trades to be involved at an early stage in the construction process, so organization according to trades chronology was not appropriate.

It became apparent that a major overhaul was required for the organization of specification sections and that a uniform system would benefit all parties in the construction process. These specifications sections were organized under general headings or divisions.

CSI/CSC *MASTERFORMAT*

Beginning with the 1978 edition of *MasterFormat,* section titles and the order of sections under each division were standardized. Also, five-digit numbers were introduced that were permanently assigned to common (Level Two) section headings. This provided a nationally recognized format for organizing construction specifications. It was endorsed by major stakeholders in design professions and construction organizations. It provided a uniform system for arranging sections under an overall 16-division format for the bound volume of specifications.

The 1995 edition of *MasterFormat* retained the 16 Divisions (Level One) headings and the five-digit (Level Two) Section numbers and titles. They are:

MASTERFORMAT LEVEL ONE NUMBERS AND TITLES

INTRODUCTORY INFORMATION (00000-00099)
BIDDING REQUIREMENTS   (00100-00499)
CONTRACTING REQUIREMENTS (00500-00999)

SPECIFICATIONS

DIVISION 1 - GENERAL REQUIREMENTS
                         (01000-01999)
DIVISION 2 - SITE CONSTRUCTION
                         (02000-02999)
DIVISION 3 - CONCRETE   (03000-03999)
DIVISION 4 - MASONRY   (04000-04999)
DIVISION 5 - METALS   (05000-05999)
DIVISION 6 - WOOD AND PLASTICS
                         (06000-06999)
DIVISION 7 - THERMAL AND MOISTURE
PROTECTION   (07000-07999)
DIVISION 8 - DOORS AND WINDOWS
                         (08000-08999)
DIVISION 9 - FINISHES   (09000-09999)
DIVISION 10 - SPECIALTIES   (11000-10999)
DIVISION 11 - EQUIPMENT   (11000-11999)
DIVISION 12 - FURNISHINGS   (12000-12999)
DIVISION 13 - SPECIAL CONSTRUCTION
                         (13000-13999)
DIVISION 14 - CONVEYING SYSTEMS
                         (14000-14999)
DIVISION 15 - MECHANICAL   (15000-15999)
DIVISION 16 - ELECTRICAL   (16000-16999)

MasterFormat — 1995 calls the 16 Divisions "Level One" headings. Under each Division are "Level Two" headings for more specific definition of content. As interpreted by the authors, these are:

MASTERFORMAT LEVEL ONE AND TWO NUMBERS AND TITLES

INTRODUCTORY INFORMATION

00001 - PROJECT TITLE PAGE
00005 - CERTIFICATIONS
00007 - SEALS
00010 - TABLE OF CONTENTS

00015 - LIST OF DRAWINGS

00020 - LIST OF SCHEDULES

BIDDING REQUIREMENTS

00100 - BID SOLICITATION

00200 - INSTRUCTIONS TO BIDDERS

00300 - INFORMATION AVAILABLE TO BIDDERS/CONTRACTOR

00400 - BID FORMS AND SUPPLEMENTS

00490 - BIDDING ADDENDA

CONTRACTING REQUIREMENTS

00500 - AGREEMENT

00600 - BONDS AND CERTIFICATES

00700 - GENERAL CONDITIONS OF THE CONTRACT

00800 - SUPPLEMENTARY CONDITIONS OF THE CONTRACT

00900 - CONTRACT MODIFICATIONS

SPECIFICATIONS

DIVISION 1 - GENERAL REQUIREMENTS

01100 - SUMMARY

01200 - PRICE AND PAYMENT PROCEDURES

01300 - ADMINISTRATIVE REQUIREMENTS

01400 - QUALITY REQUIREMENTS

01500 - TEMPORARY FACILITIES AND CONTROLS

01600 - PRODUCT REQUIREMENTS

01700 - EXECUTION REQUIREMENTS

01800 - FACILITY OPERATION

01900 - FACILITY DECOMMISSIONING

DIVISION 2 - SITE CONSTRUCTION

02050 - BASIC SITE MATERIALS AND METHODS

02100 - SITE REMEDIATION

02200 - SITE PREPARATION

02300 - EARTHWORK

02400 - TUNNELING, BORING AND JACKING

02450 - FOUNDATION AND LOAD-BEARING ELEMENTS

02500 - UTILITY SERVICES

02600 - DRAINAGE AND CONTAINMENT

02700 - BASES, BALLASTS, PAVEMENTS AND APPURTENANCES

02800 - SITE IMPROVEMENTS AND AMENITIES

02900 - PLANTING

02950 - SITE RESTORATION AND REHABILITATION

DIVISION 3 - CONCRETE

03050 - BASIC CONCRETE MATERIALS AND METHODS

03100 - CONCRETE FORMS AND ACCESSORIES

03200 - CONCRETE REINFORCEMENT

03300 - CAST-IN-PLACE CONCRETE

03400 - PRECAST CONCRETE

03500 - CEMENTITIOUS DECKS AND UNDERLAYMENT

03600 - GROUTS

03700 - MASS CONCRETE

03900 - CONCRETE RESTORATION AND CLEANING

DIVISION 4 - MASONRY

04050 - BASIC MASONRY MATERIALS AND METHODS

04200 - MASONRY UNITS

04400 - STONE

04500 - REFRACTORIES

04600 - CORROSION-RESISTANT MASONRY

04700 - SIMULATED MASONRY

04800 - MASONRY ASSEMBLIES

04900 - MASONRY RESTORATION AND CLEANING

DIVISION 5 - METALS

05050 - BASIC METAL MATERIALS AND METHODS

05100 - STRUCTURAL METAL FRAMING

05200 - METAL JOISTS

05300 - METAL DECK

05400 - COLD-FORMED METAL FRAMING

05500 - METAL FABRICATIONS

05600 - HYDRAULIC FABRICATIONS

05650 - RAILROAD TRACK AND ACCESSORIES

05700 - ORNAMENTAL METAL

05800 - EXPANSION CONTROL

05900 - METAL RESTORATION AND CLEANING

DIVISION 6 - WOOD AND PLASTICS

06050 - BASIC WOOD AND PLASTIC MATERIALS AND METHODS

06100 - ROUGH CARPENTRY

06200 - FINISH CARPENTRY

06400 - ARCHITECTURAL WOODWORK

06500 - STRUCTURAL PLASTICS

06600 - PLASTIC FABRICATIONS

06900 - WOOD AND PLASTIC RESTORATION AND CLEANING

DIVISION 7 - THERMAL AND MOISTURE PROTECTION

07050 - BASIC THERMAL AND MOISTURE PROTECTION MATERIALS AND METHODS

07100 - DAMPPROOFING AND WATERPROOFING

07200 - THERMAL PROTECTION

07300 - SHINGLES, ROOF TILES AND ROOF COVERINGS

07400 - ROOFING AND SIDING PANELS

07500 - MEMBRANE ROOFING

07600 - FLASHING AND SHEET METAL

07700 - ROOF SPECIALTIES AND ACCESSORIES

07800 - FIRE AND SMOKE PROTECTION

07900 - JOINT SEALERS

DIVISION 8 - DOORS AND WINDOWS

08050 - BASIC DOOR AND WINDOW MATERIALS AND METHODS

08100 - METAL DOORS AND FRAMES

08200 - WOOD AND PLASTIC DOORS

08300 - SPECIALTY DOORS

08400 - ENTRANCES AND STOREFRONTS

08500 - WINDOWS

08600 - SKYLIGHTS

08700 - HARDWARE

08800 - GLAZING

08900 - GLAZED CURTAIN WALL

DIVISION 9 - FINISHES

09050 - BASIC FINISH MATERIALS AND METHODS

09100 - METAL SUPPORT ASSEMBLIES

09200 - PLASTER AND GYPSUM BOARD

09300 - TILE

09400 - TERRAZZO

09500 - CEILINGS

09600 - FLOORING

09700 - WALL FINISHES

09800 - ACOUSTICAL TREATMENT

09900 - PAINTS AND COATINGS

DIVISION 10 - SPECIALTIES

10100 - VISUAL DISPLAY BOARDS

10150 - COMPARTMENTS AND CUBICLES

10200 - LOUVERS AND VENTS

10240 - GRILLES AND SCREENS

10250 - SERVICE WALLS

10260 - WALL AND CORNER GUARDS

10270 - ACCESS FLOORING

10290 - PEST CONTROL

10300 - FIREPLACES AND STOVES

10340 - MANUFACTURED EXTERIOR SPECIALTIES

10350 - FLAGPOLES

10400 - IDENTIFICATION DEVICES

10450 - PEDESTRIAN CONTROL DEVICES

10500 - LOCKERS

10520 - FIRE PROTECTION SPECIALTIES

10530 - PROTECTIVE COVERS

10550 - POSTAL SPECIALTIES

10600 - PARTITIONS

10670 - STORAGE SHELVING

10700 - EXTERIOR PROTECTION

10750 - TELEPHONE SPECIALTIES

10800 - TOILET, BATH AND LAUNDRY SPECIALTIES

10880 - SCALES

10900 - WARDROBE AND CLOSET SPECIALTIES

DIVISION 11 - EQUIPMENT

11010 - MAINTENANCE EQUIPMENT

11020 - SECURITY AND VAULT EQUIPMENT

11030 - TELLER AND SERVICE EQUIPMENT

11040 - ECCLESIASTICAL EQUIPMENT

11050 - LIBRARY EQUIPMENT

11060 - THEATER AND STAGE EQUIPMENT

11070 - INSTRUMENTAL EQUIPMENT
11080 - REGISTRATION EQUIPMENT
11090 - CHECKROOM EQUIPMENT
11100 - MERCANTILE EQUIPMENT
11110 - COMMERCIAL LAUNDRY AND DRY CLEANING EQUIPMENT
11120 - VENDING EQUIPMENT
11130 - AUDIO-VISUAL EQUIPMENT
11140 - VEHICLE SERVICE EQUIPMENT
11150 - PARKING CONTROL EQUIPMENT
11160 - LOADING DOCK EQUIPMENT
11170 - SOLID WASTE HANDLING EQUIPMENT
11190 - DETENTION EQUIPMENT
11200 - WATER SUPPLY AND TREATMENT EQUIPMENT
11280 - HYDRAULIC GATES AND VALVES
11300 - FLUID WASTE TREATMENT AND DISPOSAL EQUIPMENT
11400 - FOOD SERVICE EQUIPMENT
11450 - RESIDENTIAL EQUIPMENT
11460 - UNIT KITCHENS
11470 - DARKROOM EQUIPMENT
11480 - ATHLETIC, RECREATIONAL AND THERAPEUTIC EQUIPMENT
11500 - INDUSTRIAL AND PROCESS EQUIPMENT
11600 - LABORATORY EQUIPMENT
11650 - PLANETARIUM EQUIPMENT
11660 - OBSERVATORY EQUIPMENT
11680 - OFFICE EQUIPMENT
11700 - MEDICAL EQUIPMENT
11780 - MORTUARY EQUIPMENT
11850 - NAVIGATION EQUIPMENT
11870 - AGRICULTURAL EQUIPMENT
11900 - EXHIBIT EQUIPMENT

DIVISION 12 - FURNISHINGS

12050 - FABRICS
12100 - ART
12300 - MANUFACTURED CASEWORK
12400 - FURNISHINGS AND ACCESSORIES
12500 - FURNITURE
12600 - MULTIPLE SEATING
12700 - SYSTEMS FURNITURE
12800 - INTERIOR PLANTS AND PLANTERS
12900 - FURNISHINGS RESTORATION AND REPAIR

DIVISION 13 - SPECIAL CONSTRUCTION

13010 - AIR-SUPPORTED STRUCTURES
13020 - BUILDING MODULES
13030 - SPECIAL PURPOSE ROOMS
13080 - SOUND, VIBRATION AND SEISMIC CONTROL
13090 - RADIATION PROTECTION
13100 - LIGHTNING PROTECTION
13110 - CATHODIC PROTECTION
13120 - PRE-ENGINEERED STRUCTURES
13150 - SWIMMING POOLS
13160 - AQUARIUMS
13165 - AQUATIC PARK FACILITIES
13170 - TUBS AND POOLS
13175 - ICE RINKS
13185 - KENNELS AND ANIMAL SHELTERS
13190 - SITE-CONSTRUCTED INCINERATORS
13200 - STORAGE TANKS
13220 - FILTER UNDERDRAINS AND MEDIA
13230 - DIGESTER COVERS AND APPURTENANCES
13240 - OXYGENATION SYSTEMS
13260 - SLUDGE CONDITIONING SYSTEMS
13280 - HAZARDOUS MATERIAL REMEDIATION
13400 - MEASUREMENT AND CONTROL INSTRUMENTATION
13500 - RECORDING INSTRUMENTATION
13550 - TRANSPORTATION CONTROL INSTRUMENTATION
13600 - SOLAR AND WIND ENERGY EQUIPMENT
13700 - SECURITY ACCESS AND SURVEILLANCE
13800 - BUILDING AUTOMATION AND CONTROL
13850 - DETECTION AND ALARM
13900 - FIRE SUPPRESSION

DIVISION 14 - CONVEYING SYSTEMS

14100 - DUMBWAITERS
14200 - ELEVATORS

14300 - ESCALATORS AND
MOVING WALKS

14400 - LIFTS

14500 - MATERIAL HANDLING

14600 - HOISTS AND CABLES

14700 - TURNTABLES

14800 - SCAFFOLDING

14900 - TRANSPORTATION

DIVISION 15 - MECHANICAL

15050 - BASIC MECHANICAL MATERIALS
AND METHODS

15100 - BUILDING SERVICES PIPING

15200 - PROCESS PIPING

15300 - FIRE PROTECTION PIPING

15400 - PLUMBING FIXTURES AND
EQUIPMENT

15500 - HEAT-GENERATION EQUIPMENT

15600 - REFRIGERATION EQUIPMENT

15700 - HEATING, VENTILATING, AND AIR
CONDITIONING EQUIPMENT

15800 - AIR DISTRIBUTION

15900 - HVAC INSTRUMENTATION AND
CONTROLS

15950 - TESTING, ADJUSTING AND
BALANCING

DIVISION 16 - ELECTRICAL

16050 - BASIC ELECTRICAL MATERIALS AND
METHODS

16100 - WIRING METHODS

16200 - ELECTRICAL POWER

16300 - TRANSMISSION AND
DISTRIBUTION

16400 - LOW-VOLTAGE DISTRIBUTION

16500 - LIGHTING

16700 - COMMUNICATIONS

16800 - SOUND AND VIDEO

USING *MASTERFORMAT*

Until publication of the 1978 edition of *Master-Format,* it was accepted practice to assign an alphanumeric code to each section of the specifications. For example, a code such as "3A" was used for the first Section in Division 3 - Concrete or "9D" was used for the fourth section in Division 9 - Fin-

ishes. These alphanumeric codes varied from project to project. For example:

	Project A		Project B
Section 9A	Non-Load-Bearing Wall Framing Systems	Section 9A	Gypsum Board
Section 9B	Gypsum Board	Section 9B	Ceramic Tile
Section 9C	Resilient Flooring	Section 9C	Resilient Flooring
Section 9D	Paints, Stains and Coatings	Section 9D	Carpeting
		Section 9E	Paints, Stains and Coatings

This system worked, but it made finding information in the specifications difficult, especially for those who were scanning the specifications to find out whether certain products were used. Also, it made filing of product information inconsistent and filing of Specifications master documents difficult.

As *MasterFormat* developed, fixed numbers and titles for Sections were assigned. For example:

	Project A		Project B
Section 09110	Non-Load-Bearing Wall Framing Systems	Section 09250	Gypsum Board
Section 09250	Gypsum Board	Section 09310	Ceramic Tile
Section 09610	Resilient Flooring	Section 09610	Resilient Flooring
Section 09900	Paints and Coatings	Section 09650	Carpet
		Section 09900	Paints and Coatings

As the example illustrates, "gypsum board" is always "09250 - Gypsum Board." Also, Sections are omitted or included according to the requirements of the project. If all wall framing is with wood studs, Section 09110 - Non-Load Bearing Wall Framing is not included.

Through the 1988 edition of *MasterFormat,* a concept was included to address varying degrees of detail in information in Specifications. The terms "broad scope," "mediumscope" and "narrowscope" were used. These represented lesser to greater degrees of detail. In the current (1995) *MasterFormat,* these have been replaced by specifications "levels." For example:

Level One: The 16 Divisions
Level Two sections (formerly broadscope): Broad categories of work with the widest latitude in describing a unit of work.

08100 - METAL DOORS AND FRAMES

08200 - WOOD AND PLASTIC DOORS

08300 - SPECIAL DOORS

08400 - ENTRANCES AND
STOREFRONTS

08500 - WINDOWS

08600 - SKYLIGHTS

08700 - HARDWARE

08800 - GLAZING

08900 - GLAZED CURTAIN WALL

Level Three sections (formerly mediumscope): Covers units of work of more limited scope.

08300 - SPECIAL DOORS (Level Two)

08310 - ACCESS DOORS (Level Three)

08320 - DETENTION DOORS AND
FRAMES

08330 - COILING DOORS AND GRILLES

08340 - SPECIAL FUNCTION DOORS

08350 - FOLDING DOORS AND GRILLES

08360 - OVERHEAD DOORS

08370 - VERTICAL LIFT DOORS

08380 - TRAFFIC DOORS

08390 - PRESSURE-RESISTANT DOORS

Level Four sections (formerly narrowscope): For use in covering extremely limited and very specific elements of work.

08300 - SPECIAL DOORS   (Level Two)

08320 - DETENTION DOORS AND FRAMES
(Level Three)

08321 - DETENTION DOOR FRAMES (Level Four)

08322 - DETENTION DOORS

08330 - COILING DOORS AND GRILLES
(Level Three)

08311 - COILING COUNTER DOORS   (Level Four)

08313 - OVERHEAD COILING DOORS

08317 - OVERHEAD COILING GRILLES

08318 - SIDE COILING GRILLES

08360 - OVERHEAD DOORS   (Level Three)

08361 - ALUMINUM SECTIONAL OVERHEAD
DOORS   (Level Four)

08362 - STEEL SECTIONAL OVERHEAD DOORS

08363 - WOOD SECTIONAL OVERHEAD DOORS

MasterFormat does not list the Level Four section numbers and titles above. These must be interpolated by the specifier from suggested headings under Level Three headings.

The specifier may use a combination of all or some of the levels in a set of specifications. Specification may include Level One (division) headings plus Level Three and Level Four sections but no Level Two sections. Or a specification may use mostly Level Two sections with a few Level Three sections. It is a matter of professional judgment.

Using more narrowly focused section numbers and titles can facilitate data filing and make it easier to find information in specifications that include many types of products with diverse requirements for quality assurance, installation materials, and installation methods. Keeping information separated can reduce the chance that improper associations between materials and methods will occur. However, it adds redundancy, as common information is repeated in similar sections.

One way to reduce redundancy is to use the "basic materials and methods" sections at the beginning of a division. In these sections, information applicable to all sections in the division is presented. This works particularly well for masonry work (Division 4) and for mechanical (Division 15) and electrical (Division 16) work. It does not work for thermal and moisture protection (Division 7) and finishes (Division 9).

The advantage of using standardized formats for specifications is that it provides a standard, fixed framework for organizing specifications.

- It serves as a sequencing guide for arranging specification sections, similar in concept to the Decimal Classification System (DCS), or, as commonly known, the "Dewey Decimal System" used in libraries.

- It enables readers familiar with the format to easily locate and retrieve information.

The advantages of using Level Three or Level Four section numbers and titles to split long sections into short ones are:

- The task of assembling a Project Manual is made easier since the specifier is able to focus on limited areas of information.

- Specifications are easier to coordinate during writing.

- The contractor can exercise greater control during bidding and construction.

The disadvantages of using several similar sections are:

- Information may be repeated.

- The greater number of narrowly focused sections increases the size of the overall specifications.
- It is more difficult to cross-reference between sections.

Chapter 4

The Project Manual and Specifications Sections

THE PROJECT MANUAL CONCEPT

The AIA, through a national Committee on Specifications, in 1965 developed the "Project Manual" concept. AIA recognized that the book commonly called the "Specifications" or "Specs" normally contained more than the name implied. It was, in fact, a manual of project requirements and contract documents whose contents and functions were best implied by the title "Project Manual."

The Project Manual, in simple terms, is a bound volume or set of volumes that contains the written portion of the Bidding and Construction Contract documents. Organization of the Project Manual is typically according to *MasterFormat,* as described in the preceding chapter. Elements of the Project Manual are:

Introductory Information
- Title Page
- Certifications Page
- Table of Contents
- *Guide to Use of the Project Manual* (used by many specifiers)

Bidding Requirements
- Bid Solicitation: Advertisement/Invitation to Bid
- Instructions to Bidders
- Information Available to Bidders
- Bid Forms and Supplements

Contracting Requirements
- Agreement
- General Conditions of the Contract
- Supplementary Conditions of the Contract
- Bonds and Certificates

Specifications
- Division 01 - General Requirements
- Divisions 02 through 49 - Technical Specifications

APPENDICES

The last element, APPENDICES, is not included in *MasterFormat* but has proven to be useful for including copies of documents that are classified as "Information Available to Bidders" and that might not be formatted in compliance with the Section format, such as signage and graphics bid packages that include both written and graphic descriptions. APPENDICES may include referenced documents that the Architect or Engineer should take responsibility for, such as hazardous materials abatement program requirements and remediation procedures. Existing condition photographs and example forms for use in administration of the construction contract may also be included in the APPENDICES. Use of APPENDICES is becoming more common, especially on complex projects.

Another adaptation of the Project Manual concept is for Multiple Prime Contract projects. The overall project may have common requirements in Division 00 - Procurement and Contracting Requirements and Specifications Division 01 - General Requirements, which could be in a separately bound volume of the Project Manual. Unique requirements for each Prime Contract, including supplementary versions of Division 00 and Division 01, would be bound in separate Project Manual volumes.

Still another adaptation of the Project Manual concept is for fast-track or phased construction projects, where each increment and each phase of the project is bound into separate Project Manuals.

Even simple projects, such as an individually constructed residence or a tenant space improvement project, can use the Project Manual concept. In *MasterFormat,* especially Division 00 - Procurement and Contracting Requirements, there are many topics than can be conveniently assembled and bound in book form using the Project Manual, including the "front end" documents such as the signed Agreement form, the General Conditions of the Contract, completed insurance and bond forms, contract modification documents (change orders and construction change directives), building permits, the schedule of values, and payment applications. No project is too small for a Project Manual.

SPECIFICATIONS SECTIONS

The Project Manual can be considered a book with many chapters. The chapters are grouped as described above. Some chapters contain introductory information, Bidding Requirements and Construction Contract requirements. Other chapters concern requirements for construction to be performed under the Contract. This last group is known as "construction specifications."

Information is divided into chapters known as "Sections." Similar Sections are grouped into Divisions for convenience. All of this is reflected in *MasterFormat,* where the specifier determines which identifier (Section number) and title to give to the Section. Using the commonly recognized Section numbers and titles in *MasterFormat* makes locating information easier because of the familiarity gained through use of specifications organized according to *MasterFormat.*

Because construction projects vary in complexity and scale, dividing the specifications into Sections should reflect the relative complexity and scale of the construction. That is, large, complex projects with many types of products requiring very detailed descriptions, and stringent requirements for quality assurance require lengthy specifications. Conversely, projects of simple construction and small scale can be addressed with succinct text. As construction technology and project delivery arrangements have become more sophisticated, construction specifications have become more voluminous in order to accommodate increasingly stringent requirements.

Given the many subcontractors and trade workers, the documents for construction contracts had to become more substantial. Specifications were no longer 20 or 40 pages but became, for commercial and institutional projects, hundreds or even thousands of pages long. Obviously, these voluminous specifications required organization. That is where CSI enters the picture and CSI's formats provide the required organization. Organizing and formatting the chapters or Sections of the specifications are discussed in detail in Chapter 5.

TECHNICAL SECTIONS

In the past, Sections of the specifications were considered to be trade sections. That is, each Section described the work that a specific trade was required to accomplish. A dictionary definition of "trade" is "some line of skilled manual or mechanical work; craft" and "people engaged in a particular line of business" (*Webster's Encyclopedic Unabridged Dictionary of the English Language,* Gramercy Books, New York, 1989).

A trade could be individuals or companies that perform portions of the Work of the construction Contract. "Trade" can therefore mean a craft, such as carpentry, bricklaying, or plumbing, or it can mean a business, such as a concrete subcontractor or a plumbing and heating subcontractor. When painting is considered, the "painter" may be an individual trade worker who is an employee of the general contractor or it may be a company that specializes in painting.

Specifications have reflected the increasing specialization and sophistication of construction, as described above. Yet there is a residual concept that construction specifications are divided into units of Work or trade sections, as though one trade or subcontractor could be responsible for all the Work described in the Section. This concept breaks down when the complexity of contemporary construction, especially for nonresidential projects, is considered.

Once, concrete work was the responsibility of the project's mason. Eventually, concrete work became separate from masonry work and was performed by a concrete subcontractor. Wood forms for concrete work were once specified under carpentry but are now specified under concrete work.

Doors have been traditionally made of wood and were once specified under the carpentry Section of the specifications. Wood doors and frames were considered Work to be fabricated and installed by carpenters or a millwork subcontractor. When tempered glass doors, metal doors, and bronze doors were introduced, carpenters claimed this work as being under their jurisdiction. As new methods of work developed, they were first performed by an existing trade but eventually came under the jurisdiction of a specialty subcontractor, and new skills for manufacture and installation developed. Doors have become sophisticated building elements that have fire ratings and decorative surfaces and are manufactured according to highly technical industry-recognized standards. Trade distinctions and jurisdictions became confused and irrelevant for organizing construction information in the specifications.

It is possible that the general contractor will purchase doors from a supplier and then employ carpenters who install them. Alternatively the general contractor can subcontract both furnishing and installation of the doors to a specialty subcontractor. Although finish carpenters, employed by the specialty subcontractor, might install door hardware in the field, preparation of the doors for hardware would likely be performed in a production shop using jigs and templates for mass production. Some or all of the door hardware might be installed in a factory or shop by workers who are not members of a construction trade. Doors are now specified not according to trade (carpentry) but by construction technology (wood doors).

The complexity of construction becomes more apparent when formerly simple products, such as concrete, are considered. Virtually every design discipline and many elements of a facility involve cast-in-place concrete. Concrete is used for building foundations and for slabs on grade. That is well understood. But storm drainage structures (civil engineering) and planter walls (landscape architecture) also use cast-in-place concrete. Portland cement concrete paving is distinct from building slabs on grade, although the products used are very similar. Plumbing systems use cast-in-place concrete for thrust blocks on underground pressure lines, and the plumbing systems must penetrate concrete elements to get into, out of, and through the building. Mechanical systems such as boilers and air conditioning units require cast-in-place concrete pads, as do electrical elements such as transformers, switchgear, and parking lot light standards. Specifications for each of these elements and systems could include concrete along with the various other technologies, but accepted practice is to keep portland cement concrete related to buildings and site structures in one place in the Specifications. But that would result in a great deal of redundancy and/or even conflicting requirements in the specifications.

This does not necessarily mean that the concrete subcontractor is responsible for providing concrete for all the elements identified. It is up to the general contractor—the only Contractor with whom the Owner has a Contract—to determine how each required portion of the Work shall be provided. The general contractor may contract with a concrete subcontractor to provide all the concrete for the project or the general contractor may require each subcontractor to make arrangements for the concrete work related to his or her building elements.

So why should specifications Sections not establish the scope of work for each trade or subcontrac-

tor? Beyond the reason that the design professional who writes the specifications is not licensed as a construction contractor to establish these scopes, there are several other contractual and practical reasons why scopes of work should not be established by the specifications.

The most important reason is that the General Conditions of the Contract state that the specifications do not establish trade or subcontract jurisdictions. AIA A201 - General Conditions of the Contract for Construction and EJCDC C-700 Standard General Conditions of the Contract for Construction, in Appendices E and F, respectively, state that specifications do not establish subcontract scopes or trade jurisdictions. Paragraph 1.2.2 of AIA A201 states, "Organization of the Specifications into divisions, sections and articles, and arrangements of Drawings shall not control the Contractor in dividing the Work among Subcontractors or in establishing the extent of Work to be performed by any trade."

Perhaps the next most important reason that scopes of work should not be established by the specifications is that design professionals are not trained or licensed to make such distinctions. Professional license examinations and legislative acts do not address subcontract and trade segregation as part of a design professional's services. Owner-Architect or Owner-Engineer Agreements similarly do not require these services. For the Architect or Engineer to do so involves substantial risk of error or omission and is not in the design professional's best interests, especially when there is a general contractor who is being paid by the Owner to do this service.

Another reason is that the general contractor should be given leeway to break down required work into subcontracts and trade arrangements. There are exceptions, such as when State contracting laws and regulations for publicly funded projects require segregation of portions of the work such as plumbing, HVAC, and electrical into separate prime contracts, where the design professional is required to break down the documents according to these technologies. As a general rule, this is not required. Certainly, the design professional does not want to become involved in trade disputes between the general contractor or a subcontractor and trade union.

Regional construction practices vary, and with projects being designed by most firms for construction in various regions of North America and even on other continents, it becomes burdensome for the design professional to research and adapt to these practices. The practices are important to consider, but the party most able to make and most affected by the arrangements is the general contractor. So, let the general contractor handle this matter. The design professional should focus on coherently describing building elements and systems so that each element or system can be clearly understood.

The concept of trade Sections further breaks down when there are portions of the Work that require several trades. A Section such as "Curtain-walls" specifies a complex assembly of extruded aluminum framing, steel reinforcement and supports, glass, glazing, insulation, and joint sealers. Each of these could be a separate trade Section, but it is now generally accepted that specifying the assembly rather than establishing trade responsibilities is most important. While it can be argued that the curtainwall subcontractor is a trade contractor, what are really being specified are technical requirements for the curtainwall and related administrative and contractual requirements.

Thus, current standard practice is for Sections to specify technical requirements for products and installation and to specify related administrative requirements such as submittals, quality control, and warranties.

SECTION SCOPE

For small-scale and less complex projects, Specifications can be organized into broad categories or "broadscope" sections. As the complexity of a project increases, specifications Sections becomes more narrow in scope. Commonly, these are called "mediumscope" and "narrowscope" Sections. The 1995 edition of CSI *MasterFormat* dropped the broadscope, mediumscope, and narrowscope concepts and used four "Levels" to define Sections. Level 1, representing the then 16 Divisions of *MasterFormat,* is the broadest category for Specifications. Level 2 is the equivalent of mediumscope categories, and Levels 3 and 4 are the increasingly narrowscope categories. The Levels seem to be arbitrary and difficult to comprehend, so most specifiers still refer to broadscope, mediumscope, and narrowscope to describe specifications Sections of narrowing focus. The 2004 edition of CSI *Master-Format,* with potentially 49 Divisions, similarly follows these Levels of Section assignment.

For example, all types of wood doors may be specified in a broadscope Section titled "Wood Doors." Mediumscope Sections may be prepared for flush wood doors and for stile and rail wood doors to distinguish their requirements; flush wood doors are typically factory-manufactured, while stile and rail doors may be fabricated in a local millwork shop. Narrowscope Sections may be prepared to distinguish flush hardwood veneer doors from flush plastic laminate-faced wood doors. This is a judgment call by the specifier and depends upon the volume of text necessary to adequately specify requirements for the work.

Another reason to use more narrowly focused Sections is the scale of construction. A small project, with a relatively small quantity of a type of work, may have minimal administrative and quality control requirements that do not require lengthy text. Conversely, a large project, with only a few types of work but large quantities of products and more competitive pricing, may require more voluminous text. For example, a common portion of the work, such as cast-in-place concrete, may be specified in a single broadscope Section for a one-story wood frame structure but would be more appropriately specified in several mediumscope Sections for a competitively bid public school. Mediumscope Sections separating formwork, reinforcement, and concrete would be appropriate. Another example would be distinguishing architectural-quality concrete from structural concrete by using narrowscope Sections for each.

Many design professionals tend to accommodate the practice of general contractors to break apart the specifications by Sections and issue the portions to prospective subcontractors. This may result in incomplete information being provided to subcontractors and materials suppliers. Since the specifier does not know how the general contractor will break down the work—and since each of several general contractors bidding for the same project may choose to do the breakdowns differently—it is not feasible for each Section of the specifications to align with trade practices or trade subcontractors.

The goal of the specifier should be to break the specifications into appropriately scaled units or Sections that are organized in accordance with the widely accepted CSI *MasterFormat* Section numbers and titles. Sections become units of information for specific portions of the construction that, when taken together, become the Specifications according to the General Conditions of the Contract. As AIA A201 states in Paragraph 1.1.6, "The Specifications are that portion of the Contract Documents consisting of written requirements for materials, equipment, systems, standards and workmanship for the Work, and performance of related services."

Formats and other discussion related to internal organization of Sections follow in the next chapter.

Chapter 5

Formats for Specifications Sections

NEED FOR FORMATS FOR SPECIFICATIONS SECTIONS

The arrangement of the subject matter in an orderly, comprehensive format within a technical section is important for several reasons. The specifier, when following a definitive procedure, is less likely to overlook any item. Similarly, the contractor, estimator, materials manufacturer, and inspector will then find the information much more easily in the individual section.

A Section in the Specifications can be considered analogous to a chapter in a book; the chapter, in turn, consists of paragraphs. The material that comprises the section consists essentially of paragraphs and subparagraphs. The use of other terms to describe the breakdown of the material within the technical section, such as articles, clauses, headings, categories, or units, can lead to confusion.

The typical Section contains two categories of paragraphs, technical and nontechnical, as follows:

Technical	Nontechnical
Materials	Scope of work
Fabrication	Delivery of materials
Workmanship	Samples and shop drawings
Installation	Permits
Tests	Guarantees
Schedules	Cleaning
Preparation	Job conditions

In the years prior to the advent of a nationally promulgated Section format, the typical Section was written to include the technical and nontechnical paragraphs in the order in which they occurred chronologically, that is, in a sequence in which the Contractor would ordinarily perform the Work, and each paragraph heading was simple and self-explanatory. When the specifier followed this course, it was less likely that things would be omitted, and reliance on checklists and notes diminished accordingly.

SECTIONFORMAT

In 1969, CSI developed and promulgated the *CSI SectionFormat.* This has since been refined and updated several times and is now jointly produced by CSI and Construction Specifications Canada (CSC). This nationally approved format provides guidelines for the arrangement of information within a technical Section of the specifications, and it offers a concise, orderly method for specifiers to follow. See Exhibit 5-1 for an interpretation of *SectionFormat,* with editing comments.

Prior to the publication of *SectionFormat,* specifiers arranged the information within their technical sections in accordance with their own formulas, and in many instances without any specific method. In many cases, the lack of organization resulted in duplication and omission of information.

SectionFormat is another important step toward providing a more unified approach. It permits easier access to information by manufacturers, contractors, and inspectors. It provides a checklist for the specifier so that omission of information is minimized. It provides standardization of input that permits its use in connection with computerized specifications and information retrieval.

The benefits of using *SectionFormat* are:

- It provides an industry-accepted standard for locating information within a specification Section.

- Consistent use of *SectionFormat* reduces the chance for omissions or duplications in a project specification.

- It assists the design professional by providing a format that facilitates coordination of project documentation with a Project Manual.

- It assists specification users by consistently locating similar information in the same place in each Section of the Specifications.

- It assists the specifier by providing a standard arrangement of Articles in three parts, suitable for use in master specification systems. In writing a new Section, the standardized format helps to identify what subjects to address.

SectionFormat provides for the arrangement and presentation of information under three separate parts. The part is considered the First Level of detail in a specification Section. The three parts are:

Part 1—General

Describes administrative, procedural and temporary requirements and is an extension of Division 1 - General Requirements unique to the section. It specifies the procedures for accomplishing portions of the Work specified in the Section and the relationships with other portions of the Work or requirements in the project.

Part 2—Products

Describes, in detail, the materials, products, equipment, systems, or assemblies that are required for incorporation into the Project. It specifies the products to be incorporated into the project as specified in the Section, including their offsite fabrication.

Part 3—Execution

Describes, in detail, preparatory actions, what on-site actions are required, and how the products shall be incorporated into the project.

Under each part, *SectionFormat* provides for several standard paragraph headings following a more or less regular sequence or order. Obviously, each paragraph heading may not be pertinent for every technical section and should not be used where not applicable. In addition, where paragraph heading titles would be forced, they should be retitled to be consonant with the work intended under a specific heading. For example, under Part 3 - Execution, *SectionFormat* lists a paragraph titled "Erection, Installation, Applications, Construction." This title should be edited to reflect the content of the Section. For example, "Erection" is applicable to a structural steel Section but "Application" is not. "Installation" is applicable to wood windows but "Erection" is not. For a Section such as earthwork, use an entirely different but appropriate term, such as "Excavation."

The article and paragraph headings also should take into account certain requirements normally used in civil engineering or heavy construction specifications and in mechanical and electrical specifications. For civil engineers and those in heavy construction involved in contracts based on unit prices, a paragraph titled "Measurement and Payment" is appropriate to include under the "Summary" article because of the way the construction contract is administered and payments to the contractor are made. For mechanical and electrical engineers, an article titled "System Description" is appropriate to fully describe an involved system prior to specifying its components and their assembly and installation. Specifications for balancing the heating and ventilating system or other mechanical and electrical items should include an article titled "Adjusting." For architects, an Article titled "System Startup" is appropriate for elevators but inappropriate for ceramic tile. Use professional judgment and only include appropriate article and paragraph headings from *SectionFormat*. The headings are guidelines only; there is no obligation to use all headings in every specification Section.

It is important to note that the paragraph and subparagraph headings are appropriate when the work of technical sections can be adequately and appropriately specified thereunder. Do not use these headings when they do not apply. Introduce new headings when applicable. Deviations are proper when awkwardness would result from too close an adherence to this rule.

The 1992 *SectionFormat* was updated in 1997 to reflect and coordinate with the 1995 *MasterFormat*. *SectionFormat* is expected to be revised and updated following the publication of the 2004 edition of *MasterFormat*.

SECTION #####

SECTION TITLE

PART 1 - GENERAL

1.1 SECTION INCLUDES
****

LIST GENERALLY THE PRODUCTS SPECIFIED IN THIS SECTION. COORDINATE WITH ARTICLE AND MAJOR PARAGRAPH HEADINGS IN PART 2 - PRODUCTS. AVOID BEING OVERLY DESCRIPTIVE.

THIS ARTICLE IS OPTIONAL ACCORDING TO CSI *MASTERFORMAT.* IT IS RECOMMENDED FOR USE ONLY AS A FAMILIARIZATION AID FOR THOSE USING THE PROJECT MANUAL.
****

A. [_Element_of_Work_].

B. [_Element_of_Work_].

1.2 RELATED SECTIONS
****

LIST ONLY THOSE SECTIONS WHERE THERE IS A DIRECT RELATIONSHIP WITH THE WORK SPECIFIED IN THIS SECTION. THAT IS, IDENTIFY WORK SPECIFIED IN ANOTHER SECTION FOR WHICH THE INFORMATION CONTAINED IN THAT SECTION IS NECESSARY FOR PROPER AND COMPLETE UNDERSTANDING OF THE WORK SPECIFIED IN THIS SECTION.
****

A. Section [_Number_]: [_Description_of_related_Work_].

B. Section [_Number_]: [_Description_of_related_Work_].

1.3 ALLOWANCES
****

COORDINATE THIS SECTION WITH GENERAL REQUIREMENTS SPECIFIED IN DIVISION 1 - GENERAL REQUIREMENTS AND WITH BIDDING REQUIREMENTS, PARTICULARLY SUPPLEMENTARY INSTRUCTIONS TO BIDDERS AND THE BID FORM.
****

A. Allowance: Include in [Contract Sum] [Bid] the amount of $[_Lump_Sum_Amount_] for [_Description_of_Element_of_Work_].

1.4 UNIT PRICES
****

COORDINATE THIS SECTION WITH GENERAL REQUIREMENTS SPECIFIED IN DIVISION 1 - GENERAL REQUIREMENTS AND WITH BIDDING REQUIREMENTS, PARTICULARLY SUPPLEMENTARY INSTRUCTIONS TO BIDDERS AND THE BID FORM.
****

A. Unit Price: [State in Bid] the amount per [_Unit_of_Measure_] for [_Description_of_Element_Work_].

1.5 ALTERNATES
****

COORDINATE THIS SECTION WITH GENERAL REQUIREMENTS SPECIFIED IN DIVISION 1 - GENERAL REQUIREMENTS AND WITH BIDDING REQUIREMENTS, PARTICULARLY SUPPLEMENTARY INSTRUCTIONS TO BIDDERS AND THE BID FORM.
****

A. Alternate Bid: The Work described in this Section is affected by [an Alternate Bid item.] [Alternate Bid items.]

B. Base Bid Condition: [_Description_of_the_Work_under_Base_Bid_].

Exhibit 5-1. Example section template.

C. Alternate Bid No. [_Number_per_Bid_Form_]: [_Description_of_the_Work_under_Alternate_Bid_item].

1.6 REFERENCES

****

LIST ONLY THOSE PUBLICATIONS REFERENCED IN THIS SECTION. AVOID LISTING ASTM, ANSI AND OTHER GENERALLY KNOWN REFERENCE STANDARDS. REFER TO SECTION 01420 - REFERENCES FOR GUIDANCE.

****

A. [_Publishing_Agency_]: [_Reference_Standard_].

B. [_Publishing_Agency_]:

1. [_Reference_Standard_].
2. [_Reference_Standard_].

1.7 DEFINITIONS

****

DEFINE UNIQUE AND HIGHLY TECHNICAL TERMS ACTUALLY USED IN THIS SECTION, OR TERMS WHOSE PRECISE DEFINITION IS ESSENTIAL TO UNDERSTANDING THE WORK SPECIFIED.

****

A. [_Term_]: [_Definition_].

1.8 PERFORMANCE REQUIREMENTS

****

DESCRIBE BELOW PERFORMANCE CRITERIA TO BE MET BY COMPLETED DESIGN/BUILD CONSTRUCTION.

****

A. [_Element_of_the_Work_]: [____].

1. [_Criterion_].
2. [_Criterion_].

1.9 SUBMITTALS

****

COORDINATE THE FOLLOWING WITH DIVISION 1 - GENERAL REQUIREMENTS AND, TYPICALLY, SECTION 01330 - SUBMITTALS PROCEDURES. IT IS NOT NECESSARY TO CROSS-REFERENCE TO SECTION 01330.

****

A. Product Data: Submit [catalog data.] [_Description_].

B. Shop Drawings: Submit fabrication and installation drawings indicating [_Description_]. Shop drawings shall indicate surrounding construction as provided for the Project.

C. Manufacturer's Samples: Submit [_Description_]. Samples will [not] be returned after review and may [not] be incorporated in the Work.

D. Design Data: Submit [calculations] [and] [or] [test reports] [, signed by registered engineer] [, certified by independent testing service,] for [_Element_of_the_Work_], demonstrating conformance with [the Contract Drawings and Specifications] [specified performance requirements] [and] [applicable Code requirements].

E. Test Reports: Submit [____].

F. Certificates: Submit [____] certifying that [_Description_].

G. Instructions: Submit manufacturer's instructions and recommendations for [assembly] [application] [installation] of [_Element_of_the_Work_].

H. Field Reports: Submit reports by [____] for [_Element_of_the_Work_]

Exhibit 5-1. *(Continued)*

I. Project Record Drawings: Indicate [_Element_of_the_Work_] on project record drawings. Refer to Section 01770 - Closeout Procedures.

J. Operation and Maintenance Data: Submit for [_Element_of_the_Work_]. Comply with general requirements of Section [01770 - Closeout Procedures] [01783 - Operation and Maintenance Data] [and requirements of] [Section 15050 - Basic Mechanical Materials and Methods] [and] [Section 16050—Basic Electrical Materials and Methods].

K. Warranty Documents: Submit for all manufactured units and equipment specified in this Section. Refer to Section [01770—Closeout Procedures] [01785 - Warranties and Bonds].

1.10 QUALITY ASSURANCE
****

DESCRIBE SPECIFIC REQUIREMENTS FOR QUALITY ASSURANCE MEASURES FOR WORK SPECIFIED IN THIS SECTION. SPECIFY SHOP OR FACTORY TESTS AND INSPECTIONS IN PART 2 - PRODUCTS AND SPECIFY FIELD TESTING AND INSPECTION ACTIVITIES IN PART 3 - EXECUTION.
****

A. Qualifications: [Contractor-employed designers] [manufacturer-employed designers] [manufacturers] [fabricators] [installers] [applicators] shall have a minimum of [3] [5] [—] years full time experience [producing] [executing] work of similar scope and complexity, [and shall be certified] [by the system manufacturer] [in accordance with] [____]. Refer to Section 01450—Quality Control.

B. Regulatory Requirements: Regulatory Requirements, Comply with specific requirements of [____]. Refer to Section 01410 - Regulatory Requirements.

C. Certifications: [Applicator] [Installer] [Fabricator] [____] shall be certified [by the manufacturer] [by an independent testing service] to meet or exceed the minimum requirements specified herein.

D. Field Samples: Prepare field samples of [_Element_of_the_Work_] for [review] [and] [selection] by the [Architect] [Owner] [____] of [range of] [color] [texture] [and] [finish]. Locate field samples at [____]. Approved sample[s] shall establish standards by which the Work will be judged. Note location of field samples on project record drawings.

E. Mock-Ups: Construct full-size [working] mock-up[s] of [____] for review and approval by [Architect] [Owner] [____], showing [operation] [construction] [coordination and interface with adjoining Work]. Construct mock-ups at [____]. Approved mock-up[s] shall serve to establish standards by which the Work will be judged. Remove mock-up[s] only after Work is substantially complete and with approval of [Architect] [Owner] [____].

F. [Pre-Installation] [Pre-Application] Conference: Convene a conference at [the project site] [the Architect's office] [____], [7] [10] [____] days prior to starting [installation] [application], to review the Drawings and Specification, the reviewed submittals, [field samples,] [mock-ups], manufacturer's instructions and recommendations, sequencing and interface considerations and project conditions. Conference shall be attended by supervisory, [installation] [fabrication] [application] and quality control personnel of Contractor and all subcontractors performing this and directly related work. [Construction Manager] [Architect] [Owner] [____] will attend the conference.

1.11 DELIVERY, STORAGE AND HANDLING
****

DESCRIBE BELOW SPECIAL PROVISIONS FOR PACKING AND SHIPPING PRODUCTS SPECIFIED IN THIS SECTION.
****

A. Packing and Shipping: [____].
****

DESCRIBE BELOW SPECIAL PROVISIONS FOR ACCEPTANCE AT PROJECT SITE OF PRODUCTS SPECIFIED IN THIS SECTION.
****

B. Acceptance at Site: [____].
****

Exhibit 5-1. *(Continued)*

DESCRIBE BELOW SPECIAL PROVISIONS FOR STORAGE AND PROTECTION OF PRODUCTS SPECIFIED IN THIS SECTION.
****

C. Storage and Protection: [____].

1.12   PROJECT CONDITIONS

A. Environmental Requirements: Comply with environmental requirements and recommendations of manufacturer for proper [installation] [application] [curing] of products.

B. Temperature Criteria: Do not [install] [apply] [_Element_of_the_Work_] unless temperature is [_Criteria_].

C. Wind and Weather Criteria: Do not [install] [apply] [_Element_of_the_Work_] unless weather is [_Criteria_].

D. Field Measurements and Conditions: In addition to provisions of the Conditions of the Contract, verify dimensions and obtain field measurements prior to producing shop drawings and ordering products. Verify field conditions and condition of substrate and adjoining Work before proceding with Work specified in this Section.

1.13   SEQUENCING AND SCHEDULING
****

THIS ARTICLE IS RARELY USED. IT SHOULD BE USED WHEN THERE ARE SPECIFIC SEQUENCING AND SCHEDULING REQUIREMENTS, SUCH AS COMPLETING A PORTION OF THE BUILDING BY A CERTAIN DATE TO ALLOW FOR OWNER'S USE FOR WORK UNDER SEPARATE CONTRACT (INSTALLATION OF SPECIALIZED EQUIPMENT).
****

A. Sequencing and Scheduling, General: Refer to sequence requirements specified in Section 01100—Summary of the Project and construction progress schedule requirements specified in Section 01330 - Submittals Procedures.

B. [Sequence for [_Element_of_Work_] [Completion Schedule for [_Element_of_Work_]: Additionally, coordinate Work specified in this Section with Work specified in Section [____]—[____] [and Section][____]—[____] to properly interface the various elements.

1.14   WARRANTY
****

SPECIFY ONLY THOSE REQUIREMENTS WHICH EXCEED THE CONTRACTUAL OR STATUTORY ONE YEAR WARRANTY FROM THE CONTRACTOR. BE CAREFUL ABOUT COPYING MANUFACTURER'S WARRANTY EXCLUSIONS INTO THE CONTRACT. COORDINATE WITH SECTION 01785 - WARRANTIES AND BONDS. SEE FORMS FOLLOWING SECTION 01785.
****

A. Manufacturer's Guarantee: [_Period_], [_Conditions_].

B. [Applicator's] [Installer's] Warranty: [_Period_], [_Conditions_].

C. Warranty Bond: [_Amount_], [_Conditions_].

1.15   MAINTENANCE
****

THIS ARTICLE IS USED ONLY FOR UNIQUE SITUATIONS WHERE THE CONTRACTOR OR INSTALLER WILL ENTER INTO A SEPARATE CONTRACT WITH THE OWNER FOR MAINTENANCE SERVICE FOLLOWING SUBSTANTIAL COMPLETION OF THE WORK SPECIFIED. EXAMPLES: LANDSCAPE MAINTENANCE, ELEVATOR MAINTENANCE AND BOILER SERVICE AND MAINTENANCE.
****

A. Maintenance Service: Provide a maintenance service contract, paid in advance, covering [_Element_of_the_Work_] for a period of [_Time_] from [Substantial Completion] [Acceptance] of the Work. Such service shall be in addition to warranty service otherwise covered by the Contract and shall include [all parts and labor] [all parts, labor and consumables].

Exhibit 5-1. *(Continued)*

B. Extra Materials: Provide [_Product_] in [the amount of [_Quantity_] [_Unit_of_Measure_],] [an amount equal to [_Number_] percent of [_Element_of_Work_],] delivered to [_Person_and_Location_].

PART 2 - PRODUCTS

2.1 MANUFACTURERS

****

THIS ARTICLE IS USED ONLY WHEN THERE IS A SINGLE MANUFACTURER FOR ALL PRODUCTS SPECIFIED IN THIS SECTION. OTHERWISE, "SPECIFIED MANUFACTURER" AND "ACCEPTABLE MANUFACTURERS" ARE SPECIFIED IN EACH ARTICLE OF PART 2 - PRODUCTS, WHERE PRODUCTS ARE SPECIFIED.

THE FOLLOWING IS BASED ON USING EITHER THE "OPEN" OR "CLOSED" PROPRIETARY METHOD OF SPEC-IFYING. IF A DESCRIPTIVE, REFERENCE OR PERFORMANCE METHOD IS USED, THIS ARTICLE MAY BE DELETED.

****

A. Specified Manufacturer: [_Firm_Name_], [_City_], [_State_] ([_Telephone_No._]; local representative [_Firm_Name_], [_City_], [_State_] ([_Telephone_No._]).

B. Acceptable Manufacturers: [None identified.] [Alternate manufacturers will be considered in accordance "or equal" provision specified in Section 01620—Product Options.] [No substitutions will be considered.]

   1. [_Firm_Name_], [_City_], [_State_] ([_Telephone_No._]; local representative [_Firm_Name_], [_City_], [_State_] ([_Telephone_No._]).
   2. [_Firm_Name_], [_City_], [_State_] ([_Telephone_No._]; local representative [_Firm_Name_], [_City_], [_State_] ([_Telephone_No._]).

2.2 MATERIALS

****

SPECIFY IN THIS ARTICLE THE BASIC MATERIALS USED FOR EITHER FIELD FABRICATION OR SHOP OR FACTORY MANUFACTURE. TYPICALLY, THE PRODUCTS ARE SPECIFIED BY REFERENCE TO INDUSTRY STANDARD, SUCH AS AN ASTM STANDARD.

****

A. [_Product_]: [_Description_].

   1. [_Element_]: [_Description_].
   2. [_Element_]: [_Description_].

2.3 [MANUFACTURED UNITS] [EQUIPMENT] [COMPONENTS] [_Element_of_Work_]

****

SPECIFY IN THIS ARTICLE SHOP-FABRICATED OR FACTORY-MANUFACTURED PRODUCTS. PRODUCTS ARE TYPICALLY SPECIFIED BY IDENTIFYING THE MANUFACTURER, THE PRODUCT NAME AND CATALOG NUM-BER OR OTHER REFERENCE INDICATION.

****

A. Specified Product: [_____].

****

THE FOLLOWING IS RARELY USED BUT SHOULD BE USED WHERE THERE ARE NO TRUE EQUALS BY OTHER MANUFACTURERS.

****

B. Acceptable Products: [_____].

C. [_Attribute_]: [_Description_].

D. [_Attribute_]: [_Description_].

E. [_____].

Exhibit 5-1. *(Continued)*

2.4 ACCESSORIES

****

SPECIFY ACCESSORY PRODUCTS EITHER BY THE SPECIFIED MANUFACTURER OF THE BASIC PRODUCT OR BY OTHER MANUFACTURERS.

****

A. [_Product_]: [_Description_].

   1. [_Attribute_]: [_Description_].
   2. [_Attribute_]: [_Description_].

B. [_Product_]: [_Description_].

2.5 MIXES

****

SPECIFY MIXES (FORMULAS) TO BE USED FOR FIELD, SHOP OR FACTORY USE IN PERFORMING THE WORK. EXAMPLE: CONCRETE MIX.

****

A. [____].

2.6 FABRICATION

****

SPECIFY SHOP OR FACTORY FABRICATION. FIELD FABRICATION IS SPECIFIED IN PART 3.

****

A. Shop Assembly: [____].

B. Shop/Factory Finishing: [____].

C. Tolerances: [____].

2.7 SOURCE QUALITY CONTROL

****

SPECIFY TESTING AND INSPECTION ACTIVITIES TO BE PERFORMED DURING SHOP OR FACTORY FABRICATION.

****

A. Tests: [____].

B. Inspection: [____].

C. Verification of Performance: [____].

PART 3 - EXECUTION

3.1 EXAMINATION

****

THIS ARTICLE IS INCLUDED IN CSI MASTERFORMAT AND IS USED BY AIA MASTERSPEC. IT IS RECOMMENDED THAT IT GENERALLY *NOT* BE USED SINCE IT IS SO CLOSELY RELATED TO DIRECTING THE MEANS, METHODS, TECHNIQUES AND SEQUENCES OF CONSTRUCTION. EXCEPTIONS COULD BE SUCH MATTERS AS TESTING MOISTURE CONTENT OF SUBSTRATE AND REPORTING THE RESULT BEFORE ADHERING FINISH MATERIALS. COORDINATION OF BACKING AND BLOCKING PROVISIONS IS AN EXAMPLE OF WHAT NOT TO INCLUDE, SINCE SUCH COORDINATION IS COVERED IN DIVISION 1 - GENERAL REQUIREMENTS.

****

A. Examine Project conditions and completed Work and verify that [____].

B. Immediately correct all deficiencies and conditions which would cause improper execution of Work specified in this Section and subsequent Work.

Exhibit 5-1. *(Continued)*

C. Proceeding with Work specified in this Section shall be interpreted to mean that all conditions were determined to be acceptable prior to start of Work.

3.2 PREPARATION

****

SPECIFY ACTIVITIES IN PREPARATION FOR ERECTION, APPLICATION OR INSTALLATION OF PRODUCTS.

****

A. Protection: [____].

B. [Surface] [Substrate] Preparation: [____].

3.3 [ERECTION] [APPLICATION] [INSTALLATION]

****

SPECIFY REQUIREMENTS FOR ERECTION, APPLICATION OR INSTALLATION OF PRODUCTS. IF REQUIRE-MENTS ARE BASICALLY "IN ACCORDANCE WITH MANUFACTURER'S INSTRUCTIONS AND RECOMMENDA-TIONS," THEN SIMPLY STATE SO AND DO NOT REPEAT SUCH INSTRUCTIONS AND RECOMMENDATIONS. IF THERE ARE SPECIAL CONSIDERATIONS BEYOND WHAT THE MANUFACTURER REQUIRES OR RECOM-MENDS, THEN DESCRIBE THE REQUIREMENTS.

AVOID PRESCRIBING THE MEANS, METHODS, TECHNIQUES AND SEQUENCES OF CONSTRUCTION. DESCRIBE THE END RESULT, NOT HOW TO ACHIEVE THE RESULT.

****

A. [____].

B. [____].

C. Interface with Other Products: [____].

D. Tolerances: [____].

3.4 FIELD QUALITY CONTROL

****

DESCRIBE TESTS AND INSPECTIONS.

****

A. Field Testing and Inspection: Field inspection [and testing] will be performed as specified in Section 01450 - Quality Control.

    1. [_Test_or_Inspection_]: [_Description_].
    2. [_Test_or_Inspection_]: [_Description_].

B. Corrective Actions: Replace or repair Work to eliminate defects, deficiencies and irregularities.

3.5 MANUFACTURER'S FIELD SERVICES

A. Manufacturer's Field Services: Provide field [instruction] [inspection] services by manufacturer or authorized agent of the manufacturer in accordance with general requirements specified in Section 01610—Basic Product Requirements.

B. Schedule: Schedule site attendance by [____] manufacturer during execution of the Work.

C. Reports: Submit written reports [and certification] by manufacturer that [____] has been completed in accordance with the manufacturer's instructions and recommendations.

3.6 [ADJUSTMENT AND CLEANING

A. Labels and Coverings: Remove all labels and protective coverings from completed Work.

Exhibit 5-1. *(Continued)*

B. Adjustment: Check operation of functioning components and make adjustments for proper operation [within parameters of manufacturer] [and the Contract Drawings and Specifications]. [Refer to general requirements specified in Section [01610 - Basic Product Requirements] [01750—Starting and Adjusting]].

C. Cleaning: Thoroughly clean the Work specified in this Section and adjoining surfaces and areas affected by [application] [installation].

    1. [_Cleaning_Action_]: [_Description_].
    2. [_Cleaning_Action_]: [_Description_].

3.7 DEMONSTRATION

A. Demonstration, General: Refer to general requirements specified in Section [[01610 - Basic Product Requirements] [01750 - Starting and Adjusting]].

B. [_Equipment_] [_System_] Demonstration: Demonstrate [to] [Owner] [Architect] [Construction Manager] [____] that [equipment] [system] properly functions.

C. [____].

3.8 PROTECTION

A. Protection, General: In addition to general requirements specified in Section [01500—Temporary Facilities and Controls,] [01525—Construction Aids,] [Section 01530 - Barriers and Enclosures,] Section 01610 - Basic Product Requirements and Section 01770 - Closeout Procedures, comply also with the following requirements.

    1. [_Protective_Action_]: [_Description_].
    2. [_Protective_Action_]: [_Description_].

B. Maintenance of Protective Measures: Maintain protective devices until Work is [ready for Substantial Completion review.] [ready for Acceptance] [accepted].

C. Removal of Protective Measures: [Unless otherwise directed, remove protective devices [and complete final cleaning] [for Substantial Completion review.] [upon Acceptance of the Work].] [Protective devices will be removed by Owner.]

3.9 SCHEDULE
****

SCHEDULE PRODUCTS SPECIFIED IN THIS SECTION WHEN SCHEDULE FORMAT IS THE MOST EFFECTIVE MANNER FOR DESCRIPTION OF WORK ATTRIBUTES. THE FOLLOWING IS A GENERAL EXAMPLE ONLY.
****

A. [_System_] Fixtures:

[Element]	[Material]	[Finish]
[_Fixture_] Type 1	[_Material_A_]	[_Finish_1_]
[_Fixture_] Type 2	[_Material_A_]	[_Finish_2_]
[_Fixture_] Type 3	[_Material_B_]	[_Finish_3_]

END OF SECTION

Exhibit 5-1. *(Continued)*

PAGEFORMAT

Accompanying and closely related to *SectionFormat,* CSI *PageFormat* is a standardized presentation of text for each page of a specification Section, providing an orderly and uniform arrangement of the Articles, Paragraphs, and Subparagraphs.

- *SectionFormat* addresses the content of the Articles, Paragraphs, and Subparagraphs.
- *PageFormat* addresses the numbering of Articles, Paragraphs, and Subparagraphs.
- *PageFormat* addresses the physical arrangement on the page, such as margins, indents, headers, and footers.

The benefits of using *PageFormat* are:

- Text is presented clearly and at a density best suited for easy reading without obscuring the message or hindering rapid understanding.
- It is suitable for use in construction specifications of all types and sizes.
- It is suitable for use with most current production methods (word processing and computer-assisted specifications).

PageFormat Levels

PART 1—GENERAL (First Level)
  1.01 ARTICLE (Second Level)
    A. Paragraph (Third Level)
      1. Subparagraph (Fourth Level)
        a. Subparagraph (Fifth Level)
          1) Subparagraph (Sixth Level)

Article Numbering

Article numbers are the Second Level of detail in a specifications Section. The Article Number = PART Number + Consecutive Number. Examples:

1.04 SUBMITTALS: Article in PART 1 specifying submittals

2.01 MANUFACTURERS: Article in PART 2 specifying acceptable manufacturers

3.09 FIELD QUALITY CONTROL: Article in PART 3 specifying testing and inspections to be performed at the project site

2.11 SOURCE QUALITY CONTROL: Article in PART 2 specifying testing and inspection to be performed at on off-site production facility

1.05 QUALITY ASSURANCE: Administrative requirements governing source and field quality control activities, including field samples, mock-ups and pre-installation or pre-application conference to be convened at the project site

Paragraph Numbering

Paragraphs are the Third Level of detail. Paragraphs are the level immediately under an Article. According to *PageFormat,* Paragraphs use uppercase letters followed by a period, such as "A."

Subparagraphs are the Fourth Level of detail. Subparagraphs are all the levels under a Paragraph level. Fourth Level and all subsequent levels of detail are referred to as "subparagraphs." There are not sub-subparagraphs and sub-sub-subparagraphs.

- Fourth Level subparagraphs use an Arabic numeral followed by a period, such as "1."
- Fifth Level subparagraphs use a lowercase letter followed by a period, such as "a."
- Sixth Level subparagraphs use an Arabic numeral followed by a parenthesis, such as "2)." *(Avoid use!)*
- Seventh Level subparagraphs use a lowercase letter followed by a parenthesis, such as "d)." *(Avoid use!)*

Subsequent Subparagraph levels use lowercase letters surrounded by parentheses, Arabic numerals surrounded by parentheses, lowercase Roman numerals followed by a period (such as "iii.") and bullets such as +, –, •, ■ and □. *(Avoid use!)*

There is a great deal of concern about how to match Article numbering with word processing programs. The "leading zero" (1.03) is difficult to do with word processing programs when using the automatic paragraph numbering function. This automatic numbering function is considered by most specifiers to provide significant productivity benefit. Those who wish to hold zealously to the leading zero go to great lengths to retain it. Master guide specification publishers, such as ARCOM for *Masterspec* and CSRF for *SpecText,* deal with the problem by dropping the leading zero. Thus, "2.4" is used rather than "2.04," and when the numbering reaches "2.9" the next number is "2.10."

Another concern, dating back to when specifications were produced on typewriters, is text enhancement, that is, doing **bolding,** <u>underlining</u> and *italicizing* of text and even changing the type font and font size within a Section. Text enhancements can add to the readability and visual interest of the specifications. A note of caution for specifiers

---

FORM LETTER

FOR CONTRACTOR'S / SUBCONTRACTOR'S / MANUFACTURER'S WARRANTY
CONTRACTOR'S/SUBCONTRACTOR'S/SUPPLIER'S LETTERHEAD

SPECIAL LIMITED PROJECT WARRANTY FOR _____ WORK.

We, the undersigned, do hereby warrant that the portion of the Work described above which we have provided for [_**PROJECT_NAME**_], [_**City**_], [_**State**_] is in accordance with the Contract Documents and that all such Work as installed will fulfill or exceed all minimum warranty requirements. We agree to repair or replace Work installed by us, together with any adjacent Work which is displaced or damaged by so doing, that proves to be defective in workmanship, material, or function within a period of (*years*), commencing (*date identified in Notice of Completion, unless otherwise directed*) and terminating (*date*).

The following terms and conditions apply to this warranty (*obtain Owner's approval before submission*):

In the event of our failure to comply with the above-mentioned conditions within a reasonable time period determined by the Owner, after notification in writing, we, the undersigned, all collectively and separately, hereby authorize the Owner to have said defective Work repaired or replaced to be made good, and agree to pay to the Owner upon demand all moneys that the Owner may expend in making good said defective Work, including all collection costs and reasonable attorney fees.

Local Representative: For warranty maintenance, repair, or replacement service, contact:

(*Name*) _____

(*Address*) _____

(*City*) _____ (*State*) _____ (*ZIP*) _____

(*Phone*) _____ / _____

_____ (*signed*)   _____ (*signed*)

(*Date*) _____   (*Date*) _____

(*Typed Name*) _____   (*Typed Name*) _____

(*Title*) _____   (*Title*) _____

(*Firm*) _____   (*Firm*) _____
(*Installer, applicator, manufacturer or supplier*)   (*Contractor*)

State License No: _____

Exhibit 5-2. Example of Special Warranty Forms.

FORM LETTER

FOR CONTRACTOR'S / MANUFACTURER'S GUARANTY
CONTRACTOR'S / MANUFACTURER'S LETTERHEAD

SPECIAL LIMITED PROJECT [WARRANTY] [GUARANTY] FOR _____ WORK.

We, the undersigned, do hereby [warrant] [guaranty] that the portion of the Work described above which [we have provided] [was provided by (*Installer's or Subcontractor's Name*) for [_**PROJECT_NAME**_], [_**City**_], [_**State**_] is in accordance with the Contract Documents and that all such Work as installed will fulfill or exceed all minimum warranty requirements. We agree to repair or replace Work installed by [us,] [(*Installer's or Subcontractor's Name*)] together with any adjacent Work which is displaced or damaged by so doing, that proves to be defective in workmanship, material, or function within a period of (*years*), commencing (*date indicated in Notice of Completion, unless otherwise directed*) and terminating (*date*).

The following terms and conditions apply to this [warranty] [guaranty] (*obtain Owner's approval before submission*):

In the event of our failure to comply with the above-mentioned conditions within a reasonable time period determined by the Owner, after notification in writing, we, the undersigned, all collectively and separately, hereby authorize the Owner to have said defective Work repaired or replaced to be made good, and agree to pay to the Owner upon demand all moneys that the Owner may expend in making good said defective Work, including all collection costs and reasonable attorney fees.

Local Representative: For warranty maintenance, repair, or replacement service, contact:

(*Name*) _____

(*Address*) _____

(*City*) _____ (*State*) ____ (*ZIP*) _____

(*Phone*) ____ / _____

_____ (*signed*) _____ (*signed*)

(*Date*) _____ (*Date*) _____

(*Typed Name*) _____ (*Typed Name*) _____

(*Title*) _____ (*Title*) _____

(*Firm*) _____ (*Firm*) _____
(*Installer, applicator, manufacturer or supplier*) (*Contractor*)

State License No: _____

Exhibit 5-2. *(Continued)*

in overall charge of production of Project Manuals: there are usually several contributors to the set of specifications, and the more elaborate the format, the more difficult it can be to follow and match the format. Consulting mechanical and electrical engineers have considerable difficulty and must spend substantial time to match the various section and page formats of specifiers in a community.

Page Layout

Be adaptable and be creative in the page layout. Just as the Drawings express excellence in graphic design or are intentionally restrained graphically, for the specification Section consider the capabilities of word processing programs and go beyond the "typewriter" look of the 1960s and 1970s. Follow conventions, and for production expedience consider simple formats such as:

- Top and Bottom Margins: ½ inch; check printer limitations.
- Left and Right Margins: 1 inch (bound edge could be 1¼ inches and unbound edge could be ¾ inch); check word processing program for automatic features.
- Section Identification: Section title at the beginning of a Section and project-specific

information (see Exhibit 5-1) and section title identification on each page.

- Page Number: Five-digit *MasterFormat* number followed by consecutive numbers; conventionally located at the bottom center of each page, but alternatives are worth considering (see Exhibit 5-1).
- Section Date: Not considered in *PageFormat* but potentially very important.
- Indentations: Conventions are based on fixed-space fonts; adapt them to suit the proportional fonts commonly used with laser printers.
- Justification: Alignment of text along margins. Left margins are always justified. Justifying right margins looks professional, especially with a proportional font.
- Fonts and Font Sizes:
  - Use 10 point or 12 point, depending upon legibility.
  - The default or generic spec font is `Courier` 10 cpi (characters per inch).
  - A serif font is more legible but may be ugly.
  - A **sans serif** font is more contemporary but less readable.
  - Common fixed fonts: Courier and Letter Gothic.
  - Common proportional fonts: Times Roman and Arial/Universal/Swiss/Helvetica.

Chapter 6

Types of Specifications

METHODS FOR WRITING SPECIFICATIONS

After products have been selected to be specified, how is information about the product presented in the specifications? In Chapter 7 the subject of specifications language is addressed. Here the discussion will be about four methods for specifying products, followed by a discussion of standardized formats for specifications sections.

Four Methods of Specifying

The CSI *Manual of Practice* (The Construction Specifications Institute, Alexandria, Virginia, 1996) describes four methods of specifying, with some supplements and variations.

- *Descriptive Specifying:* Under this method, exact properties of materials and methods of installation are described in detail without using proprietary names (manufacturers' trade names).
- *Reference Standard Specifying:* Under this method, reference is made to established standards to which the specified products and processes shall comply or conform.
- *Proprietary Specifying:* Under this method, actual brand names, model numbers, and other proprietary information is specified.

- *Performance Specifying:* Under this method, required results are specified and the criteria are specified by which the performance will be verified. The Contractor is free to provide any material complying with the performance criteria.

There is a role for each of these methods in the production of typical construction specifications.

Descriptive Specifications (Exhibit 6-1)

Advantages

- Descriptive Specifications specify exactly what the design intends. If the storefront framing shall be 2 inches wide by 4½ inches deep, that is what is described. If the paint shall be "champagne metallic," then that is what is described. If the toilet seat shall have an open-front design, that is what is described.
- Descriptive specifications are applicable to all conditions, methods or situations of a project. If the project is a modernization project or new construction, or environmentally sensitive or insensitive, descriptive specifications may be used.
- Descriptive specifications permit free competition. By their very nature, descriptive specifications do not restrict the Contractor to using specific products of specific manufacturers

---

G. Gas Line Pressure Regulators: Single-stage, steel-jacketed, corrosion-resistant gas pressure regulators; with atmospheric vent, elevation compensator; with threaded ends for 2 inches and smaller, flanged ends for 2½ inches and larger; for inlet and outlet gas pressures, specific gravity, and volume flow indicated. Pressure regulators shall be suitable for outdoor installation.

B. Aggregates for Regular Weight Concrete: Fine and coarse aggregates, ASTM C 33 and as follows.

   1. Structural Concrete: Maximum size not larger than ¼ of narrowest dimension between forms, ⅓ depth of slab nor ¾ of minimum clear spacing between individual reinforcing bars; maximum aggregate size shall be ¾ inch.

   2. Other than Structural Concrete: Conform to requirements for structural concrete except maximum aggregate for concrete fill and topping shall be ⅜ inch maximum and for mass concrete shall be 1 inch.

Exhibit 6-1. Examples of descriptive specifications text.

(that is, unless the description is warped to prevent use of competing products).

- Descriptive Specifications are acceptable to all kinds of projects. This method may be used for small or large projects, for public and private projects, and for elaborate elements and for simple elements.
- Descriptive specifications provide a good basis for bidding. The description supposedly makes it clear what the result of the work shall be. For example, a description that states "¾-inch-thick basecoat" is clear, while a description that states "as necessary as a basecoat for finish coats" is not.

Disadvantages

- Descriptive specifications require the specifier to take special care in describing the design intent in order to achieve the intended results. Careful consultation and "wordsmithing" are necessary to derive a complete and understandable description.
- Descriptive Specifications tend to "bulk up" specifications with more verbiage about products than other specification methods.
- Descriptive specifications are more time-consuming to produce. They require more time and care than other methods. Leaving out important information may cause misunderstanding and misinterpretation, with unsatisfactory results.
- Descriptive specifications may require more quality control efforts. Each attribute should be verified during construction by quality assurance agents for the Owner and quality control staff for the Contractor. The Contractor must monitor the products and procedures of suppliers and subcontractors. A preengineered, proprietary elevator system requires less construction monitoring than a custom-engineered system specified by detailed descriptions of components.
- Descriptive specifications may be too elaborate for minor construction or a simple project. Consider the descriptions of gypsum board finishes in the applicable reference standard. Writing lengthy descriptions in the contract specifications can be avoided by referencing the appropriate industry standard.
- Descriptive specifications may be ambiguous. The Contractor must interpret specifications to identify and procure products available in the marketplace. The Architect/Engineer must interpret specifications when evaluating products proposed by the Contractor. It may not be the specifier who does the interpreting but a construction contract administrator who has not participated in making the design decisions and who does not know which described attributes are essential.
- Descriptive specifications are being used less often as more complete reference standards are being developed and implemented. In such cases, the reference standard, rather than the contract Specifications, describes the product.

Common Uses

Descriptive specifications are appropriate for products for which no standards exist, for products for projects where administrative restrictions prohibit identifying proprietary products, and for situations where the architect/engineer wants to exercise tight control over the specified work.

Basic Steps in Production of Descriptive Specifications

- Research available products.
- Research critical features needed.
- Determine which features to describe on drawings and which to specify ("a picture is worth a thousand words").
- Describe critical features ("lowest common denominator"?).
- Specify quality assurance measures to ensure that products comply with specifications (i.e., submittals, certifications and testing and inspection activities).

Reference Standard Specifications

Authors of Reference Standards

Many nonprofit trade associations of building product manufacturers publish reference standards. These associations are extremely knowledgeable about a particular aspect of construction technology, and they have established nonprofit associations dedicated to the promulgation of standards, among other mutual-benefit activities for their portion of the construction industry.

Examples

AWI: Architectural Woodwork Institute

BHMA: Builders Hardware Manufacturers Association

ARMA: Asphalt Roofing Manufacturers Association

METHODS FOR WRITING SPECIFICATIONS

NRCA: National Roofing Contractors Association

WSRCA: Western States Roofing Contractors Association

NEMA: National Electrical Manufacturers Association

NECA: National Electrical Contractors Association

Publishers of Standards

Trade associations, government and institutional organizations publish standards to bring diverse standards within the construction industry under a national oversight organization, such as the Steel Door Institute (SDI). Metal door and door frame standards are jointly published by SDI and the American National Standards Institute (ANSI) as an ANSI/SDI standard. Joint ANSI/ASTM standards are also published.

Examples

ASTM: ASTM International (formerly American Society for Testing and Materials)

ANSI: American National Standards Institute

Types of Reference Standard Specifications

- Basic material standard: ASTM C 33—Portland Cement
- Product standard: PS-1 - Plywood
- Design standard: SMACNA - *Architectural Sheet Metal Manual*
- Workmanship standard: Woodwork Institute - *Manual of Millwork*
- Test method standard: ASTM E 84 - Steiner Tunnel Test (flammability)
- Qualifications standard: AWS D1.1 - Welders Qualification
- Codes: *ICC Evaluation Service Research Reports*

Advantages of Reference Standard Specifications

- Widely known and accepted base for the Contractor and Owner
- Widely used materials and methods readily recognized
- Competition not limited
- Shortens specifications dramatically

Disadvantages of Reference Standard Specifications

There may be no appropriate standard to reference. Reference standards are generally available only for commonly used products.

- Referenced standards may be obsolete. Obsolescence is a problem because some standards organizations do not update regularly to keep up with developing technology.
- Reference standard specification may mean specifying to the lowest common denominator. Standards are typically consensus documents and, in order to serve the interests of the greatest number of members of the publishing organization or faction of the industry, may be more lenient than the specifier intends.
- Reference standard specifications requires research and care in use. Remember, industry associations develop and promulgate standards primarily for their own benefit.
- Reference standards must be incorporated properly, including all supplementary information.

Incorrect: "Zinc-Coated (Galvanized) Steel Sheet: ASTM A 653/A 653M." This is incomplete. The coating designation also must be specified.

Correct: "Zinc-Coated (Galvanized) Steel Sheet: ASTM A 653/A 653M, G90 (Z275) coating designation."

- Reference standard specifications are more difficult to enforce. Using them is like playing a game where the design intent is encoded into an industry standard and then decoded by the Contractor for procurement. The Contractor must interpret the specifications to identify and procure complying products available in the marketplace. Information is submitted for review by the Architect/Engineer, who must evaluate whether anything has been lost in translation and determine that the proposed product complies with both the referenced standard and the design intent.
- Reference standard specifications may require more quality control effort for the Contractor in identifying and monitoring products being provided by suppliers and subcontractors.
- Reference standard specifications may appear to be too elaborate for a minor or simple project.

Common Uses of Reference Standard Specifications

Reference standard specifications (Exhibit 6-2) are used for "commodity" products in the marketplace, where the manufacturer and brand name are not important. They are also used for products for projects where administrative restrictions prohibit identification of proprietary products.

B. Portland Cement: ASTM C 150, type as indicated on Structural Drawings for structural concrete or Type I or Type II if not indicated. For Architectural and other concrete work, provide Type II, gray color.

D. Fire-Resistive Gypsum Board: ASTM C 36, Type X (special fire-resistant), typically 48-inches wide and ⅝-inch thick (½-inch thick, proprietary fire-resistive where indicated), square cut ends, tapered sides.

H. Shop Primer: SSPC Paint 13, standard color.

Exhibit 6-2. Examples of reference standards specifications text.

Basic Steps in Production of Reference Standard Specifications:

-   Standard must be recognized as authoritative in the industry.
-   Standard must be available to all parties concerned.
-   Specifier should know the standard:
    -   Bad reference standard = bad specification.
    -   Duplication and conflict of information due to mixing of descriptive or proprietary methods with reference specifying method.
    -   Address all choices.
    -   Standards may refer to a particular trade or subcontractor.
    -   Standards may be multiple, referring to other standards for testing and materials. Conflicts and confusion can result.
-   Establish the edition date of the standard: Division 1, Section 01420 - Reference Standards and Abbreviations (or a comparable section) is commonly used to specify the applicable date generally for the specifications. For example, applicable edition date could be as of date of the bid submission or it could be the date of the building permit.
-   Incorporate the standard properly. Often in the manufacturer's literature, ASTM standards are proclaimed that look authoritative but actually only specify a test method without identifying the acceptable results.
-   Enforce the requirements of the standard. If the standard states a specific metal thickness for a certain grade of product, insist that the proposed product actually comply with that requirement.

Proprietary Specifications (Exhibit 6-3)

Advantages of Proprietary Specifications
-   Close control of product selection
-   Preparation of more detailed and complete specifications based on precise information obtained from the manufacturer's data
-   Decreases overall length of specifications
-   Reduces specifications production time
-   Simplifies bidding by narrowing the competition and removing product pricing as a major variable
-   Reflects real life by specifying actual products in the marketplace
-   Reduces the Architect/Engineer's design liability by identifying products with which the Architect/Engineer has experience and establishes a clear basis for the design intent

Disadvantages of Proprietary Specifications
-   Reduces or eliminates competition
-   May require products with which the Contractor has had little or poor experience
-   Favors certain products and manufacturers over others

Common Uses for Proprietary Specifications

Proprietary specifications are commonly used for private commercial projects. However, proprietary specifications also may be used for publicly funded projects with certain provisions. See the discussion below regarding "closed" and "open" proprietary specifications.

"Open" versus "Closed" Specifications

"Closed" Proprietary Specifying

In closed proprietary specifications, specific products of one or more manufacturers are specified, and no substitutions will be considered.

"Open" Proprietary Specifying

In open proprietary specifications, specific products of one or more manufacturers are specified, but other manufacturers will be considered ("or equal"). It is necessary to specify the process for evaluation and acceptance of products of alternative manufacturers.

A. Built-up Asphalt Roofing System: GAF Building Materials Corporation, GAF Specification N-B-4-M with GAF Type 5MB Mopped Flashing, 10-year guaranteed, hot applied, 4-ply system with ventilated base sheet, modified bitumen flashing materials, granule-surfaced cap sheet at horizontal surfaces and granule-surfaced modified bitumen flashing at parapet wall faces.

C. Toilet Compartment System, Service and Employee Areas ("Back-of-House"): Bobrick 1036 TrimLine Series, floor and ceiling anchored, particleboard core panels, stainless steel edges.

C. Synthetic Resin Polychromatic Coatings: Zolatone Series 43, inherently polychromatic coating system.

D. Color Blend: Color blend shall be as directed by Architect. Colors may be custom tints and proportions. Provide factory mixed coatings, with colors modified in field only as directed by Architect.

E. Primers: Zo-Wood 93, synthetic-based primer. Verify that coating is compatible with fire-retardant treatment of wood substrate.

F. Finishes:

   1. Base coat: Zo-Cryl Sealer 92, latex formulation based on an acrylic resin emulsion, specifically developed as part of Zolatone process.

   2. Finish coat: Zolatone 43, polychromatic terpolymer coating, consisting of a combination of separate and distinct pigmented color particles suspended in a chemically treated aqueous solution. Varied particle colors and sizes shall create the desired blend of colors and texture for the finish coat.

Exhibit 6-3. Examples of proprietary specifications text.

2.1 TOILET AND BATH ACCESSORIES

   A. Specified Manufacturer: Bobrick Washroom Equipment, Inc., North Hollywood, CA (818/764-1000).

   B. Acceptable Manufacturer: None identified. No substitutions will be considered or accepted.

   TEXT ABOVE IS FOR SPEC WITH ONLY ONE ACCEPTABLE MANUFACTURER. TEXT BELOW IS FOR SPEC WITH MORE THAN ONE ACCEPTABLE MANUFACTURER. NOTE: NO OTHER MANUFACTURERS ARE ACCEPTABLE.

2.1 TOILET AND BATH ACCESSORIES

   A. Specified Manufacturer: Bobrick Washroom Equipment, Inc., North Hollywood, CA.

   B. Acceptable Manufacturers: Equivalent products of the manufacturers listed below will be acceptable.

      1. American Specialties, Inc., Yonkers, NY.

      2. Franklin Brass Manufacturing Co., Culver City, CA.

   C. Toilet Paper Boxes: Bobrick B-272.

   D. Paper Towel Boxes: Bobrick B-263.

   E. Toilet Seat Cover Dispensers: Bobrick B-221.

   F. Grab Bars: Bobrick B-550 Series, configurations as indicated on the Drawings.

   (Text is closed proprietary because only products of three specified manufacturers may be provided.)

Exhibit 6-4. Examples of closed proprietary specifications text.

2.1 TOILET AND BATH ACCESSORIES

    A. Specified Manufacturer: Bobrick Washroom Equipment, Inc., North Hollywood, CA.

    B. Acceptable Manufacturers: Equivalent products of the following manufacturers will be accepted. Additional manufacturers will be considered in accordance with "or equal" provision specified in Section 01620—Product Options.

       1. American Specialties, Inc., Yonkers, NY.

       2. Franklin Brass Manufacturing Co., Culver City, CA.

    C. Toilet Paper Boxes: Bobrick B-272 or equal, surface-mounted, #22 gage, type 304 stainless steel, satin finish, with tumbler lock.

    D. Paper Towel Boxes: Bobrick B-263 or equal, surface-mounted, type 304 stainless steel, satin finish. Door with tumbler lock and piano hinge.

    E. Toilet Seat Cover Dispensers: Bobrick B-221 or equal, surface-mounted, type 304 stainless steel, satin finish.

    F. Grab Bars: Bobrick B-550 Series, or equal, 1½-inches o.d., 18 gage Type 304 stainless steel tubing, 1½ inches clear between grab bar and mounting surface, concealed mounting with 3-inch diameter flange secured with 4 set screws, configurations as indicated on the Drawings.

(Note: products are not restricted to specified manufacturers, but products of unidentified manufacturers may be provided in accordance with provisions in Section 01620—Product Options. Descriptive text is included to assist in evaluation of submitted products of unidentified manufacturers.)

Exhibit 6-5. Example of open proprietary specifications text.

"Nonrestrictive" Specifications

Specifications are generally written to encourage competition in bidding. It may be the policy of the Owner or the requirements of public contracting laws that specifications not restrict competition in bidding for the work. To accomplish this, specifications should be written using a method (described above) that does not restrict competition. At least two (usually three) manufacturers should be specified whose products comply with the specifications.

Specifying Method

All four of the following methods may be used, if used appropriately and correctly.

- Descriptive Specifying: Most commonly used for nonrestrictive specifications
- Performance Specifying: By nature, intentionally nonrestrictive
- Reference Specifying: By nature, nonrestrictive
- Proprietary Specifying: May be nonrestrictive by naming two or three manufacturers or by adding an "or equal" provision

De Facto Restrictive Specifying

When the specifying method is based on precise features that only one manufacturer can meet, specifications may be de facto restrictive (Exhibit 6-6). This is commonly observed in manufacturer-produced specifications that are described by the meaningless term "generic" or that have the appearance but not the true content of a "nonproprietary" (descriptive) specification. Close examination of the attributes may well reveal excessively fine tolerances, unattainable test performance (the offending manufacturer owns the only machine capable of performing the test) or unimportant dimensional and physical attributes ("cabinet 10-¼ inches wide by 29-³⁄₁₆ inches high by 6-⅞ inches deep, with swirl-polished interior surfaces").

Performance Specifications

The term "performance specifications," as used in this discussion, pertain to portions of a project specification rather than a whole-building performance specification (Exhibit 6-7). The key concept behind performance specifications, unlike other methods, is that performance specifications are not prescriptive about the products and processes to be

---

2.1 HORIZONTAL-SLIDING ACCORDION-TYPE FIRE DOORS

  A. Configurations: Single-parting and bi-parting door assemblies as indicated on Drawings.

  B. Fire Ratings: UL listings as Special Purpose Fire Doors, UL 10B and ASTM E152, without hose stream test. Time ratings shall be as indicated on Drawings.

  C. Door Construction: Two parallel walls of formed steel panels, independently suspended with no pantographs or interconnections except at lead posts. Panels shall be connected by live vinyl or formed steel hinges, according to fire rating requirements.

  D. Panels: Formed steel, 24 gage by 4½ inches, formed for maximum strength, with permanent finish.

  E. Fire Insulation: Interior surfaces of both walls shall be covered with continuous fire-resistant blanket, secured to each panel with metal clip system. Fire insulation shall extend from finished floor to tracks and shall meet at interior centers of fixed jambs and lead posts, forming an effective fire and smoke barrier.

  F. Perimeter Seal: Each wall of door assembly shall have extruded live vinyl sweeps, top and bottom, forming an effective smoke and draft seal.

  G. Track and Trolley System: Track system shall consist of two parallel tracks on 8-inch centers, of 1¾ inch by 1½ inch by 14 gage formed steel. Each panel shall be supported by ⅛-inch diameter steel hanger pin and 1¼ inch diameter, double-race ball bearing roller. Weight factor shall be maximum 9.3 pounds per lineal foot.

  H. Lead Posts: 16 gage formed steel connected to double walls by specially adapted panels.

(Not identified is the fact that the description above is verbatim from the product catalog of the only manufacturer of such a door system.)

Exhibit 6-6. Example of de facto proprietary specifications text.

---

used by the Contractor. Descriptive specifications require the Contractor to conform to specific materials, fabrication techniques, and methods of installation. Performance specifications allow the Contractor to be inventive and ingenious in complying with requirements of the construction contract. This can and perhaps should result in more efficient and economical construction.

Descriptive specifications describe what the Contractor must do to achieve the intended results, while performance specifications describe the intended result and leave it to the Contractor to determine how to achieve the result. This requires that performance specifications clearly and definitively communicate the required results while not unnecessarily limiting the products, methods or means the contractor uses to achieve those results.

Advantages of Performance Specifying
- "Design intent" (end result) only is specified, giving the Contractor latitude in selecting and applying construction products in a manner that performs as the specifier intends.
- Expedites construction through systems tech-

niques: For example, specifies a ceiling system that includes lighting, HVAC and acoustical control features.
- Encourages development of new technologies.
- Can result in shorter specifications.
- Permits free competition.
- Applicable to all types of projects.
- Delegates technical responsibilities to the construction industry. The Contractor is responsible for the results rather than the Architect/Engineer.

Disadvantages of Performance Specifying
- No firm, equal basis for bidding.
- All criteria must be clearly defined.
- More difficult to enforce.
- Delegates technical responsibilities to the construction industry. The Contractor is responsible for the results.
- The specifier is required to take special care in describing the design intent to achieve the intended results. The Contractor must interpret

specifications to identify and procure products available in the marketplace. The Architect/ Engineer must interpret specifications when evaluating products proposed by the Contractor.

- Can result in longer specifications if very detailed criteria are specified.
- Can be time-consuming to produce.
- May be too elaborate for a minor or simple project.
- "Underwhelming" success so far, but interest continues to grow.

Producing Performance Specifications

For more in-depth material on performance specifications, refer to the CSI *Manual of Practice* (1996), "Construction Documents Fundamentals and Formats Module, FF/120 Methods of Specifying" and "Construction Specifications Practice Module, SP/090 Performance Specifying."

According to the CSI *Manual of Practice,* there are four essentials in performance specifying:

- Attributes
- Requirements
- Criteria
- Tests

Attributes are the means by which performance characteristics are identified. An attribute, therefore, can be defined as a characteristic of performance.

Requirements are statements of desired results, usually in qualitative terms. More than one requirement may be defined for a single attribute.

Criteria are definitive statements of performance for a particular requirement stated in quantitative or qualitative terms. Criteria must be either measurable or observable. Several criteria may be needed to define a requirement completely and accurately.

Tests are checks for conformance with performance criteria and a measure of actual or predicted performance level. A test will be associated with each criterion and may be based on a recognized industry test method, calculation or engineering analysis, observation, or professional judgment. Test results may be evaluated by conducting the specified test or simply by submitting certified results of previous testing.

Pure performance specifications are rarely used for an entire specification section. In practice, the performance method of specifying supplements other methods or the specification uses other specifying methods to supplement the performance specification.

Common Uses of Performance Specifying

For portions of construction specification, products such as ceiling, lighting, and HVAC systems lend themselves to performance specifications. On a grand scale, entire facilities may be specified by the performance method. However, the specifications format being discussed here is not used for whole-facility performance specification. Refer to other publications by the Design-Build Institute of America (DBIA) and CSI (UniFormat™) and the computer-assisted performance specification program PerSpective™ by Building Systems Design, Inc. (Atlanta, Georgia).

Examples of Performance Specifying

Performance specifications within conventional project specifications

Selecting a Method for Specifying

The specifier should determine the following and then select an appropriate method or methods for the Specification Section.

- What does Owner require?
- What method best describes the design intent?
- What method is most appropriate for the project size and complexity?
- What method will result in the best quality of work?
- What method will result in the best price for the work?

Mixed Methods of Specifying

It is not only acceptable but recommended practice to mix methods of specifying (Exhibit 6-8). A specification section may be written using mainly the proprietary method. Real products by real manufacturers in the marketplace are specified. Then the specified product is further identified using the descriptions method and the reference standard method. Finally, performance characteristics are specified. Why the redundancy?

It is redundant to specify in such a way, but the use of the specifications and the construction context need to be considered.

C. Parking Lot Luminaires: Pole-mounted luminaires located within landscaped islands of parking lot and bracket-mounted luminaires mounted on exterior building walls providing illumination complying with the following criteria:

1. Average illumination: 2.50 maintained footcandles.
2. Minimum illumination: 1.00 maintained footcandles at all locations.
3. Maximum to minimum footcandle uniformity: 11:1 minimum to 13.1 maximum.
4. Mounting height limitation: 28 feet.
5. Lamp type and wattage: HID, high pressure sodium preferred, maximum 400 watts.

A. Ceiling System: Integrated system of suspended support structure, acoustical panels, luminaires and HVAC supply and return air fittings complying with the following criteria:

1. Noise-Reduction Coefficient: Noise-Reduction Coefficient (NRC) 0.55 (0.50–0.60), as determined in accordance with ASTM C 423, with materials tested on mounting E-400 unless otherwise noted (ASTM E 795 procedure).
2. Ceiling Sound Transmission Class: Ceiling Sound Transmission Class (STC) 35–39, as determined by manufacturer for continuous ceilings tested in accordance with AMA 1-II—Ceiling Sound Transmission Test by Two-Room Method.
3. Light Reflectance: 70–74 percent, in accordance with Fed Spec SS-S118B.

E. Panel Fire Rating: Class A Flame Spread Rating according to Fed Spec SS-S118B and ASTM E 84.

F. Luminaires: Provide minimum 100 lumens and maximum 120 lumens at work surface 28-inches above finished floor.

G. HVAC Fittings: Provide capacities for air supply and return to suit HVAC system specified elsewhere herein, with the following performance features:

1. Maximum sound pressure level shall be NC 22, based on room absorption of 8 dB (re 10^{-12} watt sound power).
2. Air diffusion distance shall provide room air velocity less than 20 fpm.

A. Temporary Connections and Fees: Contractor shall arrange for services and pay all fees and service charges for temporary power, water, sewer, gas and other utility services necessary for the Work, until Contract closeout.

D. Thermal Performance: Provide for expansion and contraction of system components over cycling temperature range of 170 degrees F, without causing detrimental effects to framing system appearance and performance.

Exhibit 6-7. Examples of performance specifications text.

Reference Standards Agencies

Inasmuch as the subject of reference specifications has been discussed earlier, it is relevant to discuss in more detail here reference standards and the organizations that produce them.

Architects, Engineers and specifiers constantly make use of reference standards in specifications, but many professionals are completely unfamiliar with the processes by which these standards are developed and promulgated. Furthermore, many are not aware of the contribution they can make in participating in the development and improvement of these standards.

Standards provide several important benefits. They reduce the number of types, sizes, and qualities of materials. They standardize methods of testing, and several provide standards on the quality of workmanship.

One major benefit is the reduction in the size of construction specifications. By incorporating a reference standard in a specification, the number of words required to specify a material and the method of testing it are reduced a hundredfold. This ensures the specifier some degree of quality since the reference standard reflects the combined knowledge and experience of the people engaged in its development.

Nevertheless, if the quality of reference standards is to be improved, there must be greater participation by users. This means affiliation of individuals and companies as members of associations producing standards. Architects, Engineers, and specifiers are particularly encouraged to participate, since their interests are more objective and less biased than those of individuals representing manufacturers and industry.

Generally, most committees producing standards

A. Vapor-Retarding Sheet Membrane: Stego Wrap 15 mil or equal, coextruded sheet of virgin resins and additives, complying with ASTM E 1745, Class A permeance and Class B puncture resistance.

  1. Vapor transmission rating: 0.006 grains/ft²/hr, when tested according to ASTM E 96 at 73.4 degrees F.
  2. Permeance rating:
    a. New material: 0.012 perms, when tested according to ASTM E 96 at 73.4 degrees F.
    b. After elevated temperature conditioning: 0.015 perms, ASTM E 154 Section 11 Method and ASTM E 96 Procedure

B.

    c. After low temperature/bending: 0.017 perms, ASTM E 154 Section 12 Method and ASTM E 96 Procedure B.
    d. After exposure to soil organisms: 0.014 perms, ASTM E 154 Section 13 Method and ASTM E 96 Procedure B.
    e. After exposure to petroleum and soil poison: 0.013 perms, ASTM E 154 Section 14 Method and ASTM E 96 Procedure B.
  3. Tensile strength: 48.6 lbf/in. (CD) and 47.8 lbf/in. (CDM), when tested according to ASTM E 154 Section 9 Method and ASTM D 828.
  4. Puncture resistance: 1970 grams when tested according to ASTM E 154 Section 10 Method and ASTM D 1709.

B. Accessory Materials:

  1. Seam tape: Stego High Density Polyethylene Tape or equal, with pressure-sensitive adhesive, minimum 4-inches wide.
  2. Pipe boots: Constructed from vapor barrier material and pressure-sensitive adhesive tape according to manufacturer's instructions.
  3. Adhesive mastic: Spray-applied contact adhesive, compatible with sheet barrier and substrate materials, with excellent water-resistive qualities, Miracle Adhesives Corporation, Bellmore, NY, Miracle 1330 Spray Adhesive, or equal.

Exhibit 6-8. Example of mixed method specifications text.

are balanced working groups representing all the interests concerned with the particular standard. Typically, they are composed of manufacturers of the basic ingredients of the material, manufacturers of the end product, suppliers, independent testing agencies, consumer groups, contractors associations, representatives of public authorities, and others who have special interests in a particular standard.

ASTM Standards

The product and test standards most widely used in both the private and public sectors of construction are ASTM standards. ASTM International (formerly the American Society for Testing and Materials) is an international private, technical, scientific, and educational society devoted, in its words, to "the promotion of knowledge of the materials of engineering and the standardization of specifications and the methods of testing."

Since 1898, this organization has conducted research on the properties of materials and has developed numerous standards concerned with the Specifications for materials, methods of testing, and definitions. An index to ASTM standards and information on membership may be obtained from the society's headquarters at 100 Barr Harbor Drive, West Conshohocken, PA 19428.

Federal Specification Standards

Among U.S. federal agencies the reference standards that were mandatory until recently were Federal Specifications (FS). ASTM committees are taking over the task of setting standards for products formerly in the FS series. Remaining Federal Specifications are in the custody of the General Services Administration. Copies of the Federal Specification Index may be obtained from the Superintendent of Documents, U.S. Government Printing Office, Washington, DC 20402.

ANSI Standards

Other widely used reference standards are ANSI standards, promulgated by the American National Standards Institute. In addition to developing standards on materials and testing procedures, ANSI has created many standards used in construction, including workmanship and installation procedures. This standards organization works closely with other technical societies engaged in developing standards, and many ANSI standards bear corresponding ASTM, AASHTO, NFPA, and CS standards numbers. An index of standards may be obtained from the association at 11 West 42nd Street, New York, NY 10036.

ACI Standards

The American Concrete Institute (ACI) is a nonpartisan organization that gathers and disseminates information about the properties and applications of concrete and promulgates recommended practices, referred to as "ACI standards." A catalog of the publications of this institute is available from P.O. Box 19150, Detroit, MI 48219.

NFPA Standards

NFPA International (formerly the National Fire Protection Association) develops fire protection standards and codes that are widely used as a basis for laws and ordinances. The more widely known standards used in construction are the National Electrical Code (NEC) and the Life Safety Code (LSC). Information on membership, technical committees, and NFPA standards may be obtained from the association at Batterymarch Park, Quincy, MA 02269.

AASHTO Standards

The American Association of State Highway and Transportation Officials (AASHTO) publishes standards on highway materials in two parts, one dealing with specifications for materials and the second with methods of testing. These AASHTO standards may be obtained from this organization at 444 N Capitol Street, Washington, DC 20001.

Standards of the National Institute of Standards and Technology (NIST)

Commercial Standards (CS) and Simplified Practice Recommendations (SPR) are voluntary standards issued by NIST and developed cooperatively with industry groups. CS establish quality requirements for products, and SPR establish sizes and classes for stock items. The NIST has consolidated these two types of standards and provided a new name, "Product Standards (PS)," to describe these new standards being developed. The list of standards may be obtained from the Superintendent of Documents, U.S. Government Printing Office, Washington, DC 20402.

Chapter 7

Specifications Writing Principles

DETERMINING CONTENT FOR THE SPECIFICATIONS

What Needs to Be Specified?

Typically, the specifier and the designers meet to determine what the design includes. The specifier may study the drawings, which tend to precede the specifications, but often it is determined early in the design documentation process that the specifier has issues for discussion that precede design development. For example, the specifier may ask whether there are toilet partitions in the design. The designer may say yes, but the material and configuration of the toilet partitions have yet to be determined. This creates an action item for the designer to follow up on and the specifier to record on a preliminary list of sections to be written.

What Are Criteria for Products to Be Specified?

The requirements of the design become more refined as the design develops and as the Owner's requirements become more apparent. As preliminary estimates of probable construction cost are produced, suitable quality and cost restraints for products become more apparent. Early in the design, criteria need to be identified for:

- Configuration Requirements: For example, shall the toilet partitions be ceiling hung, floor-mounted with overhead bracing rails, or mounted with floor-to-ceiling pilasters?
- Material Requirements: For example, shall the toilet partition panels be manufactured from metal with a painted finish, from solid high-density polyethylene, from plastic laminate-face wood fiberboard or from phenolic panels with decorative colored faces? Shall the toilet partition hardware be rated for commercial or institutional use? Will the plastic laminate on the toilet partitions be selected from the manufacturer's standard selection or will it be a custom selection from a manufacturer the partition manufacturer does not regularly buy from?

- Performance Requirements: For example, what is the shading coefficient for the glass, how many British Thermal Units per Hour (BTUH) shall the boiler produce and what is the windstorm resistance rating for the roofing system?
- Code and Regulatory Requirements: For example, what are the building code, the plumbing code, the mechanical code, and the electrical code? Are there special amendments to the Code adopted by local jurisdictions or state agencies?
- Code Jurisdiction: The building occupancy and ownership of a project may determine that different codes and authorities govern. A private school is usually governed by local codes, and a public school is usually governed by statewide Codes. One may be more restrictive than the other.
- Environmental Requirements: Regional environmental requirements, especially those regarding air quality, storm water pollution protection, waste management, haul routes through the city, and dust management may apply.
- Owner's Policies: Requirements such as sustainable design and use of recycled and low-energy or high-efficiency products may apply because of the Owner's policies, although not required by code or law.

What Are Suitable Products?

After design criteria and budget guidelines are established, suitable products need to be identified. This is a complex subject. A whole book could be written on it and, in fact, several books have been published. *Architectural Materials for Construction* by Harold J. Rosen, PE, FCSI and Tom Heineman, FCSI (McGraw-Hill Companies, New York, 1996) is one.

Answering the following questions will help define suitable products.

- What does the Owner or code authority require?

- What method best describes the design intent?
- What method is most appropriate for the project size and complexity?
- What method will result in the best quality of the project?
- What method will result in the best price for the project?

General and specific technical resources need to be utilized, along with industry product resources, to identify suitable products and to acquire technical information regarding the products. Until recently, the foremost resource for architectural product information has been *Sweets Catalog File,* published by McGraw-Hill Construction (McGraw-Hill Companies, New York, annual issues by subscription).

Competing publications are available, such as *First Source*™ by Reed Construction Data (Norcross, Georgia, annually by subscription) and *ARCAT®— The Product Directory for Architects* by ARCAT (Fairfield, Connecticut, annually by subscription). Online sources of construction information, such as "4specs.com" (www.4specs.com) have search capabilities to locate and connect directly to product manufacturer's websites. McGraw-Hill also has an online search program, "Sweets e-BuyLine" (www.sweets.com). With the decreasing size of the annual *Sweets Catalog File,* there is speculation about whether this venerable publication will be discontinued in a few years.

There are still analog sources of construction information in addition to the digital ones described above. One of the most valuable resources is networking. This includes networking within one's office, especially with experienced and technically knowledgeable staff members. A quick question to a veteran architect/engineer can save hours of research and changes in product selections. At the same time, beware of the statement that selections are not being recommended simply because "that's the way we always do it." New, superior or more appropriate products may be available.

Another resource is networking with colleagues outside of one's office. This can involve a network of people who have worked together in previous years and in other situations. It can be a network of people who have come to know each other in professional associations, such as AIA, CSI, NSPE, ASHRAE, ASPE, ASCE and ASLA. Through active participation in these organizations (not merely attending occasional meetings), professional ac-quaintances are developed with whom to network. In particular, CSI offers opportunities through national conventions, regional conferences, and local chapters to attend product exhibits and participate in professional development programs. Other associations have similar programs.

Verification of Product Attributes

After initial selection, the products should undergo additional evaluation. Issues such as the following should be addressed:

- Suitability: Do product characteristics meet all criteria of the current design? The design may have changed.
- Manufacturer: Does the manufacturer market the product in the locale of the project? Does the manufacturer have trained installers and applicators in the locale of the project? Does the manufacturer have factory-authorized technical support for construction and for warranty and maintenance service?
- Installation: Can the product be installed in the available space? Are there clearance and service access requirements that the intended product can meet?
- Cost: Is the initial cost within the budget, and are the life cycle operational and longevity costs acceptable?
- History: Does the product have a history of acceptable service and performance? Does the manufacturer have a history of acceptable support to the design professional and the owner?

Design-Build Considerations

With the increasing use of the design-build project delivery method, performance specifying is being done more often. Specific product selections may not be made for specifications issued before preparation of the design-bid proposal. Detailed performance criteria need to be developed and presented in the pre-proposal specifications. For the proposal specifications, specific products would not only need to be specified but product data would need to be presented demonstrating compliance with or deviation from specified performance criteria. These criteria may include:

- Structural serviceability
- Fire and life safety
- Habitability

- Durability
- Practicability
- Compatibility
- Maintainability
- Code acceptability
- Economics

Gathering Product-Specific Information

Once product selections have been made, detailed product information needs to be gathered. Determining the types of information needed and how to present the information is where hard-core specifications writing takes place. Without some sort of guide, this becomes a daunting task. With guidelines, it is manageable but not as easy as it might seem. The key is to follow an established format for the specifications and to know what to put in and what to leave out. These require professional judgment.

SPECIFYING

Specification Method and Language

As presented in Chapter 6 a method of specifying needs to be determined that is suitable for the products and project requirements. Determine which method—descriptive, reference standard, proprietary or performance—or combination of methods should be used. Consider the owner's requirements and policies, which may preclude use of one of the methods, such as the proprietary or "closed" proprietary method. *Note: Usually a combination of methods is used.*

Determine the type of specifications language to use: indicative, imperative, or streamlined. *Note: Often a combination of language types is used.*

Remember:

- Address only the Contractor.
- Use consistent and correct terminology. Coordinate Specifications with Drawings.
- Make correct cross references.
- Be clear, complete, correct and concise.

Specify Workmanship

Determine and specify the quality of fabrication and assembly at the source (Part 2).

Determine and specify the quality of field assembly, installation, application and finishing (Part 3).

Other Considerations

Make suitable choices based on economic considerations. Consider the project's budget and the life cycle costs of maintenance and consumption.

Consider sustainable design and construction, whether mandated or as an ethical matter. Determine what is required by authorities having jurisdiction, and determine whether the facility is intended to achieve a certain level of sustainable design and construction (LEED certification).

Considerations for trades and subcontracts: While not specifically recommended, be realistic but do not violate the provisions of the General Conditions or assume responsibilities belonging to the Contractor. Consider whether specifying waterproofing as a component of the ceramic tile flooring assembly is advantageous for quality assurance, as compared to specifying the product in a distinct section where surface quality could become a matter of dispute between two subcontractors.

Related Documents

Many specifications sections include a statement at the very beginning, in an article titled "Related Documents," placed before the "Summary" article. Typically, it reads, "Drawings and general provisions of the Contract, including General and Supplementary Conditions and Division 1 Specification Sections, apply to this Section." Well, of course they do, and the Contractor knows this very well. Why is it included? For the benefit of the Contractor when the specifications are broken up and distributed to subcontractors and suppliers for bidding. But, as discussed in other chapters, the Specifications are only written as from the Owner to the Contractor, and provisions for subcontracting are not included. The "Related Documents" article is unnecessary and redundant.

Scope of Work

Frequently, a specification section will begin with one or two articles that identify the content of the section. As indicated on CSI *SectionFormat,* the first article is titled "Summary." Under it are several paragraph headings, including paragraphs titled "Section Includes" and "Related Sections." It is understandable that the section should open with descriptive statements about what is written in the section. This helps the reader to find pertinent information and to distinguish between similar products in various sections. The problem comes when these statements are misused or misinterpreted to a scope of work for a trade or subcontractor.

SectionFormat does not help the matter by also including paragraphs titled "Products Supplied But Not Installed Under This Section" and "Products Installed But Not Supplied Under This Section." The only purpose of such titles is to indicate that the section is intended to be used to define the scope of subcontracts or trade jurisdiction, which is contrary to what is stated in typical General Conditions of the Contract. For example, under Section 08710—Door Hardware, the Summary article states under the paragraph "Products Supplied But Not Installed Under This Section" that door hardware shall be supplied for installation under Section 06200 - Finish Carpentry. Likewise, under Section 06200 - Finish Carpentry, the paragraph "Products Installed But Not Supplied Under This Section" states that door hardware is furnished under Section 08710—Door Hardware. Why split the products from the installation? Because it is intended that the door hardware supplier furnish the hardware to the finish carpenter to install. As discussed in preceding chapters, this practice is not recommended, and can lead to problems and claims against the design professional.

Assuming that the above problem is resolved, there is a problem when the description of Work includes superfluous and redundant language. For example, the description may include a statement such as "furnish and install at locations indicated on the drawings all labor, materials, equipment, tools, consumables, services, temporary utilities, tests and inspections for fire extinguisher cabinets." All of these are covered by the General Conditions of the Contract and by specifications in Division 1 - General Requirements. Simply state what is specified in the section if the Summary article is used. For example:

1. Built-up SEBS modified-bitumen roofing system.

Don't write a description of all the components of the roofing system. (There's another article in *SectionFormat* that is precisely for this purpose, titled "System Description.")

As soon as attempts are made to have all-inclusive statements at the beginning of the specification section, something will be omitted. These statements are redundant and time-consuming to produce. The Contractor is already obligated by the General Conditions to provide complete assemblies and systems. The specifier should spend time on other matters.

A review of current manufacturers' guide specifications, which are often the worst at using scope statements, indicated that the use of all-encompassing scope statements has lessened greatly. Even a review of public agency guide specifications, which also tended to use wordy scope statements, indicated the use of concise statements merely to introduce the products specified in the section.

Perhaps attention to this can be reduced but with caution because specialty consultants, such as roofing and waterproofing consultants, often include the wordy descriptions in order to make it "clear" what the subcontractor shall do and what the General Contractor shall do. Not stated is how such complex statements "prove" that the work is complicated and requires the expertise of the consultant to make the parties live up to their obligations.

If used, the paragraph titled "Related Sections" should only list other sections that specify requirements directly related to the work in the subject section. The key is to add a short description after the section number and title that states what is in the "related" section that is actually related. For example:

1. Section 05090 - Anchors and Fasteners: General requirements for anchors and fasteners to building substrates
2. Section 03100 - Concrete Formwork: Requirements for setting embedded products in cast-in-place concrete
3. Section 09970 - Coatings for Exterior Steel: Primer requirements applicable to shop-fabricated components

Broadscope, Mediumscope, Narrowscope Terms

The terms "broadscope," "mediumscope," and "narrowscope" for specifications sections at various levels of detail for scope have been superseded by the terms "Level Two," "Level Three," and "Level Four." These refer to the level of product detail covered by the section. For example, the broad-scope title "Section 07500 - Membrane Roofing" (Level Two title) does not suggest what type of material is used for the membrane. It is overly broad for practical use as a specification section title. The medium-scope title "Section 07530 - Elastomeric Membrane Roofing" is more clear and certainly differentiates this roofing from the more common built-up bituminous membrane roofing. It might be even clearer and beneficial to roofers who might want to bid on the Project if the narrowscope title "Section 07534 - Ethylene Propylene Diene Terpolymer (EPDM) Roofing" is used.

Grandfather Clauses

Individuals who are not properly grounded in the principles of specifications writing habitually fall back on general and all-inclusive language, which often results in what are termed "grandfather clauses" by specifiers and "murder clauses" by contractors—clauses that embrace everything yet fail to be specific. A typical example of a grandfather clause might read as follows: "The Contractor shall furnish and include everything necessary for the full and complete construction of the building, whether shown or specified or not shown or described." When an architect or engineer is incompetent, he or she takes cover behind such a series of clauses, which may be interpreted to mean anything or nothing. In their failure to be specific, these clauses will, during the course of construction, require interpretations by the architect or engineer that may be difficult to enforce.

A clause such as "Concrete floors shall be finished level as approved by the Architect" without stating a tolerance is interpreted by the Contractor as "Guess what I will make you do." An instruction to a Contractor by means of a drawing or a specification must be specific, and no architect or engineer should expect a contractor to fulfill nonspecific requirements.

Residuary Legatee

Where several different kinds or classes of similar materials are used, they should be described in a manner that permits some material to be specified for every part of the building. This technique has been borrowed from the legal profession and is known as the "residuary legatee." To illustrate, let us assume that in preparing a will an individual wishes to leave the bulk of his estate to his wife, but wishes to make several minor bequests to his children or to relatives. He first enumerates his minor bequests and then states, in substance, "The residue of my property I bequeath to my wife." She is then known as the *residuary legatee.*

In applying this principle to specifications writing, the materials occurring in the smallest quantity or in the fewest places should be listed first, and the material occurring in the remaining places becomes the residuary legatee and can be covered by some such phrase as "the rest of the building."

As examples of this technique the following samples are offered:

1. In specifying glass, one can list the following:
   a. Obscure glass: locker room windows
   b. Tempered glass: nonrated doors and side lights
   c. Laminated glass: acoustical windows
   d. Fire-rated laminated glass: fire-rated doors
   d. Clear float glass: all other interior locations
   f. Tinted float glass: all other exterior locations

2. In specifying paint:
   a. Plaster surfaces in toilets: semigloss enamel
   b. Plaster surfaces in kitchens: gloss enamel
   c. Plaster surfaces in bedrooms: flat enamel
   d. All other plaster: latex emulsion paint

3. In specifying concrete:
   a. 2500 psi concrete: concrete foundations
   b. 3000 psi concrete: concrete pavements
   c. 3500 psi concrete: all other concrete work

If this method is followed, some material will always be specified for every part of the building, whereas any other plan obliges the specifier to check all listings most carefully for fear of not including some minor portion.

Duplication-Repetition

In Chapter 2 it was noted that the necessary information for the construction of a building is communicated to a contractor in two forms, graphic (the drawings) and written (the specifications), and that these documents should complement one another. If this information overlaps, there can be duplication, which may lead to a difference in instructions and disagreements as to which is the proper document to follow.

If this duplication were exact in each instance and remained so, it might be harmless at best; but too often, the information presented on the drawings and in the specifications either does not agree in the first place or, owing to last-minute changes, errors, and differences develop that create entirely new meanings. Repetition in the Contract Documents is always dangerous and should be avoided.

Technically, duplication is an exact repetition of a sentence or a paragraph in a specification, or else it is an exact repetition of a detail on a drawing. For example, a steel ladder might be detailed on a drawing, giving the size of the side members and the diameter and spacing of the rungs. The specification should describe the quality of the material and how the rungs are let into the side members, but it should not repeat the sizes and spacing since the drawing may be altered by the draftsman, with a resulting conflict in the two documents. The unnecessary expense involved in writing and reproducing statements that merely repeat may be minor in comparison to the ultimate cost to the Owner for mistakes in specification interpretation.

An exact duplication in the specification or drawing should cause no misunderstanding. However, we seldom see exact duplication. In most cases, the specifier attempts to avoid duplication or repetition by stating in different words what has been stated elsewhere in order to amplify. But it is precisely in attempting to amplify or reiterate in different words that conflict and ambiguity occur. It is therefore good practice to make a statement only once; if it is not satisfactory, it should be discarded and rewritten rather than amplified or explained in other terms.

Chapter 8

Bidding Requirements

THREE BASIC BIDDING DOCUMENTS

Bidding requirements consist of documents that are used in the solicitation of bids by an owner or an agency. The documents are directed to bidders who might be interested in submitting bids for a project. These documents consist of three essential forms: (1) an invitation dealing with advertising or notifying interested bidders of the existence of a proposed project; (2) instructions pertaining to the submission of a proposal or bid; and (3) the sample form on which the bid is to be executed by a bidder.

The three documents are *Invitation to Bid, Instructions to Bidders,* and *Bid Form.* Because of the varying practices of individual specifiers and the lack of order and terminology for the material preceding the technical specifications, some chaos and nonuniformity in the arrangement and nomenclature of these documents existed in the past.

Bidding requirements are neither specifications nor Contract Documents. The basic difference lies in the fact that bidding requirements apply to a bidder prior to awarding the Contract, whereas the Contract Documents apply to the Contractor and obligations of the Contractor after execution (signing) of the Agreement (Contract). Generally speaking, certain information contained in the bidding requirements that is pertinent to a Contractor's obligations—such as time for completion, base bid, alternates, and unit prices—should be entered into the Agreement Form after awarding the contract to ensure its fulfillment by the Contractor.

Invitations to Bid are generally circulated in the case of private work to certain selected bidders, and in the case of public agencies, they are advertised in local newspapers. In any event, the Invitation to Bid, along with the Instructions to Bidders and a sample copy of the Bid Form, should be bound in the Project Manual.

Invitation to Bid

Other terms have been used for "Invitation to Bid," but they are used somewhat incorrectly as the heading for this document. These include "Advertisement to Bid" (sometimes mandated for use in public work construction for public advertising), "Notice to Bidders," and "Notification to Contractors." The term "Invitation to Bid" is preferred since it best describes the intent of this document. The purpose of an Invitation to Bid is to attract bidders in sufficient numbers to ensure fair competition and to notify all parties who might be interested in submitting proposals. It should be limited to information that will tell a prospective bidder whether the work is in his or her line, whether it is within that bidder's capacity, and whether he or she will have the time to prepare a bid prior to opening (Exhibit 8-1). It should be brief, simple, and free from subject matter not consistent with its purpose. It should consist of the following elements:

1. *Project Title.* State the name of the project, its location, and the project number if any.
2. *Identification of Principals.* State the name and address of the architect or issuing agency, together with the date of issue.
3. *Time and Place for Receipt of Bids.* State the time and place where bids will be received and whether they will be publicly opened. If they will be opened privately, indicate whether prime bidders can attend.
4. *Project Description.* Provide a brief but adequate description of the project, including the size, height, and any unusual features so that the bidder will be in a position to determine whether he or she has the financial and technical ability to undertake the construction of the project.
5. *Type of Contract.* State whether bids are being solicited for a single or segregated contract and on what basis.
6. *Examination and Procurement of Documents.* State where the contract documents can be examined and when and where they can be obtained. Indicate whether a deposit or a

```
*************************************************************************************************************
```

> USE ARCHITECT'S LETTERHEAD. EDIT TEXT TO SUIT PROJECT REQUIREMENTS. CONSULT WITH OWNER
> AND OWNER'S COUNSEL FOR SPECIFIC REQUIREMENTS. COORDINATE WITH INSTRUCTIONS TO BIDDERS
> AND SUPPLEMENTARY INSTRUCTIONS TO BIDDERS.

```
*************************************************************************************************************
```

[_Month_] [_Day_], 20[_#_]

[_Invited_General_Contractor_]
[_Street_Address_]
[_Mailing_Address_]
[_City_], [_State_] #####

Subject: Invitation to Bid, [_Project_Name_]
        [_Architect's_Project_No._#####.##_]

Gentlemen:

INVITATION

[_Owner's_Name_] hereby invites you to bid on a General Contract, including sitework, mechanical and electrical work, to construct a [_Brief_Project_Description_]. Project areas, type of construction and general nature of the Work is described in more detail on Drawing [1] [A1] [____], located in the Drawings, and in Specification Section 01100 - Summary of the Project, located in the Project Manual.

Project site is located at [____], in [_City_], [_State_] as shown approximately on the Vicinity Map in the Drawings.

OWNER

Owner is [_Owner's_Name_], [_Mailing_Address_], [_Street_Address_], [_City_], [_State_].

ARCHITECT

Architect is [_Architect's_Name_], [_Mailing_Address_], [_City_], [_State_] [_#####-####_].

BASIS OF BIDS REQUIRED

All bids shall be on a [lump sum basis, without qualifications.] [[____] basis.] Segregated Bids will not be accepted.

Each bid shall be made in accordance with the Bidding Documents on file at the office of the [Owner] [Architect] [Construction Manager].

TIME OF COMPLETION

[Project shall be completed within [300] [____] calendar days from the date of Award of the Contract for Construction.] [Project shall be completed within time nominated by Bidder and stated by Bidder on Bid Form. Time of Completion will be considered in the selection of Contractor by Owner.] [Time shall not exceed [300] [____] calendar days from the date of Award of the Contract for Construction.]

BID OPENING

Owner will receive bids until [10:00 A.M.] [12:00 noon] [2:00 P.M.] [____] Pacific [Standard] [Daylight] Time on [_Month_] [_Day_], 20[____], at the office of [_Owner_] [_Architect] [_Construction_Manager_] [____], [_Street_Address_], [_City_],

Exhibit 8-1. Sample Invitation to Bid.

[_State_]. Bids received after this time will not be accepted. [Bids will be reviewed privately and Bidders notified in writing.] [Bids will be opened and announced publicly.]

BIDDING DOCUMENTS

Bidding Documents may be examined at the office of the [Owner] [Architect] [Construction Manager]. [Bidding Documents will be issued to [invited] [pre-qualified] General Contractors in accordance with the Instructions to Bidders and Supplementary Instructions to Bidders, contained in the Project Manual in the Bidding Documents.] [Copies of Bidding Documents may be purchased from [_Reprographic_Company_] [_Street_Address_], [_City_], [_State_], [_Telephone_No._] at Bidder's expense, for the cost of reproduction.]

BIDDING PROCEDURES

Bids shall be prepared and submitted in accordance with [Document 00100 - Instructions to Bidders and Document 00120 - Supplementary Instructions to Bidders, contained in the Project Manual in the Bidding Documents] [instructions provided separately by the [Owner] [Construction Manager]].

BID SECURITY

[A bid security must accompany each Bid in accordance [in the amount of [10] [____] percent of the Bid price,] [in the amount, form and subject to conditions] as specified in [Document 00100—Instructions to Bidders and Document 00120—Supplementary Instructions to Bidders, contained in the Project Manual in the Bidding Documents.] [instructions provided separately by the [Owner] [Construction Manager].] [A bid security is not required] [____].

BIDDER'S QUALIFICATION

Bidding will be open only to [invited] [pre-qualified] General Contractors, [experienced in work comparable to that required for the Project,] licensed in the State of [California] [_State_] [and 100 percent bondable to perform the Work required]. Owner will pre-qualify General Contractors prior to issuance of Bidding Documents. General Contractors desiring to bid must complete and submit to Owner, AIA Document A305 - CONTRACTOR'S QUALIFICATION STATEMENT, 1986 Edition, and include all listed supplementary information.

RIGHT TO REJECT BIDS

Owner reserves the right to waive irregularities and to reject any and all bids.

Sincerely,

[_Name_]
[_Title_]
[_Architect's_Firm_Name_]
for the Owner

Exhibit 8-1. *(Continued)*

charge will be required for procurement of the documents and whether there will be any refunds.

7. *Bid Security.* State whether a Bid Bond or other type of bid guarantee will be required to ensure the execution of an Agreement between the Owner and the Contractor.

8. *Guarantee Bonds.* State whether Performance Bonds and Labor and Materials Payment Bonds will be required to ensure the completion of the contract.

Instructions to Bidders

The Instructions to Bidders have also been identified by other terms, such as "Information for

Bidders" and "Conditions of Bid." The purpose of the Instructions to Bidders is to outline the requirements necessary to prepare and submit a bid properly. As such, they are truly detailed instructions to a bidder; they guide the bidder in soliciting information concerning discrepancies in the contract documents and provide him or her with all the information necessary to execute the bid form. The AIA document *A701 Instructions to Bidders,* has been developed by the AIA, and a sample is included in Appendix E.

The Instructions to Bidders consists of the following elements:

1. *Form of Bid.* Identify the form of the bid and indicate the number of copies to be submitted.

2. *Preparation of Bid.* Describe which blank spaces in the Bid Form are to be filled in by the bidder, including base bids, alternates, unit prices, and so on.

3. *Submission of Bid.* State how bids are to be sealed, addressed, and delivered.

4. *Examination of Documents and Site.* Instruct the bidder to examine the contract documents and the site of the proposed project in order to become familiar with all aspects of the project.

5. *Interpretation of Documents.* State how discrepancies in contract documents discovered by bidders will be interpreted and resolved by the architect.

6. *Withdrawal and Modification of Bids.* State how bids can be withdrawn or modified prior to bid opening.

7. *Award of Contract.* Describe the procedure under which the award of the contract will be made.

8. *Rejection of Bids.* State the conditions under which the bids can be rejected.

9. *Other Instructions to Bidders.* State whether certain information relative to financial status, subcontractor, and substitutions are to be submitted with the Bid Form.

If preprinted Instructions to Bidders are used, such as the AIA document *A701 Instructions to Bidders,* modifications are necessary to suit specific project requirements. The document describing these modifications is titled *Supplementary Instructions to Bidders.* An example of a *Supplementary Instructions to Bidders* is shown in Exhibit 8-2.

Bid Form

The Bid Form, sometimes termed the "Proposal Form" or "Form of Proposal," is a document prepared by the Architect/Engineer or issuing agency in order to ensure similarity in the preparation and presentation of bids by all bidders and to obtain a uniform basis for comparison. When only the forms prepared by the issuing agency are used, the Owner is assured that all bidders are submitting proposals on an equal basis.

The Bid Form is prepared in the form of a letter from the bidder to the Owner. It contains the necessary blank spaces for the bidder to fill in proposed prices, as well as spaces for the required signatures and addresses (Exhibit 8-3).

The Bid Form consists of the following essential elements:

1. *Addressee.* State the name and address of the individual receiving bids.

2. *Name and Address of Bidder.* State the name of the organization and the address of the bidder.

3. *Project Identification.* State the name of the project.

4. *Acknowledgment.* Provide an enumeration of the documents and a statement that the site has been visited and examined.

5. *Bid Schedule.* Provide a bid list of all the major bid proposals.

6. *Alternates.* Provide a list of all alternate prices. A description of the alternates should be set forth under Division 1, General Requirements.

7. *Unit Prices.* Provide a list of unit prices and their description.

8. *Time of Completion.* Establish the time of completion or permit the bidder to insert his or her own time of completion.

9. *Acknowledgment of Addenda.* Provide spaces for acknowledgment of the receipt of addenda by bidders.

10. *Agreement to Accept Contract.* State the conditions under which the bidder agrees to enter into a formal contract within a specified time.

11. *Signature and Address of Bidder.* Provide spaces to be filled in by the bidder for his or her signature, address, and seal where necessary.

DOCUMENT 00120

SUPPLEMENTARY INSTRUCTIONS TO BIDDERS

I. DOCUMENT INCLUDES

A. The following supplement or modify American Institute of Architects Document A701 - INSTRUCTIONS TO BIDDERS, referenced in Document 00100 of the Project Manual. Where a portion of the Instructions to Bidders is modified or deleted by these Supplementary Instructions, the unaltered portions of the Instructions to Bidders shall remain in effect.

II. SUPPLEMENTARY INSTRUCTIONS

〰〰〰

THE FOLLOWING ARE EXAMPLES OF SUPPLEMENTARY INSTRUCTIONS. SUPPLEMENTARY INSTRUCTIONS ARE NECESSARY TO TAILOR THE PROVISIONS OF STANDARD AIA DOCUMENT A701 TO THE PARTICULAR REQUIREMENTS OF A PROJECT.

REVIEW THESE PROVISIONS WITH THE OWNER AND THE OWNER'S LEGAL COUNSEL. EDIT THEM TO SUIT THE PROJECT.

〰〰〰

A. Article 1, DEFINITIONS, Paragraph 1.8.

Delete Paragraph 1.8 and substitute the following:

"1.8 A Bidder is a person or entity who submits a Bid. Bidders shall only be qualified General Contractors invited by Owner to submit Bids. Qualified Bidders are required to be prequalified before receiving Bidding Documents for the Project. Instructions for prequalification are in Document 00420 - Contractor's Qualification Form in the Project Manual."

B. Article 3, BIDDING DOCUMENTS, Paragraph 3.1 COPIES.

Refer to Subparagraph 3.1.2. Add new Clauses 3.1.2.1 and 3.1.2.2 as follows:

"3.1.2.1 Bidding Documents will be issued directly to Sub-bidders or others upon request from an invited Bidder to the Architect. Bidding Documents issued directly to Sub-bidders or others will not contain referenced pre-printed AIA Documents. AIA Documents may be obtained separately, if requested by Sub-bidders or others, for a nominal additional, non-refundable charge from the office of the Architect or from most local Chapter offices of the American Institute of Architects. Deposit provisions designated in the Invitation to Bid shall apply to Bidding Documents issued to Sub-bidders and others.

"3.1.2.2 Partial sets of Bidding Documents may be obtained as designated in the Advertisement or Invitation to Bid."

C. Article 3, BIDDING DOCUMENTS, Paragraph 3.3, SUBSTITUTIONS.

Add new Subparagraph 3.3.5 as follows:

"3.3.5 The provisions specified in Section 01600 - Product Requirements of the Specifications shall apply to proposals for substitutions."

D. Article 4, BIDDING PROCEDURES, Paragraph 4.1, FORM AND STYLE OF BIDS.

Delete Subparagraph 4.1.1 and substitute the following:

"4.1.1 Bids shall be submitted in duplicate on unaltered forms identical to the Bid Form provided in the Project Manual under the heading of Document 00300 - Bid Forms. Each copy of submitted Bid Form shall be signed with orig-

Exhibit 8-2. Sample supplementary instructions to bidders.

inal, wet-ink signature of party authorized by Bidder to enter into Contract for Construction, with name and title of signing party typed below signature."

E. Article 4, BIDDING PROCEDURES, Paragraph 4.1, FORM AND STYLE OF BIDS.

Refer to Subparagraph 4.1.1. Add new Clauses 4.1.1.1, 4.1.1.2 and 4.1.1.3 as follows:

"4.1.1.1 Each Bid shall include all Work described in the Bidding Documents. Failure to comply may be cause for rejection of Bid. Segregated Bids or assignments will not be considered.

"4.1.1.2 Bidder shall indicate on the Bid Form the percentages for overhead and profit mark-ups for changes in Work which Owner may order.

"4.1.1.3 Bidder shall indicate on the Bid Form the number of calendar days required from date of commencement of Work of Contract, as defined by the Agreement, to complete Work ready for Acceptance by Owner (Substantial Completion). Such number of calendar days shall include consideration of normal adverse conditions due to weather which can be reasonably anticipated for the Project location. Owner will evaluate and consider proposed time from commencement of Work to Substantial Completion as well as sum of Base Bid and Alternates, if any, and Unit Prices, if any, in selection of Contractor. Time of Completion stated in Invitation to Bid, if any, is the maximum number of days acceptable to Owner."

F. Article 4, BIDDING PROCEDURES, Paragraph 4.1, FORM AND STYLE OF BIDS.

After Subparagraph 4.1.7, add the following new Subparagraph 4.1.8:

"4.1.8 Attach to each copy of the Bid a Preliminary Subcontractor List, as instructed in Document 00430—Subcontractor List, except percentage of Work to be performed by listed subcontractors and suppliers need not be less [five] [ten] [____] percent of Base Bid Sum. Final Subcontractor List shall be submitted within [three] [seven] days of notice by Owner to Bidder and shall be prepared in conformance with Document 00430 - Subcontractor List, contained in the Project Manual, including stated minimum percentage of Contract Sum (Base Bid plus Alternate Bids and Unit Prices as Owner may direct) or dollar amount required for listing."

G. Article 4, BIDDING PROCEDURES, Paragraph 4.2, BID SECURITY.

Refer to Subparagraph 4.2.1. Add the following new Clauses 4.2.1.1 and 4.2.1.2:

"4.2.1.1 Bid security is required. Bid security shall accompany each Bid submitted and shall be in the amount of [five] [____] percent of Bid for Base Bid.

"4.2.1.2 Bid security shall be in the form of either a certified check made payable to Owner or a bid surety bond issued by surety licensed to conduct such business in State where Project is located."

H. Article 4, BIDDING PROCEDURES, Paragraph 4.3, SUBMISSION OF BIDS.

Refer to Subparagraph 4.3.1. Add new Clauses 4.3.1.1 and 4.3.1.2 as follows:

"4.3.1.1 The party receiving bids shall be as stated on the Bid Form.

"4.3.1.2 The mailing address for submission of bids shall be as stated on the Bid Form."

Refer to Subparagraph 4.3.2. Add new Clause 4.3.2.1 as follows:

"4.3.2.1 The designated location for deposit of bids shall be as stated on the Bid Form."

I. Article 4, BIDDING PROCEDURES, Paragraph 4.4, MODIFICATION OR WITHDRAWAL OF BID.

Refer to Subparagraph 4.4.1. Add new Clause 4.4.1.1 as follows:

Exhibit 8-2. *(Continued)*

"4.4.1.1 In submitting a Bid, each Bidder shall agree that such Bid will not be modified, canceled or withdrawn by Bidder for a period of [thirty] [forty-five] [sixty] [____] days following time and date for receipt of Bids."

J. Article 5, CONSIDERATION OF BIDS, Paragraph 5.2, REJECTION OF BIDS.

Change name of Paragraph 5.2 to "REJECTION AND DISQUALIFICATION OF BIDS."

Add new Subparagraph 5.2.2 as follows:

"5.2.2 Owner reserves the right to disqualify any Bid, before or after opening, upon evidence of collusion with intent to defraud or other illegal practices by Bidder."

K. Article 7, PERFORMANCE BOND AND PAYMENT BOND, Paragraph 7.1, BOND REQUIREMENTS.

Refer to Paragraph 7.1. Delete Subparagraphs 7.1.1, 7.1.2 and 7.1.3 and substitute new Subparagraphs 7.1.1 and 7.1.2 as follows:

"7.1.1 Owner reserves the right to require, prior to execution (Award) of Contract, selected Bidder to furnish bonds covering 100 percent of Contract Sum covering faithful performance of the Contract and payment of all obligations arising thereunder.

"7.1.2 Costs for such bonds and the name of the issuing surety shall be entered where designated on the Bid Form. Surety shall be subject to acceptance by Owner and, if not acceptable, shall be replaced with a surety acceptable to Owner at no change in cost to Owner."

L. Article 7. PERFORMANCE BOND AND PAYMENT BOND, Paragraph 7.2, TIME OF DELIVERY AND FORMS OF BONDS.

At line two of Subparagraph 7.2.1, delete "not later than three days following" and substitute "promptly."

Add new Subparagraph 7.2.5 as follows:

"7.2.5 Failure or neglect to deliver such bonds shall be considered as having abandoned Contract and the bid security shall be retained as liquidated damages."

M. Article 8, FORM OF AGREEMENT BETWEEN OWNER AND CONTRACTOR, Paragraph 8.1, FORM TO BE USED.

Refer to Paragraph 8.1. Add new Subparagraphs 8.1.2, 8.1.3 and 8.1.4 as follows:

"8.1.2 Notwithstanding any delay in the preparation and execution of the Agreement, each Bidder shall be prepared, upon receipt of written notice of Bid acceptance to commence Work within [ten] [fourteen] calendar days following such notice or on the date of commencement stated in the Agreement, whichever is later.

"8.1.3 Upon notice of Bid acceptance, selected Bidder shall assist and cooperate with Owner in preparing Agreement and other Contract Documents, including submission of all required or requested data concerning Bidder and Sub-bidders, and provision of all information necessary for Architect to prepare, if directed by Owner, conformed Drawings and Specifications incorporating all Addenda, substitutions and other relevant data.

"8.1.4 Within [five] [____] calendar days of presentation of prepared Agreement and, if produced, conformed Contract Drawings and Specifications, selected Bidder shall execute the Agreement unless delay is mutually agreed upon by Owner and selected Bidder."

END OF DOCUMENT

Exhibit 8-2. *(Continued)*

[BID] [PROPOSAL] FORM

[(TO BE COMPLETED IN [DUPLICATE] [TRIPLICATE])]

GENERAL COVENANTS

DATE: _____

PROJECT: [_PROJECT NAME_]
           [_City,_State_]

           [_OWNER_] Project No. [#####.##]
           [_ARCHITECT_] Project No. [#####.##]

BID TO: [_OWNER'S_NAME_]
           [_Mailing_Address_]
           [_Street_Address_]
           [_City_], [_State_]

BIDDER: _____

           _____

           _____

THE WORK

Construction and services required by Bidding Documents, whether completed or partially completed, and including all labor, materials, equipment and services provided by or to Bidder for construction of [_BRIEF_PROJECT_DESCRIP-TION_] as described in the Bidding Documents, consisting of a Project Manual dated [____], containing Bidding and Contract Requirements and Contract Specifications and Drawings as listed in the Project Manual in Document 00015 - List of Drawings. The Work may constitute the whole or part of the Project.

ACKNOWLEDGEMENTS

1. Bidder hereby acknowledges that Bidding Documents have been received, consisting of Project Manual titled "PROJECT MANUAL For The Construction of [_PROJECT_NAME_]," dated [##/##/##], containing Bidding and Contract Requirements and Specifications, and Drawings as listed in Document 00015 - List of Drawings in the Project Manual, have been received.

2. Bidder hereby acknowledges receipt of the following Addenda:

   _____

   _____

3. Bidder hereby acknowledges that Bidder has examined the site and the conditions under which Work is to be performed, that Bidder has examined and is familiar with contents of Bidding Documents and that all costs associated with the site, the conditions under which Work is to performed and all requirements of the Bidding Documents, including all Addenda, are included in the Bid.

4. Bidder has carefully checked the figures submitted below and understands that the Owner will not be responsible for any errors or omissions on the part of the Bidder in making this Bid.

Exhibit 8-3. Sample Bid Form.

BIDDER OBLIGATIONS

1. Bidder hereby agrees to not modify, cancel or withdraw Bid for a period of [thirty] [forty-five] [sixty] [_____] days following time and date established for receipt of Bids by Owner.

2. Bidder hereby agrees to enter into and execute a Contract for Construction, if awarded, on the basis of this Bid.

3. Bidder hereby agrees to submit all data requested by the [Owner] [Construction Manager] and stated in Document 00420 - Contractor's Qualifications Form in the Project Manual, to determine Bidder's qualifications to perform Work.

4. Bidder hereby agrees to identify subcontractors and suppliers Bidder intends to utilize for Work and to submit documentation as stated in Document 00430 - Subcontractor List in the Project Manual. [Bidder further agrees to assist Owner in determining qualifications of subcontractors and suppliers to perform Work.]

BID SECURITY

****

EDIT THE FOLLOWING TO SUIT OWNER'S REQUIREMENTS. TYPICAL REQUIREMENTS ARE INDICATED.

****

1. Accompanying this Bid is _____ (Insert the words, "Bidders Bond," "Cashiers Check," or "Certified Check") in an amount equal to at least [five percent (5)] [ten percent (10)] of the total Bid amount, payable to [_OWNER'S_NAME_].

2. The undersigned deposits the above named security as a Bid security and agrees that it shall be forfeited to [_OWNER'S_NAME_], as liquidated damages in case this proposal is accepted by [_OWNER'S_NAME_] and the undersigned fails to execute a contract with [_OWNER'S_NAME_], as specified in the Contract Documents.

3. In addition, should [_OWNER'S_NAME_] be required to engage the services of an attorney in connection with the enforcement of this Bid, Bidder promises to pay [_OWNER'S_NAME_] reasonable attorney's fees, incurred with or without suit.

4. Bidder hereby agrees to disposition of bid security described in the Instructions to Bidders contained in the Project Manual.

BASIS OF BIDS

1. Contract Time: Bidder hereby proposes that Contract Time described in the Bidding Documents, including all addenda, be (show amount in words and figures; in case of discrepancy, the amount shown in words shall govern) _____ (_____) calendar days from date of commencement of Work to achieving of Substantial Completion of the entire Work. Such number of calendar days shall include consideration of normal adverse conditions due to weather which can be reasonably anticipated for the Project location.

   a. Bidder further proposes to partially complete that portion of the building [_Description_], as described in Section 01010 - Summary of the Work in the Project Manual, in (show amount in words and figures; in case of discrepancy, the amount shown in words shall govern) _____ (_____) calendar days from the date of commencement of the Work. Such number of calendar days shall include consideration of normal adverse conditions due to weather which can be reasonably anticipated for the Project location.

   b. Bidder further proposes to partially complete that portion of the building [_Description_], as described in Section 01010 - Summary of the Work in the Project Manual, in (show amount in words and figures; in case of discrepancy, the amount shown in words shall govern) _____ (_____) calendar days from the date of commencement of the Work. Such number of calendar days shall include consideration of normal adverse conditions due to weather which can be reasonably anticipated for the Project location.

Exhibit 8-3. *(Continued)*

2. Base Bid Sum: Bidder hereby proposes to perform Work described in Bidding Documents, [including Allowances stated in Specifications Section 01020 - Allowances in the Project Manual and] including all Addenda thereto, for the Sum indicated (show amount in words and figures; in case of discrepancy, the amount shown in words shall govern):

Sum of $_____

_____ Dollars

a. Allowance No. 1 - [_TITLE_]: Sum of $[####.##] ([_AMOUNT_IN_WORDS_] Dollars)

****

ADD ADDITIONAL ALLOWANCES TO SUIT.

****

3. Alternate Bid Sums: Bidder hereby proposes to perform all Work described in the following Alternate Bids and in Specifications Section 01030 - Alternates in the Project Manual, for the Sums indicated (show amount in words and figures; in case of discrepancy, the amount shown in words shall govern; indicate whether sum is Add or Deduct):

a. Alternate Bid No. [_#_]: [_Brief_Description_].

Add / Deduct (indicate) Sum of $_____

_____ Dollars

****

ADD ADDITIONAL ALTERNATE BIDS TO SUIT.

****

4. Overhead and Profit Factors: Bidder hereby proposes to perform all Work which may be ordered by Owner for the percentages indicated:

a. For Work performed by Contractor under Change Order, Overhead and Profit will not exceed (show amount in words and figures; in case of discrepancy, the amount shown in words shall govern):

_____ Percent (____%)

b. For Work performed by a subcontractor under Change Order, Overhead and Profit will not exceed (show amount in words and figures; in case of discrepancy, the amount shown in words shall govern):

_____ Percent (____%)

ATTACHMENTS

Bidder hereby acknowledges attachment to [Bid] [Proposal] Form of the following:

1. CONTRACTOR'S QUALIFICATION Form, as referenced in Document 00420 - Contractor's Qualification Form.

2. Preliminary SCHEDULE OF VALUES, as referenced in Document 00425 - Schedule of Values in the Project Manual.

3. Preliminary SUBCONTRACTOR LISTING, as referenced in Document 00430 - Subcontractor List in the Project Manual.

4. Preliminary CONSTRUCTION PROGRESS SCHEDULE, prepared in accordance with requirements specified in Specifications Section 01310 - PROGRESS SCHEDULES, in the Project Manual.

5. Bid security.

Exhibit 8-3. *(Continued)*

```
****************************************************************************************
```

ADD OTHER DOCUMENTS AS REQUIRED BY OWNER, SUCH AS NON-COLLUSION AFFIDAVIT, LIST OF MINORITY- AND WOMEN-OWNED SUBCONTRACTORS, ETC.

```
****************************************************************************************
```

BIDDER'S INFORMATION

Firm Name: _____

Address: _____

_____

Telephone No. _____/_____

FAX Telephone No. _____/_____

Type of Organization: _____
(Individual, Partnership, Corporation)

State in which incorporated: _____
(if applicable)

State Contractors License No.: _____

License Classification _____   Expiration Date: _____

SUBMITTED BY:

I declare under penalty of perjury that the foregoing is true and correct and that I am authorized to sign this proposal on behalf of the Bidder as an agreement to the provisions herein.

Seal of Corporation:

Signature: _____   (   )
                                                (   )
Typed Name: _____   (   )
                                                (   )
Title: _____   (   )

END OF BID FORM

```
****************************************************************************************
```

VERIFY PAGE COUNT AND EDIT FOOTER A ACCORDINGLY. AFTER EDITING THE BID FORM, ADD ADDITIONAL COPIES AS REQUIRED FOR SUBMISSION FOLLOWING THE HARD PAGE BREAK, BELOW.

```
****************************************************************************************
```

Exhibit 8-3. *(Continued)*

Chapter 9

General Conditions of the Contract

CONTRACTUAL RELATIONSHIPS

The Conditions of the Contract define basic rights, responsibilities, and relationships of the entities involved in the performance of the Contract (Agreement). The Conditions of the Contract are an inherent part of the Owner-Contractor Agreement and are considered to be the "general clauses" of the Agreement. There are two types of Conditions of the Contract, typically:

1. General Conditions of the Contract
2. Supplementary Conditions of the Contract

The General Conditions will be addressed in this chapter. The next chapter will consider the Supplementary Conditions.

The General Conditions are the provisions that establish the legal responsibilities and relationships among the parties involved in the Work of the Contract. These parties are formally termed the Contractor and the Owner, but also include the former's subcontractors and the latter's consultants, such as legal and insurance counsels and marketing and leasing agents. Other parties are mentioned in the General Conditions only as needed to set forth and clarify responsibilities and relationships in more complex contractual arrangements. If there is to be a bond, the surety will play a part; if there is a Construction Manager, his or her duties will be included; if there will be multiple contracts and contractors, the basic responsibilities and relationships among them will be stated.

The Architect or Engineer, whichever is the prime design professional, is identified in the General Conditions. Although not signers of the Owner-Contractor Agreement, duties and responsibilities of the Architect or Engineer are stated in the General Conditions.

General Conditions of the Contract are often preprinted documents published by professional and other industry associations and by public agencies. In the cases where the Owner is a public agency, the General Conditions may be modified to suit specific project conditions, and there are no Supplementary Conditions.

General and Supplementary Conditions should be closely coordinated with related documents such as the Owner-Architect/Engineer Agreement, the Owner-Contractor Agreement, and Division 1 - General Requirements in the Specifications. Duties and responsibilities prescribed in one of these documents must be consistent with the other documents. For example, if Division 1 - General Requirements states that the Architect will review mock-ups at the project site but the Owner-Architect Agreement does not provide for construction phase services by the Architect, then either the review will not be performed or the Architect, who prepared Division 1, has said that the Architect will perform the review without compensation. Similarly, if the Owner-Contractor Agreement states that the Engineer will review the layout and routing of piping and valves but the Owner-Engineer has not contracted with the Owner for such services, then the Contractor and Owner may have a difficult-to-resolve dispute if, after the piping is installed, the layout conflicts with other building components and systems.

In one case, an overzealous architect agreed in the Owner-Architect Agreement that the architect's construction phase services would be as stated in the General Conditions accompanying the Owner-Contractor Agreement. Unfortunately for the Architect, the Owner-Contractor Agreement and its General Conditions were not written when the architect signed the Owner-Architect Agreement. The architect was required to perform additional site visits, with very substantial labor charges and travel expenses.

Owners—those who undertake to build buildings and engineering works—are very diverse. Manufacturers need plants, counties want roads and bridges, a church congregation needs a new sanctuary, the township needs a new school, a developer wants to turn 1500 acres of farmland into an office park, a university needs a new co-generation plant, the Veterans Administration needs an addition for new medical diagnostic equipment and an outpatient clinic, and the Johnson family wants to build their dream house just like the one on the Saturday afternoon television program.

Despite this diversity in owners and building types, there are some consistent needs. Provisions must be established for progress payments by the Owner to the Contractor. Obligations for preparation, submission and review of shop drawings and sample reviews need to be established. Provisions for dispute resolution and contract termination or suspension need to be included in the contract. Each of these is a subject for the General Conditions and, in fact, each is addressed in standard General Conditions published by professional and industry associations.

Over 90 years ago, it was recognized that standard General Conditions would be beneficial for everyone involved in the construction contract. Owners, with their diverse building programs but similar interests, would benefit from standard General Conditions, especially if they regularly engaged in building or facility construction. Contractors would benefit from standard General Conditions because as they became familiar with the provisions—like them or not—there would be clear expectations about what the Contractor was and was not responsible for. The Architect/Engineer would also benefit from standardized documents because of their familiarity, as with the Contractor, and also because use of standardized, coordinated documents would mean that the Architect/Engineer and Owner will have clear expectations and responsibilities during construction.

Architect General Conditions

In the absence of a unified voice for the owners, the AIA offered prospective owners of construction a set of contract conditions as early as 1888, then called the "Uniform Contract." This was a printed contract form designed for use as the contractual agreement between the owner and the contractor. It defined many responsibilities and set forth some administrative procedures. In 1911 the Uniform Contract was divided into two parts that are familiar to us today: Standard Form of Agreement between Owner and Contractor, now called AIA A101, and General Conditions of the Contract for Construction, called AIA A201. (See Appendix E for a review-only copy of *AIA A201*.)

Engineer General Conditions

Similarly, for many years various engineering societies prepared sample agreements and contract conditions for the guidance of practicing engineers and their clients. In 1963, the Professional Engineers in Private Practice, now a division of the National Society of Professional Engineers (NSPE/PEPP), formed a contract documents committee, which published a number of contract forms. In the 1970s, the NSPE joined with two other engineering societies, the American Council of Engineering Companies (ACEC) and the American Society of Civil Engineers (ASCE), to form the Engineers Joint Contract Documents Committee (EJCDC).

EJCDC is responsible for preparing engineering-related construction contract documents similar to those AIA produces for architectural documents. EJCDC contract documents are available from the member associations. Examples of EJCDC documents are:

- *EJCDC C-520 Suggested Form of Agreement Between Owner & Contractor, Stipulated Price* (2002)
- *EJCDC C-525 Standard Form of Agreement Between Owner & Contractor, Cost Plus* (2002)
- *EJCDC C-700 Standard General Conditions of the Construction Contract* (2002)
- *EJCDC C-800 Guide to the Preparation of Supplementary Conditions* (2002)
- *EJCDC C-200 Guide to Preparation of Instructions to Bidders* (2002)

(See Appendix F for a reference-only copy of *EJCDC C-700*.)

Differences Between AIA and EJCDC General Conditions

The architecture and engineering societies appear to have avoided needless disagreement in preparing their General Conditions documents. By working from court cases at all levels, through similar experience in planning and building, and by inviting review by the Associated General Contractors of America (AGC), they have given the building community two sets of documents that accurately distinguish the North American tradition of building from the European and British traditions.

The engineering documents, of course, add or amplify topics needed for engineering projects, such as provisions for unit price work, availability of lands, underground conditions and reference points.

But the AIA and EJCDC documents are very similar.

Beyond these necessary differences lie some topics in which the EJCDC *C-700 Standard General Conditions* differs from the AIA *A201 General Conditions*. For example:

- "The Work," for engineers, is the entire completed construction; for architects, it may be completed and partially completed construction.
- For engineers, substitutions may be requested only after the Agreement is signed, and then according to detailed rules for submission and assumption of responsibility by the Contractor; for architects, these procedures are defined in the Supplementary Conditions.
- Contract closeout provisions vary. The correction ("punch") list is made by the Engineer in the EJCDC documents, while in the AIA documents the punch list is made by the Contractor and reviewed and supplemented by the Architect.
- What is called a Work Change Directive by EJCDC is called a Construction Change Directive" by AIA.
- The responsibility for the means and methods of construction are more emphatically placed on the Contractor in the EJCDC documents than in the AIA documents.

These differences may seem subtle, but they can be substantial under the pressure of a construction contract dispute. Architects and engineers should be sensitive to the differences and check the general conditions for carryover of inappropriate provisions. For example, if the owner's representative has an engineering background, then engineering-type provisions may be inappropriately applied to a building project. Similarly, architectural-type provisions may be applied inappropriately to general engineering construction such as storm drainage or rail transport projects.

Use of Standardized AIA and EJCDC General Conditions

Led by the AIA's documents, the pattern set by the professional societies has generally been adopted by the construction industry and its clients. Although in theory the General Conditions of each owner's contract are written by the owner's legal counsel, in fact the AIA General Conditions have been overwhelmingly adopted.

This wide use has led to testing of the General Conditions in the courtroom, in courthouse corridors, in arbitration, and, most important, in that arena where most disputes are settled: the jobsite trailer. This generally successful track record for the AIA and EJCDC documents over many years is an important argument in favor of using them wherever possible and with as few modifications as possible.

It appears that AIA *A201* has not only stood up well in court but has been deemed a fair document by owners and contractors as well. Since important contractor objections were worked out to their satisfaction in the 1915 edition of the *AIA General Conditions,* there has generally been a history of cooperation between architects and contractor groups in preparing each new edition. The notable exception was the 1966 edition, which was withdrawn and reissued in 1967 to satisfy contractors' complaints about the indemnification clause. It is notable that AIA A201 is endorsed by the Associated General Contractors of American (AGC).

AIA has revised its General Conditions many times since 1911 and now holds to roughly a 10-year cycle of restudy and rewriting. This allows AIA and AGC to adjust to new realities of practice while sparing users the confusion of a document that is being tinkered with incessantly. Legal decisions are reflected in updated versions, and risk management considerations are also incorporated.

For the small project, AIA *A201* is considered by many owners and architects to be too voluminous and unwieldy. AIA has developed a combination agreement form and general conditions, known as the *AIA A107 Abbreviated Standard Form of Agreement Between Owner and Contractor for Construction Projects of Limited Scope.* A copy of this form is included in Appendix E. For simple and small projects, such as tenant space improvements and residential construction, this brief document may be adequate.

Other Private-Sector Sources for General Conditions

The AGC issues three series of General Conditions and agreement documents that are of use in special methods of project delivery advocated by AGC.

- For design-build projects, the 400 Series provides documents that define the role of each prime contractor under the overall control of the design-build entity.
- The 500 Series contains documents that handle construction management in a different way from AIA *A201/CM. AGC 500* allows the construction manager to do work with his or her own forces and provides for a guaranteed maximum price (GMP) option. *AGC 510* provides construction management, with the owner awarding all trade contracts.
- In the 600 Series, which primarily covers subcontract agreements, there is *AGC 645,* which

is for a negotiated agreement between the owner and contractor.

Although the *AGC General Conditions* use AIA provisions as a starting point, there are distinct differences. Just as AIA *A201* has been criticized by contractors and owners for its attention to risk management for the architect's benefit, the AGC documents may tend to be disadvantageous to the Architect/Engineer. Those unfamiliar with the AGC documents should have them reviewed by the Architect/Engineer's legal and insurance (risk management) counsels, especially regarding differing treatment of contract administration, warranty and claims.

The American Association of State Highway and Transportation Officials (AASHTO) devotes Section 100, the initial section in its *Guide Specifications for Highway Construction,* to General Provisions, which precede sections on earthwork, the several types of highway construction and highway materials. These General Provisions are not conceived as part of the Project Manual concept; there are no separate bidding requirements, supplementary conditions, or a general requirements division. Since highway material quantities cannot be predicted precisely, the work is carried out on a unit price basis. The AASHTO general provisions reflect this mode of project delivery by including detailed measurement and payment language. As would be expected with roadwork, there is suitable text covering force account work, inspection, field laboratories, traffic maintenance, erosion and sedimentation control, and protecting vegetation. Although there is a value engineering article, virtually no provision is made for substitutions, claims or warranty of the work beyond final acceptance.

Federal Government General Conditions

For agencies in the executive branch of the U.S. federal government, the Construction Contract Clauses serve as General Conditions. The GPO Superintendent of Documents publishes and distributes the Federal Acquisition Regulation (FAR). FAR is available in the Code of Federal Regulations (48 CFR), in looseleaf bound form and in a CD-ROM. Periodic updates are available from the Government Printing Office and the Department of Commerce Clearing House. FAR governs construction as well as other procurement activities of the federal government.

Construction for the military services, Veterans Administration (VA), National Aeronautics and Space Administration (NASA) and General Services Administration (GSA) is governed by FAR contract clauses selected from CFR Title 48 and from supplements that each agency is allowed to add for its own unique purposes. Two bodies, the Defense Acquisition Regulatory Council (DAR Council) and the Civilian Agency Acquisition Council (CAA Council), prepare the contract clauses, frequently in duplicate to suit the differing needs or preferences of the Defense and Civilian Agency Councils. The work of the two councils and the secretariat is reviewed by the Office of Management and Budget (OMB).

FAR is administered for construction by three entities:

1. General Services Administration
2. Department of Defense
3. National Aeronautics and Space Administration

Each service or agency assembles clauses that are suitable for its construction projects and publishes that compendium for its own use. Form 3506 is the current designation for the GSA's primary construction contract clauses compendium. Slightly different compendia, for construction inside the United States, construction outside the United States, fixed-price contracts, unit price contracts, and so forth, become the boilerplate at the beginning of each Project Manual.

The private-sector general conditions have been tested in the courts, and are written and interpreted in the light of the law as it applies to everyone. With federal projects, however, the courts may not be involved. One does not sue the government unless the government gives permission. However, there are avenues of appeal from a Contracting Officer's decisions during construction.

Since the government knows that even it can make mistakes, and since it is in its interest to be fair with contractors and others who suffer damage working for the government, Congress has set up various appeals boards that sit as a court of claims. Over the years, FAR and its predecessors have been tested in bodies such as the Armed Services Board of Contract Appeals, the General Services Board of Contract Appeals, the NASA Services Board of Contract Appeals, the DOT Services Board of Contract Appeals and the VA Contract Appeals Board. The procedures are much like those of the courts, and decisions handed down are published, to become precedents in subsequent cases.

Many of the features of the general conditions of professional societies have their counterpart in the federal contract clauses. However, the design professional is almost totally absent from the federal documents: the contract recognizes the government and the contractor only. In federal contracts, much attention is given to subcontractors and employment practices, as well as to audits that reveal the actual prices paid by the contractor for materials and labor. Claims by either party are decided by the Contracting Officer within 60 days and are final, although subject to appeal.

When working under federal contract clauses, one must be aware of the many differences from private-sector documents. Specifications take precedence over drawings in some instances; correction of work and warranty provisions are mingled; there is no concept of substantial completion, only final acceptance; and quality control procedures by the Contractor are expected.

Supplements are often made to the contract clauses in such matters as labor standards and value incentive clauses. Special clauses having the effect of supplementary conditions are often added. These vary according to the service or agency writing the contract and may cover such subjects as liquidated damages, salvage, use of explosives, testing, scheduling, safety, acceleration of work, and accommodations and meals for inspectors working far from the home office.

Architects and engineers performing professional services for federal agencies should request instructions and documents from the project's Contracting Officer. Familiarization with FAR and with the agency's general provisions for the construction contract is very important.

OTHER PUBLIC AGENCY AND CORPORATE GENERAL CONDITIONS

General Conditions of states, municipalities, and corporations, prepared by their attorneys for their building programs, appear in most cases to have been based on some edition or of the AIA documents. Some of these tailored general conditions reflect a building climate that existed decades ago, where the administrator of the contract had powers that would be considered dangerous today, and where contractors were expected to do a complete job even if everything was not shown or specified. City solicitors and corporate counsel have added provisos that make some of these general conditions barely recognizable to a contractor who is used to working with current AIA and EJCDC documents.

Almost every general conditions document used in the nonfederal building construction sector until 1989 was based on an obsolete edition of the *AIA General Conditions.* For the legal protection of all parties, and to recognize the changes that have occurred in the building process, it is time for many of these dog-eared documents, some of which have attained the status of sacred scrolls, to be brought up-to-date. They should be revised to follow more closely current national professional society general conditions. AIA and EJCDC documents have been revised by market forces to reflect court decisions and changes in law, and to stay current with the way building is done today.

Supplementary Conditions

It is always necessary for the owner to modify AIA or EJCDC General Conditions to suit the unique requirements of a specific project. These modifications, in the form of additions, deletions, and substitutions to the General Conditions, are called "supplementary conditions."

Terms such as "special conditions," "special provisions," "general provisions" and "supplementary general conditions" are sometimes used. These terms are incorrect. There are General Conditions and there are Supplementary Conditions—at least according to the publishing associations. It should be left to the owner's legal counsel, who is responsible for preparation of the conditions of the contract, to establish the titles that will be used.

A new development has occurred with standard AIA General Conditions. Instead of publishing separate Supplementary Conditions, AIA offers computerized versions of their documents, particularly the AIA *A201 General Conditions,* with a publishing license, that allows the Owner's legal counsel and the Architect to make revisions to the AIA *A201* text and print the modified document. Additions, deletions and revisions are graphically indicated in the text. Contact the AIA for information and a demonstration disk for AIA's Electronic Format Documents for more information and availability.

See Chapter 10 for further discussion of Supplementary Conditions of the Contract.

Location of the General Conditions

The General Conditions should always be bound into (signed contract copies of) the Project Manual.

See Chapters 3 and 4 for discussion of the organization and content of the Project Manual.

Duties and Responsibilities of the Architect/Engineer

The Architect/Engineer is *not* a signer of the Owner-Contractor Agreement. However, certain duties and responsibilities are described in the General Conditions for the Architect/Engineer, which generally include the following:

- Acting as the Owner's representative for matters concerning the Work (projects not under the construction management delivery method)
- Routinely visiting the construction site to observe and evaluate progress (coordinated with the Owner-Architect/Engineer Agreement)
- Validate the progress of the Work for the purpose of recommending to the Owner that progress payments be made
- Preparing Change Orders and other Contract modification documents
- Clarifying and interpreting the Contract Documents
- Exercising approval and disapproval authority on submittals
- Reviewing shop drawings and samples
- Rejecting defective and noncomplying Work
- Determining dates of Substantial Completion and Final Completion

Preprinted General Conditions

To obtain standard printed General Conditions and related documents, the following associations may be contacted for precise titles, dates of issue, ordering information and price. A note of caution: These documents are copyrighted, and some publishing organizations aggressively protect the copyright by pursuing actions against those who make unauthorized copies.

American Association of State Highway and Transportation Officials (AASHTO)
444 North Capitol Street NW, Suite 249
Washington, DC 20001
202/624-5800
www.aashto.org

American Council of Engineering Companies (ACEC): EJCDC documents
1015 15th Street NW, 8th Floor
Washington, DC 20005-2505
202/347-7474
www.acec.org

American Institute of Architects (AIA)
1735 New York Avenue, NW
Washington, DC 20006
202/626-7300 800/242-3837
www.aia.org
Also most local chapter offices of AIA

American Society of Civil Engineers (ASCE): EJCDC documents
1801 Alexander Bell Drive
Reston, VA 20191
800/548-2723
www.asce.org

Associated General Contractors of America (AGC)
333 John Carlyle Street, Suite 200
Alexandria, VA 22314
703/548-3118
www.agc.org

National Society of Professional Engineers (NSPE): EJCDC documents
1420 King Street
Alexandria, VA 22314-2715
703/684-2862
www.nspe.org

Chapter 10

Supplementary Conditions of the Contract

SUPPLEMENTARY CONDITIONS

It is always necessary to modify preprinted General Conditions of the Contract to suit specific project requirements. Both AIA and EJCDC recognize this, and both provide documents to guide the process. AIA publishes *AIA Document A511—Guide for Supplementary Conditions,* and EJCDC publishes *EJCDC C-800 Guide to the Preparation of Supplementary Conditions.* In addition, contact AIA for its electronic format documents, including software to produce a customized version of *AIA A201 General Conditions of the Contract for Construction,* eliminating the need for separate General Conditions and Supplementary Conditions. The topics and principles for modifications to the General Conditions remain the same.

Modifications to the General Conditions take the form of additions, deletions and substitutions and are called Supplementary Conditions. They are not called "special conditions," "special provisions," or "supplementary general conditions." They are the Supplementary Conditions.

In the past, the Supplementary Conditions were larded with many topics that are, in essence, work of a general nature to be performed by the Contractor in order to construct the building or facility. These specific work topics went beyond the broad contractual provisions of the contract that are the proper scope of General and Supplementary Conditions. Out-of-place work topics in some Supplementary Conditions have included administrative matters such as who shall prepare meeting agendas and publish meeting minutes. Matters regarding temporary construction, such as the size of temporary worksite offices and the types of telecommunications services and equipment on the worksite, do not belong with provisions prescribing general responsibilities for signing parties of the contract for construction.

There is a well-meaning but misguided tendency for construction managers, who have responsibility for preparation of the General Conditions and Supplementary Conditions, to address important construction matters in the documents they control. This creates conflicts between the construction specifications, which more appropriately address detailed administrative and temporary construction matters, and the broad and authoritative General and Supplementary Conditions.

Until the advent of the CSI *MasterFormat,* there was little choice but to address administrative and temporary construction matters in the Supplementary Conditions or to create another layer to the hierarchy of documents called "special conditions." Now, with CSI *MasterFormat,* there is a place for these detailed requirements, in Division 1 of the specifications, titled General Requirements. (Division 1 - General Requirements is discussed in detail in Chapter 12.)

To aid in determining which topics are best inserted in the General Conditions, in the Supplementary Conditions, and in Division 1 - General Requirements, AIA and EJCDC have jointly published the *Uniform Location of Subject Matter.* See Exhibit 10-1 for sample pages from this document. The *Uniform Location of Subject Matter* is a compendium of practically every general subject that can come up in preparing a set of contract documents. It advises which topics are best located in the Invitation to Bids, the Instructions to Bidders, the Bid Form, the Agreement, the General Conditions, the Supplementary Conditions, and Division 1 of the specifications.

Modifying the General Conditions

Although this book is limited to subjects of technical construction specifications and cannot presume to offer expertise on legal (contractual) matters, it can forewarn architects and engineers about the use of General Conditions and their modifications. The most important reason for having an owner's document that covers only the legal responsibilities and relationships of the parties is that designers are not in the business of practicing law. Under no circumstances should design professionals draw up or modify legal (contractual) provisions unless, upon doing so, these forms and modifications are forwarded to the owner and the owner's legal and insurance counsels. To protect the design professional, responsibility for these

AIA Document A521 / EJCDC Document 1910-16
Uniform Location of Subject Matter

Information in Construction Documents

Prepared by

Engineers' Joint Contract Documents Committee

Issued and Published Jointly by

Professional Engineers in Private Practice
a practice division of the
National Society of Professional Engineers

American Consulting Engineers Council

American Society of Civil Engineers

With Participation and Joint Publication by

The American Institute of Architects

Construction Specifications Institute

Exhibit 10-1. Sample pages from Uniform Location of Subject Matter

Introduction

An examination of the bidding and construction documents published for the construction industry indicates that there are varying approaches to where in these documents particular subject matter is located. It is apparent that parties both familiar and unfamiliar with these documents will benefit from a reference list and location of subject matter.

With this in mind, the Engineers' Joint Contract Documents Committee (EJCDC), with concurrence of the AIA Documents Committee, has prepared the attached directory of information for those assembling these documents. The bidding and construction documents considered here are those customary to a construction project: Advertisement or Invitation, Instructions to Bidders, Bid, Agreement, General Conditions, Supplementary Conditions, General Requirements and Specifications. An understanding of the decisions reflected in the Uniform Location of Subject Matter may prove helpful.

The Contract Documents include the Agreement, General Conditions, Supplementary Conditions, General Requirements and Specifications. The Bidding Documents (Advertisement or Invitation and Instructions) are not part of the Contract Documents. Their substance pertains to relationships existing before the construction agreement or contract is signed. The Bid is only an offer to perform the Work. Performance and payment bonds and other instruments of surety, as well as Addenda and drawings, are not addressed here.

The Agreement should contain the contract price, terms of payment and retainage. In a negotiated contract, some of the terms of the Agreement may be confidential. For example, when the parties to the contract decide that the Agreement will not be made available to subcontractors or suppliers, important information about retainage and other subjects must be covered elsewhere, as in the Supplementary Conditions.

The General Conditions contain the basic provisions affecting the rights and duties of the parties involved, and should be altered only where mandated by the specific requirements of a given project and following the advice of an attorney. There are many provisions of the General Conditions that need to be supplemented, such as the procedural arrangements for the processing of **Shop** Drawings as set forth in Division 1 of the Specifications, or the specific requirements for property and liability insurance coverage that should appear in the Supplementary Conditions.

Significant efforts have been made by the AIA and EJCDC committees to develop guidelines for preparation of Supplementary Conditions. While such documents are primarily intended for use when modifying or supplementing the basic contractual relationships, the division of subject matter between the Supplementary Conditions and General Requirements in Division 1 of the Specifications is more difficult and may appear arbitrary at times. Division 1 pertains to performance of the Work and applies to the other divisions of the Specifications. For example, language supplementing the General Conditions on Shop Drawing processing and the requirement to maintain a current marked-up set of record documents at the site will appear in Division 1, whereas amounts of property and liability insurance will appear in the Supplementary Conditions. This concept appears in CSI's *Spectext* (master specifications guide), Construction Sciences Research Foundation's *Manual of Practice*, AIA's *Masterspec*, and by AIA and EJCDC in their guidelines to the preparation of Supplementary Conditions; all of which are identified below.

This listing is intended as a guide, not only for design professionals, but also for owners, attorneys, contractors, subcontractors, lenders, sureties and others who work with these documents.

1

Exhibit 10-1: *(Continued)*

Code Symbols

The following code symbols have been used in the Uniform Location of Subject Matter:

P = The principal or primary statement of the topic involved that includes the basic principles of the relationships to the parties.

S = A further statement to supplement or amend the principle statement, to add details, procedural requirements or specific data applicable to the basic statement, to add language required by law or regulation, or to adapt the basic statement of principle to the requirements of the particular contract.

X = Cross reference to where the basic or primary statement may be located.

§ = The term or subject matter referred to is not used or dealt with in the standard Contract Documents prepared by EJCDC.

= The term or subject matter referred to is not used or dealt with in the standard Contract Documents prepared by AIA.

In some instances, a particular matter may be coded with a "P" in two places. This is intentional, and is used to recognize that, in certain instances, repetition is acceptable (i.e., where the statement will appear in both the Agreement and the Bid), or to indicate a choice.

Bibliography

The following is a bibliography of the documents prepared by AIA, EJCDC and CSI to which this Uniform Location of Subject Matter applies.

The American Institute of Architects

Instructions to Bidders—A701

Owner-Contractor Agreement Form, Stipulated Sum—A101

Owner-Contractor Agreement Form, Cost Plus a Fee—A111

Abbreviated Owner-Contractor Agreement Form, Stipulated Sum—A107

Abbreviated Owner-Contractor Agreement Form, Cost Plus a Fee—A117

General Conditions of the Contract for Construction—A201

Guide for Supplementary Conditions—A511

MASTERSPEC

MASTERSPEC—Division 1 (General Requirements)

The Engineers' Joint Contract Documents Committee

Guide to the Preparation of Instructions to Bidders—No. 1910-12

Suggested Bid Form Commentary for Use—No. 1910-18

Standard Form of Agreement Between Owner and Contractor on the Basis of a Stipulated Price—No. 1910-8-A-1

Standard Form of Agreement Between Owner and Contractor on the Basis of Cost-Plus—No. 1910-8-A-2

Standard General Conditions of the Construction Contract—No. 1910-8

Guide to the Preparation of Supplementary Conditions—No. 1910-17

Commentary for Procurement Documents—No. 1910-26-E

Instructions to Bidders for Procurement Contracts—No. 1910-26-D

Standard Form of Procurement Agreement Between Owner and Contractor—No. 1910-26-A

Procurement General Conditions—No. 1910-26-B

Guide to the Preparation of Procurement Supplementary Conditions—No. 1910-26-C

The Construction Specifications Institute

CSI Manual of Practice

Construction Documents Fundamentals and Formats Module

FF/030	Construction Documents (CD)
FF/040	Construction Contracts (CO)
FF/050	Bidding Requirements (BR)
FF/060	Conditions of the Contract (CC)
FF/070	Specifications (SP)
FF/100	Division 1—General Requirements (GR)
FF/140	Allowances, Unit Prices, and Alternatives (AL)
FF/150	Substitutions (SB)
FF/160	Warranties (WA)

Construction Specifications Practice Module

SP/010	The Agreement (AG)
SP/020	Construction Bonds (CB)
SP/030	Construction Insurance (CI)
SP/080	Procurement Specifying (PR)
SP/130	Design-Build (DB)

MasterFormat (CSI/CSC)

Construction Sciences Research Foundation (CSRF)

SPECTEXT—Division 1 (General Requirements)

2

Exhibit 10-1: *(Continued)*

The American Institute of Architects is pleased to provide this sample copy of an AIA Contract Document for education purposes. Created with the consensus of contractors, attorneys, architects and engineers, the AIA Contract Documents represent over 110 years of legal precedent.

KEY OR CODE

P = Primary Statement of Subject Matter
S = Supplementary Statement
X = Cross Reference to Primary Statement
§ = Not used by EJCDC
# = Not used by AIA

Subject Matter	Advertisement or Invitation	Instructions	Bid	Agreement	General Conditions	Supplementary Conditions	General Requirements Division 1	Specifications Division 2-16	Notes
A. Abbreviations—definition of those used in specs.							P		
Acceptance of Work					P				
Access to the Work by A/E and others (see Work)									
Addenda—definition of		X			P				
Addresses for Notices				P			S		
Adjustment Prices			S	S			P	X	#
Agreement—definition of		X			P				#
Agreement—execution of		P							
Allowances—cash		X	X			S	P	X	
Alternates/Alternatives—description of		S	X	X			P		
Alternates/Alternatives—for grouping of parts of Work in bid		S	P				S		
Alternates/Alternatives—for materials and equipment in bid		S	P	X			S	S	
Application for Payment—definition of (see Payment)									
Application for Payment—form or contents of (see Payment)									
Application for Final Payment (see Payment)									
Application for Progress Payment (see Payment)									
Approval (Recommendation) of Payments (see Payment)									
Arbitration					P	S			
Architect—definition of					P				§
Architect—identification of		S		P					§
Architect—responsibilities and limitations of					P	S			§
Architect—status during construction					P	S			§
Architect—supplemental instructions					P				§ For term used by EJCDC, see Field Order.
Architect—visits to site					P	S			§
Assignment of Other Contracts to General Contractor						P	S		
Assignment by Contractor				P					
Availability of Lands or Site					P	S	S		
Award—basis of		P							
Award—Notice of, defined (see Notice of Award)									
Award—Notice of, timing and procedure for giving (see Notice of Award)									
B. Before Starting Construction—Contractor's responsibilities (see Contractor)									
Before Starting Construction—Owner's responsibilities (see Owner)									
Bid—definition of		P			P				
Bid—evaluation		P							
Bid—opening procedure		P							
Bid—Owner's discretion to accept, reject or waive		P							
Bid—requirements for preparation and submission of		P							
Bid Opening—time and place	P								
Bid Security—in general and detailed requirements for	S	P							All requirements pertain to time before execution of Agreement.
Bidder—authority to sign		P		S					
Bidder—qualification of	X	P							Reference to specific laws or regulations sometimes required.
Bidding Documents—deposit refund	X	P							
Bidding Documents—obtaining	P	X							
Bonds and Insurance—in general		X			P	S			
Bonds, Bid (see Bid Security)									
Bonds, Performance and Payment—definition of					P				

3

WARNING: Unlicensed photocopying violates U.S. copyright laws and will subject the violator to legal prosecution.

Exhibit 10-1: *(Continued)*

KEY OR CODE

P = Primary Statement of Subject Matter
S = Supplementary Statement
X = Cross Reference to Primary Statement
§ = Not used by EJCDC
# = Not used by AIA

SUBJECT MATTER	Advertisement or Invitation	Instructions	Bid (May be attached as Exhibit to Agreement and become one of Contract Documents)	Agreement	General Conditions	Supplementary Conditions	General Requirements Division 1	Specifications Division 2-16	NOTES
Bonds, Performance and Payment—delivery of		P							
Bonds, Performance and Payment—requirements for	S	X			P	S			See Bond Forms
C. Cash Allowances (see Allowances)									
Changes and Supplements to Agreement and General Conditions—not specifically allocated herein to General Requirements, Division 1.						P			
Change Order—definition of					P				
Changes in the Work					P	S	S		
Claims for Additional Cost or Time					P	S			
Claims for Damages					P				§
Claims, Waiver of					P				
Clarifications and Interpretations—by A/E of Bidding Documents		P							
Clarifications and Interpretations—by A/E of Contract Documents					P				
Cleaning—before construction							P		
Cleaning—during construction					P		S	S	
Cleaning—final					P		S	S	
Closing Out Project—procedures							P		
Commencing Work at the Site					P	S			
Communications During Construction					P	S			
Communications Prior to Opening of Bids		P							
Completion—Final					P	S			
Completion—Substantial (see Substantial Completion)									
Conferences and Meetings—after award but before construction					P	S			
Conferences and Meetings—during construction						P			
Conferences and Meetings—pre-award		P							#
Conferences and Meetings—pre-bid	X	P							#
Conferences and Meetings—progress							P		
Conflicts, Errors or Discrepancies—reporting					P				
Construction—by other contractors					P	S			
Construction—by Owner					P	S			
Construction Change Directive					P				§ See Work Change Directive
Construction—Coordination					P	S			
Construction—Future					P				§
Construction Equipment Furnished by Contractor					P	S			
Construction Manager—identity of	S			P					§ See also AIA CM Series Documents:
Construction Manager—representative of Owner and responsibilities of						P			§ See also AIA CM Series Documents:
Construction Schedule					P	S			
Construction—sequences of, if specified in Contract Documents by A/E							P	S	
Coordination of Project and Work of Various Trades					P	S	S		
Continuing Work During Disputes					P				
Contract Documents—copies furnished by Owner					P	S			
Contract Documents—definition of				S	P	S			
Contract Documents—detailed listing of				P					
Contract Documents—ownership of					P				
Contract Price Breakdown by Contractor for Owner's Accounting Requirements							P		# This is not a Schedule of Values.

4

Exhibit 10-1: (Continued)

legal forms and their modifications must, in turn, be accepted in writing by the owner and the owner's counsels. Otherwise, the design professional may be accused of practicing law or insurance counseling.

By limiting the articles in the General Conditions to those having to do only with the contractual duties, responsibilities, and relationships, and having the owner take proper responsibility in regard to them, the architect or engineer can then develop other provisions in Division 1 - General Requirements, for which a design professional should be qualified by education, training and experience. The design professional is similarly cautioned with respect to the insurance provisions contained in the General Conditions. To prepare the necessary modifications and amendments to the General Conditions pertaining to insurance and bonds, the design professional must obtain specific directions and information from the owner's insurance counsel to ensure that the insurance provisions adequately protect the owner and other parties and are appropriate for the specific project.

Some architects and engineers include in the Project Manual only the Supplementary Conditions and reference standard General Conditions, such as AIA *A201* and EJCDC *C-700*. In signed contract copies of the Project Manual, it is necessary to include a copy of the preprinted, standard General Conditions. It is also advisable to include a copy of the General Conditions in all copies of the Project Manual, but AIA and EJCDC prohibit reproduction of their documents without licensing. The advantage of the recently developed and still-developing AIA electronic documents is that licensing has become easy and affordable. There is no practical reason now for omitting a copy of the standard General Conditions. The same is true for the AIA and EJCDC Instructions to Bidders documents.

Format of Supplementary Conditions

Assuming that the General Conditions document is not physically modified, such as when AIA's electronic version is used, the modification process should follow a simple concept: describe the physical change to be made to the document. Actual strike-throughs and interspersed notes on a printed copy of the General Conditions should not be done. The integrity of the original document should be preserved for two reasons: modifications of the printed copy could create copyright problems and, more important, the modifications tend to be difficult to read. Instead, Supplementary Conditions typically describe the physical changes to be made

to the General Conditions so that a clerical person could make the changes.

Using AIA *A201 General Conditions* as the example, changes should be listed in the order of the articles of the General Conditions and should be keyed to the articles and paragraphs of the original document. New articles and paragraphs should be added with care in order to preserve the integrity of the original document as much as possible. The AIA *A511 Guide for Supplementary Conditions* is set up this way and is recommended for use in making changes to AIA *A201*. Similarly, the EJCDC *C-800 Guide to the Preparation of Supplementary Conditions* should be used for EJCDC C-700 General Conditions.

Once the Supplementary Conditions are prepared, the document should be bound with the General Conditions into the Project Manual, in the proper order, and assigning the proper CSI *MasterFormat* in the table of contents.

If, for some reason, it is decided not to bind the General Conditions into the Project Manual, or to bind the document into a few copies of the Project Manual, it is recommended that the Supplementary Conditions begin with a statement such as the following: "The General Conditions of the Contract for Construction, American Institute of Architects Document A201, 15th Edition, 1997, hereinafter referred to as the General Conditions, shall be a part of the Contract Documents as fully as if bound herein."

The following are issues which should be addressed in the Supplementary Conditions.

Bonds and Insurance

AIA has published *G612 Owner's Instructions Regarding the Construction Contract, Insurance and Bonds, and Insurance Requirements*. This document is available for download at no charge from the AIA's website (*www.aia.org*). Similarly, EJCDC has published *C-050 Owner's Instructions Regarding Bidding Procedures* and *C-051 Engineer's Request for Instructions on Bonds and Insurance for Construction*. It is recommended that these documents be used, as appropriate, before construction documents production begins. The architect/engineer's project manager or principal in charge should transmit these to the owner with an appropriate cover letter early enough so that the owner's directions will be available when documents such as the Supplementary Conditions are produced.

Owners do not always respond to this request for information in a timely manner. If the owner delays returning the instructions beyond the

Project Manual's print date, the architect/engineer has no choice but to leave the specifics of insurance and bonds out of the Supplementary Conditions. It would be very unwise to make decisions of this kind on behalf of the owner even if the owner says, "Do what you usually do." Most professional liability insurance carriers specifically exclude the architect/engineer from determining bond and insurance types and amounts of coverage. State insurance officials also take exception to architect/engineers performing insurance counseling. There are laws that govern insurance agents, and it is just as inappropriate for insurance agents to design buildings as for architects and engineers to counsel others on construction bonds and insurance.

Using the completed documents recommended above, the architect/engineer may complete the Supplementary Conditions based on the owner's directions, which supposedly are based on competent counsel. If the documents are not completed and furnished to the architect/engineer, it is recommended that the Supplementary Conditions include a statement such as the following: "The extent and cost of insurance coverage shall be as negotiated by the Owner and Contractor before execution of the Agreement and shall be as stated in the Agreement." This puts the matter into the hands of the owner or, if applicable, the construction manager.

Construction Bonds

Refer to Chapter 11 for a discussion of the bonds used during construction. In the Supplementary Conditions, the General Conditions should be modified to establish requirements for, typically, a Faithful Performance Bond and a Labor and Materials Payment Bond.

Construction Insurance

There are many risks and liabilities that concern the architect or engineer in the construction of a project. The AIA and the EJCDC General Conditions each deal with basic insurance requirements, but neither includes amounts of coverage or other types of insurance coverage that may be necessary to safeguard the interests of all parties to the contract, including the designers. Obviously, the contractor bears most of the responsibility, but the owner also has responsibilities for providing insurance during construction.

Design professionals also have insurance interest during construction. To provide protection for the design professionals (the architect/engineer and his or her consultants), their names may be added to some insurance policies as "additional insured" parties for nominal additional costs but perhaps significant insurance protection. The architect/engineer should consult with his or her insurance counsel and with the owner and the owner's insurance counsel to determine insurance types and coverage limits and other conditions.

Since the General Conditions do not state specific insurance types and coverage, Supplementary Conditions should be used to include in the Contract Documents specific requirements for insurance to be furnished by the Contractor and the Owner.

In order to be sure that the required insurance is in effect for the project, the Supplementary Conditions should require certificates of insurance or copies of insurance policies to be submitted prior to start of the work. Prescribing that a specific form be used for these certificates, such as those published by AIA and EJCDC, may be difficult for the Contractor, but these forms may also ensure that necessary information is included. Also, there should be a requirement that the Contractor shall notify the Owner within a certain time period, usually brief, that changes in the insurance have occurred, including benign matters such as annual renewal of the policy but also significant matters such as cancellation of the policy or a change in the insurance carrier. These are potentially serious matters, and the architect/engineer should insist that the owner use competent insurance counsel for advice and directions in insurance matters.

Liquidated Damages

For projects where the time of completion is extremely important, the owner may desire to include a provision for liquidated damages in the Supplementary Conditions. Liquidated damages are based upon the concept that actual damages are difficult to determine should completion of the Work be delayed, and that the owner and contractor agree that an estimate of the value of the damage will be paid (deducted from amounts owed by the Owner to the Contractor) rather than the actual damage. However, by agreeing to payment of liquidated damages in the event of late completion by the Contractor, the Owner may give up the right to compensation for actual damages, which might exceed the liquidated damages amount. This is a legal concept that should be discussed and determined by the owner with the owner's legal counsel. The owner should direct the architect/engineer on whether to include or exclude liquidated damages and, if so, what the amount and terms shall be.

Typically, liquidated damages are stated in a dollar amount for each day that completion is delayed by the Contractor. The amount should be based on an approximation of actual damages. For example, if the facility is not available and the owner will incur overtime charges for movers and staff who will set up operations in the new or remodeled facility, then the amount could be several hundred dollars per day. If the owner will lose profits on operations, the amount could be very substantial. For example, on a new multiple-theater movie complex, the liquidated damages were $10,000 per day.

In requiring payment of liquidated damages, it is essential to avoid even the appearance that liquidated damages will be used to penalize the Contractor rather than to compensate the Owner for damage. The daily amount must be calculated separately for each project, using a justifiable estimate of cost, and the estimate should be retained by the owner and architect/engineer to justify the amount should it be challenged later in court or arbitration. Stock daily damage amounts, used time and again by an owner regardless of project size or need, may encourage the Contractor to challenge the stated amount.

While true damages can be collected, with the help of a court if necessary, penalties are usually unenforceable. The legal concept is that one citizen cannot punish another. Punishment is reserved for the courts. Otherwise, our society would be chaos. There is no link between liquidated damages for delay and a bonus for early completion. The two are independent. A bonus provision is not needed to make liquidated damages collectable. There is a belief that a bonus clause is needed to balance a penalty clause, but since a penalty is not within the power of the owner to exact from the contractor, the imagined balance does not exist.

Conflicts Between Documents

Much in this chapter suggests that some documents have more general application than others or that some take precedence over others. Reading the assignments of topics in the *Uniform Location of Subject Matter* (Exhibit 10-1) reinforces this notion that the more global requirements go in the Agreement and the General and Supplementary Conditions, while more particular requirements are left for documents such as the Contract Drawings and Specifications.

The idea of a strict pecking order among the Contract Documents can be dangerous. To believe that the provisions of a "higher" document nullify the provisions of a "lower" document is to be comfortable with conflicts of a type that should not occur within any contract. A "lower" document may contain particular information that must take precedence over the more general statements made in a "higher" document.

One of the many general rules of contract interpretation is that the contract shall be read as a whole. The Architect or Engineer and the Owner or Construction Manager should be aware, as the Contract Documents are prepared, that prime importance must be given to eliminating conflicts between the various portions of the documents.

A mental diagram of the way the documents interact is useful for a specifier to maintain as the table of contents for the Project Manual is assembled. But such a diagram is not a matter of pure hierarchy; phase and time also come into play. The bidding requirements are first, not because they are most important but also because they deal with the earliest phase of the project. In the AIA way of doing things, bidding requirements effectively drop out of the set of documents when the Agreement is executed (signed) and the General and Supplementary Conditions are activated. Modifications, such as Change Orders, are last in the "0" series of documents at the beginning of the Project Manual, not because they are less important than the Supplementary Conditions or bond and insurance certificates but because they occur later in the construction contract administration process.

In the case of drawings and specifications, very careful attention is necessary to avoid giving precedence to either, because each is intended to complement the other.

How, then, are conflicts to be resolved without a stated hierarchy of documents or a strict order of precedence? Construction Managers and construction contract administrators are often vehement about having a hierarchy of documents established so that interpretations and decisions are easier to make.

Most disputes are resolved by applying provisions such as the one provided in the EJCDC General Conditions: "Engineer will issue . . . clarifications or interpretations of the requirements of the Contract Documents as Engineer may determine necessary, which shall be consistent with or reasonably inferable from the overall intent of the Contract Documents." The Engineer must then do what a lawyer does to construe the meaning of a Contract. All parts must be considered to make the most reasonably inferable decision from the intent as expressed by the various requirements in the documents. The General Conditions still provide for steps that can be taken after the Engineer's decision if that decision is not acceptable to one of the parties.

Arbitration and Mediation

If a construction dispute is not resolved by the Owner and Contractor through direct negotiation by the parties, it may end up in court or in another forum. Courts are clogged with criminal matters, and civil matters such as construction-related claims take many years to be heard. Alternative dispute resolution has been developed to provide forums where construction disputes, among others, may be resolved.

AIA *A201* has two paragraphs addressing alternative dispute resolution. Paragraph 4.5 addresses mediation, and paragraph 4.6 addresses arbitration.

"Mediation" is the process whereby a mediator is engaged by the parties to hear the matter in dispute. The mediator is usually a person with knowledge and experience in construction and the particular subject in dispute.

For example, an engineer with particular expertise in chilled water systems and mediation skills would be selected by the parties to hear a dispute about why the chiller and related components do not operate within the performance criteria stated in the Specifications. The Contractor claims that the system was constructed according to the Contract Documents and that any performance shortcomings are due to errors or omissions in the design. The subcontractor claims that money should be paid by the Owner to compensate for excessive testing and balancing performed in an unsuccessful effort to make the system work as intended. The Engineer claims that construction-inefficient routing of piping and shrouding of the chiller, plus use of inferior components, are the reasons the chiller does not perform acceptably. The Owner claims 25 days of delay under the liquidated damages provisions of the Supplementary Conditions, and does not care who is responsible as long as someone else pays to fix the problem and cover the Owner's losses.

The mediator hears from all parties and assesses the problem. Based on the facts presented, the mediator may offer an opinion, based on his or her expertise, on how the mediator expects the dispute to be resolved—that is, who will pay how much. The mediator may also endeavor to get the parties to settle the dispute and avoid additional costs for litigation and damages. The mediator's decision and recommendations are not binding, and it is up to the parties to negotiate further, with the advantage of the mediator's impartial opinion, to come to a settlement of the dispute.

"Arbitration" is a quasi-legal but formal process conducted under the rules of an organization such as the American Arbitration Association (AAA) or JAMS® (formerly Judicial Arbitration and Mediation Services). An impartial panel of arbitrators hears both sides in a dispute. The parties may be represented by legal counsel, and the rules may be much like those in a trial. The parties must accept the ruling of the arbitrators as binding and, with some exceptions for misconduct on the part of the arbitrators, the ruling is enforceable subsequently in court. Because of the binding nature of the ruling and the compromise of some legal protections in the interest of expedient resolution, some attorneys advise their clients to not agree in a contract to arbitration.

This last point is very significant for architects and engineers. AIA and EJCDC General Conditions require binding arbitration, and the arbitration is required to be according to the rules of the AAA. Some attorneys may counsel their clients to agree to arbitration but under the rules of another organization such as JAMS.® The main issue, for architects and engineers, is that the owner should make an informed decision about arbitration and whether the arbitration provision should remain in the General Conditions and not be removed or modified by the Supplementary Conditions. Otherwise, if the owner does not make the decision to retain arbitration or claims not to realize that the contract required binding arbitration, the owner might claim that the architect or engineer erred in not adequately informing the owner about arbitration, and that the owner suffered monetary damage because the owner would have done better in a court hearing (lawsuit) of the case.

GUIDE FOR SUPPLEMENTARY CONDITIONS

For Supplementary Conditions based on AIA A201–*General Conditions of the Contract for Construction,* follow AIA A511–Guide for Supplementary Conditions, included in Appendix E–Sample AIA Documents.

Chapter 11

Bonds, Guaranties, and Warranties

BONDS

Bid Bond

A Bid Bond is intended to guarantee that if awarded the contract within a specified time, a bidder will enter into a contract and furnish a Performance Bond and a Payment Bond. The bidder who fails to do so is required to pay the owner the difference (not to exceed the penal sum of the bond) between the bidder's bid and a larger amount for which the owner may in good faith contract with another party to perform the work covered by the bid.

A Bid Bond is optional in private work and generally mandatory in public work. In private work, an owner may accept a Bid Bond in lieu of a certified check or a bank draft. It is recommended that the bid security be no less than 5% of the amount of the bid, and that this sum should be expressed as a specific number of dollars, not as a percentage of the bid. On public work, the amount of the bid security and its form may be specified by law or regulation, and such legal requirements will govern.

The AIA document *A310 Bid Bond* is often used in private work, and the requirement for it is set forth in the Invitation to Bid.

Surety Bonds

Surety bonds, sometimes called "guarantee bonds" or "construction bonds," are essential in construction operations. They make it possible for the contractor to provide the owner with the guarantee of a responsible surety company that the Contractor will satisfactorily perform the project at the determined price and pay the Contractor's bills. Of additional interest to the Architect/Engineer is the fact that extra professional services compensation resulting from a Contractor's default caused by delinquency or insolvency is reimbursable by the use of these bonds.

A Surety Bond is an agreement under which one party, the surety, agrees to answer to another party, the obligee, for the debt, default, or failure of a third party, the principal, to fulfill the Contractor's contract obligations. A surety is usually a corporation that underwrites or guarantees construction bonds; the obligee is usually the Owner; and the principal is the Contractor.

A Surety Bond does not impose on the surety any obligations that are separate and distinct from or additional to those assumed by the principal. Under any Surety Bond, the principal is primarily responsible, and every obligation of the surety is also that of the principal. A bond is not a substitute for the integrity, financial worth, experience, equipment, and personnel of the Contractor. Nor is such a bond an independent undertaking by the surety as long as the principal performs in accordance with the terms of the Contract Documents.

Contractor Defaults

One of the major reasons for a Contractor's default is the inadequacy of the Contractor, either on the Owner's contract or on other past or current contracts. This can derive from a variety of sources, such as deficient cost and other accounting records, unforeseen price rises, labor troubles, defaults of subcontractors, materials delays, prolonged inclement weather, and so on.

Surety Bonds provide protection against loss resulting from the failure of others to perform. Whereas the liability of the Contractor for damage may be unlimited, that of the surety is limited to a certain sum of money called the "penalty" or the "penal sum," which is set out in the bond. Such an instrument consists of an extension of credit to the Contractor, not as a loan of money but rather as an endorsement. The Performance Bond directly increases the financial responsibility of the Contractor for the benefit of the Owner by the amount of its penal sum. The architect/Engineer should instruct the Owner on the value of a Surety Bond, and should suggest that the Owner seek the advice of the Owner's legal and insurance (risk management) counsels before deciding for or against these bonds.

AIA Bond Forms

Article 11.5 of the AIA General Conditions describes a provision for furnishing of Surety Bonds by a Contractor covering the faithful performance of the contract and for the payment of obligations arising thereunder. Standard forms have been prepared by the AIA in cooperation with the surety industries, and are recommended for use in all private and public construction where a statutory form is not prescribed. These forms are in AIA Document *A312 Performance Bond and Payment Bond,* available from the AIA.

- *Performance Bond, AIA Document A312.* This bond assures the Owner that the Contractor will perform all the terms and conditions of the contract between the Contractor and the Owner and in the event of default will protect the Owner against loss up to the amount of the bond penalty.
- *Payment Bond, AIA Document A312.* This bond ensures that the surety will pay the Contractor's bills for labor and materials in the event of the Contractor's default.

Formerly, these bonds were contained in one instrument. Now they are divided into two separate documents. The AIA two-bond system has much merit. The inclusion in one instrument of the obligation to perform the contract and to pay laborers and material has given rise to certain difficulties in handling claims against the bond. These difficulties have resulted from the competing interests of the Owner, on the one hand, and laborers and material suppliers, on the other. Under the two-bond system, the surety is enabled to make payment without awaiting a determination as to the Owner's priority. These bonds are issued by the companies as a package, and there is usually no additional premium for the separate Payment Bond.

Amount of Bonds

A Performance Bond and a Payment Bond, each in the amount of 100% of the Contract sum (price), are recommended. Where a public body is the Owner, its legal counsel should obtain complete information regarding the legal requirements, amount and form of the bond, and provide this information to the Architect/Engineer for inclusion in the Project Manual.

Statutory and Nonstatutory Bonds

Surety Bonds fall into two basic categories: statutory and nonstatutory or private. Statutory Bonds are those required by law. Some states have statutory provisions relating to bonds. On private projects, counsel for the Owner may suggest special requirements. No standard form of surety bond is applicable to every project. The AIA document *A312 Performance Bond and Payment Bond* is a step toward such standardization, and their use is urged for all private and public contracts where a statutory form is not prescribed.

WARRANTIES AND GUARANTIES

Before the 1976 edition of AIA Document *A201 General Conditions,* the terms "warranty" and "guarantee" were used almost interchangeably throughout the document. Since then, only the term "warranty" appears. In the 1996 edition of the EJCDC Document *1910-8 Standard General Conditions of the Construction Contract,* Par. 6.19A, the Contractor "warrants and guarantees" that the work will be in accordance with the Contract Documents and will not be defective. This use of both "warranty" and "guarantee" by engineers and the use of "warranty" alone by architects can be disturbing to users of the two major professional society documents.

Where should the Architect/Engineer turn when it comes time to explain these differences? In this case, most general dictionaries are of help because they pattern their definitions of warranty, guaranty, and guarantee on the same thinking that governs *Black's Law Dictionary.*

"Warranty" and "guaranty" were once the same, having entered our language as *warant* and *garant,* spelled differently but pronounced almost alike in Norman French. Both spellings were related to the old Frankish *warjan,* which meant to protect or to vouch for the truth of something. Following *Black's Law Dictionary,* the meanings can be distilled to these thumbnail definitions:

1. *Warranty:* a promise that certain facts are true as represented and that they will remain so. A written warranty may also promise to repair or replace a product if it fails to meet the specification.
2. *Guaranty:* a promise, by a party called a "guarantor," to make good the mistake, debt or default of another party.

3. *Guarantee:* (1) the party to whom a guaranty is made; (2) the obligation of a guarantor.

Under the AIA General Conditions, the Contractor is required to warrant that the Contract Documents have been followed and that the Work is of good quality. That warranty can be made by the Contractor for its own work and for the work performed under its supervision by subcontractors. Engineers prefer that the Contractor warrant its own Work but offer a guarantee for the work of its subcontractors. In this, the second definition of "guarantee" is appropriately used. "Guarantee" may seem colloquial compared with "guaranty," but *Black's Law Dictionary* supports it.

What, then, is the specifier to say to the client when asked about this difference in terminology? The best course is to understand and follow the advice of the professional society that wrote the document. Despite their different wording, each society's approach is based on legal principles. The specifier is well advised to adhere to the language of the standard general conditions being used, leaving to the client and the client's attorney the responsibility for any changes in language they may choose. See the discussion regarding modification of General Conditions in Chapter 10.

The General Warranty

The General Conditions of both AIA and EJCDC, as well as the federal government's construction contract clauses require that each Contractor warrant that the Work of the Contract is:

1. Constructed according to the Contract Documents
2. Free of defects

The duration of this warranty is not limited in the AIA and EJCDC General Conditions. The warranty is open-ended and would be restricted only by statutes of limitations in the various states and by practical considerations. Practically speaking, there is difficulty in distinguishing normal wear and tear or abuse or lack of maintenance from true construction defects when the Work remains in use year after year. Although the AIA and EJCDC general warranty provisions are actually nothing more than a summary of the common law as it has developed for contracts, these provisions are very strong.

Some have confused this warranty with the one-year correction period for Work under the Contract, also required in the AIA (*AIA A201* 12.2.2) and

EJCDC documents. Reference to the one-year correction period for Work as "the 1-year guarantee" is erroneous, and it is also erroneous to take the position that after 1 year from the Contract completion date, the Contractor has no further obligations for the Work. "These arguments are incorrect and have been rejected by the courts," says the EJCDC.

For further information, consult with legal counsel and research construction claims and the Uniform Commercial Code (UCC). The subjects of latent defects and when the Statute of Limitations ends responsibility vary from state to state. These subjects are very complex.

Special Warranties

So far, the discussion of warranting and guaranteeing the Work has dealt with the overall responsibility of the Contractor or a prime contractor who has entered into a contract with the Owner. The AIA General Conditions also consider special warranties (*A201* 9.10.4.3 and 12.2.2.1). These are warranties made by subcontractors, suppliers and manufacturers for Work performed and products furnished for the Work. Special warranties extend beyond 1 year and may be 2 to 20 or even 30 years in duration, since their customary purpose is to obtain extended attention to correction of defects, much as the 1-year period for correction of Work does for the entire Work.

A special warranty is the same as a warranty, defined earlier, but in this case it does not usually refer to the Contractor's own work, although he or she is obliged to aid in enforcing it. A special warranty most frequently applies to the work of a subcontractor, although, as usual, the language of the warranty is directed to the Contractor. The requirements of the special warranty are usually written by the specifier, taking the form of the basic parts of a contract: over what period the warranty is to be effective, to what value limit, what is to be done if the work does not perform, and who is going to make good the warranty provisions. In this way, the provisions of a special warranty that is directed to subcontract work can often be controlled by the specifier through the general contractor (the Contractor).

At the other extreme, a warranty that benefits the Owner is a producer's limited or material-only warranty. This is a special warranty offered by some manufacturers in the hope that an unwary specifier will include it in the specifications. A warranty of this type usually severely limits the manufacturer's responsibility if something goes wrong. The specifier must be selective in deciding which special warranties to incorporate and which to keep out for the owner's protection. Between the harmful and beneficial extremes of special warranty provisions lie many offerings of industry, with names such as "warranty," "guarantee," and "service agreement," which can be examined by the specifier for the actual protection they offer. The conscientious specifier will take care to find out what such warranties add to the cost of the work, the stability of the offeror, their duration, exclusions that may render them inoperative, and features such as whether they cover the labor involved in repair or replacement or the prorated value of the material only.

When writing specifications, one must realize that many products are advertised as "guaranteed" and "warranted" or "warranteed." These terms may mean something or nothing. Beware of disclaimers in printed form, and of warranties that offer nothing when a failure occurs or that make the cost of replacement as great as the cost of new work. Some manufacturers go so far as to say that they will not sell their product if their self-serving warranty is not accepted by the buyer.

The most futile exercise in warranty writing is to say, "Guarantee all ceramic tile for 1 year." To begin with, this statement gives the Owner nothing that the Owner did not already have in the 1-year correction-of-work provision (*AIA A201* 12.2.2). Second, the Owner has an extended warranty against defects and nonconforming work that will cover the Owner for years if it is not compromised by a statement such as "for 1 year." In addition, the statement is vague. It should at least say "Guarantee against defects" or, if specific problems are of concern, "Guarantee against chipping, cracking, and coming loose," for example.

Note that the term "special guarantee," when used in EJCDC General Conditions, refers only to a substitution of materials or equipment.

Correction of Work

Aside from general warranty and special warranty provisions, both the AIA and EJCDC documents make provision for a 1-year service agreement by the Contractor after the work is substantially complete. During this period the Contractor is required to correct Work that is defective or otherwise not in accordance with the Contract Documents. The correction-of-work provisions require the corrective work to be done upon written notice from the Owner, and they allow the Owner to correct the work if the Contractor does not do so, the cost presumably to be borne by the

Contractor or his surety in such a case. Both the AIA and EJCDC go on to state that nothing contained in the correction-of-work provision is to limit the Contractor's obligations under his or her warranty, which extends an indefinite time beyond the date of completion.

The correction-of-work period may be extended in the Supplementary Conditions to a period longer than 1 year. Owners should realize, however, that any service agreement costs money. The correction-of-work period costs the contractor an amount that he or she must include in the price, and any extension will cost more in proportion.

The federal government handles correction-of-work differently. There the correction period runs concurrently with a 1-year warranty. Practically speaking, this has much the same effect as the warranty and correction provisions of the private sector documents. In the federal clauses, gross mistakes, latent defects, and fraud by the Contractor do not limit the government's rights.

Writing a Special Warranty

The special warranty will be written by the manufacturer or supplier who offers it as the work is performed. All the specifier does is to communicate what special warranty provisions are required. If there are industry guidelines for special warranties (whatever they are called), read them and extract as much of the industry custom as you can. After all, a special warranty that is already common with manufacturers of a product type will be easier to obtain without argument. For example, special warranties for water heaters generally will include the same points of coverage, duration, corrective action, and exclusions. With this similarity in approach, it is easy to find the small differences that offer more or less benefit to the Owner.

Manufacturer-generated warranty text often contains many exclusions. Analyze the exclusions: if they make sense, and many do, include them in your text. For instance, if stock warranty text withdraws responsibility for a roof that has been altered by someone other than the original roofer, consider saying that you will allow such an exclusion: it is common and it is fair. If you want protection by means of a special warranty, do not invent new and unheard-of requirements and expect distant roofing manufacturers two tiers below the Contractor to jump to your request on a $20,000 order. You will find it difficult to get a piece of paper to deliver to your client. Your time may be better spent keeping an eye on the roof as it is installed.

Remember that the Owner's contract is with the

contractor (general contractor) only. Involve the Contractor in a subcontractor's or supplier's warranty as much as possible. Use the traditional elements of a Contract to spell out what is required: what, what value, how long, for whom, and by whom. Here are examples of very brief special warranty text, taking several different approaches to ensuring performance:

1. *SPECIAL WARRANTY:* The Contractor hereby warrants that the waterproofing is free from defective materials and execution and will remain so for 3 years after the date of Substantial Completion. Upon notification of defects within that special warranty period, the Contractor shall make repairs and replacements at no cost to the Owner, according to AIA General Conditions 12.2.

This example does not require that a written warranty be submitted since it states that "the Contractor hereby warrants" and so on. By his signature on the Owner-Contractor Agreement, the Contractor has already furnished the necessary written special warranty. Note also that the text follows the form of the dictionary definition of warranty: First, there is the promise that the facts are true as represented and that they will remain so. Second, there is the promise to repair or replace the product if it fails to meet the specification.

2. *SPECIAL WARRANTY:* The Contractor hereby warrants that the waterproofing shall not leak or delaminate for 3 years after the date of Substantial Completion. Upon notification of defects within that special warranty period, the Contractor shall make repairs and replacements to leaking or delaminating waterproofing at no cost to the Owner, in accordance with AIA General Conditions 12.2.

Like Example 1, this special warranty, which follows the classic formula, is already signed into effect by the Contractor. It differs in that it is written in performance rather than prescriptive terms. It will be binding only if the specification states at some point, "Provide a waterproofing assembly that will not leak or delaminate."

3. *SPECIAL WARRANTY:* Before final payment the Contractor shall furnish to the Owner a 3-year written warranty of products and execu-

tion provided in waterproofing the Work. The special warranty shall promise that the waterproofing is free from defective products and execution and that it shall remain so for 3 years after the date of Substantial Completion, and that the Contractor shall make repairs and replacements at no cost to the Owner during the special warranty period, in accordance with AIA General Conditions 12.2.

Example 3 is used in those instances where the Owner wants the assurance of a written special warranty separate from any promise made in the contract documents. The administrative cost to the Owner and Architect/Engineer in gaining physical possession of such a piece of paper should not be underestimated.

4. *SPECIAL WARRANTY:* The Contractor hereby warrants that the waterproofing is free from defective materials and execution and will remain so for 3 years after the date of Substantial Completion. Upon notification of defects within that special warranty period, the Contractor shall make repairs and replacements at no cost to the Owner, in accordance with AIA General Conditions 12.2. The Contractor shall deliver to the Owner a 3-year warranty bond of a surety company approved by the Owner, guaranteeing that the Contractor or his surety will repair or replace defective waterproofing at no cost to the Owner.

Example 4 is used where the Owner wants still further assurance that a third party will undertake to answer for the performance of the Contractor (and therefore the subcontractor).

The law regarding warranties, guarantees, and sureties varies from state to state. Although the specifier may draft a realistic set of special warranty requirements for consideration by the Owner, all warranty text should be discussed with, reviewed by, and approved by the Owner or his or her legal counsel. As with all of the legal-fiscal-managerial portions of the contract documents, especially the agreement, the conditions of the contract, bidding, warranties, bonds, and insurance, the Architect/Engineer may prepare the draft from customary forms and the owner's instructions, but the final legal form of the documents should be reviewed by an attorney for the Owner who is familiar with construction and construction law.

Chapter 12

Division 1 - General Requirements

SCOPE OF DIVISION 1

The general requirements of the specifications consist of certain sections listed under Division 1 of the *MasterFormat* (see Chapter 3). The sections recommended by the *MasterFormat* for inclusion under Division 1 are shown in Exhibit 12-1.

The recommendations originally suggested under the *CSI Format,* which were published in 1963 for inclusion under Division 1 - General Requirements, consisted of alternates, alterations, inspections, tests, allowances, and temporary facilities. With subsequent updates, Division 1 has undergone various philosophical changes, depending on the views of the CSI committee charged with the responsibility for updating the *MasterFormat* at each revision. Since the inception of the 1963 *CSI Format,* the various revisions of Division 1 have seen the number of sections in Division 1 expand exponentially and their order change frequently.

The prime purpose of the general requirements in the original format was to provide a place for the nonlegal (noncontractual) and administrative requirements for construction of the project. All those general requirements not suitable for inclusion under the sections of Divisions 2 through 16 were also expected to be set forth here by the early proponents of the *CSI Format.* As stated in Chapter 10, the Supplementary Conditions had become the catchall document in which were specified tempo-

rary utilities, temporary facilities, and a host of other requirements not of a legal (contractual) nature. The intention was that the establishment of a Division titled "General Requirements" would create a convenient place for instructions to the Contractor that could not logically be placed anywhere else.

Division 1 in *MasterFormat* 1995

Division 1 of the 1995 edition of *MasterFormat* hardly resembles the original intent and scope of the general requirements of 1963. As noted in the Introduction to the 1995 edition of *MasterFormat,* Division 1 "was completely rearranged to clarify that general requirements may apply to products and execution as well as administration" and that the "new arrangement is more closely aligned with *SectionFormat.*"

It is somewhat disingenuous to state that it may apply to products without qualification. Products per se belong in Divisions 2 through 16. Topics of general information concerning products—namely, product handling, storage, and substitution requirements—that are germane to all products used in the project are subjects to be established in Division 1 as procedural elements.

Execution is likewise specified. The installation of products, their application, erection, and integration into the project, are specified in Divisions 2

DIVISION 1 - GENERAL REQUIREMENTS

Section No. - TITLE

01100 - SUMMARY
01200 - PRICE AND PAYMENT PROCEDURES
01300 - ADMINISTRATIVE REQUIREMENTS
01400 - QUALITY REQUIREMENTS
01500 - TEMPORARY FACILITIES AND CONTROLS
01600 - PRODUCT REQUIREMENTS
01700 - EXECUTION REQUIREMENTS
01800 - FACILITY OPERATION
01900 - FACILITY DECOMMISSIONING

Exhibit 12-1. Division 1 - General Requirements.

through 16. Basic requirements for all products in regard to examination and preparation of surfaces to receive them and final cleaning and protection are appropriately spelled out in general terms in Division 1 as procedural elements.

Facility decommissioning is also an area that is not truly the province of a specifier or a design firm. It is related primarily to needs of facility owners and managers. With the tendency of construction specifications to be used throughout the life cycle of a facility, these topics are being included in *MasterFormat*. Indeed, one of the major controversies about the radical reconstruction of the pending, revised *MasterFormat* 2004 is inclusion of "life cycle" and "maintenance" categories in the format. Some question whether these are specifications issues, and the publishers of master guide specifications have not indicated that these topics will be addressed in the foreseeable future.

Recommended Division 1 Sections

The authors, having applied the original concept of Division 1 information in projects ranging from $1 million to $1 billion, suggest that the information in the examples shown in the exhibits will suffice for the vast majority of projects without using the more than 50 section titles contained in Division 1 of the *MasterFormat*. From the standpoint of usage, the value of some of the section titles identified in the *MasterFormat* has yet to be demonstrated.

Sample sections delineating the scope and content of Division 1, without use of the 3-part section format, are presented in Appendix C.

RELATIONSHIP OF DIVISION 1 SECTIONS TO OTHER DOCUMENTS

Division 1 Relationships to Bidding Requirements

Standard published General Conditions of the Contract state that bidding requirements are not part of the Contract Documents. Consequently, provisions stated *only* in the bidding requirements are not enforceable during construction contract administration. Therefore, Instructions to Bidders should cross-reference Division 1 sections to direct bidders to relevant information but should not repeat Division 1 provisions. Division 1 topics of concern to bidders include:

• Substitution, alternate, and unit price procedures

• Use of the site
• Phasing of the Work
• Owner occupancy (early or continuing)
• Owner-furnished products (OFCI)
• Definition of scope of separate prime contracts

Relationship of Division 1 to the Conditions of the Contract

Refer to the exhibits at the end of this chapter, titled:

• "Comparison Between Bidding and Contract Requirements and Division 1 - General Requirements" (Exhibit 12-2)
• "Relationship Between Conditions of the Contract and Division 1 - General Requirements" (Exhibit 12-3)
• "Comparison Between Division 1 - General Requirements and Typical Spec Section Articles" (Exhibit 12-4)

Conditions of the Contract

The General Conditions and the Supplementary Conditions are inherent parts of the Agreement. With the Agreement, they govern the content of the entire Contract. They contain contractual principles applicable to most projects with supplements for the particular project.

• General Conditions of the Contract
  ◦ Are broad contractual conditions.
  ◦ Contain "constants." Relatively static content allows use of published standard documents from project to project (typically, AIA *A201* or EJCDC *C-700*).
• Supplementary Conditions of the Contract
  ◦ Modify contractual conditions stated in the General Conditions.
  ◦ Take precedence over the General Conditions.
  ◦ Modify the "constants" of the General Conditions to suit specific requirements for the project.
  ◦ Must be written separately for each project.
• Division 1
  ◦ Is an inherent part of the specifications.
  ◦ Contains specific administrative and procedural requirements for administering the contractual principles of the General Conditions and Supplementary Conditions.
  ◦ Also, administratively governs all Division 2 through 16 specification sections, carrying

COMPARISON BETWEEN BIDDING AND CONTRACT REQUIREMENTS AND DIVISION 1 - GENERAL REQUIREMENTS

Bidding and Contract Requirements

("00" file numbers) . . . Example Documents:

00100 - BID SOLICITATION

00110 - ADVERTISEMENT FOR BIDS
00120 - INVITATION TO BID
00140 - REQUEST FOR PROPOSAL
00150 - REQUEST FOR QUALIFICATIONS
00155 - PRE-QUALIFICATION FORMS

00200 - INSTRUCTIONS TO BIDDERS

00201 - INSTRUCTIONS TO BIDDERS (General Construction)
00202 - INSTRUCTIONS TO PROPOSER (General Construction)
00203 - INSTRUCTIONS TO BIDDERS (FF&E)
00204 - INSTRUCTIONS TO PROPOSER (FF&E)
00211 - SUPPLEMENTARY INSTRUCTIONS TO BIDDERS (Gen'l Construction)
00212 - SUPPLEMENTARY INSTRUCTIONS TO BIDDERS (FF&E)
00213 - SUPPLEMENTARY INSTRUCTIONS TO PROPOSER (Gen'l Const)
00214 - SUPPLEMENTARY INSTRUCTIONS TO PROPOSER (FF&E)
00221 - SCOPES OF BIDS (Multiple Contracts)
00222 - SCOPES OF PROPOSALS (Multiple Contracts)
00223 - SCOPES OF BIDS (FF&E)
00224 - SCOPES OF PROPOSALS (FF&E)
00251 - PRE-BID CONFERENCE
00252 - PRE-PROPOSAL CONFERENCE
00253 - PRE-BID SITE VISITS
00254 - PRE-PROPOSAL SITE VISITS

00300 - INFORMATION AVAILABLE TO BIDDERS

00310 - PRELIMINARY SCHEDULE
00320 - GEOTECHNICAL DATA
00330 - EXISTING CONDITIONS
00340 - ENVIRONMENTAL ASSESSMENT INFORMATION
00350 - PROJECT FINANCIAL INFORMATION
00360 - PERMIT APPLICATION PROCEDURES

00400 - BID FORMS AND SUPPLEMENTS

Division 1 - General Requirements

("01" file numbers) . . . Example Sections:

01100 SUMMARY

01001 - BASIC REQUIREMENTS (Simple Construction)
01100 - SUMMARY OF THE PROJECT

01200 - PRICE AND PAYMENT PROCEDURES

01210 - ALLOWANCE PROCEDURES
01230 - ALTERNATE [BID] [PROPOSAL] PROCEDURES
01250 - CONTRACT MODIFICATION PROCEDURES
01270 - UNIT PRICE PROCEDURES
01290 - MEASUREMENT AND PAYMENT PROCEDURES

01300 ADMINISTRATIVE REQUIREMENTS

01311 - PROJECT COORDINATION
01312 - PROJECT MEETINGS
01321 - CONSTRUCTION PROGRESS SCHEDULES
01322 - CONSTRUCTION PROGRESS REPORTS
01323 - CONSTRUCTION PHOTOGRAPHS
01330 - SUBMITTALS PROCEDURES
01350 - SPECIAL PROJECT PROCEDURES

01400 QUALITY REQUIREMENTS

Exhibit 12-2. Comparison Between conditions of the contract and Division 1 - General Requirements.

00405 - PROPOSAL FORM (Design/Build)
00411 - BID FORM (General Construction - Construction Management)
00412 - BID FORM (General Construction - Cost plus Fee)
00413 - BID FORM (General Construction - Stipulated Sum)
00414 - BID FORM (General Construction - Unit Price)
00416 - PROPOSAL FORM (General Construction - Construction Management)
00417 - PROPOSAL FORM (General Construction - Cost plus Fee)
00418 - PROPOSAL FORM (General Construction - Stipulated Sum)
00419 - PROPOSAL FORM (General Construction - Unit Price)
00421 - BID FORM (FF&E - Construction Management)
00422 - BID FORM (FF&E - Cost plus Fee)
00423 - BID FORM (FF&E - Stipulated Sum)
00424 - BID FORM (FF&E - Procurement)
00426 - PROPOSAL FORM (FF&E - Construction Management)

00427 - PROPOSAL FORM (FF&E - Cost plus Fee)
00428 - PROPOSAL FORM (FF&E - Stipulated Sum)
00429 - PROPOSAL FORM (FF&E - Procurement)
00431 - BID SECURITY
00432 - SCHEDULE OF VALUES
00433 - SUBCONTRACTORS LIST
00450 - CONTRACTOR'S QUALIFICATION FORM

00490 - BIDDING ADDENDA

00500 - AGREEMENT

00521 - AGREEMENT [FORM] (General Construction - Construction Management)
00522 - AGREEMENT [FORM] (General Construction - Cost plus Fee)
00523 - AGREEMENT [FORM] (General Construction - Stipulated Sum)
00524 - AGREEMENT [FORM] (General Construction - Unit Price)
00525 - AGREEMENT [FORM] (General Construction - Procurement)
00526 - AGREEMENT [FORM] (FF&E - Construction Management)
00527 - AGREEMENT [FORM] (FF&E - Cost plus Fee)
00528 - AGREEMENT [FORM] (FF&E - Stipulated Sum)
00530 - AGREEMENT AND GENERAL CONDITIONS (Simple Construction)
00550 - NOTICE TO PROCEED

01410 - REGULATORY REQUIREMENTS
01420 - REFERENCE STANDARDS AND ABBREVIATIONS
01450 - QUALITY CONTROL
01465 - CUTTING AND PATCHING PROCEDURES

01500 TEMPORARY FACILITIES AND CONTROLS

01500 - TEMPORARY FACILITIES AND CONTROLS
01510 - TEMPORARY UTILITIES (to be developed)
01520 - CONSTRUCTION FACILITIES (to be developed)
01550 - VEHICULAR ACCESS AND PARKING (to be developed)
01560 - TEMPORARY BARRIERS AND ENCLOSURES (to be developed)
01570 - TEMPORARY CONTROLS (to be developed)
01580 - PROJECT IDENTIFICATION AND SIGNAGE

01600 PRODUCT REQUIREMENTS

01610 - BASIC PRODUCT REQUIREMENTS
01620 - PRODUCT OPTIONS
01630 - PRODUCT SUBSTITUTION PROCEDURES
01640 - OWNER-FURNISHED PRODUCTS
01650 - PRODUCT DELIVERY REQUIREMENTS
01660 - PRODUCT STORAGE AND HANDLING REQUIREMENTS

01700 EXECUTION REQUIREMENTS
01710 - EXAMINATION REQUIREMENTS
01720 - PREPARATION REQUIREMENTS
01722 - FIELD ENGINEERING
01730 - EXECUTION REQUIREMENTS
01740 - CLEANING REQUIREMENTS
01750 - STARTING AND ADJUSTING PROCEDURES
01760 - PROTECTION OF CONSTRUCTION
01770 - CONTRACT CLOSEOUT PROCEDURES
01781 - FINAL SITE SURVEY
01782 - MAINTENANCE CONTRACTS
01783 - OPERATION AND MAINTENANCE DATA
01785 - PRODUCT WARRANTIES AND BONDS
01789 - PROJECT RECORD DOCUMENTS

Exhibit 12-2. *(Continued)*

01800 FACILITY OPERATION

01810 - COMMISSIONING
01820 - DEMONSTRATION AND TRAINING
01830 - OPERATION AND MAINTENANCE
01890 - RECONSTRUCTION

01900 FACILITY DECOMMISSIONING

01901 - DECOMMISSIONING SYSTEMS AND EQUIPMENT
01902 - FACILITY DEMOLITION AND REMOVAL
01903 - HAZARDOUS MATERIALS ABATEMENT
01904 - HAZARDOUS MATERIALS REMOVAL AND DISPOSAL
01905 - PROTECTION OF DEACTIVATED FACILITIES

00600 - BONDS AND CERTIFICATES

00611 - PERFORMANCE AND PAYMENT BONDS
00621 - CERTIFICATES OF INSURANCE
00622 - APPLICATION FOR PAYMENT CERTIFICATES
00623 - CERTIFICATES OF COMPLIANCE
00625 - CERTIFICATE OF COMPLETION
00641 - RELEASE OF LIEN CERTIFICATES
00642 - CONSENT OF SURETY FOR FINAL PAYMENT
00650 - STATUTORY DECLARATION FORMS

00700 - GENERAL CONDITIONS OF THE CONTRACT

00700 - GENERAL CONDITIONS OF THE CONTRACT (Project-Specific)
00701 - GENERAL CONDITIONS OF THE CONTRACT (Std - General Construction)
00702 - GENERAL CONDITIONS OF THE CONTRACT (Std - Const Management)
00703 - GENERAL CONDITIONS OF THE CONTRACT (Std - FF&E)

00800 - SUPPLEMENTARY CONDITIONS OF THE CONTRACT

00801 - SUPPLEMENTARY CONDITIONS OF THE CONTRACT (General Const)
00802 - SUPPLEMENTARY CONDITIONS OF THE CONTRACT (Const Mgt)
00803 - SUPPLEMENTARY CONDITIONS OF THE CONTRACT (FF&E)
00830 - WAGE DETERMINATION SCHEDULE
00850 - SPECIAL CONDITIONS (Modification to Std Suppl Conditions)
00890 - PERMITS FOR CONSTRUCTION

00900 - ADDENDA AND MODIFICATIONS

00910 - ADDENDA
00931 - REQUESTS FOR INTERPRETATION (RFI)
00932 - REQUESTS FOR PROPOSAL (RFP)
00933 - CONSTRUCTION BULLETINS
00941 - ARCHITECT'S SUPPLEMENTAL INSTRUCTIONS
00942 - CONSTRUCTION CHANGE DIRECTIVES
00943 - CHANGE ORDERS

Exhibit 12-2. (Continued)

RELATIONSHIP BETWEEN CONDITIONS OF THE CONTRACT AND DIVISION 1 - GENERAL REQUIREMENTS

Conditions of the Contract ("00" file numbers)
An inherent part of the Agreement.

With the Agreement, govern the content of the entire Contract Documents (Drawings and Specifications; Other bidding and contract requirements such as bonds, insurance, information available to Bidders).

With the Agreement, the contents stand alone. No specific coordination is required in products specifications.

Contain principles which are applicable to most projects nationwide.

Conditions of the Contract are in two documents:

General Conditions	Supplementary Conditions
Are broad contractual conditions.	Modify broad contractual conditions to apply to particular project.
Contain the "constants."	Modify the "constants" to apply to a specific geographic region, project type, Owner's requirements, method of contracting, A/E's practices.
	Take precedence over General Conditions by specific reference.
Content changes infrequently, providing standard practices for common understanding by all parties.	Content always unique, written specifically for each project, easy to determine variations from standard practices.
Written by legal and construction experts. For each project, A/E typically uses preprinted forms (AIA A201 and EJCDC C-700 are examples) and incorporates them into Project Manual.	A/E typically prepares these *in consultation with Owner's legal and insurance counsels* for each project and incorporates them into Project Manual.

Division 1 - General Requirements ("01" file numbers)
An inherent part of the Specifications.

Administratively govern product specifications (Divisions 2 through 16).

Describe administrative procedures for the Contract.

Mutually interdependent with the content of products specifications (Divisions 2 through 16).

Example: Section 01710 - Final Cleaning specifies cleaning generally; Section 08800 Glazing specifies how to clean glass.

Division 1 - General Requirements, individual Sections

Contain specific administrative and procedural requirements.

Contain "variables" directly applicable to a specific project.

Content unique to each office, project type, contracting type and Owner's requirements.

Written by A/E for each project.

Exhibit 12-3. Relationship Between conditions of the contract and Division 1 - General Requirements.

General Conditions

Are Contract Documents.
Are **NOT** Specifications.

Note: If preprinted General Conditions are used, then Supplementary Conditions must also be used. If unique General Conditions are used (i.e., Owner produces General Conditions for the project), then Supplementary Conditions are typically omitted.

Supplementary Conditions

Are Contract Documents.
Are **NOT** Specifications.

Division 1 - General Requirements

Are Contract Documents.
Are Specifications.

Exhibit 12-3. *(Continued)*

**COMPARISON BETWEEN DIVISION 1 - GENERAL REQUIREMENTS
AND TYPICAL SPEC SECTION ARTICLES**

Division 1 - General Requirements   **Typical Specification Section Articles**

01100 SUMMARY

    01001 - BASIC REQUIREMENTS (Simple Construction)   1.1 SECTION INCLUDES
    01100 - SUMMARY OF THE PROJECT   1.2 RELATED SECTIONS

01200 PRICE AND PAYMENT PROCEDURES

        1.3 ALLOWANCES
    01210 - ALLOWANCE PROCEDURES   1.4 UNIT PRICES
    01230 - ALTERNATE [BID] [PROPOSAL] PROCEDURES   1.5 ALTERNATES
    01250 - CONTRACT MODIFICATION PROCEDURES
    01270 - UNIT PRICE PROCEDURES
    01290 - MEASUREMENT AND PAYMENT PROCEDURES

01300 ADMINISTRATIVE REQUIREMENTS

        1.6 REFERENCES
        1.7 DEFINITIONS
        1.8 DESIGN REQUIREMENTS
        1.9 PERFORMANCE REQUIREMENTS
        1.14 SEQUENCING AND SCHEDULING
    01311 - PROJECT COORDINATION
    01312 - PROJECT MEETINGS
    01321 - CONSTRUCTION PROGRESS SCHEDULES
    01322 - CONSTRUCTION PROGRESS REPORTS
    01323 - CONSTRUCTION PHOTOGRAPHS   1.10 SUBMITTALS
    01330 - SUBMITTALS PROCEDURES
    01350 - SPECIAL PROJECT PROCEDURES

01400 QUALITY REQUIREMENTS   1.11 QUALITY ASSURANCE

    01410 - REGULATORY REQUIREMENTS   2.7 SOURCE QUALITY CONTROL
    01420 - REFERENCE STANDARDS AND ABBREVIATIONS   3.4 FIELD QUALITY CONTROL
    01450 - QUALITY CONTROL   3.5 MANUFACTURER'S FIELD SERVICES
    01465 - CUTTING AND PATCHING PROCEDURES

01500 TEMPORARY FACILITIES AND CONTROLS   1.13 PROJECT CONDITIONS

    01500 - TEMPORARY FACILITIES AND CONTROLS
    01510 - TEMPORARY UTILITIES (to be developed)
    01520 - CONSTRUCTION FACILITIES (to be developed)
    01550 - VEHICULAR ACCESS AND PARKING (to be developed)
    01560 - TEMPORARY BARRIERS AND ENCLOSURES (to be developed)
    01570 - TEMPORARY CONTROLS (to be developed)
    01580 - PROJECT IDENTIFICATION AND SIGNAGE

01600 PRODUCT REQUIREMENTS

    01610 - BASIC PRODUCT REQUIREMENTS   2.1 MANUFACTURERS
    01620 - PRODUCT OPTIONS   2.2 MATERIALS
    01630 - PRODUCT SUBSTITUTION PROCEDURES   2.3 [MANUFACTURED UNITS]
            [EQUIPMENT]
            [COMPONENTS]

Exhibit 12-4. Comparison Between Division 1 - General Requirements and typical spec section articles.

01640 - OWNER-FURNISHED PRODUCTS	2.4 ACCESSORIES
	2.5 MIXES
	2.6 FABRICATION
01650 - PRODUCT DELIVERY REQUIREMENTS	1.12 DELIVERY, STORAGE AND HANDLING
01660 - PRODUCT STORAGE AND HANDLING REQUIREMENTS	

01700 EXECUTION REQUIREMENTS

01710 - EXAMINATION REQUIREMENTS	3.1 EXAMINATION
01720 - PREPARATION REQUIREMENTS	3.2 PREPARATION
	3.3 [ERECTION] [APPLICATION] [INSTALLATION]
01722 - FIELD ENGINEERING	3.9 SCHEDULE
01730 - EXECUTION REQUIREMENTS	3.6 [ADJUSTMENT AND] CLEANING
01740 - CLEANING REQUIREMENTS	
01750 - STARTING AND ADJUSTING PROCEDURES	3.8 PROTECTION
01760 - PROTECTION OF CONSTRUCTION	
01770 - CONTRACT CLOSEOUT PROCEDURES	1.16 MAINTENANCE
01781 - FINAL SITE SURVEY	1.15 WARRANTY
01782 - MAINTENANCE CONTRACTS	
01783 - OPERATION AND MAINTENANCE DATA	
01785 - PRODUCT WARRANTIES AND BONDS	
01789 - PROJECT RECORD DOCUMENTS	

01800 FACILITY OPERATION

01810 - COMMISSIONING	
01820 - DEMONSTRATION AND TRAINING	3.7 DEMONSTRATION
01830 - OPERATION AND MAINTENANCE	
01890 - RECONSTRUCTION	

01900 FACILITY DECOMMISSIONING

01901 - DECOMMISSIONING SYSTEMS AND EQUIPMENT
01902 - FACILITY DEMOLITION AND REMOVAL
01903 - HAZARDOUS MATERIALS ABATEMENT
01904 - HAZARDOUS MATERIALS REMOVAL AND DISPOSAL
01905 - PROTECTION OF DEACTIVATED FACILITIES

Exhibit 12-4. *(Continued)*

out the specification writing principle of "say it once."
○ Contain specifics directly applicable to the particular project.
○ Must be written separately for each project.

• Divisions 2 through 16

○ Specify administrative requirements applicable only to a separate specification section.
○ Administrative requirements are specified in PART 1 - GENERAL in Division 2 through 16 specification sections.

Examples—Testing
• General Conditions of the Contract: "Owner will pay for testing."
• Division 1 - General Requirements: "Owner will engage an independent testing and inspection agency" plus information regarding authority of the agency.
• Divisions 2 through 16: Specific tests to conduct, such as test for water or air infiltration.

Examples—Submittals
• Submittals requirements in the General Conditions of the Contract (Exhibit 12-5)

3.12 SHOP DRAWINGS, PRODUCT DATA AND SAMPLES

3.12.1 Shop Drawings are . . .

3.12.2 Product Data are . . .

3.12.3 Samples are . . .

3.12.4 Shop Drawings, Product Data and Samples are not Contract Documents. The purpose of their submittal is to demonstrate for those portions of the Work for which submittals are required by the Contract Documents the way by which the Contractor proposes to conform to the information given and the design concept expressed in the Contract Documents. . . .

3.12.5 The Contractor shall review for compliance with the Contract Documents, approve and submit to the Architect Shop Drawings, Product Data, Samples and similar submittals required by the Contract Documents . . .

Exhibit 12-5. Sample General Conditions text regarding submittals.

- An example from the Supplementary Conditions of the Contract, modifying the General Conditions of the Contract (Exhibit 12-6)
- An example from Division 1 - General Requirements, Section 01330 - Submittal Procedures, PART 1 - GENERAL, in which general administrative procedures are specified (Exhibit 12-7)
- An example from a product specification section, Division 5, Metals, Section 05210 - Steel Joists, in which specific requirements for steel joist submittals are specified (Exhibit 12-8)

Relationship of Drawings and Division 1

The Contract Drawings graphically define certain Division 1 subjects, including the scope of Work and related work that may or may not be part of the Contract. There are other elements of the Work requiring close coordination between the Drawings and Division 1, including:

- Use of the site: Owner occupancy or use of the facility during construction
- Phased construction
- Multiple prime contract construction

Examples: Drawings show project limits, such as access ways, areas to be maintained clear for continuing use of the facility by the Owner, and location of temporary construction such as the construction fence and project identification signage.

WRITING DIVISION 1 SECTIONS

Division 1 Style and Format

Division 1 sections are organized and written in the same outline style and three-part section format as the sections in Division 2 through 16. For most Division 1 sections, Part 2 - Products and Part 3 - Execution do not apply and are noted "Not applicable to this section" or a similar phrase is used. Division 1 sections specify general requirements applicable to all of the Work and are written to specify requirements broadly enough to cover the content of all sections in Divisions 2 through 16.

Organization

The re-organization of Division 1 in the 1995 edition of *MasterFormat* resulted in a more consistent and logical arrangement of sections, which is also related more closely to the organization of PART 1 in the specification sections of Divisions 2 through 16. Many public agencies and corporations have resisted changing their Division 1 section numbers and titles. Looking at a comparison between the 1995 and 1988 versions, in the parallel comparison below (Exhibit 12-9), the superior arrangement of the 1995 version is apparent. Those who use the outdated format should update.

As a comparison to the broadscope (Level Two) headings above, Exhibit 12-10 is a list of Division 1 section numbers and titles from specifications for a large, complex project that illustrates the range of topics that may be covered in Division 1.

K. Add Subparagraph **3.12.9** as follows:

3.12.9 Procedures for submittals are further prescribed in Division 1 - General Requirements, Section 01330 - Submittals Procedures, in the Specifications.

Exhibit 12-6. Sample Supplementary Conditions text regarding submittals.

1.08 SHOP DRAWINGS

A. Copies: Submit one reproducible and one print, minimum sheet size 22 inches by 17 inches or a multiple of 8½ inches by 11 inches.

B. Preparation: Shop drawings shall be original drawings prepared for submittal review, fabrication and execution of Work. Direct copies and modified reproductions of Contract Drawings will not be accepted for review.

C. Coordination: Show all field dimensions and relationships to adjacent or critical features of Work.

Exhibit 12-7. Sample Divisions 1 - General Requirements text regarding submittals.

1.05 SUBMITTALS

A. Shop Drawings: Detailed drawings, indicating layout of joists, joist girders, special connections, joining and accessories. Include mark, number, type, location and spacing of joists and bridging. Use AWS symbols for all welds.

Exhibit 12-8. Sample specifications section text regarding submittals.

COMPARISON: DIVISION 1 - 1995 VS. 1998 MASTERFORMAT

1995 MasterFormat	1988 MasterFormat
01100 SUMMARY	01010 SUMMARY OF WORK
	01020 ALLOWANCES
01200 PRICE AND PAYMENT PROCEDURES	01025 MEASUREMENT AND PAYMENT
	01030 ALTERNATES
	01035 MODIFICATION PROCEDURES
01300 ADMINISTRATIVE REQUIREMENTS	01040 COORDINATION
	01070 IDENTIFICATION SYSTEMS
	01090 REFERENCES
	01060 REGULATORY REQUIREMENTS
	01100 SPECIAL PROJECT PROCEDURES
	01200 PROJECT MEETINGS
01400 QUALITY REQUIREMENTS	01300 SUBMITTALS
01500 TEMPORARY FACILITIES AND CONTROLS	01400 QUALITY CONTROL
	01500 CONSTRUCTION FACILITIES AND TEMPORARY CONTROLS
01600 PRODUCT REQUIREMENTS	
01700 EXECUTION REQUIREMENTS	01600 MATERIAL AND EQUIPMENT
	01700 CONTRACT CLOSEOUT
01800 FACILITY OPERATION	01050 FIELD ENGINEERING
01900 FACILITY DECOMMISSIONING	01800 MAINTENANCE

Exhibit 12-9. Comparison: Division 1 - 1995 versus 1988 *MasterFormat*.

```
DIVISION 1 - GENERAL REQUIREMENTS (COMPLEX PROJECT)

Section No. - TITLE

SECTION 01110 - SUMMARY OF THE PROJECT
SECTION 01140 - WORK RESTRICTIONS
SECTION 01210 - ALLOWANCE PROCEDURES
SECTION 01230 - ALTERNATE PROPOSAL PROCEDURES
SECTION 01250 - CONTRACT MODIFICATION PROCEDURES
SECTION 01270 - UNIT PRICES
SECTION 01290 - PAYMENT PROCEDURES
SECTION 01310 - CONSTRUCTION PROGRESS SCHEDULE
SECTION 01311 - PROJECT COORDINATION
SECTION 01312 - PROJECT MEETINGS
SECTION 01315 - MECHANICAL AND ELECTRICAL COORDINATOR
SECTION 01330 - SUBMITTALS PROCEDURES
SECTION 01340 - REQUESTS FOR INTERPRETATION (RFI)
SECTION 01351 - HAZARDOUS MATERIAL PROCEDURES
SECTION 01352 - INTERIM LIFE SAFETY MEASURES
SECTION 01353 - INFECTION CONTROL MEASURES
SECTION 01410 - REGULATORY REQUIREMENTS
SECTION 01420 - REFERENCE STANDARDS AND ABBREVIATIONS
SECTION 01450 - QUALITY CONTROL
SECTION 01455 - FIELD MOCK-UPS
SECTION 01458 - TESTING LABORATORY SERVICES
SECTION 01459 - PERFORMANCE AND OPERATIONAL TESTING
SECTION 01510 - TEMPORARY UTILITIES
SECTION 01520 - CONSTRUCTION FACILITIES
SECTION 01540 - CONSTRUCTION AIDS
SECTION 01550 - VEHICULAR ACCESS AND PARKING
SECTION 01560 - TEMPORARY BARRIERS AND ENCLOSURES
SECTION 01570 - TEMPORARY CONTROLS
SECTION 01580 - PROJECT IDENTIFICATION AND SIGNAGE
SECTION 01610 - BASIC PRODUCT REQUIREMENTS
SECTION 01630 - PRODUCT SUBSTITUTION PROCEDURES
SECTION 01640 - OWNER-FURNISHED PRODUCTS
SECTION 01650 - PRODUCT DELIVERY REQUIREMENTS
SECTION 01660 - PRODUCT STORAGE AND HANDLING REQUIREMENTS
SECTION 01710 - EXAMINATION REQUIREMENTS
SECTION 01720 - PREPARATION REQUIREMENTS
SECTION 01722 - FIELD ENGINEERING
SECTION 01730 - EXECUTION REQUIREMENTS
SECTION 01732 - CUTTING AND PATCHING REQUIREMENTS
SECTION 01740 - CLEANING REQUIREMENTS
SECTION 01770 - CONTRACT CLOSEOUT PROCEDURES
SECTION 01783 - OPERATION AND MAINTENANCE DATA
SECTION 01785 - PRODUCT WARRANTIES AND BONDS
SECTION 01789 - PROJECT RECORD DOCUMENTS
SECTION 01820 - DEMONSTRATION AND TRAINING
```

Exhibit 12-10. Sample Division 1 - General Requirements (complex project).

Caution: Division 1 administrative and procedural matters and requirements for temporary facilities are cost items, the same as products and other Work specified in Divisions 2 through 16. Overspecifying Division 1 items can increase construction costs unnecessarily and create excessive obligations for the firm administering the contract.

See the example short-form versions of Division 1 specifications included with the example Short-form Specifications in Appendix B: Sample Division 1 - General Requirements.

Chapter 13

Modifications

MODIFICATIONS TO BIDDING AND CONTRACT DOCUMENTS

Modifications to the bidding and Contract Documents will be necessary despite efforts to avoid or minimize them. The issue is how they will be documented. There are several types of documents that modify bidding and construction Contract Documents:

- Addendum
- Change Order
- Work Directive Change
- Construction Change Directive
- Field Order
- Architect's Supplemental Instructions

ADDENDA

The dictionary definition of an *addendum* is "a thing to be added; an addition" *Webster's Encyclopedic Unabridged Dictionary of the English Language* (Gramercy Books, New York, 1989). Each addendum is a document added to a previously prepared and issued set of bidding documents during the bid period. It becomes a part of the contract documents, such as defined by AIA *A201 General Conditions,* Article 1.1.1.

Addenda are written or graphic instruments issued to clarify, revise, add to, or delete information from original bidding documents or previous addenda. AIA *A701 Instructions to Bidders* describes an addendum as including "interpretations, corrections and changes in the Bidding Documents" (3.2.3) and states that addenda will be issued prior to receipt of bids. Similarly, EJCDC *C-700* defines addenda as modification documents published before bid opening. Addenda communicate changes in the bidding documents.

Following the AIA *A701* process for addenda:

- Addenda will be transmitted to all who are known by the issuing office to have received a complete set of bidding documents. (This is

why it is important to maintain an accurate list of planholders.)
- Addenda will be made available for inspection at the location where other bidding documents are on file.
- Addenda will be issued no later than 4 days before the date set for receipt of bids, unless the addendum withdraws the Invitation for Bids, or if the addendum postpones the date bids will be received.

Purposes of Addenda

In Chapter 8, under the heading "Instructions to Bidders," there is a discussion regarding interpretation of documents. The primary purposes of addenda are to clarify, in writing, the meaning of the Drawings and Specifications and respond to inquiries from prospective bidders regarding discrepancies, omissions, and conflicts in the bidding documents. In addition, an addendum may be used to make additional information part of the bidding documents. This information can take any of the following forms:

- Corrections of errors and omissions
- Clarifications of ambiguities
- Additions or deletions to increase or reduce the scope of the proposed work
- Other information that can affect the bid prices
- Change in the time and place for receipt of bids
- Change in the quality of the work
- Listing of additional names of qualified "or equal" products (see the discussion of Substitutions and Product Options in Chapter 16)

For example, if the drawings show sheet vinyl flooring but the specification describes vinyl composition tile flooring, one of these documents must be changed. If vinyl composition tile is actually required, the addendum should state the Drawing containing the information regarding sheet vinyl flooring, and should give the instruction to delete the term "sheet vinyl flooring" and substitute the term "resilient tile flooring." If sheet vinyl flooring is actually required, the Specifications should be altered by the addendum with instructions to change specific paragraphs by deletions, additions, and substitutions of text, up to and including wholesale deletion and substitution of the entire section.

Precautions

- Do not issue a complex addendum that requires considerable work on the part of the bidders unless there is still sufficient time before the bid due date. If time is insufficient for assimilating such a change, postpone the bid due date.
- Do not respond verbally to a bidder's telephone inquiries. Instruct the Bidder to submit the inquiry in writing and, if a clarification is in order, answer the inquiry by a written addendum issued to all bidders so that every bidder is informed in the same manner, with the same information, and the responses are clearly documented. A recommended technique is to require, in the Instructions to Bidders, that inquiries be submitted only by facsimile (fax). Thus, there will be no chance of verbal misunderstanding.
- While an urgent change may be issued by fax, formal issuance of the addendum should still occur through established methods. With changes in communication through facsimile documents, e-mail, and intranet sites, the dissemination process is changing. The enforceability of addenda other than written ones delivered by the U.S. Postal Service or courier service is yet to be definitively determined.

In private construction work, where changes are often negotiated with the successful bidder before award of contract, changes should be summarized and incorporated into an addendum that is issued prior to execution (signing) of the Owner-Contractor Agreement form. This applies to projects under AIA contract documents. There is no such provision under EJCDC documents.

All addenda should be prepared, controlled, and issued by one party who has intimate knowledge of all the Contract Documents and who serves as the clearinghouse for gathering and arranging all the information. Consultants, such as structural, mechanical, and site engineers, should not issue addenda since they may inadvertently assign wrong addenda numbers or issue instructions that conflict with other instructions contained elsewhere in the bidding and Contract Documents.

Itemize each instruction or change within the addendum by number for future reference during construction and correspondence.

Describing Changes in Bidding Documents

Changes typically are described in addenda by one of two methods:

- Narrative Method: A narrative description of changes is presented that should be followed to make physical changes to the documents, in the form of strikeouts to delete text, notations, and taped text segments for revised text. The narrative should be brief and give only enough information to make the change clear. Avoid oversimplification, but be clear by repeating enough of the previous text to make each change self-explanatory.
- Revised Page Method: With word processing, changes and reprinting of documents are easy. By revising the section or document, changes are effectively communicated. With functions such as "track changes" in Microsoft Word and "redlining" in WordPerfect, changes may be noted and graphically indicated. If these word processing functions are not used, the revised page method may still be used, with a brief summary or narrative description as above.

This describes changes to the documents and Specifications in the Project Manual. For the Drawings, similar methods may be used, including issuing new details to be cut and pasted over existing details, or narrative descriptions for revisions, additions, and deletions to notes on the Drawings, or reissuing of an entire sheet of the drawings if the changes are extensive. Drawings revisions are typically surrounded by a "cloud" with a numeral in a triangle ("delta" or different) that is keyed to the revision box in the Drawing title block.

If there are conflicts in the Drawings, in which one detail shows one arrangement and another detail shows another arrangement, describe in the addendum the deletion of the inappropriate detail rather than make a statement that one detail shall govern over the another or that one detail is preferred over another.

Similarly, for conflicts within the Specifications, delete the inappropriate material by addendum. Do not explain that one specification is preferred and the other should be ignored. Make the revision by describing physical changes to be made to the Drawings and Specifications. Because paper is cheap, addendum revisions to the Specifications are best handled by deleting an entire section and substituting a revised section with a different publication date.

Procedures for Addenda

Addenda should be prepared and issued in a timely manner. Bidders must be given corrections and additional information in time for actual use in preparing their bids. A minor addendum can be issued as late as 5 days before bids are due without imposing hardship on bidders. It might be possible to issue a minor addendum within even a shorter time if bidders are alerted that a late addendum will be issued.

Preserve the bid date if possible. However, if a critical question arises, an addendum should be issued even if it means delaying the bid opening. This is not always possible, such as when bids must be received and reviewed prior to a scheduled meeting of the board of a public agency, which will authorize acceptance of the bid and direct that a contract be awarded to the selected bidder. Delay could mean losing several weeks or a month until the next scheduled board meeting.

Number addenda consecutively. For multiple prime contracts, provide a separate series of addenda numbers for each contract. Where feasible, use a simple system of numbering for items within an addendum to permit future cross referencing.

Each addendum should be arranged in an orderly sequence. Following a prepared format helps to ensure that all required elements of the addendum are included. The format should follow the same sequence as the Project Manual and the Drawings as follows:

1. Introduction
2. Addendum number
3. Name of issuing party: Owner, Construction Manager, Architect, or Engineer
4. Project identification
5. Date of addendum
6. Opening remarks and instructions
7. Addendum changes, in sequence
8. Changes to prior Addenda
9. Changes to the Project Manual
10. Changes to introductory documents
11. Project Manual cover
12. Certifications
13. Table of contents
14. Changes to bidding requirements
15. Invitation to Bid
16. Instructions to Bidders
17. Bid Forms
18. Other bidding requirements
19. Changes to the Owner-Contractor Agreement and other Contract forms
20. Changes to the Owner-Contractor Agreement form
21. Change to General Conditions and Supplementary Conditions
22. Change to other Contract forms
23. Changes to Specifications, in sequence of Section numbers
24. Changes to Appendices (if included in the Project Manual)
25. Changes to Drawings, in sequence of the Index of drawings

References to addenda, the method of issue, and other pertinent facts concerning addenda are included in the Instructions to Bidders, the Bid Form, and the agreement. See Exhibit 13-1 for an example of an addendum.

CHANGE ORDERS

Change Order Format

Change orders are documents describing changes to the contract documents and are issued after execution (signing) of the agreement (contract). They are written instructions to the contractor, signed by the owner, architect/engineer, and contractor, authorizing and ordering an addition, deletion, or revision to the work, including, as applicable, an adjustment to the Contract Sum or the Contract Time.

In terms of descriptions of the changes, a Change Order follows an organization and production procedure similar to those of an addendum, except that the topic of the Change Order is usually much more limited. Only one or a few topics are included in a change order. In fact, the fewer the topics, the easier it is to trace the process of the Change Order should there be a subsequent dispute or performance problem.

Change Order Procedures

Preparing and authorizing a change early in construction allows time for consideration of proposals and determination of a mutually acceptable change

---

ADDENDUM TO BIDDING DOCUMENTS

DATE: September 12, 2004

FROM: City of Richmond
             Parks, Recreation and Community Services Division
             316 E. Broadway, Room 120
             Richmond, CA 94805

TO: All Prospective Bidders.

PROJECT: RICHMOND CIVIC AUDITORIUM - REHABILITATION AND UPGRADES
             (City of Richmond Specifications No. 2621-A and 2569-A)

SUBJECT: Addendum No. 1

NOTICE: This Addendum modifies the original Bidding Documents and shall become part of the Contract Documents.
             Make the following modifications and place this Addendum in the Project Manual following the introductory
             document, ADDENDA under the heading of BIDDING REQUIREMENTS AND FORMS - OVERALL PRO-
             JECT. Acknowledge receipt of this addendum on the Bid Form.

             This Addendum consists of two pages and the following enclosures:

                 Enclosure 1: Introductory Document titled "ADDENDA," dated 9/12/04.
                 Enclosure 2: Section 02150 - Shoring and Underpinning, dated 9/12/04.
                 Enclosure 3: Section 11150 - Parking & Revenue Control System, dated 9/12/04.
                 Enclosure 4: Sketch SK-E9-1-1, dated 8/25/04.

             The Bid date, time and location for receipt of Bids are unchanged by this Addendum.

INTRODUCTORY DOCUMENTS - PROJECT MANUAL

TABLE OF CONTENTS - PROJECT MANUAL

1. Refer to TABLE OF CONTENTS dated 8/16/04. Under heading "BIDDING REQUIREMENTS AND FORMS - OVERALL
   PROJECT," add the following at the end of the list:

       "ADDENDA 9/12/04"

2. Refer to TABLE OF CONTENTS dated 8/16/04. Make the following corrections:

   a. Change date of Section 11150 - Parking & Revenue Control System to 9/12/04.
   b. Change date of Section 02150 - Shoring and Underpinning to 9/12/04.

CONTRACT REQUIREMENTS AND SPECIFICATIONS

SPECIFICATIONS

SECTION 02150 - SHORING AND UNDERPINNING

1. Delete Section 02150 - Shoring and Underpinning dated 8/16/04 and add revised Section 02150 - Shoring and Underpinning dated 9/12/04. The revised Section adds requirements for installation, maintenance and removal of shoring and
   underpinning.

Exhibit 13-1. Sample addendum.

SECTION 02620 - SUBDRAINAGE SYSTEM

1. Delete Article 2.2 in its entirety and substitute the following. Article 2.2 has been revised to require subsurface drainage mat to be approved by the below-grade waterproofing manufacturer.

"2.2 SUBSURFACE DRAINAGE MAT

    A. Specified Manufacturer: Carlisle Coatings & Waterproofing Incorporated (CCW), Wylie, TX.

    B. Acceptable Manufacturers: Equivalent products of the manufacturers listed below will be acceptable in accordance with the "or equal" provision specified in Section 01630 - Product Options and Substitutions.

        1. JDR Enterprises, Inc., Alpharetta, GA.
        2. Solutia, Inc., St. Louis, MO.

    C. Subsurface Drainage Composite: CCW MiraDRAIN 6000 or 6200, prefabricated in-plane wall drainage composite, as part of overall foundation drainage system, to function as drainage system and protection for below-grade waterproofing specified in Section 07130 - Sheet Membrane Waterproofing.

    D. Drainage Core: Manufacturer's standard three-dimensional, non-biodegradable, plastic material designed to effectively conduct water to foundation drainage system under maximum soil pressures.

    E. Filter Fabric, Vertical Applications: CCW MiraDRAIN 140-N, standard non-woven geotextile fabric of polypropylene or polyester fibers, or a combination of these fibers.

    F. Filter Fabric, Horizontal Applications: CCW MiraDRAIN HP 500, FW 40/10 or FW 70/20, to suit site-specific soil and fill materials, woven geotextile fabric of polypropylene or polyester fibers, or a combination of these fibers."

SECTION 07110 - SHEET MEMBRANE WATERPROOFING

1. Delete Paragraph 2.1.A. and substitute the following:

  "A. Carlisle Coatings & Waterproofing Incorporated (CCW), Wylie, TX"

2. Delete Paragraph 2.1.C. and substitute the following:

  "C. Sheet-Applied Waterproofing: CCW MiraDRI 860, self-adhering, cold-applied, composite sheet waterproofing, consisting of 56 mil layer of rubberized asphalt integrally bonded to cross-laminated, 4 mil high density polyethylene sheeting."

SECTION 11150 - PARKING & REVENUE CONTROL SYSTEM

1. Delete Section 11150 - Parking & Revenue Control System dated 8/16/04 and substitute revised Section 11150 - Parking & Revenue Control System dated 9/12/04. The revised Section clarifies Work under the Contract and "future" (not in Contract) Work.

DRAWINGS

DRAWING C-4

1. Add Note No. 4, as follows:

  "4. ALL PEDESTRIAN RAMPS SHALL HAVE A GROOVED BORDER 12″ WIDE AT THE LEVEL SURFACE ALONG THE TOP OF ALL RAMPS AND EACH SIDE AT CURBS AND RAMPS."

Exhibit 13-1. *(Continued)*

DRAWING C-5

1. Replace Demolition Note No. 11 with the following:

   "11. REMOVE EXISTING FLAGPOLE AND CONCRETE FOUNDATION. SALVAGE FLAGPOLE AND DELIVER TO OWNER AT CITY STORAGE YARD, AS DIRECTED."

2. Replace Demolition Note No. 15 with the following:

   "15. REMOVE EXISTING LAMPPOST, CONCRETE FOUNDATION AND ELECTRICAL CONDUITS. SALVAGE LAMP-POST AND DELIVER TO OWNER AT CITY STORAGE YARD, AS DIRECTED."

DRAWING E9-1

1. Delete conduit and wiring to the four landscape step lights along the north walkway.

2. Add one additional Type "PC" pole light fixture and foundation and revise landscape lighting circuiting as indicated on enclosed Sketch SK-E9-1-1.

3. Revise landscape lighting circuiting at southeast corner of structure and add circuiting for landscape lighting at southwest corner of structure as indicated in enclosed Sketch SK-E9-1-2.

END OF DOCUMENT

Exhibit 13-1. *(Continued)*

in Contract Time or Contract Sum (if any) without affecting Work in progress. This is not always possible, but it is a goal. Change Order procedural requirements should be specified in Division 1 - General Requirements and should be coordinated with the General Conditions and Supplementary Conditions. Change Order procedures should be included in Division 1 - General Requirements in a section such as Section 01250, Contract Modification Procedures.

- Only the Owner has authority to direct a change to be made. However, there are several ways that a change order may be initiated.
- Owner-initiated change: With the Owner's authorization, the Architect/Engineer prepares and issues a Request for Proposal (RFP) to the Contractor.
- The RFP should include a detailed description of a proposed change, with supplementary or revised drawings and Specifications as appropriate.
- The RFP may include an estimate of additions or deductions in Contract Time and Contract Sum for executing the change, and may include stipulations regarding overtime work and the period of time the requested response from the Contractor shall be considered valid.

- In response to the RFP, the Contractor prepares and submits a written proposal within a specified time period.
- The Owner, Architect/Engineer, and Contractor review the proposal and negotiate the change in Contract Time and Contract Sum as applicable. *Note:* Both time and money need to be considered. Time is money, and extension of the contract completion date may result in a claim by the Contractor for additional overhead and contract administration costs.
- When the scope of the proposed change, the change in Contract Time, and the change in Contract Sum are agreed upon by the Owner and Contractor, the Architect/Engineer prepares the formal Change Order document, typically using a preprinted document, such as AIA *G701 Change Order* or EJCDC *C-941 Change Order,* or a custom document as required by the Owner. The Change Order should reference and, as practical, include attached revised Contract Drawings and Specifications.
- Change Orders are signed by the Owner, Architect/Engineer, and Contractor.

Contractor-Initiated Change

The Contractor may propose a change by submitting a request for a change to the Architect/

Engineer, describing the proposed change and its full effect on the Work, with a statement describing the reason for the change and a full description of its effects on the Contract Sum, Contract Time, related Work, and work being performed under separate contracts. Such proposed changes may be requests to substitute products or assemblies, or to change the agreed-upon sequence of construction represented in the approved construction progress schedule, or to address unforeseen conditions such as utility lines in locations other than where previous project record drawings ("as-built" drawings) indicate.

- To keep the administrative process consistent, contractor-initiated changes may be informal in nature and lead to preparation and issuance by the Architect/Engineer of an RFP as described above.
- When the change requires urgent action, the initiation by the Contractor may be in the written form of a memorandum or letter to the Owner, with preliminary supporting information such as will be included in the Contractor's response to the eventual RFP.
- Urgent action would likely take the form of a Construction Change Directive or a Work Directive Change, described below.
- The Owner should respond to the Contractor-initiated change with preparation and issuance by the Architect/Engineer of a formal RFP.

Change orders have the following basic elements:

- Change Order number and date of issue.
- Project name and job number.
- Architect/Engineer's name and address.
- Owner's name and address.
- Contractor's name and address.
- Original Contract completion date.
- Statement that this Change Order modifies the original Contract.
- Description of revisions, including the cost of each. Reference can be made to detailed descriptions in previously issued bulletins (Architect's Supplemental Instructions or Construction Change Directive [AIA] or Work Directive Change [EJCDC].).
- Revised Contract Drawings and Specifications, referenced or, if practical, attached to the Change Order.

Construction Changes Based on Unit Costs or Quantities: When the scope of a change in the Work cannot be accurately determined in advance, a Construction Change Directive should be executed based on mutually acceptable quantities and predetermined unit prices. Actual costs should be determined after completion of the Work, and a Change Order for this amount should be executed.

Construction Changes Based on Time and Material Costs: When the scope of a change in the Work cannot be accurately determined in advance, a Construction Change Directive should be executed based upon an agreement that the Owner will adjust the Contract Sum and the Contract Time based on the actual costs and time expended by the Contractor in performance of the change.

Substantiating Data for Changes in Contract Sum and Contract Time: The Contractor should provide full information required for evaluation of proposed changes and to substantiate the costs of changes in the Work.

1. Document each quotation for a change in Contract Sum and Contract Time, with sufficient data to allow evaluation of the quotation.
2. Upon request, provide additional data to support computations:
   a. Quantities of products, labor, and equipment
   b. Taxes, insurance, and bonds
   c. Overhead and profit
   d. Justification for change in Contract Time, if claimed
   e. Credit for deletions from the Contract, similarly documented

Cost and Time Resolution: If the amounts for changes in contract sum and contract time cannot be agreed upon by the Owner and Contractor, they shall be resolved in accordance with provisions of the Conditions of the Contract for resolution of disputes and the following:

1. The Contractor should keep accurate records of time, both labor and calendar days, and of the cost of materials and equipment.
2. The Contractor should prepare and submit an itemized account and supporting data after completion of the changed Work within the time limits indicated in the conditions of the contract.

3. The Contractor should provide full information as required and requested for the Owner and Architect/Engineer to evaluate and substantiate proposed costs and time for the change in the Work.

4. When the Owner and Contractor determine mutually acceptable amounts for changes in the Contract Sum and Contract Time, a Change Order should be executed for these amounts.

5. The Owner should have the right to audit the Contractor's invoices and bid quotations to substantiate the costs for Change Orders.

Reconciliation of Change Orders

• Schedule of Values: The Contractor should promptly revise the Schedule of Values and Application for Payment forms to record each authorized Change Order as a separate line item and adjustment to the Contract Sum.

• Construction Schedules: The Contractor should promptly revise the progress schedules to reflect changes in Contract Time, revising subschedules to adjust the time for other items of Work that may be affected by the change. The Contractor should submit revised schedules at the next Application for Payment following approval and acceptance of the Change Order.

The Owner will sign all Change Orders, Construction Change Directives, or Work Directives because each is an extension or modification of the Contract between the Owner and Contractor.

OTHER MODIFICATIONS

There are modifications in addition to Addenda and Change Orders.

Architect's Supplemental Instructions

The Architect's Supplemental Instructions are for contracts using AIA General Conditions and are used for minor written instructions or interpretations from the Architect/Engineer to the Contractor that do not involve change orders and do not change the Contract Sum or Contract Time.

Construction Change Directives

Construction Change Directives are for contracts using AIA General Conditions and are written authorizations directing the Contractor to proceed with described changes, subject to subsequent adjustments in the Contract Sum or Contract Time through execution of a Change Order. In accordance with provisions of the *AIA A201* General Conditions, the Owner may direct the Contractor to proceed with a change in the Work prior to formal preparation, review, and agreement of a Change Order in order to not delay construction.

The Architect prepares a Construction Change Directive that, when signed by the Owner and the Architect, instructs the Contractor to proceed with a change in the Work for subsequent inclusion in a Change Order.

Should the Construction Change Directive result in disputed costs and time adjustments, the dispute will be resolved in accordance with the dispute resolution provisions of the General Conditions and Supplementary Conditions.

Construction Change Directives should follow procedures specified above for preparation of Change Orders, except that the Contractor should immediately proceed with the change upon receipt of the signed Change Directive and not wait until completion of the proposal and negotiation process.

Work Directive Changes

Work Directive Changes are for contracts using EJCDC General Conditions and are written directives to the Contractor to make changes in the Work, which may or may not affect the Contract Sum but are evidence that changes will be incorporated in a Change Order once value of the Work is established. The Work Directive Change operates similarly to the Construction Change Directive. EJCDC publishes document *C-940 Work Change Directive* for this purpose.

Field Order

A Field Order is for contracts using EJCDC documents only. It is an antiquated term for contracts using AIA documents (therefore, it is not a valid term). A Field Order is defined as an authorization directing the Contractor to perform minor variations in the Work when such changes do not change the Contract Sum or Contract Time. EJCDC publishes document *C-942 Field Order* for this purpose.

Chapter 14

Specifications Language

GENERAL PRECEPTS

It is not intended, nor indeed is it possible, for this chapter to be a primer on English grammar and readable writing. Rather, the chapter presents precepts and examples of proper specification language.

In order to communicate with proper language, the specifier must sufficiently master the tools of specifications language, including grammar, vocabulary, spelling, use of abbreviations and symbols, punctuation, capitalization, sentence structure, and the unique considerations of "streamlined" writing and specifications detail. The specifier must not only follow the rules of language but must understand the subtleties of language. It is much like carpentry. Sometimes rough carpentry, performed with power saws and pneumatic nail guns, is adequate. Sometimes finish carpentry is called for, with hand crafting and fine finishing.

Imagine that each statement in the specifications carries a dollar sign, whether it is concerned with specifying materials, instructing the Contractor on installation procedures, or describing workmanship. The Contractor expects compensation for each requirement in the Contract Documents, and the Contractor's bid theoretically reflects every statement made in the specifications and noted on the drawings. Specification language should be precise. Vague, ambiguous language indicates that the specifier may want something but is unsure about demanding it. Statements such as "tests shall be performed unless waived," "additional shop drawings and samples may be required," and "uneven surfaces may be cause for rejection" are examples of equivocation that plague the Contractor and lead to disputes between the Owner and Contractor.

Precise specifications can be enforced. Vague specifications are open to multiple interpretations and invariably result in the Contractor's choosing the least restrictive, most expedient, and least costly outcome. Vague specifications are difficult to enforce and cost the Owner additional money because the Contractor will claim that the most restrictive, least expedient, and most costly out-come is not what the bid was based upon, and additional time and money must be added to the Contract if a different interpretation of the specifications is enforced. It is difficult to contradict this assertion without precisely specified requirements.

The essential requirement for writing specifications, aside from knowledge of the technical content, is the ability to express oneself in proper English, or at least in the English that is conventionally used for writing construction specifications. Specifications English is not the same as the English used in reports, narratives, and business correspondence. It is certainly not the slang-laden, abbreviated English expediently used in e-mail. Specifications English is a technical style of writing, suitable for the outline format of specifications, which often relies upon terse sentences and an even, no-nonsense tone.

Construction specifications writing is the preparation of written English documents that become contractual (legally enforceable) documents. Despite this "legal" aspect of specifications, legal phraseology is not necessary and should be avoided. Likewise, eloquence and creative expressions are inappropriate for construction specifications. Specifications should state requirements in clear, concise, and correct terms in plain English.

There are four important Cs in specifications writing:

1. *Be Clear:* Avoid ambiguity. Consider the reading ability of the reader. Choose precise words that convey exact meanings.

2. *Be Correct:* Present ideas and explanations accurately and precisely. Specifications should be correct technically and grammatically. Proper terminology is important.

3. *Be Complete:* Do not leave out anything that is important. Brevity at the expense of completeness should be avoided.

4. *Be Concise:* Eliminate unnecessary words but not at the expense of clarity, correctness, or completeness. Typically, delete the articles "a," "an," and "the" where clarity is not diminished.

Eliminate verbiage. A well-crafted specification is one containing the fewest words that can be used to complete the description and make sense. Verbosity and repetition can lead to ambiguity rather than avoiding it.

Consider those who use the specifications. Specifications must be clearly written so that they can be understood by someone whose vocabulary and English comprehension are limited. A clearly written specification in English should be clear enough even for the mechanics on the job. If a mechanic cannot understand and interpret specifications, he or she will not be able to follow them.

Specified requirements should be definite, unequivocal, and understandable to the reader, whether he or she is the owner, the contractor's superintendent, a cost estimator, an architect checking shop drawings, a purchasing agent placing an order for materials, or (perish the thought) an attorney for the opposing side in a construction dispute.

Since Specifications are instructions to the Contractor, they should be definite and mandatory. To be mandatory, they must be indicative. Therefore, use the indicative "shall" with reference to the Work of the Contractor; never use the vague and indefinite "will" or "to be." The proper place to use "will" is in a statement describing the acts to be performed by the Owner or the Architect.

Language is a means of communication. Unlike graphic communication, where symbols and graphic indications such as crosshatching have representative meanings, written words must be carefully selected for their precise meanings. Unlike spoken words, where inflections and gestures add subtle nuances and emphasis, written language relies solely upon the printed characters on a page. Because words may have subtle connotations, they should be selected carefully to minimize misinterpretation.

Consider the word "smooth." The dictionary defines it as "having an even surface; devoid of surface roughness." The term "smooth" has been employed in specifications as follows: "Bituminous road surfaces shall be smooth." Yet the preferred texture for a road surface, to reduce skidding, is a rough texture. The word "smooth" means something different for road surfaces than for concrete floors. Concrete floors may be specified to have a smooth, wood float finish, or a smooth, rubbed finish, or a smooth, troweled finish. However, in each case the degree of smoothness varies significantly, and it is necessary for the specifier to understand the differences.

It might seem preferable to specify the tool that will accomplish the intended result and rely on it to achieve the desired surface finish. For example, "Finish concrete floors with wood float," or "rub concrete steps with carborundum stone," or "burnish concrete floors with steel trowel" is grammatically preferable specifications writing. However, none of these adequately describes how smooth the concrete should be. Should a few swipes with a wood float, a carborundum stone, or a steel trowel be sufficient? No. Better descriptions, with measurable qualities, are needed.

Technical adequacy needs to be considered as well as proper and preferred language. It is necessary to understand the technology, specifications language, and construction practices.

APPLICATIONS

Grammar

The rules of grammar apply to specifications writing.

Subject and Verb Agreement

Single subjects require single verbs and plural subjects require plural verbs. Avoid long, complicated sentences because they make it more difficult to keep the subject and verb in agreement.

Incorrect: Two beads of sealant is applied before securing cover.

Correct: Two beads of sealant are applied before securing cover.

Preferred: Apply two beads of sealant before securing cover.

The incorrect example uses the plural subject "beads" with the singular verb "is." The correct example uses the plural subject "beads" with the plural verb "are." The preferred example uses a different sentence structure, described in more detail below, and lessens the chance of disagreement should the sentence be edited further and the number of "beads" be changed. The verb "apply" works with both singular and plural subjects.

Parallel Construction

Use the same style in both portions of a sentence with a compound subject or predicate. Also, use identical style when there is a series of nouns, adverbs, or prepositional phrases.

Incorrect: Inspections shall be conducted to determine quality of welds and verifying of compliance with specified tolerances.

Correct: Inspections shall be conducted to determine quality of welds and to verify compliance with specified tolerances.

Preferred: Conduct inspections to determine quality of welds and verify compliance with specified tolerances.

Inappropriate Terms

Do not use phrases with missing objects such as "as allowed" (by whom?), "as appropriate" (according to what?), "as approved" (by whom?), "as directed" (by whom?), "as indicated" (where?), "as required" (according to what or whom?), and "as necessary" (according to what or whom?). The last phrase, "as necessary," may be appropriate if a definition is specified that establishes the criteria for determining the necessity.

Avoid Legal-Sounding Adverbs

Do not try to write "legalese." Avoid adverbs such as "hereinafter," "hereinbefore," "herewith," and "wherein." They do not make the text more authoritative.

Avoid Certain Articles

Do not use "Any or all" (what does it mean and how is it enforced?), and avoid making an article out of "such" ("such accessories shall be silver plated"). Avoid making articles or pronouns out of certain words ("Polish said floor with wax"). "Do not use "same" either as an article ("Polish same floor with wax") or as a pronoun ("Mop floor and polish same with wax").

Avoid Certain Words and Expressions

Do not use "etc." (what precisely are the "etcetera," and how is a specification with this ambiguous term enforced?), "as per" (it is great-sounding spec lingo but grammatically incorrect), "in a workmanlike manner" (workmanlike is an ambiguous and unenforceable term), "to the satisfaction of the architect/engineer" (it is subjective, subject to abuse, and ambiguous), and "shall function as intended" (does this mean that the contractor must hire a psychic to read the mind of the architect/engineer?).

Avoid Unnecessary Words

- Eliminate superfluous words. The definite article "the" and the indefinite articles "a" and "an" need not be used in most instances. However, where these articles enhance clarity or readability, use them.
  Acceptable practice: "The Contractor shall prepare a Construction Progress Schedule."

  Preferred practice: "Contractor shall prepare Construction Progress Schedule."
- *All:* The use of "all" is usually unnecessary. The General Conditions of the Contract usually establish that the adjective "all" is implied, especially if AIA or EJCDC preprinted documents are used.
  Acceptable practice: Store all millwork under shelter.
  Preferred practice: Store millwork under shelter.
- *Contractor:* Avoid using "Contractor" as the subject of the sentence. Specifications are written to the Contractor; therefore, using "Contractor" is unnecessary, especially if "streamlining," discussed below, is used.
- *Which:* "Which" and other relative pronouns such as "who" and "that" should be used sparingly, if at all.
  Acceptable practice: Install bathroom accessories which are provided by Owner.
  Preferred practice: Install bathroom accessories provided by Owner.

Prepositional Phrases

Sentences should be kept short in specifications. An easy way to eliminate verbiage is to use modifiers in place of prepositional phrases.

Acceptable grammar: Apply vinyl wallcovering of selected pattern and color in remodeled classrooms of Building B.

Preferred grammar: Apply selected pattern and color vinyl wallcovering in Building B remodeled classrooms.

Acceptable phrase: "surfaces of concrete."

Preferred phrase: "concrete surfaces."

Acceptable phrase: "within temperature limits recommended by manufacturer."

Preferred phrase: "within manufacturer's recommended temperature limits."

Include Key Headings

This is not the same as streamlined writing, but it appears similar. Key headings, stated at the beginning of a paragraph express the topic of the paragraph and end with a colon (:). They greatly enhance the readability of specifications because the reader is able to scan the page and focus on the key words in order to find the topic of interest or concern, as in the following examples:

A. Corrosion Protection: Protect galvanized and nonferrous metal surfaces from corrosion or galvanic action by heavy coat of bituminous coating on surfaces which will be in contact with concrete or dissimilar metals.

B. Environmental Requirements: Condition interior casework and trim products to building environment. Maintain temperature and humidity at completed Work in accordance with requirements for storage.

C. Loose Segments Repair: Remove all loose concrete segments, tile and other subfloor elements and repair subfloor with patching compound to provide a solid base for underlayment.

Sentence Structure

Imperative Mood

This is recommended for instructions covering the installation of products and equipment. The verb that clearly defines the action becomes the first word in the sentence. The imperative sentence is concise and readily understandable, as in the following examples:

- "Spread adhesive with notched trowel."
- "Install equipment plumb and level."
- "Apply two coats of paint to each exposed surface."

Indicative Mood

The traditional language of specification sentences is the indicative mood, passive voice. This requires the use of "shall" in nearly every statement. This sentence structure can cause unnecessary wordiness and monotony, as in the following:

"Adhesive shall be spread with notched trowel."

"Equipment shall be installed plumb and level."

"Two coats of paint shall be applied to each exposed surface.

Avoid: "Contractor shall install all equipment plumb and level."

Vocabulary

Develop and Use Proper Terminology

This is a by-product of learning construction technology and construction contracting. One way to avoid conflicts in terminology is to define the terms. Consider the following terms and their definitions.

1. *Do not use "must" and "is to."* Substitute the word "shall" or use the imperative mood (described below). This prevents the inference of different degrees of obligation.

   Poor practice: Each joint must be filled solid with mortar.

   Poor practice: Each joint is to be filled solid with mortar.

   Better practice: Each joint shall be filled solid with mortar.

   Preferred practice: Fill each joint solid with mortar. (Imperative mood)

2. *Do not use "any" when a choice is not intended.* Because "any" implies a choice, it should not be used when a choice is not intended, as in the following examples:

   Poor practice: Any materials rejected shall be removed. (This gives the contractor the unintended choice of removing some but not necessarily all materials.)

   Better practice: Materials rejected shall be removed.

   Preferred practice: Remove rejected materials. (Imperative mood)

3. *Do not use "either" when a choice is not intended.* The word "both" should be substituted for "either" when no choice is intended.

   Poor practice: Glass panels shall be installed on either side of main entrance.

   Better practice: Glass panels shall be installed on both sides of main entrance.

   Preferred practice: Install glass panels on both sides of main entrance. (Imperative mood)

4. *Do not use "same" as a pronoun.*

   Poor practice: If materials are rejected, the Contractor shall replace same at no additional cost.

   Better practice: Replace rejected materials at no additional cost.

5. *Do not use "said" as an adjective.*

   Poor practice: Said materials shall be replaced at no additional cost.

   Better practice: Replace rejected materials at no additional cost.

6. *Do not use "and/or."* This is a stilted legal expression. The word "or" or "both" should be used in place of "and/or."

   Poor practice: Brick shall be made of clay and/or shale.

Better practice: Brick shall be made of clay, shale, or a combination of both.

7. *Do not use "etc."* Placed at the end of a list of items, "etc." shows that the specification writer obviously does not know what comprises the complete list or is too lazy to write it out. The use of "etc." is vague and results in ambiguous specification. It puts unnecessary responsibility on the contractor and therefore should not be used. As one specification writer puts it, "It is better to be definite even if you are wrong; then, at least, there is a firm basis for negotiating the corrections."

Poor practice: All standing trim, running trim, etc., shall be painted.

Better practice: Paint exposed millwork. (Imperative mood)

8. *Do not use the phrase "furnish and install."* Since it is established by the general conditions that the contractor shall provide and pay for all materials, labor, water, tools, equipment, light, power, transportation, and other facilities, unless otherwise stipulated, for the execution and completion of the work, it is redundant to use the phrase in other sections. Define the term "provide" and use it to mean "furnish and install." (See "Defined Terms and Expressions" below.)

Poor practice: Contractor shall furnish and install standard size face brick.

Better practice: Face brick shall be standard size.

9. *Do not use the phrase "to the satisfaction of the Architect"* and similar phrases, such as "as the Architect may direct," "acceptable to the Architect," and "in the opinion of the Architect." Instead, specify exactly what the Architect's directions are, or definitely what would be satisfactory or acceptable to the Architect. Do not leave the Contractor guessing and at the mercy of the Architect's future decisions.

Poor practice: Brick shall be laid to the satisfaction of the Architect.

Better practice: Brick shall be laid plumb and true with all joints completely filled with mortar.

10. *Do not use the phrase "a workmanlike job"* and similar phrases, such as "a high-class job" or "a first-class job." These are undefined and ambiguous expressions. Instead, the type of workmanship expected should be described in detail.

Poor practice: Brick shall be laid in a workmanlike manner.

Better practice: Brick shall be laid plumb and true with all joints completely filled with mortar.

11. *Use "shall" in connection with acts of the Contractor.* Do not use "will." Use of the simple imperative mood is even better.

Poor practice: Brick will be laid in running bond.

Better practice: Brick shall be laid in running bond.

Preferred practice: Lay brick in running bond. (Imperative mood)

Defined Terms and Expressions: In Division 1 - General Requirements, define certain terms and phrases that have specific meaning in the specifications. Examples:

- *And/or:* If used, shall mean that either or both of the items so joined are required. (This is a concession to those who prepare only a portion of the specifications and who insist on using "and/or." It results in ambiguity but gives some leverage to the Architect/Engineer.)

- *Applicable:* As appropriate for the particular condition, circumstance, or situation.

- *Approved:* This term, when used in conjunction with the Architect/Engineer's action on the Contractor's submittals, applications, and requests, is limited to the Architect/Engineer's duties and responsibilities as stated in the Conditions of the Contract.

- *Approve(d):* Limited to duties and responsibilities of the Architect/Engineer stated in the Conditions of the Contract, for actions performed in the professional judgment of the Architect/Engineer, in conjunction with submittals, applications, and requests. Approvals shall be valid only if obtained in writing and shall not apply to matters regarding the means, methods, techniques, sequences, and procedures of construction. Approval shall not relieve the Contractor from responsibility to fulfill Contract requirements.

- *Directed:* Limited to duties and responsibilities of the Architect/Engineer stated in the conditions of the contract, meaning as instructed by the Architect/Engineer or the owner, in writing, regarding matters other than the means, methods, techniques, sequences, and procedures of construction. Terms such as "directed," "requested," "authorized," "selected," "approved," "required," and "permitted" mean "directed by the Architect/Engineer," "requested by the Architect/Engineer," and similar phrases. No implied meaning shall be interpreted to extend the Architect/Engineer's responsibility to the Contractor's supervision of construction.

- *Equal or equivalent:* As determined by the Architect/Engineer as being equivalent, considering such attributes as durability, finish, function, suitability, quality, utility, performance, and aesthetic features.

- *Furnish:* Means Contractor shall procure indicated products or perform indicated services. Where used regarding products, the term 'furnish' is understood and intended to mean delivery of products to site of the Work but is not intended to include the installation, application or other action to incorporate of products, either temporarily or permanently, into the Work.

- *Indicated:* Refers to graphic representations, notes, or schedules on the Drawings, or other paragraphs or schedules in the Specifications, and similar requirements in the Contract Documents. Terms such as "shown," "noted," "scheduled," and "specified" are used to help the reader locate the reference. There is no limitation on location. Refer also to *General Conditions of the Contract (AIA A201),* Subparagraph 4.2.12.

- *Install:* Means Contractor shall receive, unload, transport and temporarily store products at the site of the Work and to perform assembly, fitting, installation, application, erection and similar actions as necessary to incorporate products complete in place and ready for use, including furnishing of necessary labor, materials, tools, equipment and transportation. The term "install" is also understood and intended to include testing and inspection that is necessary for proper installation, application, erection, and similar actions and for verification of the quality of the work, as provided in the contract documents.

- *Installer:* Refers to the Contractor or an entity engaged by the Contractor as an employee, subcontractor, or sub-subcontractor for performance of a particular construction activity, including installation, erection, application, and similar operations. Installers are required

to be experienced in the operations they are engaged to perform. *Experienced installer:* When used with "installer," the term "experienced" means having a minimum of five previous projects similar in size to this project, knowing the precautions necessary to perform the work, and being familiar with the requirements of the authorities having jurisdiction over the work.

- *Jobsite:* Same as *Site,* defined below.

- *Necessary:* With due consideration of the conditions of the project and as determined in the professional judgment of the Architect/Engineer as being necessary for performance of the work in conformance with the requirements of the Contract Documents, but excluding matters regarding the means, methods, techniques, sequences, and procedures of construction.

- *Noted:* Same as *Indicated.*

- *Products:* Material, system, or equipment.

- *Project site:* Same as *Site,* defined below.

- *Proper:* As determined by the Architect/Engineer as being proper for the Work, excluding matters regarding the means, methods, techniques, sequences, and procedures of construction, which are solely the Contractor's responsibility to determine.

- *Provide:* Means Contractor shall both 'furnish' and 'install' indicated products, as defined above. This definition applies equally to future, present and past tenses, except the word "provided" may mean "contingent upon" where such is the context.

- *Regulation:* Includes laws, ordinances, statutes, and lawful orders issued by authorities having jurisdiction, as well as rules, conventions, and agreements within the construction industry that control performance of the Work.

- *Required:* Necessary for performance of the Work in conformance with the requirements of the Contract Documents, excluding matters regarding the means, methods, techniques, sequences, and procedures of construction, such as:
  1. Regulatory requirements of authorities having jurisdiction
  2. Requirements of referenced standards
  3. Requirements generally recognized as accepted construction practices of the locale
  4. Notes, schedules, and graphic representations on the Drawings

5. Requirements specified or referenced in the Specifications
6. Duties and responsibilities stated in the bidding and contract requirements

- *Scheduled:* Same as *Indicated.*
- *Selected:* As selected by the Architect/Engineer or Owner from the full national product selection of the manufacturer, unless otherwise specifically limited in the contract documents to a particular quality, color, texture, or price range.
- *Shown:* Same as *Indicated.*
- *Site:* Same as *Site of the Work* or *project site;* the area or areas or spaces occupied by the project and including adjacent areas and other related areas occupied or used by the Contractor for construction activities, either exclusively or with others performing other construction on the project. The extent of the site is shown on the Drawings and may or may not be identical with the description of the land upon which the project is to be built.

Additional Terms

Words and terms not otherwise specifically defined on the Drawings and in the Specifications should be as customarily defined by trade or industry practice, by reference standard, and by specialty dictionaries such as the *Dictionary of Architecture and Construction,* third edition, by Cyril M. Harris (McGraw-Hill Book Companies, 2000).

Spelling

Simply stated, spelling matters. Misspelled words may cause misunderstanding of the intended meaning. Misspelled words undermine the credibility of the specifier. Misspelling may be avoided by using the spell-check feature of most word processing programs. Use an unabridged dictionary of English and a specialized dictionary, such as the *Dictionary of Architecture and Construction,* to check spelling and proper use of terminology.

Many technical terms are not included in the spell-check vocabulary of word processing programs. However, these programs have the capability of adding user-supplied words. Of course, the added words must be correctly spelled.

Occasionally, there are two spellings for a word— for example, "caulk" and "calk," "fascia" and "facia," "moulding" and "molding," and "gauge" and "gage." The preferred practice is to use the shorter spelling. However, avoid the use of improper yet brief spellings, such as "thru" for "through."

Punctuation

Construct sentences so that the misplacement or elimination of punctuation marks will not change the meaning or cause confusion. Follow conventional rules of punctuation. A good source for these rules is a basic English textbook, from a college-level English class, or a style manual.

Use commas to separate elements in a sentence that might otherwise seem to run together and cause confusion about the meaning. Examples:

- *Confused sentence:* After installing water heater piping insulation shall be installed.
- *Less confused sentence:* After installing water heater piping, insulation shall be installed. (Is insulation installed on both water heater and piping or only on piping?)
- *Clarified sentence:* After installing water heater, piping insulation shall be installed.
- Use commas after each item in a series to enhance readability. Placing a comma after the item preceding a conjunction is optional according to many grammarians. Breaking a sentence into phrases set off by commas enhances readability. *Example:* "Deliver manufactured material to the job site in original, unopened and undamaged containers, with name of the product and name of manufacturer clearly identified, and with manufacturer's seals and labels intact."
- Avoid semicolons (;).
- Use a colon instead of "shall" or "shall be." (See "Streamlined Writing" below.)
- End sentences with a period (.).

One way to avoid punctuation errors is to use short sentences and phrases. Long sentences offer opportunities for confusion and error. Break up long blocks of text by using subparagraphs.

Poor practice: "Tests: Materials used in this Work shall be tested by the manufacturer before shipping. Drainage and vent piping shall be tested before fixtures are installed by capping or plugging the openings, filling the entire system with water, and allowing it to stand thus filled for 3 hours. Water supply piping and hot-water tanks and heaters inside the building shall be tested by capping or plugging the openings, connecting up a test pump, filling the system with water, and applying a hydrostatic pressure of 150 psi. Water piping may be tested before fixtures or faucets are connected. Each fixture shall be tested for soundness, stability of support, and satisfactory operation of all its parts. After fixtures have been installed, all traps shall be filled and a smoke test shall be applied to expose leaks in the fixtures or connections. Piping shall have tight seals when tested. Screwed and soldered piping not tight under test shall be removed and reconstructed. Joints in cast iron piping not tight under test shall be replaced with new heaters and tanks. Certified test reports delivered to the Architect before Substantial Completion review."

Preferred practice:

A. Tests: Conduct tests as follows:
   "1. Materials used in this Work shall be tested by the manufacturer before shipping.
   "2. Drainage and vent piping shall be tested before fixtures are installed by capping or plugging the openings, filling the entire system with water, and allowing it to stand thus filled for 3 hours.
   "3. Water supply piping and hot-water tanks and heaters inside the building shall be tested by capping or plugging the openings, connecting up a test pump, filling the system with water, and applying a hydrostatic pressure of 150 psi. Water piping may be tested before fixtures or faucets are connected.
   "4. Each fixture shall be tested for soundness, stability of support, and satisfactory operation of all its parts.
   "5. After fixtures have been installed, all traps shall be filled and a smoke test shall be applied to expose leaks in the fixtures or connections. Piping shall have tight seals when tested. Screwed and soldered piping not tight under test shall be removed and reconstructed. Joints in cast iron piping not tight under test shall be replaced with new heaters and tanks.
   "6. Certified test reports delivered to the Architect before Substantial Completion review."

Use of parenthetical phrases should be minimized or avoided. They add complexity and make the specifications less readable.

Capitalization

The general rule for capitalization is to capitalize the first letter of proper names and defined terms, especially the names and terms used in Agreement and the General Conditions. For example:

1. *Contract Terms:* Defined terms in the Agreement and General Conditions. These include parties to the Contract, such as the Owner and Contractor, and parties defined in the General Conditions of the Contract, including the Architect, Engineer, and Subcontractor. Also, the term "Work" should be capitalized when referring to Work as defined in the General Conditions. (See the definition of the term "Work" in Chapter 2 of this book.) Terms such as "work day" do not require capitalization.

2. *Contract Documents:* As identified in the General Conditions of the Contract, including the Agreement, General Conditions, Drawings and Specifications, Change Order, and Construction Change Directive. Shop drawings, however are not Contract Documents and should not be capitalized.

3. *Spaces of the Building:* Principal's Office, Auditorium, Library, Teachers' Lounge, Lobby, Clinic, and Mechanical Room are examples.

4. *Grades of Materials:* B and Btr southern pine, Intermediate Heat Duty fire clay brick, Standard Grade ceramic tile, and Type I Regular Core hardwood plywood are examples. Note that terms such as "southern pine," "douglas fir," and "portland cement" are not capitalized because they are not proper names.

5. *Portions of the Specifications:* Within a specification section and when referencing portions of a specification section, capitalize PART and ARTICLE titles, Article, Paragraph, and Subparagraph in a specification section. (See the discussion of the format of a specification section in Chapter 3 of this book.)

Capitalize the first letter of the first word in sentence. Capitalize the first letter of the first word following a colon (:).

Abbreviations, Acronyms and Symbols

The principle to follow regarding abbreviations is to ensure that the reader understands what the abbreviation stands for. Often abbreviations are used when the writer does not know how to spell a word. Sometimes the use of abbreviations is a carryover from the Drawings where space is limited (or was limited when drawings were produced by manual drafting). Stop this practice when writing specifications.

Establish and enforce office standards for use of abbreviations, acronyms and symbols. Two comprehensive sources for industry-recognized abbreviations and symbols are *Uniform Drawing System* (*UDS*), Module 5, titled "Terms and Abbreviations," and Module 6, titled "Symbols," published in the *National CAD Standard* (*NCS*), available from The Construction Specifications Institute (CSI). A comprehensive source for acronyms is *Encyclopedia of Associations,* published by Gale Research Company.

Coordinate abbreviations and symbols between Drawings and Specifications.

While a list of common abbreviations is usually specified in a Division 1 section of the specifications, explain the abbreviation in the text to enhance clarity. For example, at the first occurrence of an abbreviation or acronym, write out what the abbreviation stands for, such as "barrel (bbl)" or "Association of Home Appliance Manufacturers (AHAM)." After this, in the same specification section, use the abbreviation or acronym.

Common Abbreviations (from John Regener's office master specifications)

AC or ac	Alternating current or air conditioning (depending upon the context)
AMP or amp	Ampere
BTU	British thermal units
BTUH	British thermal units per hour
C	Celsius
CFM or cfm	Cubic feet per minute
CM or cm	Centimeter
CY or cy	Cubic yard
DC or dc	Direct current
DEG or deg	Degrees
F	Fahrenheit
FPM or fpm	Feet per minute
FPS or fps	Feet per second
FT or ft	Foot or feet
Gal or gal	Gallons
GPM or gpm	Gallons per minute
IN or in	Inch or inches
Kip or kip	Thousand pounds
KSF or ksf	Thousand pounds per square foot
KSI or ksi	Thousand pounds per square inch
KV or kv	Kilovolt
KVA or kva	Kilovolt amperes
KW or kw	Kilowatt
KWH or kwh	Kilowatt hour
LBF or lbf	Pounds force
LF or lf	Lineal foot
M or m	Meter
MM or mm	Millimeter
MPH or mph	Miles per hour
PCF or pcf	Pounds per cubic foot
PSF or psf	Pounds per square foot
PSI or psi	Pounds per square inch
PSY or psy	Per square yard
SF or sf	Square foot
SY or sy	Square yard
V or v	Volts

With the advent of word processing programs, symbols are easier to add to text. Macros (recorded instructions or key strokes) make insertion of symbols easy. More than the symbols on the keyboard are available. There are several ASCII (American Standard Code for Information Interchange) symbol libraries that can be accessed with commercial office-grade word processing programs. These include commonly used trademark (™), registered trademark (®), copyright (©) and plus-minus. (±). Refer to instructions for word processing software for additional information.

Numbers

There are many accepted practices for use of numbers in specifications. Selection of the practice should be guided by the principles of making specifications unambiguous and readable. There may be tension between these two principles. Typically, it is recommended to use Arabic numerals rather than words for numbers.

Some recommend using both Arabic numerals and numbers expressed in words—for example, "sixteen (16) fasteners per panel" rather than "16 fasteners per panel." The reasoning is based on the misbelief that "sixteen (16)" is more "legal" than simply "16." And, there is a misbelief that there would be less of a problem if the two did not agree, such as "twelve (16) fasteners per panel." Which is the correct interpretation: twelve or sixteen (12 or 16)?

Use Numerals Instead of Words

Generally, use numbers instead of words. Consider that the Drawings, which are Contract Documents, use numbers exclusively. There is no less risk of error by using numbers in specifications than there is on the drawings.

An exception to this principle is when numbers are used to state both size and quantity. In this case, state the quantity in words and the quality (dimension or size) in numerals—for example, "three

½-inch bolts," "six 2 × 4 studs," and "two 4-position hold-opens." The last example, if stated in numbers only, could be misread as "2 4-position hold-opens" (twenty-four positions).

When including dimensions in the text of specifications, the following conventions are recommended:

1. Spell out feet when no inches are used—for example, "8 feet."

2. Spell out inches when no feet are used—for example, "8 inches."

3. When both feet and inches are used, the CSI *Manual of Practice* recommends use of symbols—for example, 8′-8″ or 8′-2-½.″ Others, however, recommend using "8-feet 9-inches" because missing punctuation can cause very significant confusion. Consider readability and use professional judgment to choose a convention for dimensions, and then follow the convention consistently.

4. A complete dimension should appear on one line. This requires careful proofreading and understanding of the word processing program. There are several symbols for a dash; some are interpreted like hyphens that break words at the end of a line. Understand the difference between typewritten text and text produced with a word processor.

5. Spell out numbers ten and below. Use numerals for 11 and above.

   Example: "four coats of paint," not "4 coats of paint" (except in schedules).

   Example: "12 samples of each brick unit type and color to show range of color variations."

Use Numerals for Dimensions, Degrees of Temperature, Percent, and Money

Examples: "8-inches on center," "2-½ inch diameter pipe," "45 degrees F (7 degrees C)," "85 percent," and "$250 per day."

Use Numerals for Dates and Time

Examples: "2:00 pm on February 14, 2005." Note that it is not written "February 14th." Omit the "th." Exceptions are the times "noon" and "midnight." Which is 12:00 am? Properly, they should not be "12 noon" or "12 midnight." However, common practice seems to be to state "12:00 noon" and "12:00 midnight" for consistency with other expressions of time in numerals and for clarity.

Decimals

Express decimals in numerals—for example, "5.75-inches." When the quantity is less than zero, include a leading "0." For example, "0.08-inch" or "0.16-foot."

Fractions

Type out fractional dimensions and quantities. Although word processing programs have the capability of using special characters for fractions, such as ¼, ½, and ¾, do not use them. This means changing the default setting in the word processing program. This is a carryover from the time when typewriters could not produce these special characters, but now the reason is to avoid conflicts in converting between word processing programs. Some programs may not recognize the special character, and a converted file will lose these fractions. Thus, "3-½" translates as "3- ."

Examples: "3-1/2 inches" and "3/4-inch."

Zeros

Omit zeros if clarity and consistency are not compromised. For example, "$400" is as clear as "$400.00" but "10 a.m." may not be as clear or readable as "10:00 a.m.," especially when the context includes other expressions of time such as "10:15 a.m."

Streamlined Writing

Consider the use of *streamlined specifications.* In *Pencil Points Magazine* in August 1939, there appeared an article entitled "Streamlined Specifications" by Horace W. Peaslee, FAIA, proposing writing specifications in an outline form without the use of complete sentences. The CSI *Manual of Practice* recommends and even encourages the use of streamlined specifications. After decades of practice, the authors also recommend using streamlined specifications. Streamlined specifications resolve many grammatical issues of specifications writing and help to produce clear and concise specifications. Consider these points:

1. Streamlining is the most important factor in concise specifications.

2. Streamlining is not proper English, but it is proper specifying.

3. Streamlining is not an excuse for not knowing proper English—but it definitely helps.

The key to streamlined specifications is definition of the colon (:) to mean "shall" or "shall be." It is very helpful to include an explanatory note in Division 1 or in the supplementary conditions regarding this:

> These Specifications are written in the imperative mood and streamlined form. This imperative language is directed to the Contractor, unless specifically noted otherwise. The words "shall be" shall be included by inference where a colon (:) is used within sentences or phrases.

Examples of Streamlined Sentences

Adhesive: Spread with notch trowel.

Equipment: Installed plumb and level.

Portland Cement: ASTM C 150, Type 1.

Aggregate: ASTM C 33.

Air-entraining Agent: More-Air Brand, More-X Manufacturing Co.

Specifications Detail

Specification detail should not be confused with complexity of language. Specification detail pertains to the magnitude of specified information relative to the scope and complexity of the project. One size of specifications does not fit all projects, nor does it fit portions of a single project. Professional judgment is needed to determine the appropriate level of detail for project-specific construction specifications.

Specification detail should be commensurate with the complexity and required quality of the project. A speculative warehouse shell building requires less specification detail than a corporate headquarters office building. A tract house requires less specification detail than a mansion-like custom residence. Although similar in construction quality, a multiunit residential complex requires greater specification detail than a single tract house; the multiplied effect of small details in the specifications significantly affects the overall cost.

At the same time, there may be an unintended cost impact of bulky specifications to consider. There is a prejudice at work in the construction industry that says, the bigger the specs, the more expensive the project. This may be a reaction by those who are uncomfortable with written documents or who simply are functionally illiterate. However, specifications are considered well crafted when they cover all important details without elaborate and unnecessary language. This is the main argument for shortform specifications.

This, then, brings the discussion back to the beginning of this chapter and the four Cs in specifications writing:

1. Be Clear.
2. Be Correct.
3. Be Complete.
4. Be Concise.

These principles are in dynamic tension in the specifications. There needs to be balance between completeness and conciseness. There needs to be balance between conciseness and clarity; sometimes more information is necessary, and sometimes brevity causes no significant consequence. And there needs to be correctness, which requires care in crafting the specifications so that expedience, often represented in conciseness, does not inhibit proper selection and description of products and their application.

Chapter 15

Specifications Resources

OVERVIEW

The scope of constructions specifications is extremely broad, encompassing the spectrum of materials from acoustical ceilings through zinc coatings. No individual specifier can possibly have a complete and intimate knowledge of all building products, nor can the specifier possibly keep up with the constant changes in building products. Even when the specifier has a focused design discipline, such as one of the engineering professions, or a construction technology, such as door hardware and waterproofing, change in technology is still daunting. Information overload due to the ever-increasing complexity of construction technology is a great challenge.

Knowing where to look for information, and how to identify and extract the applicable information, is half of the battle. Knowing how to apply the information according to recommended specifications organization, and writing principles and procedures, is another major component of the equation. All of this takes place in a construction project delivery environment that is rapidly evolving and a business environment that is increasingly competitive. To make the situation more manageable, the specifier must acquire and use resources that enable information to be located, extracted, and recorded expediently and correctly.

Due to rapid and fundamental changes in how construction information is recorded, identified, and retrieved, it is impossible to describe with certainty what the future of specifying will be. It is the authors' opinion that (1) current trends in building products and project delivery methods will continue to lead to significant changes in specifying and (2) the traditional principles and procedures for specifications will endure. That is, computer-assisted information management programs will be used with computer-assisted drafting (CAD) for specifications production, but it will be essential to understand and apply the fundamental principles of construction contracts and construction technology to the process in order for competent specifications to be produced.

In *Alice's Adventures in Wonderland* and *Through the Looking Glass* by Lewis Carroll (1865), Alice was traveling and came to a fork in the road. She asked the Cheshire Cat, "Would you tell me, please, which way I ought to go from here?" In response, the Cheshire Cat said, "That depends a good deal on where you want to go." Alice replied, "It doesn't much matter where." The cat responded aptly, "Then I suppose it doesn't matter what road you take."

Mark Kalin, FAIA, FCSI, CCS, uses the term "fake specs" to describe construction specifications prepared without proper thought and understanding. The specifier who does not know what the end result (the specifications) shall be and does not understand construction technology, product evaluation, contractual duties and responsibilities, terminology, and the subtleties of specifications language will be like Alice trying to decide what to do.

To paraphrase the above passage, if you don't know where you are going or what you are doing, it doesn't matter where you go or what you do. Concerning construction information management, project delivery methods, construction contracts, and production of construction drawings and specifications, too often those who produce construction specifications merely "go through the motions." They do not understand the consequences of choices, and therefore they don't understand which choices to make.

It is essential that specifiers develop suitable resources for preparing construction specifications in order to be informed and able to make proper choices.

CONSTRUCTION SPECIFICATIONS RESOURCES

In previous editions of this book, at this point publications, businesses, associations, and other resources were listed. The problem is, these lists are immediately out of date. Change is happening that fast. Industry associations not only relocate but also change their names. Building code publishers are joining together under new association names, and new alternatives are being published. International considerations are affecting the scope of the construction industry. The North America market includes the United States, Canada, and Mexico in increasingly equal measures, with impacts of different languages and construction practices.

The key to accessing resources in the future will be the Internet. This is not a visionary statement but a statement that represents current mainstream practices in construction specifications production. Architectural and engineering firms that do more than light construction projects have high-speed Internet service. Most independent construction specifiers have high-speed Internet service through cable service or DSL (digital subscriber line). Design professionals are not only wired but intensely using these communication technologies. Reference books and building product catalogs may still sit on shelves, but they are only occasionally needed for understanding and resolving issues of design and construction. With high-speed Internet service and powerful search engines, information access is restricted only by the quality of the information that is available.

Building product manufacturers, standards associations, code authorities, and publishers of construction specifications–related documents are being accessed by specifiers regularly on the Internet. Therefore, the lists of resources below will emphasize current Internet sources and some printed publications and associations. While it is still possible to prepare construction specifications using print resources—indeed, print resources have some advantages—the predominant resources are found through online communications of the Internet.

General Resources for Construction Specifications

There are several Internet sites for locating and acquiring current construction information. Perhaps the most comprehensive and suitable one for specifiers is titled "Construction Specifiers Library" and is found on the website of Building Systems Design, Inc. (BSD): *www.bsdsoftlink.com/library/speclibrary-principles.htm*. This library includes links to numerous businesses, organizations, and specifications-related resources. It also includes a comprehensive glossary and links to building product search engines. Even competing guide specifications publishers are listed, and there are complete listings of federal agency sites where construction specifications are available.

BSD's website, *www.bsdsoftlink.com*, also includes pages addressing general specifying and design topics, including:

- Whole building performance specifying
- Energy and environment
- Design-build
- Agreements and General Conditions

ARCOM has a competing website at *www.arcomnet.com*, portions of which are accessible by nonsubscribers to *Masterspec*. Find the page titled "Resource Links" for general specifications resources.

CSI has many resources available at its website *www.csinet.org*, including (for CSI members) access to archives of CSI's magazine, *The Construction Specifier*. CSI also has a bookstore that sells specifications, construction- and design-related publications.

The Internet has not abolished books. Indeed, the Internet has made finding and acquiring books easier. Book publishers, such as John Wiley & Sons, have corporate websites that include the capability of locating and purchasing books by the publisher (*www.wiley.com/WileyCDA*). One of the most notable e-commerce firms is Amazon.com, which is best known for book sales (*www.amazon.com*). Using the search engine of a book publisher's website or the generic book sales website, current publications can be identified and purchased for expedient delivery directly from the warehouse to the buyer's door. Even used books can be found, usually for lower costs, and out-of-print books can be located.

Until recently, the foremost resource book has been CSI's *Manual of Practice* (1996 edition). The *Manual of Practice* (CSI publication number MP-1) consists of four modules plus appendices. The four modules are:

- Fundamentals and Formats
- Construction Specifications Practice
- Construction Contract Administration (with a separate CD of forms)
- Construction Product Representation

Appendices to the CSI *Manual of Practice* include CSI's *MasterFormat* (1995 edition), *SectionFormat* (1997 edition), and *PageFormat* (1999 edition). These documents have been recognized for many years as industry standards for organizing and preparing construction specifications.

As this is written, CSI's *Manual of Practice* is in the process of major revision and renaming to reflect a broader purpose. The new CSI *Project Resource Manual—Manual of Practice,* will be published in hardbound form rather than as a looseleaf publication that can be updated and expanded. Contact CSI:

The Construction Specifications Institute
99 Canal Street, Suite 300
Alexandria, VA 22314
703/684-0300 OR 800/689-2900
www.csinet.org

Construction Specifications Books

The search for books related to construction specifications reveals many books in addition to this one and the CSI *Manual of Practice* noted above. John Wiley & Sons has an extensive catalog of architectural, engineering, and building construction titles. Go to *www.wiley.com/WileyCDA* and search under book headings for "Architecture & Design," "Building Design," and "Engineering."

The following list of books is not exhaustive, and no opinion is given regarding their content or suitability. These books cover general topics of construction specifications, specialized construction specifications (such as interiors construction and engineering projects), construction technology, construction contracts, and construction terminology. The list includes out-of-print books that may be available through used-book sources. See *www.amazon.com* and search for the book title to identify potential sources of used books. Some of these books focus on specifications for light construction and address construction specifications in less comprehensive ways that may be suitable for outline and shortform specifications.

The Specifications Writer's Handbook by H. Leslie Simmons (John Wiley & Sons, 1985) ISBN: 0-471-88615-7

Spec Writers' Handbook: A Textbook and Reference Book for Writing Architectural Specifications, 2nd ed., by John A. Weyl, RA (Pacific Odyssey, Inc., 2000) ISBN: 0-962-32930-4

Specifications for Architecture, Engineering and Construction, 2nd ed., by Chesley Ayers, P.E. (McGraw-Hill Text, 1984) ISBN: 0-070-02642-4

The Practical Specifier: A Manual of Construction Documents for Architects by Walter Rosenfeld, AIA, CSI (McGraw-Hill Text, 1985) ISBN: 0-070-53779-8

Specifications Writing for Architects and Engineers, by Donald A. Watson (McGraw-Hill Publishing Co., 1964) ISBN: 0-706-84731-8

Specifying Interiors: A Guide to Construction and FF&E for Commercial Interiors Projects by Maryrose McGowan (John Wiley & Sons, 1996) ISBN: 0-471-10619-4

Specifications for Commercial Interiors by S. C. Reznikoff (Whitney Library of Design, 1989) ISBN: 0-823-04893-4

A Guide to Writing Successful Engineering Specifications by David C. Purdy (McGraw-Hill Text, 1990) ISBN: 0-07050-999-9

Engineering Construction Specifications: The Road to Better Quality, Lower Cost, Reduced Litigation by Joseph Goldbloom (Van Nostrand Reinhold, 1989) ISBN: 0-442-22994-1

Construction Specifications: Managing the Review Process by William T. Lohmann (Butterworth Architecture, 1992) ISBN: 0-750-69148-4

Specifying Buildings: A Design Management Perspective by Stephen Emmitt and David T. Yeomans (Butterworth-Heinemann, 2001) ISBN: 0-750-64849-X

Dunham and Young's Contracts, Specifications, and Law for Engineers, 4th ed., by Joseph T. Bockrath (McGraw-Hill Book Company, 1986) ISBN: 0-070-18237-X

Building Construction Specifications by I. T. Rathbun (McGraw-Hill Text, 1972) ISBN: 0-070-51209-4

Construction Specifications Handbook, 4th ed., by Hans W. Meier (Builder's Book.) *www.buildersbook.com*

Library of Specifications Sections, 6th ed., by Hans W. Meier (Builder's Book) (*www.buildersbook.com*)

Construction Specifications Portable Handbook by Fred A. Stitt (McGraw-Hill Professional, 1999) ISBN: 0-071-34103-X

Easyspec: Construction Specifications, 2nd ed., by Bni Building News (*www.buildersbook.com;* 2000) ISBN: 1-557-01338-1

Construction Reference Books

The following dictionaries cover a broad range of construction types, including historical building materials, archaic terminology, slang terms, and specialized construction terms.

The Construction Dictionary, 9th ed., by the Greater Phoenix Chapter of the National Association of Women in Construction (NAWIC) (*www.constructiondictionary.com*)

Dictionary of Architecture and Construction, 3rd ed., by Cyril M. Harris (McGraw-Hill Book Company, 2000) ISBN: 0-07-135178-7

Construction Glossary: An Encyclopedic Reference and Manual, 2nd ed., by J. Stewart Stein (John Wiley & Sons, 1993) ISBN: 0-471-56933-X

A Visual Dictionary of Architecture by Francis D. K. Ching (John Wiley & Sons, 1996) ISBN: 0-471-28821-7

Means Illustrated Construction Dictionary by Howard Chandler (R. S. Means Company, 1991) ISBN: 0-876-29218-X

Construction Dictionary Illustrated by William D. Mahony, PE (editor) (Building News, 1997)

Craftsman's Illustrated Dictionary of Construction Terms by James T. Frane (Craftsman Book Co., 1994)

The following books cover construction contracts and construction contract administration:

The Architect's Handbook of Professional Practice, 13th ed., by The American Institute of Architects (John Wiley & Sons, 2001, with updates) ISBN: 0-471-41969-9

The Building Professional's Guide to Contract Documents, 3rd ed., by Waller S. Poage (R.S. Means Co., 2000) ISBN: 0-876-29577-4

Construction Contracting, 6th ed., by Richard H. Clough and Glenn A. Sears (John Wiley & Sons, 1994) ISBN: 0-471-30968-0

Smith, Currie & Hancock LLP's Common Sense Construction Law: A Practical Guide for the Construction Professional, 2nd ed., by Robert B. Ansley, Jr., Thomas J. Kelleher, Jr., and Anthony D. Lehman (general editors) (John Wiley & Sons, 2000) ISBN: 0-471-39090-9

Construction Graphics: A Practical Guide to Interpreting Working Drawings by Keith A. Bisharat (John Wiley & Sons, 2004) ISBN: 0-471-21983-5

Construction Claims: Prevention and Resolution, 3rd ed., by Robert A. Rubin, Virginia Fairweather and Sammie D. Guy (John Wiley & Sons, 1999) ISBN: 0-471-34863-5

Construction Contract Law by John J. P. Krol (John Wiley & Sons, 1993) ISBN: 0-471-57414-7

Construction Administration: An Architect's Guide to Surviving Information Overload by Patrick C. Mays and B. J. Novitski (John Wiley & Sons, 1997) ISBN: 0-471-15419-9

Construction Contract Administration by Hans W. Meier, AIA, FCSI (Builder's Book, Inc., *www.buildersbook.com*)

Construction Project Administration, 7th ed., by Ed Fisk (Prentice Hall, 2002) ISBN: 0-130-98472-8

A Guide to Successful Construction: Effective Contract Administration by Arthur F. O'Leary

Construction Operations Manual of Policies and Procedures, 3rd ed., by Andrew M. Civitello, Jr. (McGraw-Hill Professional, 2000) ISBN: 0-071-35495-6

Contractor's Guide to Change Orders: The Art of Finding, Pricing, and Getting Paid for Contract Changes and the Damages They Cause by Andrew M., Civitello, Jr. (Prentice Hall Trade, 1987) ISBN: 0-131-71588-7

The following books cover architectural and engineering design and construction technology in general and specific terms:

Americans with Disabilities Act Accessibility Guidelines Manual (*ADAAG*) by the U.S. Access Compliance Board (*www.access-board .gov/adaag/html/adaag.htm*) *Note:* ADAAG has been superseded by UFAS, below.

Uniform Federal Accessibility Standards (*UFAS*) by the U.S. Access Compliance Board (*www .access-board.gov/ufas/html/ufas.htm*)

Time-Saver Standards for Building Types, 4th ed., by Joseph De Chiara, Michael J. Crosbie and Mike Crosbie (McGraw-Hill Professional, 2001) ISBN: 0-070-16387-1

Time-Saver Standards for Interior Design and Space Planning, 2nd ed., by Joseph De Chiara, Julius Panero and Martin Zelnik (McGraw-Hill Professional, 2001) ISBN: 0-071-34616-3

Architectural Graphic Standards, 10th ed., by Charles George Ramsey, Harold Reeve Sleeper and John Ray Hoke, Jr. (editor-in-chief) (John Wiley & Sons, 2000) ISBN: 0-471-34816-3

Architectural Graphic Standards for Residential Construction by The American Institute of Architects (John Wiley & Sons, 2003) ISBN: 0-471-24109-1

Construction Principles, Materials, and Methods, 7th ed., by H. Leslie Simmons (John Wiley & Sons, 2001) ISBN: 0-471-35640-9

Fundamentals of Building Construction: Materials and Methods, 4th ed., by Edward Allen and Joseph Iano (John Wiley & Sons, 2003) ISBN: 0-471-21903-7

Building Materials: Dangerous Properties of Products in MasterFormat Divisions 7 and 9 by H. Leslie Simmons and Richard J. Lewis, Sr. (John Wiley & Sons, 1997) ISBN: 0-471-29084-X

Architectural Materials for Construction by Harold J. Rosen, PE, FCSI and Tom Heineman, FCSI (McGraw-Hill Professional, 1995) ISBN: 0-070-53741-0

Green Building Materials: A Guide to Product Selection and Specification by Ross Spiegel and Dru Meadows (John Wiley & Sons, 1999) ISBN: 0-471-29133-1

Mechanical and Electrical Equipment for Buildings, 9th ed., by Ben Stein and John S. Reynolds (John Wiley & Sons, 1999) ISBN: 0-471-15696-5

The Engineering Resources Code Finder for Building and Construction by Dennis Phinney

Carpentry and Building Construction, 5th ed., by John Louis Feirer, Gilbert R. Hutchings and Mark D. Feirer (Glencoe McGraw-Hill, 1997) ISBN: 0-028-38699-X

Materials Standards

Standards for materials have been devised and issued in the United States by national technical associations and by the federal government to provide uniform criteria, grades, and tests.

The U.S. federal government is privatizing the standards-making process. By executive order, the former Federal Specifications are not to be cited if an adequate private standard exists. FS, CS, SPR, IS, and PS documents are no longer to be developed by the federal government and are to be withdrawn in time. Specifications and standards by private-sector organizations, such as ASTM International (formerly the American Society for Testing and Materials) and the American National Standards Institute (ANSI), are superseding the former Federal Specifications and standards. ASTM and ANSI have responded with many new standards, organized similarly to the former federal standards, using identical product types and classes in many cases. Confusion has thus been minimized as the standards-setting process is taken over by industry and consumer-oriented organizations.

Originally called the National Standards System Network (NSSN), this organization has become NSSN—A National Resource for Global Standards (see *www.nns.org*). From this site, "QuickLinks" provides direct linkage to more than 200 federal, state, and local government agency websites focused on standards, regulations, legislation, and technical issues. The site is operated by the ANSI.

The NSSN site is the current source of standards and specifications for federal agencies. From this site, documents may be downloaded with restrictions. The primary restriction is that the party downloading documents must have a contract with a federal government agency that requires the acquisition of the standard or specification. General access and downloading is not possible.

Federal specifications were once the primary reference in construction specifications for General Services Administration, Department of Defense and National Aeronautics and Space Administration projects. With a few exceptions, these federal specifications have been rendered obsolete for construction purposes. They have been replaced by non-government standards (NGS), which are familiar in commercial and nonfederal institutional construction.

NGS specifications and standards are the standards of ASTM International and ANSI, as noted above. Other association standards, such as those of the Architectural Woodwork Institute, Tile Council of America, and Steel Door Institute (SDI), have also been adopted. Sometimes the adopted standard is a joint standard, such as ANSI 250.8, which is SDI 100.

Similarly, Commercial Standards, (CS), Product Standards (PS), Simplified Practice Recommendations, (SPR), and Industry Standards, (IS), formerly issued by the Department of Commerce, have been withdrawn and are no longer indexed or sold. Exceptions are two Product Standards (PS) for plywood and softwood lumber (PS 1 and PS 20, respectively), still available from the National Institute of Standards and Technology (NIST). Also remaining is Consumer Products Safety Commission (CPSC) Standard 16CFR1201 for safety glazing.

Those requiring specifications and standards applicable to federal agencies may also access these documents through the Construction Criteria Base (CCB) at *www.ccb.org*. The CCB is an extensive electronic library of construction guide specifications, manuals, standards, and many other essential criteria documents published on the Internet and on a set of eight CD-ROM disks or one DVD by the nonprofit National Institute of Building Sciences. Updated continuously at *www.ccb.org* and on disk twice a year in April and October, CCB contains the complete unabridged, approved, current electronic equivalents of over 10,000 documents and executable programs, direct from 18 federal agencies and over 100 industry organizations. CCB is an effective tool for finding and using current, approved U.S. construction criteria. Subscription information can be found at *www.ccb.org/ subscribe.html.*

The former American Society for Testing and Materials is now an international organization with a name to suit: ASTM International. Founded in 1898, ASTM International is a nonprofit organization that provides a global forum for the development and publication of voluntary consensus standards for materials, products, systems and services. Over 30,000 individuals from 100 nations are members of ASTM International, including producers, users, consumers, and representatives of government and academia.. In over 130 varied industry areas, ASTM standards serve as the basis for manufacturing, procurement and regulatory activities. ASTM International provides standards that are accepted and used in research and development, product testing, quality systems and commercial transactions around the globe.

ASTM standards, individually or in *ASTM Standards in Building Construction,* may be obtained from:

ASTM International
100 Barr Harbor Drive
West Conshohocken, PA 19428-2959
610/832-9585

ANSI is a private, nonprofit organization (501(c)3) that administers and coordinates the U.S. voluntary standardization and conformity assessment system. ANSI's mission is to enhance both the global competitiveness of U.S. business and the quality of American life by promoting and facilitating voluntary consensus standards and conformity assessment systems and safeguarding their integrity. ANSI standards may be obtained online at *http://webstore.ansi.org/ansidocstore/default.asp* or by contacting ANSI at:

American National Standards Institute (ANSI)
25 West 43rd Street, 4th Floor
New York, NY 10036
212/642-4900

In some cases, synopses of standards are available from the publishing organization's website. Try the website prior to purchasing very expensive and seldom-read documents.

Master Guide Specifications

Master guide specifications serve as an aid in the preparation and development of construction specifications. These are being supplanted by true computer-assisted specifications, but in the meantime, subscriptions to portions or all of the libraries of guide specifications and supporting documents may be obtained from the publishers. Master guide specifications will be discussed further in Chapter 18. Major sources for master guide specifications are:

MasterSpec
ARCOM, Inc.
332 E. 500th Street
Salt Lake City, UT 84111
801/521-9162 or 800/424-5080
www.arcomnet.com

SpecText
Construction Sciences Research Foundation,
    Inc. (CSRF)
The CSRF Support Center
P.O. Box 926
Bel Air, MD 21014-0926
410/838-7525 or 877/SPECTXT (773-2898)
www.spectext.com

UFGS—Unified Facilities Guide Specifications
    (Federal agency specifications)
Construction Criteria Base (CCB)
National Institute for Building Sciences (NIBS)
1090 Vermont Avenue NW, Suite 700
Washington, DC 20005
202/289-7800 or 877/222-5667
www.ccb.org/docs/ufgshome/UFGSToc.htm

GreenSpecs and Master Shortform Specifications
Kalin Associates
154 Wells Avenue
Newton, MA 02459
617/964-5477 or 800/565-2546
www.kalinassociates.com

Automated Construction Specifications Programs

Automated or computer-assisted specifications will be discussed in Chapter 19. Resources are:

BSD SpecLink®
Building Systems Design, Inc.
3520 Piedmont Road NE, Suite 415
Atlanta, GA 30305
404/365-8900 or 888/273-7638
www.bsdsoftlink.com

Specware (Masterworks™ and Linx™
ARCOM, Inc.
332 E. 500th Street
Salt Lake City, UT 84111
801/521-9162 or 800/424-5080
www.arcomnet.com

EditSpec
Construction Sciences Research Foundation,
   Inc. (CSRF)
The CSRF Support Center
P.O. Box 926
Bel Air, MD 21014-0926
410/838-7525 or 877/SPECTXT (773-2898)
www.spectext.com

e-Specs
InterSpec LLC
100 Commercial Street
Portland, ME 04101
207/772-6135 or 888/507-7327

SpecsIntact (based on Unified Facilities Guide
                    Specifications)
Kennedy Space Center, FL 32899
321/867-8800
http://specsintact.ksc.nasa.gov/

Journals and Periodicals

Construction industry periodicals occasionally have construction specifications–related articles. However, articles on topics such as those addressed in this book are infrequently published. An exception is CSI's *The Construction Specifier,* which after several years of neglecting its core constituency—construction specifiers—regularly publishes articles on Division 1 - General Requirements specifications issues and construction technology. Subscription information may be found at the CSI's website, *www.csinet.org,* or at *www.constructionspecifier.com.* A subscription to *The Construction Specifier* is included with CSI membership.

General construction industry publications sometimes contain articles on construction specifications. Typically, however, they focus on construction technology and architectural and engineering design. Examples:

Architectural Record
877/876-8093
http://archrecord.construction.com

Building Design & Construction
2000 Clearwater Drive
Oak Brook, IL 60523
www.bdcmag.com

Consulting-Specifying Engineer
2000 Clearwater Drive
Oak Brook, IL 60523
www.csemag.com

Engineering News-Record
877/876-8208
www.enr.com

Other architecture and engineering magazines have occasional articles on construction contract documents and construction specifications. Contact industry associations, code associations and professional societies for information regarding subscriptions and editorial content.

Newsletters with a focus on construction specifications are published by several corporate and professional associations and are available for free downloading. These include:

- *KnowHow* by Specifications Consultants in Independent Practice (SCIP): *www.scip.com*
- *Creating a Common Language* by The Construction Sciences Research Foundation (CSRF), publisher of SPECTEXT: *www.csrf.org*
- *SpecPress* by ARCOM, Inc. (publisher of *Masterspec, Masterworks,* and *Linx*): *www.arcomnet.com*
- *Building Science Newsletter* by the National Institute of Building Sciences
- *Canadian Building Digest* by the National Research Council

Building Codes and Ordinances

There are codes and ordinances promulgated by authorities having jurisdiction over public agencies. These agencies include cities, counties, states, and special districts. The purpose of the codes and ordinances is to protect health, life, and property. Included are zoning regulations, building codes, fire and life safety codes, plumbing codes, mechanical (HVAC) codes, and electrical codes. These codes should be consulted to ensure that the project design and specified requirements comply with applicable regulations. Where appropriate, references to these codes and ordinances should be incorporated in specifications. Often plancheckers require citation of code requirements in the specifications to ensure compliance during construction.

Occasionally, these codes may cover only minimum standards and types of construction. The Architect/Engineer or specifier may exercise professional judgment and determine that construction shall be of higher quality than that required under the codes.

Buildings and other facilities have been governed by model codes published by regional code associations and adopted as law by governing authorities. Major cities such as New York City, Los Angeles, and San Francisco and states such as Massachusetts and California have adopted their own codes, often based on model codes but with significant amendments. Other unique codes are found in Florida, such as the Metro-Dade South Florida Building Code.

An effort has been made to consolidate, standardize, and simplify building codes. Statewide codes have been adopted, with some local amendments to suit unique requirements. With the merging of several code associations, state-to-state standardization has occurred.

ICC Codes and Standards

In 1994, Building Officials and Code Administrators International, Inc. (BOCA), the International Conference of Building Officials (ICBO), and Southern Building Code Congress International, Inc. (SBCCI) formed the International Code Council (ICC). In 1997, the Council of American Building Officials (CABO) also joined. This resulted in the 1999 publication of a series of international codes covering most aspects of construction and other facilities, including:

- International Building Code
- International Energy Conservation Code
- International Existing Building Code
- International Fire Code
- International Fuel Gas Code
- International Mechanical Code
- ICC Performance Code
- International Plumbing Code
- International Private Sewage Disposal Code
- International Property Maintenance Code
- International Residential Code
- International Zoning Code
- ICC Electrical Code

This means that the former BOCA National Building Code, CABO One and Two Family Dwelling Code, Standard Building Code and Uniform Building Code and their related plumbing, mechanical and other codes no longer apply. According to ICC, more than 95% of U.S. cities, counties and states adopt building and safety codes published by the ICC.

While this merger would seem to indicate that architects, engineers and specifiers will only need to refer to one series of codes from one model code association (ICC), practical experience suggests that there will be numerous local amendments to address matters such as disaster mitigation (windstorm and earthquake), energy conservation, accessibility, innovative technology and fire protection. Responsible design professionals need to confirm the exact codes, including special amendments, that apply to the project and comply with these special requirements.

In California, where all jurisdictions were mandated to adopt the California Building Code, which at the time of publication of this book is the 1997 Uniform Building Code (UBC) with 2001 State of California Amendments, the standardization effort did result in a great deal of commonality in the codes. However, the cities of San Francisco and Los Angeles, and Los Angeles County, each adopted additional amendments to the California Building Code that amended the UBC. Further, for projects under state jurisdiction for planchecking and permitting, such as public school projects under the Division of the State Architect, (DSA) and medical facilities under the Office of Statewide Health Planning and Development, (OSHPD), additional amendments for accessibility and structural safety (seismic design) were adopted.

Building codes are being standardized; Some are more standard than others. For further information, contact the following organizations:

Headquarters (FC)
International Code Council
5203 Leesburg Pike, Suite 600
Falls Church, VA 22041
703/931-4533
www.iccsafe.org

Birmingham District Office (BIR)
International Code Council (ICC)
900 Montclair Road
Birmingham, AL 35213-1206
205/591-1853

Chicago District Office (CH)
4051 W Flossmoor Road
Country Club Hills, IL 60478-5795
800/214-4321

Los Angeles District Office (LA)
5360 Workman Mill Road
Whittier, CA 90601-2298
800/284-4406

NFPA Codes and Standards

Competing with ICC is NFPA. NFPA, now officially known only by its initials, is an international nonprofit organization founded in 1896 and best known for codes and standards related to fire and life safety. NFPA has published over 300 safety codes and standards. The most prominent ones are:

- *NFPA 1*, Fire Prevention Code™: Provides the requirements necessary to establish a reasonable level of fire safety and property protection in new and existing buildings.
- *NFPA 13*, Installation of Sprinkler Systems: Rules that apply to the full range of fire sprinkler systems, from concept to installation.
- *NFPA 54*, National Fuel Gas Code: Safety requirements for fuel gas installations.
- *NFPA 70*, National Electrical Code®: A widely used and accepted code for electrical installations.
- *NFPA 101*, Life Safety Code®: Establishes minimum requirements for new and existing buildings to protect building occupants from fire, smoke and toxic fumes.
- *NFPA 5000™, Building Construction and Safety Code™: A building code developed through an open, consensus-based process that is accredited by ANSI. NFPA 5000 will be a cornerstone of the Comprehensive Consensus Codes (C3) set.*

For further information, contact:

NFPA
1 Batterymarch Park
P.O. Box 9101
Quincy, MA 02169-7471
617/770-3000 or 800/344-3555
www.nfpa.org

At the time of publication of this book, ICC and NFPA are aggressively competing for adoption of their codes and standards. Architects, engineers, and specifiers must monitor developments in codes and standards and remain current through continuing education and professional development activities. In practice, too many design decision makers have barely a superficial understanding of codes and standards. Taking a basic course on building codes, especially building code interpretation, is highly recommended.

Evaluation Reports

A final source of codes and standards information, applicable to ICC Codes and standards, are the publications of the ICC Evaluation Service (ICC-ES). Concurrent with the establishing of the ICC, ICC-ES assimilated the evaluation services of the four model code associations that formed ICC. These were CABO National Evaluation Service, BOCA Evaluation Services, ICBO Evaluation Service, and SBCCI Public Service Testing and Evaluation Services. Through "legacy" evaluation services, ICC-ES has a history that goes back more than 70 years.

At time of publication of this book, ICC-ES is both adopting new ICC-ES evaluation reports and retaining "legacy reports" originally issued under the rules of one of the legacy evaluation services. Eventually, all reports will be reevaluated and issued as ICC-ES evaluation reports. ICC-ES reports may be viewed and downloaded online from *http://www.icc-es.org/Evaluation_Reports/ index.shtml*. For further information, contact:

ICC Evaluation Service
5360 Workman Mill Road
Whittier, CA 90601-2298
562/699-0543
www.icc-es.org

Materials Investigations

A valuable service that has been performed by some governmental agencies and national technical associations is the laboratory investigation of properties of building materials and the structural elements of buildings, as well as the performance of mechanical equipment for buildings. Many of these reports have also been compiled on the basis of the experience record of many individuals who have been closely associated with certain materials. Examples of these resources are:

Building Science Series (NIST BSS)
National Institute of Standards and Technology
100 Bureau Drive
Gaithersburg, MD 20899-3460
301/975-6478
www.nist.gov

Council Notes and Research Reports
Building Research Council
School of Architecture, University of Illinois at
    Urbana-Champaign
1 East St. Mary's Road
Champaign, IL 61820
http://brc.arch.uiuc.edu/Pubcatalog.htm

The Wood Handbook FPL-GTR-113
Forest Products Laboratory
USDA Forest Service
One Gifford Pinchot Drive
Madison, WI 53726-2398
608/231-9200
www.fpl.fs.fed.us

Approval Guide
FM Global (Factory Mutual Insurance Co.)
1301 Atwood Avenue / P.O. Box 7500
Johnston, RI 02919
401/275-3000
www.fmglobal.com

UL Building Materials, Roofing Materials and Systems, Fire Protection Equipment, Fire Resistance Directory
Listing Information Services—5152XGNK
Underwriters Laboratories, Inc.
333 Pfingsten Road
Northbrook, IL 60062-2096
847/664-2899 or 800/704-4050
http://ulstandardsinfonet.ul.com

Association Standards

Other pertinent architectural and engineering information can be found in standards issued by various manufacturing, contracting and technical associations. Some frequently referenced sources are:

ACI Manual of Concrete Practice
American Concrete Institute / ACI International
38800 Country Club Drive
Farmington Hills, MI 48331
248/848-3700
www.aci-int.org

Aluminum Curtain Wall Design Manual and Metal Curtain Wall Design Manual; ANSI/AAMA/NWWDA 101 / IS 2-97 Voluntary Specifications for Aluminum, Vinyl (PVC) and Wood Windows and Glass Doors; Anodic Finishes / Painted Aluminum
American Architectural Manufacturers Association (AAMA)
1827 Walden Office Square, Suite 550
Schaumberg, IL 60173-4268
847/303-5664
www.aamanet.org

Standard Specifications for Transportation Materials and Methods of Sampling and Testing

American Association of State Highway and Transportation Officials (AASHTO)
444 North Capitol Street NW, Suite 249
Washington, DC 20002
202/624-5800
www.aashto.org

North American Specification for the Design of Cold-Formed Steel Structural Members
American Iron and Steel Institute (AISI)
1140 Connecticut Avenue, Suite 705
Washington, DC 20036
202/452-7100
www.steel.org

AISC Steel Construction Manual
American Institute of Steel Construction
One East Wacker Drive, Suite 3100
Chicago, IL 60601-2001
312/670-2400 or 800/644-2400
www.aisc.org

Architectural Woodwork Quality Standards Illustrated
Architectural Woodwork Institute (AWI) and Architectural Woodwork Manufacturers Association of Canada (AWMAC)
1952 Isaac Newton Square
Reston, VA 20190
703/733-0600 or 800/449-8811
www.awinet.org

ASHRAE Standards and Guidelines
American Society of Heating, Refrigerating and Air-Conditioning Engineers (ASHRAE)
1791 Tullie Circle NE
Atlanta, GA 30329
404/636-8400 or 800/527-4723
www.ashrae.org

AWWA Standards
American Water Works Association (AWWA)
666 Quincy Avenue
Denver, CO 80235
303/794-7711 or 800/926-7337
www.awwa.org

Certified Product Listings
CSA International (CSA) (Formerly: IAS—International Approval Services)
178 Rexdale Boulevard
Toronto, Ontario, Canada M9W 1R3
416/747-4000 or 866/797-4272
www.csa-international.org

GANA Glazing Manual
Glass Association of North America (GANA)
2945 SW Wanamaker Drive, Suite A
Topeka, KS 66614
785/271-0208
www.glasswebsite.com

Technical Notes—Brick and Tile
Brick Industry Association (BIA)
11490 Commerce Park Drive
Reston, VA 22091
703/620-0010
www.bia.org

Architectural Sheet Metal Manual and numerous HVAC-related documents
Sheet Metal and Air-Conditioning Contractors
    National Association (SMACNA)
4201 Lafayette Center Drive
Chantilly, VA 20151-1209
703/803-2980
www.smacna.org

Manual of Millwork
Woodwork Institute (formerly Woodwork Institute of California)
P.O. Box 980247
West Sacramento, CA 95798
916/372-9947
www.wicnet.org

Handbook for Ceramic Tile Installation
Tile Council of America, Inc.
100 Clemson Research Boulevard
Anderson, SC 29625
864/646-8453
www.tileusa.com

Manufacturers' Product Data

Manufacturers' catalogs are other specification reference sources. However, the suggested specifications in these catalogs should be used with caution. Some catalogs and websites include manufacturers' specifications that are appropriately written, while others are vague and written so as to exclude competitors' products. Therefore, they often fail to provide precise, informative, and clear subject matter.

When using specifications produced by manufacturers, it is absolutely essential to be discriminating in using the text. Do not use clauses as written unless every statement is clearly understood. Modify the language where necessary to ensure competition and complete understanding.

Sources of manufacturers' product data include the following. Those known to include access to construction specifications in word processing formats are noted by the publisher.

ARCAT, The Product Directory for Architects
(including access to library of manufacturers' specifications in word processing formats)
ARCAT, Inc.
1275 Post Road
Fairfield, CT 06824
203/256-1600
www.ARCAT.com

First Source (including Spec-Text data format and Manu-Spec guide specifications)
Reed Construction Data
30 Technology Parkway South, Suite 100
Norcross, GA 30092
770/417-4000 or 800/906-3406
www.FirstSourceONL.com

Sweet's Catalog File—Architects, Engineers & Contractors Edition
McGraw-Hill Construction Sweets
Two Penn Plaza, 10th Floor
New York, NY 10121
800/442-2258
www.sweets.com

4specs.com (comprehensive online directory of links to manufacturers' websites, with discussion forums on construction specifications topics)
www.4specs.com

Specifications published by ARCAT, ARCOM (*Product MASTERSPEC®*, available to *Masterspec* subscribers and issued by the manufacturer), *Manu-Spec* (issued by First Source and BSD (accessible by *SpecLink+* subscribers) conform to CSI *SectionFormat* and *PageFormat* and comply with the specifications writing criteria of the *Manual of Practice*.

General References

Additional reference sources for materials, workmanship, standards, tests and general information are contained in the publications of various associations of manufacturers, technical societies and contractors associations. Names and addresses of these associations are updated annually in the *Encyclopedia of Associations*, published by Gale Research Co., Detroit, available at most libraries. For convenience, see the current listings available from the Building Systems Design, Inc. (BSD) website at *www.bsdsoftlink.com.*

American Institute of Architects Documents

These documents are available from the American Institute of Architects (AIA) (*www.aia.org*), including most local chapter offices. Contact the national office at the following address or a local office for a current list of documents and ordering instructions.

American Institute of Architects (AIA)
1735 New York Avenue, NW
Washington, DC 20006
202/626-7300 or 800/365-2724
www.aia.org/documents

The following documents may be suitable for reference and for inclusion with the bidding and construction contract documents in the Project Manual:

- *A101™-1997,* Standard Form of Agreement Between Owner and Contractor Stipulated Sum (minimum order of five copies)
- *A101™/CMa-1992,* Standard Form of Agreement Between Owner and Contractor-Stipulated Sum, Construction Manager—Adviser Edition
- *A105™-1993,* Standard Form of Agreement Between Owner and Contractor for a Small Project, and *A205™-1993,* General Conditions of the Contractor for Construction of a Small Project (two-document set)
- *A107™-1997,* Abbreviated Standard Form of Agreement Between Owner and Contractor for Construction Projects of Limited Scope Stipulated Sum
- *A111™-1997,* Standard Form of Agreement Between Owner and Contractor Cost of the Work Plus a Fee, With a Negotiated Guaranteed Maximum Price (GMP)
- *A114™-2001,* Standard Form of Agreement Between Owner and Contractor where the basis of payment is the Cost of the Work Plus a Fee without a Guaranteed Maximum Price (GMP)
- *A121™ CMc-2003* Standard Form of Agreement Between Owner and Construction Manager Where the Construction Manager is also the Constructor (also AGC Document 565)

- *A131™ CMc-2003,* Standard Form of Agreement Between Owner and Construction Manager Where the Construction Manager is also the Constructor and Where the Basis of Payment is the Cost Plus a Fee and there is no Guarantee of Cost" (also AGC Document 566)
- *A175™ ID-2003,* "Standard Form of Agreement Between Owner and Vendor for Furniture, Furnishings and Equipment where the basis of payment is a Stipulated Sum
- *A191™-1996,* Standard Form of Agreement Between Owner and Design/Builder
- *A201™-1997,* General Conditions of the Contract for Construction (minimum order of five copies)
- *A201™ Commentary* (free for downloading at *www.aia.org/documents*)
- *A201™ 1987 to 1997 Comparison* (free for downloading at *www.aia.org/documents*)
- *A201™/CMa-1992,* General Conditions of the Contract for Construction, Construction Manager-Adviser Edition (minimum order of five copies)
- *A201™/SC-1999,* Federal Supplementary Conditions of the Contract for Construction
- *A275™ ID-2003* General Conditions of the Contract for Furniture, Furnishings, and Equipment
- *A305™-1986,* Contractor's Qualification Statement (minimum order of five copies)
- *A310™-1970,* Bid Bond (minimum order of five copies)
- *A312™-1984,* Performance Bond and Payment Bond
- *A401™-1997,* Standard Form of Agreement Between Contractor and Subcontractor
- *A491™-1996,* Standard Form of Agreement Between Design/Builder and Contractor
- *A501™-1995,* Recommended Guide for Competitive Bidding Procedures and Contract Awards for Building Construction
- A511™-2001, Guide for Supplementary Conditions (free for downloading at *www.aia.org/documents*)
- *A511™/CMa-1993,* Guide for Supplementary Conditions, Construction Manager—Adviser Edition
- *A521™-1995,* Uniform Location of Subject Matter
- *A701™-1997,* Instructions to Bidders (minimum order of five copies)

- *A775™ ID-2003* Invitation and Instructions for Quotation for Furniture, Furnishings and Equipment
- *G601™-1994,* Request for Proposal Land Survey
- *G602™-1993,* Request for Proposal Geotechnical Services
- *G605™-2000,* Notification of Amendment to the Professional Services
- *G606™-2000,* Amendment to the Professional Services Agreement (50-pack)
- *G607™-2000,* Amendment to the Consultant Services Agreement (50-pack)
- *G612™-2001,* Owner's Instructions Regarding the Construction Contract, Insurance and Bonds, and Bidding Procedures (free for downloading at *www.aia.org/documents*)
- *G701™-2000,* Change Order (50-pack)
- *G701™/CMa-1992,* Change Order, Construction Manager—Adviser Edition (50-pack)
- *G702™-1992,* Application and Certificate for Payment (50-pack or continuous roll)
- *G702™/CMa-1992,* Application and Certificate for Payment, Construction Manager—Adviser Edition (50-pack)
- *G703™* Continuation Sheet (50-pack or continuous roll)
- G704™-2000, Certificate of Substantial Completion (50-pack)
- *G704™/CMa-1992,* Certificate of Substantial Completion, Construction Manager-Adviser Edition (50-pack)
- *G706™-1994,* Contractor's Affidavit of Payment of Debts and Claims (50-pack)
- *G706A™-1994,* Contractor's Affidavit of Release of Liens (50-pack)
- *G707™-1994,* Consent of Surety to Final Payment (50-pack)
- *G707A™-1994,* Consent of Surety to Reduction in or Partial Release of Retainage (50-pack)
- *G709™-2001,* Proposal Request (50-pack)
- *G710™-1992,* Architect's Supplemental Instructions (50-pack)
- *G711™-1972,* Architect's Field Report (50-pack)
- *G712™-1972,* Shop Drawing and Sample Record (50-pack)
- *G714™-2001,* Construction Change Directive (50-pack)
- *G714™/CMa-1992,* Construction Change Directive, Construction Manager—Adviser Edition (50-pack)
- *G715™-1991,* Instruction Sheet and Attachment for ACORD Certificate of Insurance (50-pack)
- *G722™/CMa-1992,* Project Application and Project Certificate for Payment, Construction Manager—Adviser Edition and *G723™,* "Project Application Summary, Construction Manager—Adviser Edition (50-pack)
- *G804™-2001,* Register of Bid Documents (50-pack)
- *G805™-2001,* List of Subcontractors (50-pack)
- *G806™-2001,* Project Parameters Worksheet (50-pack)
- *G807™-2001,* Project Team Directory (50-pack)
- *G808™-2001,* Project Data
- *G809™-2001,* Project Abstract (50-pack)
- *G810™-2001,* Transmittal Letter (50-pack)

Engineers Joint Contract Documents Committee (EJCDC) Documents

The following EJCDC documents are forms, commentaries and other documents useful to those engaged in the preparation of bidding and contract documents for engineering projects. They are offered through the American Consulting Engineers Council, the American Society of Civil Engineers, and the National Society of Professional Engineers.

American Consulting Engineers Council (ACEC)
1015 15th Street, 8th Floor, NW
Washington, DC 20005-2605
202/347-7474
www.acec.org/shoppingcart/

American Society of Civil Engineers (ASCE)
1801 Alexander Bell Drive
Reston, VA 20191-4400
703/295-6300 or 800/548-2723
www.asce.org/ejcdc/

National Society of Professional Engineers (NSPE)
1420 King Street
Alexandria, VA 22314
703/684-2800
www.nspe.org/ejcdc/home.asp

EJCDC documents may be suitable for inclusion with the bidding and construction contract documents in the Project Manual. Prospective documents are:

- *C-050* "Owner's Instructions Regarding Bidding Procedures and Construction Contract Documents" (formerly 1910-29)
- *C-051* "Engineer's Letter to Owner Requesting Instructions Concerning Bonds and Insurance for Construction" (formerly 1910-20)
- *C-052* "Owner's Instructions to Engineer Concerning Bonds and Insurance for Construction" (formerly 1910-21)
- *C-200* "Guide to the Preparation of Instructions to Bidders" (formerly 1910-12)
- *C-410* "Suggested Bid Form for Construction Contracts" (formerly 1910-18)
- *C-430* "Bid Bond" (Penal Sum Form) (formerly 1910-28-C)
- *C-435* "Bid Bond" (Damages Form) (formerly 1910-28-D)
- *C-510* "Notice of Award" (formerly 1910-22)
- *C-520* "Standard Form of Agreement Between Owner and Contractor on the Basis of a Stipulated Price" (formerly 1910-8-A-1)
- *C-525* "Standard Form of Agreement Between Owner and Contractor on the Basis of Cost-Plus" (formerly 1910-8-A-2)
- *C-550* "Notice to Proceed" (formerly 1910-23)
- *C-610* "Construction Performance Bond" (formerly 1910-28-A)
- *C-615* "Construction Payment Bond" (C-615, formerly 1910-28-B)
- *C-620* "Application for Payment" (formerly 1910-8-E)
- *C-625* "Certificate of Substantial Completion" (C-625, formerly 1910-8-D)
- *C-700* "Standard General Conditions of the Construction Contract" (formerly 1910-8)
- *C-800* "Guide to the Preparation of Supplementary Conditions" (formerly 1910-17)
- *C-940* "Work Change Directive" (formerly 1910-8-F)
- *C-941* "Change Order" (formerly 1910-8-B)
- *C-942* "Field Order

  "Focus on Shop Drawings" (1910-9C)

  "Indemnification by Engineers: A Warning" (1910-9G)

  "Limitation of Liability in Design Professional Contracts" (1910-9E)

  "Uniform Location of Subject Matter" (1910-16)

  "Commentary on Agreements for Engineering Services and Construction-Related Documents" (1910-9)

Chapter 16

Products Selection

PRODUCT SELECTION FACTORS

The selection of products for the design of a facility is the responsibility of the Architect or Engineer of Record. There are many factors to consider when selecting and evaluating materials, equipment, components, and systems for use in construction. Some are mandated, and others are determined by applying architecture and engineering principles. Still others are ethical, economic, and political. The architect or engineer also is responsible for selecting products and designing assemblies that result in an integrated design for the project. These factors combine, requiring the architect or engineer to understand the technology represented in the design and appropriately address the various product selection factors, appropriately using the methods for specifying the products discussed in Chapter 5.

MANDATED REQUIREMENTS

The first and perhaps most obvious factor in product selection is what is mandated. Sources of those mandates include the applicable building code, regulations, laws, and ordinances, and the stated requirements, policies and preferences of the owner. Especially on institutional projects (educational, correctional and medical facilities), mandated requirements are abundant.

Fire and Life Safety Requirements

Review of the drawings at the Design Development phase of a project should reveal the fire and life safety requirements. When these are determined, products complying with the fire and life safety requirements should be listed according to the CSI *MasterFormat* section number. For example, if fire-resistive gypsum board is needed for walls, partitions and ceilings, this should be noted to be included in the gypsum board section. If doors in fire-rated walls and partitions are needed, this should be noted to be included in the steel doors and frame section and (if used) in the wood doors section. If access panels are needed in fire-rated walls and ceilings, this should be noted to be included in the access doors and panels section. If ducts pass through fire-rated walls and floors/ceilings, then it should be noted that fire dampers are needed on ductwork and the specifications should include a section for firestopping and smoke seals.

Carefully going through the drawing plans, sections, and details as they develop will enable the project architect or engineer to identify for the specifier what needs to be specified. Handing a set of drawings to the specifier without consultation with the architect or engineer, who is very knowledgeable about the project, is inefficient and leads to false assumptions, overspecifying, and underspecifying.

The more clearly the drawings indicate fundamental fire and life safety information, including ratings of walls, partitions, and ceilings, plus exitways and room occupancies, the easier it will be for the specifier (and planchecker) to understand the project design and select appropriate products.

Structural Safety Requirements

It is not only the structural engineer who must address seismic design issues. As lessons are learned from analyses of major earthquakes, the more seismic safety provisions become mandated by code. The 1994 earthquake in Northridge, California, involved heavy shaking and damage to building structural elements, which was not surprising. What was surprising was the damage to mechanical and electrical systems. Sprinkler piping was frequently severed, and systems were rendered useless; leakage caused flooding and extensive damage as well. HVAC equipment and ductwork were put out of commission due to failed anchorages. Emergency power was not available at one hospital because the violent shaking of the earthquake damaged the internal components of the emergency generator's switchgear. The generator started, but the electrical load could not be shifted to it.

Mechanical and electrical engineers, as well as architects and structural engineers, must address the seismic requirements in revised building, mechanical, and electrical codes. This is not a phenomenon limited to California. Alaska, the Pacific Northwest, Missouri, and other locations with a high seismic risk will require attention to structural safety.

Storm Damage Resistance

Areas subject to windstorm damage, including Pacific islands such as Guam, the Gulf and Atlantic coasts of the United States, the Caribbean islands, and Midwestern areas subject to tornados all require attention to structural safety requirements. Architects and engineers need to comply not only with lateral bracing and roof structural anchorages requirements but also with strengthened glass framing and missile-resistant glass requirements.

The consequences of hurricane damage can cost billions of dollars, as evidenced by Hurricane Andrew (1992; $26.5 billion), Hurricane Hugo (1989; $7 billion), and Hurricane Floyd (1999; $4.5 billion). Building codes have been updated and will continue to be updated to require facilities to be more resistant to storm damage. These requirements will be reflected in the selection and specification of building products.

Accessibility

Access compliance in facility designs is not a building code matter. It is a civil rights matter. This makes it even more serious for architects and engineers. Federal and federally funded facilities require compliance with accessibility regulations, most commonly the U.S. federal *Uniform Facility Accessibility Standards (UFAS)* (which superseded the Americans with Disability Act Accessibility Guidelines for Buildings and Facilities [AADAG]). These regulations are usually the basis of state/province and local accessibility regulations.

Accessibility regulations address not only architectural barriers and matters such as toilet room grab bars and lever handles on doors, but also many other matters including those affecting civil, landscape, plumbing, and electrical components. For example, slip resistance and gradients of exterior walking surfaces, plumbing fixtures, and electrical outlet locations are governed by accessibility regulations. Product selections must consider use by the physically challenged.

Sustainable Design and Construction

Government entities at federal, state/province, and local levels are mandating sustainable design

of facilities, which includes selection of products used in construction. Even corporations are mandating sustainable designs, as evidenced by the LEED 2.0 certified corporate headquarters for the Premier Automotive Group (Ford Motor Company) in Irvine, California, and the LEED registered South Campus Office Development for Toyota Motor Sales, U.S.A., in Torrance, California (a 624,000-square-foot complex constructed at a cost comparable to that of a conventional commercial office building). Both projects were designed by LPA, Inc., of Irvine (specifications consultant: David E. Lorenzini, FCSI, CCS, Associate AIA of Architectural Resources Company, Leesburg, Virginia).

Sustainable design involves far more than selecting products with recycled materials content. It means designing from the project's inception to meet energy conservation and waste management criteria.

Energy Conservation

This is a major topic for architects and mechanical and electrical engineers. States such as Massachusetts and California have highly developed and restrictive energy conservation regulations affecting thermal insulation, air and water vapor barriers, mechanical equipment efficiency, lighting fixtures, and building controls such as energy management and control systems. Energy conservation is now an integral part of the design of all types of buildings, from individual residences to schools, hospitals, commercial office buildings, hotels, restaurants, and even warehouses. Product selections must be made with sensitivity to energy conservation factors.

Waste Management

This is an increasing issue for municipalities, and local waste management regulations must be determined and followed. The city of Vancouver, British Columbia, is a leader in the development of waste management specifications for Division 1 - General Requirements. For LEED-certified projects, construction waste not only is required to be separated for recycling, but offsite disposal shall be minimized.

LEED Certification

The LEED (Leadership in Energy and Environmental Design) Green Building Rating System is a voluntary, consensus-based national standard for developing high-performance, sustainable buildings by the U.S. Green Building Council (USGBC). USGBC represents segments of the building industry and has developed (or is in the process of developing) LEED standards for:

- New construction and major renovation projects (LEED-NC)
- Existing building operations (LEED-EB)
- Commercial interiors projects (LEED-CI)
- Core and shell projects (LEED-CS)
- Homes (LEED-H)

LEED provides a complete framework for assessing building performance and meeting sustainability goals. Based on well-founded scientific standards, LEED emphasizes state-of-the-art strategies for sustainable site development, water savings, energy efficiency, materials selection, and indoor environmental quality. LEED recognizes achievements and promotes expertise in green building through a comprehensive system offering project certification, professional accreditation, training, and practical resources.

DESIGN CRITERIA

Occupancy and Use

Consider the occupancy and use of the facility. Institutional occupancies have more restrictive criteria than commercial occupancies. Commercial occupancies put building elements to heavier use than residential occupancies. Products suitable for heavy-duty use may be overdesigned for a residential use, with commensurate overpricing. Conversely, residential-grade materials may be unsuitable for heavier loads, greater frequencies of use, and resistance to soiling of commercial and institutional occupancies.

There are occasions when the design challenge is to build products that appear residential in character while performing at institutional quality. An example is congregate living or senior housing, especially assisted-living facilities. Products in living units may be satisfactory if they are of residential quality, but common area finishes, hardware, and fire performance require the same criteria as for an institutional setting such as a dormitory or hospital. Even hospitals sometimes require residential-appearing products, such as in alternative birthing rooms that resemble hotel rooms or residential bedrooms but have provisions for rapid access to medical equipment and are in compliance with hospital standards for infection control.

The designer must understand these criteria in order to select appropriate products. The architect, engineer, owner, and construction manager must understand that heavier-duty products need higher construction budgets.

Life Expectancy

Related to the above discussion is the criterion of life expectancy. For a hotel, guest room finishes might have a life expectancy of only a few years. For a tenant space in a commercial office building, the washability of paint on a wall might be less than that for a residence. The tenant space may have an expected life of a few years before remodeling. Public schools might be built with a life expectancy of 20 or 30 years, which is what building codes tacitly assume is the life of a building. Products need to be selected based on the expected life of the building.

In Southern California, two projects represent extremes in life expectancy. A new campus for Soka University, a private university constructed in 2002, has products selected for an expected life of 100 years for the facility. In Los Angeles, Our Lady of Angels Cathedral is constructed with products and details for a 400-year expected life.

Meanwhile, public schools constructed with the typical 20- to 30-year life in the 1950s and 1960s are 40 to 50 years old without major reconstruction. Modernization and rehabilitation of existing schools and construction of new schools are being performed with economy (lowest initial cost) as the primary design factor.

Designers should understand the life expectancy of products when making selections and should endeavor to inform the owner of options in product selections that balance cost and quality over the expected life of the facility.

Geography and Environment

Architects and engineers are designing projects throughout North America and on other continents. Geophysical factors vary widely throughout the United States and Canada. Designers need to understand these factors when making product selections.

Corrosion considerations at a coastal location are generally more important than those in an inland or desert location. Galvanized sheet metal may be adequate for most locations but not in a tropical environment such as that of Hawaii or Puerto Rico. Alclad (aluminum allow sheet coated with pure aluminum) may be necessary in high-humidity and salt-air environments. Stainless steel door frames and doors may be necessary in the corrosive environment of a sewage treatment plant, and the stainless steel may even need to be coated for additional corrosion resistance (stainless steel is technically corrosion-resistant steel).

A residential complex designed for the relatively benign coastal environment of Southern California will almost self-destruct under the snow loads and freeze-thaw conditions of a mountain ski resort. This has unfortunately been proven, with large financial settlements after extended litigation.

Energy consumption in a desert environment such as Phoenix, Arizona, or in the severe winter cold of Buffalo, New York, warrants selection of products that are very energy efficient. Aluminum windows and storefront framing, and glass and glazing as well, in extremely hot and cold environments require greater value to be placed on thermal attributes. In coastal environments such as South Florida and along the Atlantic seaboard, resistance of window and storefront framing and glass to extremely high winds and airborne objects is mandated by code as well as proper design.

The designer and specifier need to identify these factors and apply them when making product selections.

- Proposed life span of the structure, the geographic location, the environment, and the proposed occupancy
- Professional judgment of the responsible design professional

Using Performance Characteristics

Exhibit 16-1 is provided to aid the designer and specifier in analyzing the performance characteristics of products. Each product under consideration should have applicable factors from the "Performance Characteristics" list addressed and performance criteria assigned. This way, the specifier will be less likely to overlook an essential attribute.

An attribute is an inherent characteristic. The use of an analytical approach, such as suggested by Exhibit 16-1, will trigger other essential inquiries that might not otherwise surface since this is a coherent interrogation. Although the list may be long, the specifier may not need to examine every attribute, since a particular material intended for use in a particular portion of the building may not be subjected to the specific performance.

In using Exhibit 16-1, which is a list of attributes sought in a product, there are two additional elements that must be considered to determine whether the product meets the needs of the attribute: (1) criterion and (2) test. For example, if the attribute "flame spread" is an essential attribute for a ceiling material in an exit corridor, most fire codes and building codes establish a flame spread not to exceed 25. The material to be specified

would be required to meet this criterion, measured by test method ASTM E 84. So, the performance could be expressed as:

- Attribute: Flame spread.
- Criterion: Not to exceed 25
- Reference test standard: ASTM E 84

The accomplished specifier will usually have no difficulty in assessing traditional building materials for specific geographic and environmental conditions. Certain materials have a long history of performance under known conditions. However, with the recent developments in building materials, particularly those that are the products of modern chemistry, the same long-term performance behavior patterns cannot be applied.

Even traditional materials will experience faster degradation when used in new geographical environments or when the environment changes. For example, Brownstone, quarried in arid climates in the American West, fares poorly when used for steps and facades of "brownstones" in New York City that are subject to substantial abrasion, moisture, freeze-thaw cycles, and de-icing chemicals. As another example, marbles that withstood several millennia in structures on the Acropolis in Athens have shown marked deterioration in less than 100 years when exposed to the products of combustion generated by automobile fumes and acid rain.

Performance characteristics of products need to be evaluated under the conditions of use in order to foresee problems such as those described above.

New Products Considerations

For new products, there are two major areas that involve materials evaluation. The first deals with the development of a product or material to fit a particular situation created by specific requirements. The second involves an evaluation of the properties of a new material or product to determine whether the manufacturer's claims match its test results, thus warranting the use of its product.

For development of a product to meet a specific requirement, the specifier must establish the conditions of use and the criteria for testing and acceptance. For example, if a floor will be subjected to unusual hazards, such as moisture, acid spillage, hot jet fuels, or printers' ink, a standard commercial flooring material would probably not satisfy all the criteria. Specific design criteria need to be established. The specifier would have to research and determine which fluids would be likely to spill on

PERFORMANCE CHARACTERISTICS

Structural Serviceability
Natural forces
Wind
Seismic
Strength
Compression
Hardness
Indentation
Modules of rupture
Shear tension
Torsion

Fire Safety
Fire resistance
Flame spread
Smoke development
Toxicity

Habitability
Acoustic properties
Sound absorption
Sound reflectance
Sound dispersion
Sound transmission
Noise reduction coefficient
Hygiene, comfort, safety
Air infiltration
Mildew resistance
Slip resistance
Toxicity
Vermin infestation
Thermal properties
Thermal expansion
Thermal shock
Thermal transmittance
Thermal resistance

Water permeability
Moisture expansion
and drying shrinkage
Water absorption
Water vapor
transmission

Durability
Adhesion of coatings
Blistering
Delamination
Dimensional stability
Expansion
Shrinkage
Volume change
Mechanical properties
Resistance to bursting
Resistance to fatigue
Resistance to splitting
Resistance to tearing
Resistance to wear
Abrasion
Scratching
Scrubbing
Scuffing
Weathering
Bactericidal
Chemical fumes
Fading
Freeze-thaw
Ozone
Ultraviolet (UV) radiation

Compatibility
Chemical interaction
Differential thermal movement
Galvanic interaction

Exhibit 16-1. Example of performance characteristics.

the floor and to what extent the proposed flooring should resist the effects of such spillage.

The specifier would also need to take into account resistance to abrasion, slip resistance, indentation, hardness, heat resistance, and similar factors. Then the specifier could establish the overall criteria and select certain ASTM test procedures by which these characteristics would be measured. After determining which test procedure to use, the specifier could set minimum and maximum values for test results and either identify products that meet these criteria or develop criteria that a manufacturer would be required to meet with a custom-made product. Unless a specific composition is specified, the end product from a manufacturer

could be an epoxy, neoprene, polyester, acrylic, or urethane formulation. The specific basic ingredients are not important to the specifier. The end result (or the performance characteristics determined by the materials evaluation) is all that counts.

New products are developed by manufacturers either to fill a specific need or to improve existing products. For the most part, manufacturers, rather than architects, have taken the lead in developing new products. After they are developed, the manufacturer brings the items to the attention of architects and specifiers. Where the new products are referenced by the manufacturer to a reference standard, such as a federal, ASTM, or ANSI specification, there is no major problem in evaluating them. However, many new products are specifically designed by manufacturers to keep ahead of their competition.

In these cases, the physical and chemical properties are not referenced to known standards. A specifier investigating these products finds them difficult to evaluate without normal standards of comparison. Sometimes the manufacturer develops its own test methods, and the results have no correlation with standard test procedures.

What procedures does a specifier follow in evaluating new products? The specifier must take several factors into account. One is the integrity of the manufacturer. Has it had a successful record of developing good products? Has it field tested the new product? Is there any correlation between its field tests and its laboratory tests? Has it tested the significant properties of the product?

The reliability of the source of the information and its authenticity should be investigated. Check with other architects and engineers, if they are given as references, to determine whether the condition of use is similar to that proposed for your project. Demand additional test data if necessary. Suggest specific properties to be tested.

Review the problems to be encountered in the field in the handling and installation of a new product. Will there be an adequate, fully trained corps of tradespeople who understand how to handle the product? Are there franchised applicators? Are there any special precautions to be observed with respect to volatile solvents, flammable materials, or staining of adjacent surfaces?

The evaluation of new or untried materials for possible use should include discussions with the manufacturer to obtain long-term guarantees to ensure additional safeguards for the client and the design professional.

For a more comprehensive treatise on materials and their evaluation and selection, the reader is directed to *Architectural Materials for Construction* by Harold J. Rosen, PE, FCSI and Tom Heineman, FCSI (McGraw-Hill Professional, 1995; ISBN: 0-070-53741-0).

SUBSTITUTIONS AND PRODUCT OPTIONS

The construction side of the project team has a strong incentive to reduce costs. As a result, general contractors, subcontractors, and product suppliers usually want to substitute less costly products for those specified. This is not necessarily an indict-

ment of architects and engineers for "gold plating" the facility design. It means that there is always pressure to construct cheaper and faster. In lay terms, it means shopping at Wal-Mart rather than Nordstrom or Neiman-Marcus.

Under competitive bidding for public projects, the substitution process is intense and aggressive. Residential projects, especially multiunit housing for low- and middle-income occupants, also involve intense pressure for cost reduction and substitutions. Sometimes a process called "value engineering" is used to apply reason to the substitution process where the life cycle cost and performance of products and assemblies are considered. Experience indicates, however, that this process is applied too late in the design process (such as immediately prior to and during bidding, rather than during the Design Development phase), so that the only factor is reduction of the initial cost.

There are built-in disincentives when the Owner's representative or Construction Manager is required only to construct the facility on time and within budget. Quality considerations are ignored, and the consequences are deferred to operation and maintenance of the facility. Meanwhile, the responsibilities of the architect or engineer are unchanged, and poor performance or premature failure of products is blamed on the responsible design professional. It is strange that design professionals are willing to assume additional financial risk by accepting inferior products, while others benefit financially from decisions for which they do not have to take responsibility.

This discussion emphasizes how serious substitutions and product options can be.

When proprietary products are used as the basis of design, it is almost impossible to identify and name all the competitive products in the marketplace. With international trade increasing, the problem has become staggering. To make the process manageable, a "specified manufacturer" is usually identified and "acceptable manufacturers" are listed whose equivalent products will be acceptable. This invites or ensures competitive pricing while realistically specifying known products available in the marketplace. The brand name products of the specified manufacturer, including catalog numbers, are specified. Attributes of the specified products, such as reference standards compliance and physical descriptions of the products, are included to make submittals reviews and reviews of substitution requests easier.

When acceptable products are not known, the specifications typically state, "Acceptable manufacturers: None identified" or "Acceptable manufacturers: Unknown." To keep the competitive process in force, the infamous term "or equal" is used. See Chapter 6 for a discussion of the "or equal" provision, as well as the following.

No phrase in specifications has been subject to more severe criticism than "or equal." The problems that have arisen from its use, the countless seminars that have been held to discuss alternative approaches, and the many magazine and newsletter articles that have appeared over the years attest to the fact that use of this term is not satisfactory in controlling the selection of products to be used for construction. No more satisfactory solution has developed to address the need to allow multiple sources of construction products in order to keep pricing competitive.

Use of "or equal" has often led to conflicts about who shall determine what is an "equal" product. Is it the architect/engineer? Is it the owner's representative or construction manager? Is it the general contractor?

To make the "or equal" provision and the general subject of substitutions manageable, clearly established procedures are needed for determining the equivalence of products. In Division 1 - General Requirements in the specifications, "or equal" needs to be defined, and the criteria and procedures for determination of acceptability of products under the "or equal" provision or other well-defined criteria for substitutions should be specified. These criteria and procedures are usually specified in Section 01600 - Product Requirements or a combination of Section 01610 - Product Options and Section 01630 - Product Substitution Procedures. See the example of a shortform Section 01600 - Product Requirements in Appendix B. See Exhibit 16-2 for example text for product options and Exhibit 16-3 for example text for substitution provisions and procedures.

Using Product Information Resources

In Chapter 15, resources for construction specifications are identified and discussed. The following supplement that discussion.

ICC-ES Evaluation Reports

Determining code compliance for products is made easier by evaluation and research reports. As discussed in Chapter 15, major model building code associations have combined to form the International Code Council (ICC). ICC Evaluation Service publishes evaluation reports for building products, with recommendations and conditions of acceptance by building officials. Current reports may be found at the ICC-ES website: *http://www.icc-es.org/Evaluation_Reports/index.shtml.*

Evaluation reports were also published by the four code associations that formed ICC. These legacy reports are being superseded by ICC-ES reports, but in the interim they are available through the ICC-ES website above.

ICC-ES evaluation reports are an excellent resource for specifiers who wish to use the reference standard method of specifying and minimize the text in a specification section. By referencing an ICC-ES report, a tremendous amount of information is incorporated into the contract documents. In addition to being an expedient way to comply with building code requirements, this also saves a great deal of research time for the architect, engineer and specifier.

Product Evaluations

Subscribers to *Masterspec,* published by ARCOM for the AIA, receive *Masterspec Evaluations.* Even subscribers to the short language version of *Masterspec* receive the *Evaluations.* In addition to providing an introduction and overview of products specified in a section, *Masterspec Evaluations* provide product information that enables the specifier to compare and select products.

Construction Specifications Canada's *TEK-AID* program of technical guides also provides guidance for product selections. Reed Construction Data (*www.reedconstructiondata.com*) publishes *Build-Select Product Data* for the Canadian construction market.

For the U.S. construction market, Reed Construction Data publishes *Spec-Data,* which presents construction information in a 10-part format: product name, manufacturer contact data, product description, technical data, installation recommendations, product availability and costs, warranty, maintenance, technical services, and filing systems.

The foremost unpublished resource for product evaluations is building product representatives. Ethical, knowledgeable and responsive representatives are invaluable to specifiers. Over time, specifiers get to know building product representatives who are reliable and whose recommendations can be counted on to be fair. These representatives tend to represent the leading products in the marketplace. Their products may not be the least expensive, but they usually serve as the basis of design because of their quality and the available resource of a local product representative.

1.3 GENERAL PRODUCT REQUIREMENTS

A. Products, General: "Products" include items purchased for incorporation in the Work, whether purchased for the Project or taken from previously purchased stock, and include materials, equipment, assemblies, fabrications and systems.

1. Named Products: Items identified by manufacturer's product name, including make or model designations indicated in the manufacturer's published product data.
2. Materials: Products that are shaped, cut, worked, mixed, finished, refined or otherwise fabricated, processed or installed to form a part of the Work.
3. Equipment: A product with operating parts, whether motorized or manually operated, that requires connections such as wiring or piping.

B. Specific Product Requirements: Refer to requirements of Section 01450 - Quality Control and individual product Specifications Sections in Divisions 2 through 16 for specific requirements for products.

C. Minimum Requirements: Specified requirements for products are minimum requirements. Refer to general requirements for quality of the Work specified in Section 01450 - Quality Control and elsewhere herein.

D. Product Selection: Provide products that fully comply with the Contract Documents, are undamaged and unused at installation. Comply with additional requirements specified herein in Article titled "PRODUCT OPTIONS."

E. Standard Products: Where specific products are not specified, provide standard products of types and kinds that are suitable for the intended purposes and that are usually and customarily used on similar projects under similar conditions. Products shall be as selected by Contractor and subject to review and acceptance by the Architect.

F. Product Completeness: Provide products complete with all accessories, trim, finish, safety guards and other devices and details needed for a complete installation and for the intended use and effect.

G. Code Compliance: When applicable Codes products to comply with prescribed attributes, all products, other than commodity products prescribed by Code, shall have a current ICC Evaluation Service, Inc. (ICC ES). Refer to additional requirements specified in Section 01410 - Regulatory Requirements.

H. Interchangeability: To the fullest extent possible, provide products of the same kind from a single source. Products required to be supplied in quantity shall be the same product and interchangeable throughout the Work. When options are specified for the selection of any of two or more products, the product selected shall be compatible with products previously selected.

I. Product Nameplates and Instructions:

1. Except for required Code-compliance labels and operating and safety instructions, locate nameplates on inconspicuous, accessible surfaces. Do not attach manufacturer's identifying nameplates or trademarks on surfaces exposed to view in occupied spaces or to the exterior.

2. Provide a permanent nameplate on each item of service-connected or power-operated equipment. Nameplates shall contain identifying information and essential operating data such as the following example:

Name of manufacturer
Name of product
Model and serial number
Capacity
Operating and Power Characteristics
Labels of Tested Compliance with Codes and Standards

3. For each item of service-connected or power-operated equipment, provide operating and safety instructions, permanently affixed and of durable construction, with legible machine lettering. Comply with all applicable requirements of authorities having jurisdiction and listing agencies.

Exhibit 16-2. Example of product options text.

1.4 PRODUCT OPTIONS

A. Product Options: Refer to General Conditions of the Contract, Articles 3.4.2 and 3.5.1. Provisions of [*applicable_procurement_law_*] shall apply, as supplemented by the following general requirements.

B. Products Specified by Description: Where Specifications describe a product, listing characteristics required, with or without use of a brand name, provide a product that has the specified attributes and otherwise complies with specified requirements.

C. Products Specified by Performance Requirements: Where Specifications require compliance with performance requirements, provide product(s) that comply and are recommended by the manufacturer for the intended application. Verification of manufacturer's recommendations may be by product literature or by certification of performance from manufacturer.

D. Products Specified by Reference to Standards: Where Specifications require compliance with a standard, provided product shall fully comply with the standard specified. Refer to general requirements specified in Section 01420 - Reference Standards and Abbreviations regarding compliance with referenced standards, standard specifications, codes, practices and requirements for products.

E. Products Specified by Identification of Manufacturer and Product Name or Number:

  1. "Specified Manufacturer": Provide the specified product(s) of the specified manufacturer.
     a. If only one manufacturer is specified, without "acceptable manufacturers" being identified, provide only the specified product(s) of the specified manufacturer.
     b. If the phrase "or equal" is stated or reference is made to the "or equal provision," products of other manufacturers may be provided if such products are equivalent to the specified product(s) of the specified manufacturer. Equivalence shall be demonstrated during submission of required submittals.
  2. "Acceptable Manufacturers": Product(s) of the named manufacturers, if equivalent to the specified product(s) of the specified manufacturer, will be acceptable in accordance with the requirements specified herein in the Article titled " 'OR EQUAL' PRODUCTS."
  3. Unnamed manufacturers: Products of unnamed manufacturers will be acceptable only as follows:
     a. Unless specifically stated that substitutions will not be accepted or considered, the phrase "or equal" shall be assumed to be included in the description of specified product(s). Equivalent products of unnamed manufacturers will be accepted in accordance with the "or equal" provision specified herein, below.
     b. If provided, products of unnamed manufacturers shall be subject to the requirements specified herein in the Article titled " 'OR EQUAL' PRODUCTS."
  4. Quality basis: Specified product(s) of the specified manufacturer shall serve as the basis by which products by named acceptable manufacturers and products of unnamed manufacturers will be evaluated. Where characteristics of the specified product are described, where performance characteristics are identified or where reference is made to industry standards, such characteristics are specified to facilitate evaluation of products by identifying the most significant attributes of the specified product(s).

F. Products Specified by Combination of Methods: Where products are specified by a combination of attributes, including manufacturer's name, product brand name, product catalog or identification number, industry reference standard, or description of product characteristics, provide products conforming to all specified attributes.

G. "Or Equal" Provision: Where the phrase "or equal" or the phrase "or approved equal" is included, product(s) of unnamed manufacturer(s) may be provided as specified above in subparagraph titled "Unnamed manufacturers."

  1. The requirements specified herein in the Article titled " 'OR EQUAL' PRODUCTS" shall apply to products provided under the "or equal" provision.
  2. Use of product(s) under the "or equal" provision shall not result in any delay in completion of the Work, including completion of portions of the Work for use by Owner or for work under separate contract by Owner.
  3. Use of product(s) under the "or equal" provision shall not result in any costs to Owner, including design fees and permit and plancheck fees.

Exhibit 16-2. *(Continued)*

4. Use of product(s) under the "or equal" provision shall not require substantial change in the intent of the design, in the opinion of the Architect. The intent of the design shall include functional performance and aesthetic qualities.
5. The determination of equivalence will be made by the Architect and such determination shall be final.

H. Visual Matching: Where Specifications require matching a sample, the decision by the Architect on whether a proposed product matches shall be final. Where no product visually matches but the product complies with other requirements, comply with provisions for substitutions for selection of a matching product in another category.

I. Selection of Products: Where requirements include the phrase "as selected from manufacturer's standard colors, patterns and textures," or a similar phrase, selections of products will be made by indicated party or, if not indicated, by the Architect. The Architect will select color, pattern and texture from the product line of submitted manufacturer, if all other specified provisions are met.

1.5 "OR EQUAL" PRODUCTS

A. "Or Equal" Products: Products are specified typically by indicating a specified manufacturer and specific products of that manufacturer, with acceptable manufacturers identified with reference to this "or equal" provision. If Contractor proposes to provide products other than the specified products of the specified manufacturer, provisions of General Conditions of the Contract and [applicable_procurement_law_] shall apply. Contractor shall submit if and when directed by Architect, complete product data, including drawings and descriptions of products, fabrication details and installation procedures. Include samples where applicable or requested.

1. Submit a minimum of 4 copies. Form and other administrative requirements shall be as directed by the Owner's Representative.
2. Include appropriate product data for the specified product(s) of the specified manufacturer, suitable for use in comparison of characteristics of products.
   a. Include a written, point-by-point comparison of characteristics of the proposed substitute product with those of the specified product.
   b. Include a detailed description, in written or graphic form as appropriate, indicating all changes or modifications needed to other elements of the Work and to construction to be performed by the Owner and by others under separate contract with the Owner, that will be necessary if the proposed substitution is accepted.
3. "Or Equal" product submissions shall include a statement indicating the substitution's effect on the Construction Schedule. Indicate the effect of the proposed products on overall Contract Time and, as applicable, on completion of portions of the Work for use by Owner or for work under separate contract by Owner.
4. "Or Equal" product submissions shall include signed certification that the Contractor has reviewed the proposed products and has determined that the products are equivalent or superior in every respect to product requirements indicated or specified in the Contract Documents, and that the proposed products are suited for and can perform the purpose or application of the specified product indicated or specified in the Contract Documents.
5. "Or Equal" product submissions shall include a signed waiver by the Contractor for change in the Contract Time or Contract Sum due to the following:
   a. "Or equal" product failed to perform adequately.
   b. "Or equal" product required changes in other elements of the Work.
   c. "Or equal" product caused problems in interfacing with other elements of the Work.
6. If, in the opinion of the Architect, the "or equal" product request is incomplete or has insufficient data to enable a full and thorough review of the proposed products, the proposed products may be summarily refused and determined to be unacceptable.

B. Product Substitutions: For products not governed by the "or equal" provision, comply with substitution provisions of the General Conditions of the Contract and requirements specified in Section 01630 - Product Substitution Procedures.

Exhibit 16-2. *(Continued)*

1.2 SUBSTITUTION OF MATERIALS AND EQUIPMENT

A. Substitutions, General: Catalog numbers and specific brands or trade names are used in materials, products, equipment and systems required by the Specifications to establish the standards of quality, utility and appearance required. Alternative products which are of equal quality and of required characteristics for the purpose intended may be proposed by Contractor, subject to provisions of the General Conditions of the Contract, Paragraphs 3.4.2, 3.5.1 and 7.3.7 and subject to the following:

1. See Section 01610 - Basic Product Requirements, for requirements regarding product options.
2. Substitutions will only be authorized by properly executed Change Order or Construction Change Directive.

B. Substitution Provisions:

1. Documentation: Substitutions will not be considered if they are indicated or implied on shop drawing, product data or sample submittals. All requests for substitution shall be by separate written request from Contractor. See paragraph below for documentation required for submission of request for substitution.
2. Cost and Time Considerations: Substitutions will not be considered unless a net reduction in Contract Sum or Contract Time results to Owner's benefit, including redesign costs, life cycle costs, plancheck and permit fees, changes in related Work and overall performance of building systems.
3. Design Revision: Substitutions will not be considered if acceptance will require substantial revision of the Contract Documents or will substantially change the intent of the design, in the opinion of the Architect. The intent of the design shall include functional performance and aesthetic qualities.
4. Data: It shall be the responsibility of the Contractor to provide adequate data demonstrating the merits of the proposed substitution, including cost data and information regarding changes in related Work.
5. Determination by Architect: Architect will determine the acceptability of proposed substitutions and Owner's Representative will notify Contractor in writing of acceptance or rejection. The determination by the Architect regarding functional performance and aesthetic quality shall be final.
6. Non-Acceptance: If a proposed substitution is not accepted, Contractor shall immediately provide the specified product.
7. Substitution Limitation: Only one request for substitution will be considered for each product.

C. Request for Substitution Procedures:

1. Contractor shall prepare a request for substitution and submit the request to Architect through Owner's Representative for review and recommendation for acceptance or rejection. Formal acceptance or rejection of substitutions shall be by Owner's Representative, based on recommendation by Architect.
   a. Submit a minimum of 4 copies.
   b. Present request for substitution using form provided by Owner's Representative.
   c. Comply with other administrative requirements as directed by Owner's Representative.
2. Substitution requests shall included complete product data, including drawings and descriptions of products, fabrication details and installation procedures. Include samples where applicable or requested.
3. Substitution requests shall include appropriate product data for the specified product(s) of the specified manufacturer, suitable for use in comparison of characteristics of products.
   a. Include a written, point-by-point comparison of characteristics of the proposed substitute product with those of the specified product.
   b. Include a detailed description, in written or graphic form as appropriate, indicating all changes or modifications needed to other elements of the Work and to construction to be performed by the Owner and by others under separate contracts with Owner, that will be necessary if the proposed substitution is accepted.
4. Substitution requests shall include a statement indicating the substitution's effect on the Construction Schedule. Indicate the effect of the proposed substitution on overall Contract Time and, as applicable, on completion of portions of the Work for use by Owner or for work under separate contracts by Owner.
5. Except as otherwise specified, substitution requests shall include detailed cost data, including a proposal for the net change, if any, in the Contract Sum.
6. Substitution requests shall include signed certification that the Contractor has reviewed the proposed substitution and has determined that the substitution is equivalent or superior in every respect to product requirements indi-

Exhibit 16-3. Example substitution provisions and procedures text.

cated or specified in the Contract Documents, and that the substitution is suited for and can perform the purpose or application of the specified product indicated or specified in the Contract Documents.

7. Substitution requests shall include a signed waiver by the Contractor for change in the Contract Time or Contract Sum because of the following:
   a. Substitution failed to perform adequately.
   b. Substitution required changes in other elements of the Work.
   c. Substitution caused problems in interfacing with other elements of the Work.
   d. Substitution was determined to be unacceptable by authorities having jurisdiction.

8. If, in the opinion of the Architect, the substitution request is incomplete or has insufficient data to enable a full and thorough review of the intended substitution, the substitution may be summarily refused and determined to be unacceptable.

D. Contract Document Revisions:

1. Should a Contractor-proposed substitution or alternative sequence or method of construction require revision of the Contract Drawings or Specifications, including revisions for the purposes of determining feasibility, scope or cost, or revisions for the purpose of obtaining review and approval by authorities having jurisdiction, revisions will be made by Architect or other consultant of Owner who is the responsible design professional, as approved in advance by Owner's Representative.

2. Services of Architect, other responsible design professionals and Owner for researching and reporting on proposed substitutions or alternative sequence and method of construction shall be paid by Contractor when such activities are considered additional services to the design services contracts of Architect or other responsible design professional under contract with Owner.

3. Costs of services by Architect, other responsible design professionals and Owner shall be paid, including travel, reproduction, long distance telephone and shipping costs reimbursable at cost plus usual and customary mark-up for handling and billing.

4. Such fees shall be paid whether or not the proposed substitution or alternative sequence or method of construction is ultimately accepted by Owner and a Change Order or Construction Change Directive is executed.

5. Such fees shall be paid from Contractor's portion of savings if a net reduction in Contract Sum results. If fees exceed Contractor's portion of net reduction, Contractor shall pay all remaining fees unless otherwise agreed in advance by Owner's Representative.

Exhibit 16-3. *(Continued)*

Sweet's Catalog File—Architects, Engineers & Contractors Edition, published by McGraw-Hill Construction Sweets, is a multivolume set of manufacturers' catalogs conveniently bound in green books. Several versions of *Sweets* are published. Unfortunately, participation by building product manufacturers in *Sweets* is steadily declining. Other sources of product information, including manufacturers' websites and product search websites such as *Sweets* own (*www.sweets.com*), are replacing printed product data.

Online building product information should be more current and substantial than that published in *Sweets Catalog File* and other printed forms such as manufacturers' product binders. Manufacturers are coming to realize the shift by architects, engineers and specifiers to online resources and are developing substantial websites containing not only product information but also automated selection programs.

Resources for locating building product manufacturers' websites, in addition to *Sweets,* include:

- *4specs.com: www.4specs.com,* with very comprehensive listings of manufacturers' websites organized according to *MasterFormat* section numbers and titles.

- *ARCAT™, The Product Directory for Architects: www.ARCAT.com;* synopses of building products and manufacturers listed by *MasterFormat* section numbers, including listings of manufacturers' websites.

- *First Source™: www.FirstSourceONL.com;* synopses of building products and manufacturers listed by *MasterFormat* section numbers, including listings of manufacturers' websites. *Spec-Data* are available through this website.

COMMUNICATING AND RECORDING PRODUCT SELECTIONS

Conscientious project architects and engineers, and their designers and job captains, often send product information to specifiers. This information is usually just a start for product research and selection. Due to lack of familiarity with what is needed by specifiers (and drafters who need similar information), the information is often superficial and lacks an important consideration: detailed selection of capacities, sizes, finishes, and other options.

For example, a project designer unfamiliar with loading dock equipment sent copies of manufacturers' catalog pages for a specific series of loading dock leveler to the specifier. However, this literature contained a table of sizes, capacities, and operating range of the leveller and a highlighted list of options. None of these were marked, highlighted, or otherwise indicated for selection. This information was of limited value to the specifier without the necessary detailed design decisions. A time-consuming interchange of e-mail and telephone calls was necessary to resolve the matter because the project designer did not understand all the detailed design decisions that must be made for all products.

Checklists, "action item" lists, and memoranda are helpful for identifying and recording design decisions that must be made and that have been made. Preliminary Project Descriptions are useful tools, as are Outline Specifications; see Chapters 20 and Chapter 21, respectively, for further discussion of these documents.

Whenever product information is discussed in e-mail or memoranda, it is very helpful to categorize the documentation in the "subject" heading of the e-mail or memorandum using the CSI *MasterFormat* section number and title; discussions of products can be structured by this information. It is easier to find pertinent information in an action item list if the list is organized according to *MasterFormat*. This is not as burdensome as it might seem. Familiarity with common *MasterFormat* section numbers and titles comes quickly.

Organized selection and communication processes greatly aid the designers and specifiers to communicate and ensure that appropriate products are specified.

Chapter 17

Specifications Writing Procedures

GETTING STARTED

How does one write construction specifications?

The uninitiated architect or engineer, faced with the task of writing construction specifications for a project, does what other novices with no basic understanding of the principles and procedures of specifications writing have done. With an urgent need to produce specifications, the specifier often succumbs to the expedient practice of taking similar specifications from a (hopefully) similar project or (worst of all) copies verbatim specifications produced by product manufacturers. Recognizing that some changes are unavoidable, the novice specifier proceeds to cut and paste from various versions of specifications and manufacturers' product data, writes in requirements to reduce the design professional's risk and control the bidding process, and deletes requirements that are not understood or appreciated. The resulting patchwork is published and issued for bidding, planchecking, or both. During bidding and planchecking, the specifications are coordinated with the drawings, which may have been produced in a similar manner using graphic details cut, pasted, and adapted from other projects. This cynical description is unfortunately too representative of actual practices.

Neutralizing the cynicism, it must be said that there is value in using previous project specifications. There is constant learning in the specifications writing process. What is learned needs to be carried forward to new projects. This is the "corporate memory" of the firm and represents many design decisions. It is appropriate to build on past experience, but only if past work is used as a resource for a current project and documents are not thoughtlessly copied.

PRODUCT SELECTIONS

Where does one start when producing a set of construction specifications? Rarely does the specifier start with a blank page. Start production of construction specifications by making design decisions for products, assemblies, and systems. See Chapter 16 for a discussion of the product selection process.

Construction specifications should be written according to an organized process that is complete but not burdensome. Just as a good drafter develops systematic methods of laying out drawings and follows drafting conventions, the specifier should be disciplined about acquiring, recording, and applying product information. A systematic approach also expedites production of the specifications because more timely and complete decisions are made. Because similar projects usually use similar products, decisions from previous projects can quickly be used to develop specifications for the current project; these decisions must be reviewed and validated. It is a waste of time and energy to act on decisions that were erroneous and find out that products must be revised, added, or deleted. While it is not advisable to blindly copy construction specifications from other projects, building on experience by considering and recording past project product information is wise. The product information must be validated, just as for new product selections.

Product selections begin during the Schematic Design and Design Development phases of the design of a project (to use AIA terminology for convenience). In these phases, design decisions are made and recorded in the Preliminary Project Descriptions (PPDs) and Outline Specifications. Other communication devices, such as memoranda and checklists, may also be used to prompt and record design decisions and product selections. In Chapters 20 and 21, the topics of Preliminary Project Descriptions (PPDs) and Outline Specifications are discussed in detail.

PPDs and Outline Specifications provide the advantage of product selections prior to drawings production. For this reason, they may be used to guide detailing of the drawings. Drafters are informed of what to detail and can make appropriate decisions. Yet, by their nature, PPDs and Outline Specifications are preliminary, and many decisions about specific products must be made during the Construction Documents (CD) phase of the design. During the CD phase, project architects and engineers and job captains need to work with the specifier in order to make informed, coordinated decisions. The earlier basic decisions are made and recorded, the more efficiently drawings and specifications can be produced.

With a substantial PPD and set of Outline Specifications, the specifier is prepared to review the developed design with the project architect or engineer, project designers, job captains, manufacturers' representatives, and the Owner or Construction Manager. A Project Manual Checklist and checklists for construction specifications are effective aids in the process of identifying and recording detailed decisions about products, assemblies, and systems beyond the superficial descriptions of the PPD and Outline Specifications.

Conduct a review of project requirements using the outline specifications, which are based on the same 16-division *MasterFormat* as the construction specifications. Mark the decisions on a set of the Outline Specifications and work from these specifications until the first draft of the full construction specifications is published. If there are no Outline Specifications, use a comprehensive table of contents from the office master specifications sections. Another choice is to go through *MasterFormat* itself, noting sections to be included. *MasterFormat* provides brief descriptions of what is included in each section.

Sometimes the specifier is simply provided a set of preliminary drawings and told to make some specifications to accompany them. This is a poor way to communicate. In addition, there are undoubtedly design decisions and products yet to be identified and included in the set of drawings. No effective program for telepathic communication between the specifier and other project team members is known. Project architects and engineers must take the time to make clear, complete design decisions and communicate them to the specifier in a timely manner.

Keynotes included in the drawings greatly aid in identifying products to be specified. The specifier can rapidly go through the keynotes and list these products. However, there are occasions when keynotes backfire. This often occurs when the "office master" list of keynotes is included in the set and extraneous notes result. Products get specified because they are in the keynotes but ultimately are not used. Conversely, products that are unique to the project but not included in the office master keynotes may be overlooked. For example, it is a waste of time to try to find out where metal louvers are located in the drawings when an extraneous keynote about louvers is included. Also, time is wasted seeking direction from the project architect or engineer regarding materials and finishes needed to write the specifications when in fact the product is not used.

PROJECT INFORMATION

A Project Manual Checklist should be used to identify basic information that is needed for preparation of bidding and construction contract documents and construction specifications. Download from the AIA website document *G612,* Owner's Instructions to the Architect Regarding the Construction Contract and use it as a guide to obtaining basic information regarding the project. For engineering projects using EJCDC construction contract documents, obtain *EJCDC C-050,* "Owner's Instructions Regarding Bidding Procedures and Construction Contract Documents" (formerly *1910-29*), *EJCDC C-051,* "Engineer's Letter to Owner Requesting Instructions Concerning Bonds and Insurance for Construction" (formerly *1910-20*), and *EJCDC C-052,* "Owner's Instructions to Engineer Concerning Bonds and Insurance for Construction" (formerly *1910-21*). See Chapter 15 for information regarding sources for these documents.

Additionally, there is basic information regarding the project that should be assembled, recorded and disseminated, including:

- Project Directory: Firm or agency names, contact persons, addresses, telephone numbers, and e-mail addresses. For example:
  ○ Owner
  ○ Owner's consultants: Geotechnical engineer, hazardous materials remediation consultant, testing and inspection agency, specialty designers under direct contract with the Owner (graphics designer and interior furnishings designer)
  ○ Construction Manager or Program Manager (if applicable)
  ○ Prime design professional (Architect or Engineer)
  ○ Prime design professional's consultants: Sustainable design (LEED) consultant, civil engineer, landscape architect, acoustician, door hardware consultant, food service equipment designer, laboratory designer, medical equipment designer, theatrical equipment designer, audio/visual systems designer, mechanical engineers (plumbing and HVAC), electrical engineer, and low-voltage systems engineers
  ○ Authority having Jurisdiction (AHJ): Build-

ing department, fire department, health department, public works department, air quality control district, and waste management authorities
  ◦ Serving Utilities: Storm drainage, domestic water, landscape water, sewer, gas, steam, power, cable television, and telecommunications
  ◦ Contractor or Prime Contractors (if known)
  ◦ Major Subcontractors (if known)
• Codes and Regulations
  ◦ Building codes: Model codes plus amendments applicable to the locale of the project
  ◦ Industrial safety regulations
  ◦ Accessibility regulations
  ◦ Regional and local regulations: Air quality, waste management, and vector control
  ◦ Public safety and security regulations: Background checks of workers and airport security
• Owner's Policies and Procedures: Parking restrictions and regulations, infection control procedures, noise restrictions, exhaust fume controls, work hours restrictions, and access restrictions
  ◦ Other information appropriate to the project

Another key to starting the construction specifications is a clear, concise, complete, and correct statement of what is the project. A section such as Section 01100 - Summary is an effective tool for describing what is the work under the contract for construction and what is work to be performed outside of the contract that might affect work under the contract. Review *MasterFormat* for Section 01100 - Summary, for descriptions of Level Three headings that could be applicable.

It is recommended that this section be titled Section 01110 - Summary of the Project and that it describe the entire project. This is to clearly establish the context in which the work will be performed, including descriptions of work under separate contracts that might conflict with or affect work under the contract. This is also where requirements for coordination and management by the contractor of these other activities might be established. This puts a burden on the contractor to manage all construction, but it also provides the contractor with justification for charging in the contract sum for construction management services.

See Appendix A for a sample Project Manual checklist. Because this is written prior to CSI's pending major update of *MasterFormat,* review the current version of *MasterFormat* for section numbers and titles. It is expected that the 1995 edition of *MasterFormat* will be used for many years during transition of product data, master specifications and related data to new formats.

SPECIFICATIONS FORMAT

Each specifier has unique format requirements for the construction specifications. Although CSI *SectionFormat* and *PageFormat* may be the bases for the project construction specifications format, there will be differences in detail. Sometimes these are due to professional judgment and creativity by the specifier, sometimes to requirements and restrictions of the owner. For consultants to the architect or engineer, there are many different formats to suit various specifiers requirements.

Refer back to Chapter 5 for a discussion of specifications formats. See Exhibit 17-1 for sample specifications production standards for page format instructions.

SPECIFYING METHOD

Refer back to Chapter 6 and the four methods of specifying:

1. Descriptive specifying
2. Reference standard specifying
3. Proprietary specifying
4. Performance specifying

As noted in Chapter 6, there is a role for each of these methods in the production of typical construction specifications sections. One method, however, will be the dominant method, and this needs to be determined for the project as a whole and for individual specifications sections. The key determinants are the type of funding for the project and the requirements of the owner for competitive pricing of the work to be performed under the contract.

Typically, if the funding is from a public source, the project is required to be competitively bid and sole-source specifying of products is prohibited except under certain conditions. Again referring back to Chapter 6, review the discussion of nonrestrictive and restrictive (including de facto restrictive) specifying. For projects with public-source funding, specifications typically use descriptive or reference standard methods of specifying.

If the funding is from a private source and there is no policy against restrictive product selection and

---

SPECIFICATIONS PRODUCTION STANDARDS

WORD PROCESSING INFORMATION

Word Processing Program: Microsoft Word 2000

Printer Font: Arial 10pt.

Deliverable Specifications: Printed on white 20 lb bond paper, 1-sided plus specifications files on CD-ROM disk in Adobe Acrobat (.pdf) format suitable.

SECTION FORMAT

3-PART Format: Follow CSI *SectionFormat*™:

> PART 1 - GENERAL
> PART 2 - PRODUCTS
> PART 3 - EXECUTION
> (do not use additional PARTs)

Outline Format: Construction specifications follow the unique format established by CSI *SectionFormat*™ and *PageFormat*™ with the following exceptions and clarifications (note paragraph spacing).

PART 1 - GENERAL	(1st Level)
1.01 ARTICLE	(2nd Level)
A. Paragraph	(3rd Level)
1. Subparagraph	(4th Level)
a. Subparagraph	(5th Level)
1) Subparagraph	(6th Level)

Article Numbers: "PART" number plus consecutive number (e.g., 1.3 = third Article under PART 1; 2.11 = eleventh Article under PART 2; 3.1 = first Article under PAR 3). "Leading 0" is unnecessary (1.4 rather than 1.04).

Paragraph Numbers: As indicated above.

Schedules: At end of PART 3, by convention, rather than end of PART 2.

PAGE FORMAT

Page Margins: 0.5″ Top and Bottom; 1″ Left and 1″ Right.

Tab Settings: Relative to margin, set at –0.5,″ 0,″, 0.1″ and every 0.3″ thereafter.

Paragraph Numbering: Follow CSI *PageFormat*™ except Article numbers do not require leading "0" after period (1.4 rather than 1.04).

Headers and Footers: Each Section shall include headers and footers. Footers shall include Section number, Section name and Page number. Format shall vary for odd and even number pages for 2-sided printing, so that Page number is always on unbound edge. Header shall be constant throughout the Specifications and identify the Project and Section publication date. MS Word template file for Section styles and header text will be provided by specifications writer.

Exhibit 17-1. Sample specifications production standards.

specification, then specifications may use descriptive, reference standard, or proprietary methods of specifying.

Choose the method for specifying products that suits the requirements of the project. For example, for a publicly funded project, the reference standard method of specifying may be used as the primary method, with the descriptive method used for supplementary information to identify the most significant attributes to assist in evaluating submittals and substitution requests. Another example is specifying products by manufacturer, trade name, and catalog number, using the proprietary method of specifying as the primary method and adding reference standard and descriptive information, again to assist in evaluating submittals and substitution requests.

When the specifying method is determined, gather current and appropriate information regarding the selected products for inclusion in the specifications text.

SPECIFICATIONS DETAIL

There are two considerations regarding detail of the specifications. One is the detail level of the specifications section, and the other is the amount of information used within the section to describe the attributes of the specified products.

Refer back to the discussion of levels of detail for specifications sections in Chapter 3. Choose the level of detail to use for the project specifications for each section. Shall a Level Two (broad-scope) section number and title be used or a Level Three (medium-scope) or Level Four (narrow-scope) number and title? This is somewhat related to the level of detail used in the specifications text for the section and the method of specifying discussed above. It is also related to the draft section that is used as the basis of the project-specific specifications.

Starting with the draft, from the office master guide specifications or another source, determine whether the source document is Level Two, Level Three, or Level Four. That in itself might make the determination. If it is Level Four, then that will be probably be the level used for the project-specific section. If it is Level Two and contains several types of similar products but only one is used, then it could be edited down and retitled a Level Four section. Conversely, if several similar products are necessary for the project but the specifications will not require a great deal of written detail, the Level Four specifications could be combined into Level Three or Level Four sections. For example, if the

project includes metal toilet partitions and solid (HDPE) panel toilet partitions, but the specifications will be written using the proprietary method without much supplementary text, then the single Section 10150 - Toilet Partitions could be written rather than two separate sections: Section 1016 - Metal Toilet Partitions and Section 10170 - Solid-Plastic Toilet Partitions.

With the level of detail of the section number and title determined, the amount of detail required for the text is necessary. If the project is privately funded and what the architect specifies is what must be provided by the Contractor, without alternate manufacturers or substitutions and without substantial descriptions of the installation and quality control provisions, then the level of detail of the text may be low. If the project is publicly funded or will be competitively bid, then greater detail will be necessary. All the attributes of the products must be specified so that the quality of the products and installation is clearly stated. If there are extensive quality assurance provisions (submittals, samples, and mock-ups) and quality control provisions (source-control testing during manufacture, fabrication, field inspection, and testing), then the section will require a high level of detail.

It is essential to choose appropriate levels of detail. It is embarrassing when the specifier uses a section from a public school or hospital project, with a high level of specification detail, for a commercial office project. This also adds unnecessary cost if the excessive quality assurance and quality control provisions are followed.

ORDER OF SECTIONS TO BE PRODUCED

The order of sections to be completed—the work plan for specifications production—should be based on two concepts: (1) what needs to be decided or produced earliest and (2) what is known and can be specified.

On a building project, consulting mechanical and electrical engineers may have total responsibility for production of specifications for plumbing, fire protection, HVAC, energy management and other building controls, electrical power, lighting, and signal systems. Within these types of specifications, the second series of considerations, discussed below, apply.

There are other sections that the specifier typically provides to other consultants to edit, such as the civil engineer, landscape architect, and structural engineer. The door hardware consultant might also use a draft section from the specifier, and specialty designers such as food service equipment, gymnasium equipment, and medical and laboratory equipment designers might use the specifier's drafts. Sections used by these consultants should be prepared and transmitted to the responsible designers early in the specifications production process. These consultants typically mark up the drafts and return them to the specifier for production of the final document. This is usually the most expedient method and ensures consistency of format, language and specifications style. The specifier reviews the drafts, eliminates inconsistencies and clarifies other issues, and performs word processing.

Sections produced by the specifier are completed when information is available. There are some sections that are common to most projects, such as gypsum board and painting, and these are often completed first and set aside for review with the project architect. Other sections may be started without full information in order to get the editing process underway and issues identified. As issues are identified, they are recorded for review with the project architect or engineer.

When Outline Specifications are produced for the project, many of the specifications (product selection and installation) issues are identified and often resolved so that the specifier may proceed. Generally, the detailed technical and product issues are not identified during Design Development when the Outline Specifications are produced. They become apparent during editing of project-specific construction specifications.

EDITING AND WRITING SPECIFICATION TEXT

A novice specifier might start editing the text of a specification section at the beginning. This is not a bad idea. Start with the "Summary" or "Section Includes" article. State what is in the section. This can be extracted readily from the outline specifications. If there are no outline specifications, how is this summary information determined? It is typically determined by starting to edit the section at Part 2 - Products. For sections containing an extensive listing of equipment or other scheduled products, which *SectionFormat* locates at the end of Part 3 - Execution, begin with the scheduled products.

There is an order in the article headings of Part 2 - Products, starting with basic materials and progressing through manufactured products to shop or factory fabrication requirements and source quality control:

- Manufacturers
- Materials
- Manufactured units
- Equipment
- Components
- Accessories
- Mixes
- Fabrication
- Finishes
- Source Quality Control

Not all of these headings should be used in a specification section. Use only those that are applicable, and adapt the titles to suit the products specified. Typically, the headings "Manufactured Units," "Equipment," and "Components" are not used, but the actual product names are. See Exhibit 17-2 for examples.

Part 3 - Execution is the next portion of a specifications section to edit. It too follows a standard sequence according to CSI *SectionFormat:*

- Examination
- Preparation
- Erection/Installation/Application/Construction
- Repair/Restoration
- Field Quality Control
- Adjusting
- Cleaning
- Demonstration
- Schedules

Again, not all of these article headings are used in Part 3 of a section. See Exhibit 17-3 for examples.

Part 1 - General is the last portion of the section to edit. It too follows a standard sequence, but it not necessarily intuitive. CSI *SectionFormat* lists standard article titles as follows:

- Summary
- References
- Definitions
- System Description
- Submittals
- Quality Assurance

SAMPLE ARTICLE HEADINGS, PART 2 - PRODUCTS

SECTION 05210 - STEEL JOISTS
PART 2 - PRODUCTS
2.1   MANUFACTURERS
2.2   MATERIALS
2.3   PRIMERS
2.4   K-SERIES STEEL JOISTS
2.5   LONG-SPAN STEEL JOISTS
2.6   JOIST GIRDERS
2.7   JOIST ACCESSORIES
2.8   FABRICATION

SECTION 07710 - MANUFACTURED ROOF SPECIALTIES
PART 2 - PRODUCTS
2.1   REGLETS AND FLASHING
2.2   FORMED ALUMINUM PARAPET COPINGS
2.3   PIPE PENETRATION BOOTS
2.4   ACCESSORY MATERIALS

SECTION 16511 - INTERIOR LIGHTING
PART 2 - PRODUCTS
2.1   MANUFACTURERS
2.2   LIGHTING FIXTURES AND COMPONENTS, GENERAL REQUIREMENTS
2.3   BALLASTS FOR LINEAR FLUORESCENT LAMPS
2.4   BALLASTS FOR COMPACT FLUORESCENT LAMPS
2.5   EMERGENCY FLUORESCENT POWER UNIT
2.6   BALLASTS FOR HID LAMPS
2.7   EXIT SIGNS
2.8   EMERGENCY LIGHTING UNITS
2.9   FLUORESCENT LAMPS
2.10 HID LAMPS
2.11 LIGHTING FIXTURE SUPPORT COMPONENTS
2.12 RETROFIT KITS FOR FLUORESCENT LIGHTING FIXTURES

Exhibit 17.2. Sample article headings, Part 2 - Products.

- Delivery, Storage and Handling
- Project/site Conditions
- Sequencing
- Scheduling
- Warranty
- System Start-Up/Owner's Instructions/ Commissioning
- Maintenance

It is easy to get carried away with Part 1 in a section and forget the use of Division 1 - General Requirements. Requirements get repeated in Part 1 that are specified in Division 1. Of course, this assumes that there is a substantial Division 1 containing these requirements in general terms. Specific requirements, such as specific ambient temperature and humidity criteria for storage of products, should be specified in Part 1 of the applicable section. Specific extended warranty requirements should also be specified in Part 1.

Also, not all articles of Part 1 are necessary or even suitable for all sections. Articles such as References and System Description can be troublesome. ASTM and ANSI standards, which should be well known in the industry, do not need to be listed unless there is a policy to list in detail all references included in the section. Describing the system in Part 1 may be redundant to information in Part 2 and unnecessary. If Part 2 only specifies components of a system or assembly, then the Part 1 description is helpful and should be included. When the Part 1 description describes specific products rather than a general description of the system or assembly, it should be omitted or rewritten and Part 2 used instead.

The Summary article at the beginning of the section can be confirmed or written when the section

---

SAMPLE ARTICLE HEADINGS, PART 3 - EXECUTION

SECTION 05210 - STEEL JOISTS
PART 3 - EXECUTION
3.1   EXAMINATION
3.2   INSTALLATION
3.3   FIELD QUALITY CONTROL
3.4   REPAIRS AND PROTECTION

SECTION 07710 - MANUFACTURED ROOF SPECIALTIES
PART 3 - EXECUTION
3.1   EXAMINATION
3.2   INSTALLATION, GENERAL
3.3   REGLETS AND COUNTERFLASHING INSTALLATION
3.4   PARAPET COPING INSTALLATION
3.5   PIPE PENETRATION BOOTS INSTALLATION
3.6   CLEANING AND PROTECTION

SECTION 16511 - INTERIOR LIGHTING
PART 3 - EXECUTION
3.1   INSTALLATION
3.4   FIELD QUALITY CONTROL

Exhibit 17-3. Sample article headings, Part 3 - Execution.

editing is completed. The list of contents ("Section Includes") should not be a description of subcontract scope or trade jurisdictions. Cross references to other sections ("Related Sections") should only list those sections where there is a direct link, such as Section 09900—Painting for field finishing of products in the subject section.

SPECIFICATIONS CHECKLISTS

Specifications checklists are very helpful for identifying what needs to specified. They prompt specifiers and designers to make decisions. They can also become cumbersome to develop and maintain. Specifications checklists should be kept simple. Perhaps the checklist can be as simple as an annotated list of the office master guide specifications, with basic product types listed. For example, Section 09250—Gypsum Board could list:

- Regular (nonrated) gypsum board, ½-inch thick
- Fire-rated (Type X) gypsum board, ⅝-inch thick

- Flexible gypsum board (curved wall surfaces)
- Sag-resistant gypsum board (ceilings)
- Special fire-resistant gypsum board, ½-inch thick
- Foil-backed gypsum board
- Abuse-resistant (impact-resistant) gypsum board
- Exterior gypsum soffit board (for interior damp locations as well)

Detailed guides to product selections and decisions are better if embedded in the master guide specifications. Editing notes and optional text in the master guide specifications, addressed when the specifier is focusing on micro rather than macro matters, are appropriate in the master guide specification. Otherwise, developing and maintaining the checklist becomes too cumbersome, resulting in an incomplete or erroneous list.

Specifications editing and word processing computer programs are discussed further in Chapter 19.

Chapter 18

Master Guide Specifications

SPECIFICATION GUIDES AND INTACT MASTERS

Construction specifications are written by architects, engineers, and paraprofessional technical writers ("spec writers") who have various levels of experience, knowledge, and writing ability. Product manufacturers also write construction specifications of high to low quality. To compensate, master guide specifications are developed by competent specifiers that are both technically accurate and correctly follow established principles of construction specifications writing.

Specifications Guides

Aids have been developed for specifiers to use while producing project-specific constructions specifications. Reed Construction Data (*www.reedconstructiondata.com*) publishes *Spec-Data,* documents in a standardized format originally developed by CSI for the U.S. construction market and *BuildSelect Product Data* for the Canadian market. *Spec-Data* presents construction information in a 10-part format: product name, manufacturer contact data, product description, technical data, installation recommendations, product availability and costs, warranty, maintenance, technical services, and filing systems. *BuildSelect Product Data* also follows a 10-part format. Both are organized according to CSI *MasterFormat.* Contact Reed Construction Data for additional information.

Architectural Computer Services, Inc. (ARCOM), publishes *Evaluations* documents that accompany *Masterspec®* master guide specifications. The *Evaluations* range from fundamental information about construction products to evaluation and selection guidelines. Some building product manufacturers also publish summary documents of technical attributes of building products.

Altogether, these types of documents are known as "specifications guides." They guide the specifier's editing of a separate specification section document. The separate document is an outline of the section with some fundamental and typical techni-

cal information plus bracketed blanks in which the specifier writes specific descriptions, numerical values, and other attributes and requirements. CSRF's *SpecText®* is typically written in this manner.

Intact Masters

Another approach is to include technical information and editing instructions embedded in the specification master. These types of masters are known as "intact masters." Intact masters have optional text set off by brackets, which are predetermined to be likely choices for the specifier. Some bracketed blanks are usually included for the specifier to add custom text, as for the specification guides discussed above. The text is usually comprehensive, and the editing process is deductive (edit-by-delete) rather than additive. ARCOM's *Masterspec* is typically written in this manner.

Public agency master guide specifications, such as the *Unified Facilities Guide Specifications* (*UFGS*), are typically written as intact masters, with complete text written using descriptive and reference standard methods of specifying. Little or no proprietary product information is included, and bracketed blanks for user-written text is minimized. Again, the intent is to edit-by-delete.

PUBLIC AGENCY MASTER GUIDE SPECIFICATIONS

U.S. Federal Agency Master Guide Specifications

Public agencies have long recognized the value of master guide specifications. Guide specifications and master specifications have long been associated with U.S. federal agencies. The U.S. Army Corps of Engineers (COE) and the Naval Facilities Engineering Command (NAVFAC) were early developers of master specifications. Other federal agencies, such as the General Services Administration (GSA), the Veterans Administration (VA), and the National Aeronautics and Space Administration (NASA), also have developed extensive libraries of master specifications. These are necessary in order to maintain consistency in construction quality and procedures for the billions of dollars of construction performed annually by the federal government. It should be noted that these agencies have a culture that highly values conformity, and they have financial assets and personnel to develop and maintain these masters.

Except for public works (general engineering) construction for infrastructure works, state and local agencies have not developed master specifications. Those infrastructure specifications that have been developed rarely follow CSI formats and construction specifications writing procedures. Public agencies are changing, however, and many are developing sets of standard specifications for buildings and some general engineering construction. Some of these are in the form of Outline Specifications, with enough information to describe product requirements but not complete enough for bidding and construction documents. Other public agencies, such as large metropolitan public school districts and state general services administration agencies, as well as private organizations, such as universities and medical systems, have developed sets of full-length master specifications that are expected to be used almost verbatim. These agencies and organizations have recognized the value of master guide specifications to produce consistent and higher-quality construction specifications.

Federal agencies no longer publish agency-specific master guide specifications. They have moved to the *UFGS,* a very comprehensive library of standard specifications. These specifications require use of a special word processing program, *SpecsIntact.* Refer to the discussion below of SpecsIntact for additional commentary on *UFGS* and *SpecsIntact.*

Canadian National Master Specifications

In Canada, there are two major national master specifications systems for public and private construction. For Government of Canada organizations, there are the *National Master Specifications* (*NMS*). For other organizations, there are *CSC Master Specifications,* which are part of Construction Specifications Canada's *TEK-AID* program of technical guides and master specifications. These specifications also rely upon standardization to ensure quality. For more information on these Canadian masters, contact:

Construction Specifications Canada (CSC)
120 Carlton Street, Suite 312
Toronto, Ontario, Canada M5A 4K2
416/777-2198
www.csc-dcc.ca

COMMERCIAL MASTER GUIDE SPECIFICATIONS

Beginning in the mid-1960s, large architecture and engineering firms recognized the value of master guide specifications and began to develop their own sets of these masters. These were based on the firms' own standards and practices. Developments in document production, especially computer-based word processing programs, enabled firms to record specifications decisions and criteria in standard documents: the office master specifications.

The AIA recognized the value of master guide specifications for architects and developed *Masterspec.* Based on annual subscriptions, *Masterspec* has grown to several libraries of specifications for various design disciplines. Also, *Masterspec* versions have been developed in full-length, short language and outline versions. ARCOM has taken over development and marketing of *Masterspec,* under license from AIA.

CSI similarly developed master guide specifications through the Construction Sciences Research Foundation (CSRF). Now independent of CSI, CSRF continues to develop and market *SpecText* (full-length), *SpecText II* (abbreviated version) and *Outline* specifications.

Masterspec and *SpecText* both offer comprehensive sets of master guide specifications. Each offers specifications in printed form and as electronic files in Microsoft Word and Corel WordPerfect word processing file formats. Each also formats its specifications to use add-in editing enhancements to both word processing programs. These tools should not be confused with true computer-assisted specifications programs, discussed in Chapter 19.

The quality of these commercial master guide specifications is high. They conform well to CSI recommended formats and specifications writing principles. Both *Masterspec* and *SpecText* include substantial numbers of sections in Division 2 - Site Construction, Division 13 - Special Construction (building automation and fire protection), Division 15 - Mechanical and Division 16 - Electrical that are suitable for engineering firms.

The primary purpose of commercial guide specifications is not to provide an editing draft for project-specific specifications. This might be a surprise, but the primary purpose is to develop office master guide specifications. Commercial guide specifications, such as *Masterspec* and *SpecText,* are starting points for editing into master guide specifications that are customized for the standard principles and practices of a specific architecture or engineering firm.

ARCOM *Masterspec*

The AIA has recognized the value of master guide specifications for architects. Since 1969, AIA has published, first through its Production System for Architects and Engineers and now under license through ARCOM, the comprehensive library of master guide specifications called *Masterspec.* These specifications conform to CSI formats and writing conventions and cover the full 16 divisions of CSI *MasterFormat.*

Based on annual subscriptions, *Masterspec* has grown to several libraries of specifications for various design disciplines. Also, *Masterspec* versions have been developed in full-length, short language, and outline versions. Each section is extensively researched, and names of manufacturers and products are included. A full-time staff of professional architects and engineers produces and updates *Masterspec* on a regular basis. *Masterspec* includes an extensive peer review process, and there is also a *Masterspec* advisory review board made up of full-time in-house and independent specification writers.

Masterspec includes Evaluation, Drawing Coordination, and Specifications Coordination documents for each section. These are substantial documents and useful resources for product selection and production of well-integrated documents.

Masterspec is an intact specification system. Except for a small number of exemplary paragraphs and schedules that must be written by the specifier, the specifier edits the section by deleting optional text and filling in numerical or specific data such as colors and finishes. This is sufficient for most commercial and industrial projects. For publicly funded projects and institutional projects requiring code-compliance information, the specifier needs to edit the text to suit the requirements of authorities having jurisdiction.

In addition to full-length *Masterspec,* ARCOM has developed and published short-form and outline versions of *Masterspec. Masterspec Small Project,* an abridged version, is published for projects that are modest in size, scope and complexity. Proj-

ects that can use these master specifications include simple residential, commercial, retail, and institutional projects. *Masterspec Outline,* designed to generate preliminary specifications to be used during the schematic and design development stages of a project, is also published.

Masterspec is available by annual subscription in several word processing software formats in nine practice-specific libraries:

- Architectural/Structural/Civil (A/S/C)
- Structural/Civil (S/C)
- Mechanical/Electrical (M/E)
- Electrical
- Landscape Architecture
- Interiors Construction
- Roofing
- Security & Detention
- General Requirements

Masterspec subscriptions also include *Masterworks* enhancements to Microsoft Word and Corel WordPerfect word processing software. For an additional subscription fee, ARCOM publishes *Linx,* an automated specifications editing program for initial editing of specifications sections. These programs are discussed in more detail in Chapter 19. For more information, contact:

ARCOM, Inc.
332 E. 500th Street
Salt Lake City, UT 84111
801/521-9162 or 800/424-5080
www.arcomnet.com

CSRF *SpecText*

In 1967, CSI created the nonprofit CSRF, whose mission is to support and develop the programs recommended by the Stanford Research Institute in the Automated Specifications Study (CSI Document STD-1). In 1978, CSRF first published *SPECTEXT Master Guide Specifications,* a library of Divisions 1–16 specification guides written in conformance to CSI *SectionFormat, PageFormat* and the CSI *Manual of Practice.* Now an independent organization from CSI, CSRF continues to develop and market three versions: *SpecText®, the SpecText® II* abridged version, and *SpecText® Outline* specifications.

SpecText is published in printed form and in word processing file formats. Editing instructions are embedded in the text, and an edit-by-delete process is used. However, there are many bracketed blanks to fill in since *SpecText* is written using the descriptive and reference standard methods of specifying. For this reason, *SpecText* is more like a specification guide. Proprietary product information and specific attributes of products must be added by the specifier. Drawing coordination information is provided for each section, but there are no product selection guides or discussion beyond embedded notes within the unedited section text.

SpecText II is an abridged version of *SpecText* composed of approximately one-third of the full *SpecText* titles. CSRF says that *SpecText II* is designed for less complex projects such as light commercial, multifamily residential and low-rise buildings and requires less editing than *SpecText.*

Included with a *SpecText* subscription is *Edit-Spec,* CSRF's enhancements to Microsoft Word and Corel WordPerfect. This program is discussed in more detail in Chapter 19.

SpecText is available in several libraries or groups of sections according to design disciplines:

- *SpecText* Facility & Building
- *SpecText* Architectural/Structural/Civil
- Mechanical/Electrical
- *SpecText* Mechanical
- *SpecText* Electrical
- *SpecText* Site/Civil
- *SpecText* Environmental
- *SpecText* Environmental Engineering Add-Ons
- *SpecText II* Facility & Building
- *SpecText II* Architectural/Structural/Civil
- *SpecText II* Mechanical/Electrical
- *SpecText II* Site/Civil
- *Outline*™ Facility & Building
- *Outline* Architectural/Structural/Civil
- *Outline* Mechanical/Electrical
- *Outline* Site/Civil

SpecText library subscriptions, in addition to the specification guides (sections), include:

- *SpecText* Speller
- *SpecText* Glossary of Terms
- *SpecText* User's Manual
- *SpecText* Glossary of Terms User's Manual
- Series 0 Documents

- Drawing Coordination Considerations (DCCs for *SpecText* Divisions 2–14)
- Automated, Combined *SpecText* and *SpecText II* Tables of Contents
- *EditSpec* Section-Editing Tools with instructions

For further information, contact:

CSRF Support Center
P.O. Box 926
Bel Air, MD 21014
410/838-7525 or 877/SPECTXT (773-2898)
www.spectext.com

Manu-Spec

Another library of commercially produced construction specifications is *Manu-Spec,* produced and distributed through First Source. Originally developed and distributed by CSI, *Manu-Spec* and the companion *Spec-Data* programs of manufacturer-produced product data and specifications are now owned by Reed Construction Data. These specifications closely follow the format of *SpecText* but are proprietary specifications written generally in proprietary method of specifying. For further information, contact:

Reed Construction Data
30 Technology Parkway South, Suite 100
Norcross, GA 30092
770/417-4000 or 800/906-3406
www.FirstSourceONL.com

Other Manufacturer (Proprietary) Specifications

ARCAT produces and distributes through its website proprietary specifications that are well written and conform to CSI formats and writing principles. These specifications may be obtained at no charge. For further information, contact:

ARCAT, Inc.
1275 Post Road
Fairfield, CT 06824
203/256-1600
www.ARCAT.com

Sources of manufacturer (proprietary) specifications may be identified from 4specs.com, a comprehensive but simple online source frequented by full-time in-house and independent specifications writers. 4specs.com provides links to building product manufacturers' websites and is developing means to identify which manufacturers offer example and master guide specifications. For further information, contact *www.4specs.com.*

SpecsIntact

As stated above, most federal government specifications are based on the *Unified Facilities Guide Specifications.* These specifications are written using a tagging mark-up system, known as SGML, developed in the 1960s by NASA for its large construction programs. Known as *SpecsIntact,* these were intact master specifications ("Specifications-Kept-Intact"), as discussed above.

After several changes of program and major revisions of text, a major revamping of *SpecsIntact* was done in 1985, with conversion of the documents from mainframe computer to personal computer (PC)–based operation. The PC-based word processing program was used to edit *SpecsIntact* master guide specifications and featured many advance features, such as computer generation of reports (required submittals, required tests and inspections, and lists of reference standards).

In 1988, *SpecsIntact* text and word processing program were issued on the CCB CD-ROM disk by the National Institute for Building Sciences (NIBS). The *SpecsIntact* word processing program was adopted by the Federal Construction Council Committee on Federal Construction Guide Specifications (FCGS) as the standard automated system for the executive departments of the federal government. Construction specifications of the COE, NAVFAC and the VA were rewritten to conform to NASA's document format so the *SpecsIntact* word processing program could be used. Now with standardization of the master specifications of these three agencies plus the GSA into the *Unified Federal Guide Specifications,* the *SpecsIntact* word processing program has taken on greater significance.

The *SpecsIntact* word processing program has continued to develop. It enables the specifier to delete and add text, move text, search and replace text, change case in text (uppercase/lowercase), check references, add missing referenced sections, check spelling and prepare quality assurance reports. It also allows the specifier to generate reports, as previously, and now the specifier may export the edited section in the widely used Microsoft Word format for further editing and publication using Word.

With standardization of all federal agencies on *UFGS,* specifiers producing construction specifications for federal projects must acquire and learn the *SpecsIntact* program or engage an independent specifications consultant who is expert in its use. There is no option; project specifications and *SpecsIntact*-generated reports must be delivered to the contracting officer in electronic format usable by the *SpecsIntact* program. For additional information, contact *http://si.ksc.nasa.gov/specsintact.*

Abbreviated Specifications

Master specifications are long. *Masterspec* and *SpecText* have developed into long documents, but out of a need for completeness rather than wordiness. This is a result of the increasing complexity of construction technology and highly competitive contracting. The trend is expected to continue.

PCs with powerful word processing software have made it easy to produce large quantities of text. Even the abridged versions of *Masterspec* and *SpecText* are voluminous for some projects. Chapter 21 addresses the use of outline specifications, which are suitable for the design development phase of a project, and shortform specifications, which are not merely abridged versions of full-length specifications but are a different style of specification that is a fraction of the length of full-length specifications but nevertheless suitable for use for bidding and construction contract documents.

OFFICE MASTER GUIDE SPECIFICATIONS

Office master guide specifications differ from commercial and public agency master guide specifications. Typically, they begin with commercial master guide specifications and then are edited to suit the requirements of an architectural or engineering firm. This is a process like pre-editing the specifications to avoid repetitive changes in the commercial master guide specifications. Considerations such as the following are addressed, and the text is edited accordingly:

- Typical geographic locations of projects
- Typical building code requirements and references for projects
- Typical product selections (PART 2 of sections)
- Typical construction contract administration requirements (PART 1 of sections)

- Typical product installation requirements (PART 3 of sections)

- Typical construction contract administration requirements (PART 1 of sections)

- Level of detail of specifications sections (e.g., use of broad-scope, medium-scope and narrow-scope sections; refer to Chapter 17)

- Level of detail within typical sections (refer to Chapter 17)

This list of tasks seems simple, but it requires substantial time and effort to accomplish for each section used in the office master guide specifications. Nevertheless, the tasks are necessary. Imagine having to go through each of these tasks, making the same revisions to a commercial master guide specification for each project. This is why office master guide specifications are essential.

The greatest challenge for office master guide specifications is maintenance of the masters. Products, standards and practices constantly change. As each project is completed, something is learned and the office master guide specifications should be updated. This can be accomplished with word processing programs by opening multiple documents and copying text from one to the other. To facilitate this, the documents can be compared with the software, and differences can be highlighted to make the updating process more thorough and expedient.

When a series of similar projects are undertaken, such as a multiphase remodeling project in a hospital or a series of modernization projects at several schools, "submasters" or prototype specifications should be created from the office master guide specifications. This advances the pre-editing process by incorporating specific requirements of the Owner into the specifications. For example,

manufacturer and series of corner guards in a hospital or markerboards in a school can be incorporated that will appear in all projects of the building program. This means maintaining another set of masters, but in the long run time is saved and profitability is enhanced. Quality assurance in design and documentation is also increased.

Because it takes so much effort to develop and maintain office master guide specifications, it is essential to subscribe to a commercial master guide specification library. The commercial masters may serve as the basis of the office masters or they may be used only as a reference resource. Because the commercial masters are updated periodically, the work of updating the office masters is reduced. By using the compare function of word processing software, the updating process can be expedited.

This discussion appears to be most appropriate for medium-sized or large firms, with extensive office master guide specifications, that produce several Project Manuals per month or even per week. In fact, the benefits of office master guide specifications are perhaps even more significant for small firms and firms producing few Project Manuals per month. The difference is the amount of office overhead budget devoted to developing and maintaining the office master guide specifications.

The need for quality in the specifications is the same, whether the office is large or small. By using commercial master guide specifications, even a small firm can produce specifications comparable in quality to those of a large firm. If the office overhead budget is too small to develop and maintain office master guide specifications, then the services of an independent specifications consultant should be considered. A resource for identifying these consultants is Specifications Consultants in Independent Practice at *www.scip.com*.

Chapter 19

Computer-Assisted Specifications

INTRODUCTION

What was most burdensome to the specifier before the advent of the computer was the specifier's inability to cope with the onrush of new building technologies after World War II. There was both a proliferation of new building materials and, simultaneously, the development of new construction techniques. However, it was the development of word processing on the computer, the mainframe at first and then the personal computer (PC), plus the development of master specifications systems, that have kept the specifier from being completely overwhelmed.

The development of the PC and its software continues. A book such as this can only present a brief history and some suggestions of what the future of computer-assisted specifications might be. This book is essentially a primer on the principles of specifications writing, which are relatively timeless, whereas the technological developments of computers for design and construction occur on a daily basis.

THE FIRST FOUR DECADES OF AUTOMATION

In the early 1960s, paper tape–driven typewriters and then typewriters with small magnetic memory devices began to help specifiers cope with the increased text needed to specify more complex buildings. Throughout the 1960s, the cost of memory typewriters came down and the capabilities of the machines improved. Not only construction specifications but also office correspondence benefited. With the advent of xerographic office copiers, the drudgery of typing messy, hard-to-correct stencils and mimeos was eliminated. Computers made it easier to assemble the rapidly thickening books of specifications and other construction contract documents. It became less necessary to cut up the last book of specifications or to compile longhand text to meet the next project deadline.

At the same time, the first uses of computers to store, alter, and print out text were being made by NASA, and some large engineering firms were already using computers for calculations. By 1965, IBM had developed Datatext, which could be used by anyone wishing to do simple text assembly. Then, dozens of architectural and engineering firms and many computer service bureaus filed away enormous blocks of retrievable construction specifications text, most of it arranged by the Divisions and Sections of CSI *MasterFormat*™. About 40 such batch processing services, many also providing specifications master text, were in operation at the beginning of the 1970s.

Altering text, whether it was generated inside the office or outside, was not always easy. The specifier was often separated from the host computer by data entry people and by remote processing, which came either from a central computer room serving the whole office or from an express-linked service bureau. If a large computer was used, the specifications generally had to be batch processed to fit in with other demands on the computer.

Two equipment advances in the 1970s began to change the picture: timesharing and the word processor. Starting in 1969, the first timeshare program with construction specification text was made available to architects and engineers: Pacific International's PIC system. It was now possible to control one's own master text and to edit it by using simple commands in one's own office, even if the mainframe processor was located the width of the continent away.

In 1972, CSI formed the nonprofit Construction Sciences Research Foundation (CSRF), contracting first with PIC and then with Bowne Information Services, to develop Comspec. Comspec could operate with homegrown or acquired master text, and it allowed interactive editing and printing at the same terminal.

By the second half of the 1970s, the first true word processors—the Laniers, Wangs, and Xeroxes—had evolved from mere memory typewriters. These word processors were actually small, limited-purpose minicomputers. They were usually standalones that could do the interactive editing and printing on demand that the timeshare systems provided at greater cost. Word processors, along with timesharing, brought the specifying process back to the specifier's workstation. This allowed flexibility in fine-tuning new and retrieved text to the needs of the project at hand.

As PCs (microcomputers) advanced in power to eclipse minicomputer-based word processors in offices, and as word processing software for PCs came to have document production capabilities that only a few experts could harness, construction specifications production for medium-sized to large design offices was almost exclusively PC-based. With high-speed laser printers and even inkjet printers that eclipse early laser printers for speed, PC-based word processing for construction specifications is usual and customary practice for architects and engineers.

To illustrate, consider the evolution of practice among one group of specifiers. Using the history of the 40 or so members of Specifications Consultants in Independent Practice (SCIP) as an indicator, a rough measure of change from 1970 to 1975 can be made. In this period the number of these specification consultants using memory typewriters grew from 2 to 10, and 1 adopted the computer for production. Another survey of SCIP, which had grown to 80+ members in the early 1990s, found about 70% using PC-based word processing programs for specifications production; the remainder used a variety of equipment and procedures that included typical electric typewriters and a few proprietary typewriter-like word processors. The proportion of SCIP members who edited text themselves directly on a PC using word processing software, rather than doing mark-ups of printed text for a clerical person to transcribe using a PC-based word processing program, went from about 50% in the early 1990s to close to 90% at the end of the decade.

TRANSFORMING THE SPECIFICATION PROCESS

Seven transformations having to do with specifications were underway simultaneously in the 1960s and 1970s and continue today. Each feeds the others and makes the others possible or necessary:

1. *The Hardware Transformation:* Commercially available equipment evolved from the punched paper tape typewriter and the vacuum tube computer to the desktop word processors and large, fast mainframes. Mainframes gave way to minicomputers and then microcomputers (PCs) that have proliferated in large and small offices. Instantaneous access to vast amounts of technical information through the Internet has been made possible by high-speed telecommunications. Production of voluminous documents has been assisted by high-speed laser printers that are affordable for even home-based offices.

2. *The Format Transformation.* Good practice was distilled into the CSI way of organizing text and ideas, from the whole Project Manual down to the subparagraph. New formats have developed to assist recording of information at early phases of project design.

3. *The Master Text Transformation.* Government agencies, large and small design firms, consultants, and text-writing organizations started to develop master text that was well researched, organized, and coordinated. These texts were comprehensive and usable, with intelligent editing, by design professionals. Strict adherence to document and computer file formats enable semiautomated production of specifications using the two major word processing computer programs.

4. *The Technology and Information Transformation.* Manufacturers continued what they had started in the 1950s, making new structural systems, claddings, waterproofings, coatings, finishes, and methods of environmental control available to designers by way of technical literature and hundreds of trained sales-engineering representatives. Availability of information through the Internet on a 24/7/365 basis greatly expedites specifications production. E-mail assists communication by eliminating "telephone tag," and "frequently asked question" (FAQ) files quickly answer inquiries without intervention by product representatives.

5. *The Responsibility Transformation.* Owners and contractors no longer accept the designer as nearly perfect in matters of technology and document preparation. Architects and engineers have had to learn to give clear, complete instructions without resorting to subjective and risk-shifting clauses or trusting that builders will provide what the architect/engineer had forgotten. The current emphasis on design-build rather than design-bid-build shifts and blurs design and construction responsibilities, without substantial understanding of the consequences.

6. *The Accountability Transformation.* Design professionals, including specifiers, have come

to realize that they live in an increasingly litigious climate. Many specifiers have become quality assurance specialists for their offices in addition to specifications producers. They have learned to select and write with concern for immediate and distant consequences.

7. *The Education Transformation.* The practical education of architects, engineers, construction managers, and contractors in specifications and construction technology has generally left the university campus. The initiative in professional development has been taken over by design firms, manufacturers, community colleges, and providers of continuing education programs—often with the support of local CSI chapters.

When these seven transformations were underway in the 1960s and 1970s, there was a remarkable unity in the methods of specification production. Large and medium-sized architectural and engineering offices adopted new hardware, new formats, and new texts. Some small offices advanced with them, but most lagged due to the cost of computerization of clerical tasks. Ultimately, by the early 1980s, most practices had adapted to computerization, primarily with CAD and office correspondence production, but also with document production for construction specifications. Integration of CAD and construction specifications production was seriously discussed, and futuristic pronouncements were made about the imminent connection of drawings and specifications.

From the 1980s on, the picture of computerization in architectural and engineering firms becomes broader, more diverse, and faster moving. Accounting fully embraced PC-based spreadsheet programs. CAD became widely available and rapidly developed so that, at the beginning of the twenty-first century, manual drafting is almost a lost art. Even professional licensing examinations require proficiency in CAD. Virtually all workstations in architectural and engineering firms have PCs, the vast majority of them are networked within the office (local area network or LAN), and all but the unenlightened are connected to the World Wide Web or the Internet for access to unlimited information and instant communication.

Technological advances in computers for architects and engineers have exceeded the advances of technology in text creation and modification of construction specifications. In the late 1980s, Corbel developed *SuperSpec*™, a combination online

(through slow modem connections) and mail/courier service for computer-assisted specifications. *SuperSpec* was based on the specifier's making decisions about the text, recorded on a printed checklist or in a PC-based program, which were sent to SuperSpec's home office for downloading into Corbel's large, high-speed mainframe computer and high-speed reproduction equipment designed for preparation of custom documents for the pension and insurance industries. The Internet had not yet taken hold, and modem connections were pitifully slow by current standards. In the early 1990s, Corbel ceased to provide computer-assisted construction specifications services.

A program competing with *SuperSpec,* called *SweetSpec,* was offered in the late 1980s by McGraw-Hill/Sweets. It was also PC-based and provided assistance to specifiers in producing customized specifications based on SweetSpec's fixed text. The specifier went through a checklist for each section, and the computer worked through a decision tree to eliminate text that was not applicable. This edit-by-delete process yielded a construction specification that was not completely satisfactory and was difficult to modify further using the specifier's PC-based word processing program. SweetSpec was taken over by ARCOM and renamed *Masterspec Q&A,* but it was discontinued in the early 1990s.

The federal government, through NASA, developed computer-assisted text production for construction specifications with the program *SpecsIntact,* discussed in Chapter 18. *SpecsIntact* was highly proprietary and required substantial training to use. Eventually adopted for the U.S. Army Corps of Engineers and the Naval Facilities Engineering Command for production of construction specifications, and now the basis for production of most federal government construction specifications under the *Unified Facilities Guide Specifications* (*UFGS*) program, *SpecsIntact* is not a true computer-assisted specifications program because its automation features are used only for generation of reports and other documents from the edited specifications text. Text creation is not computer-assisted.

Today, true computerized specifications, first introduced in the mid-1990s, are beginning to evolve into not only useful but perhaps essential tools for production of construction specifications. The remainder of this chapter, along with the discussion in Chapter 18 of master guide specifications, addresses the many elements of the still developing practice of computer-assisted specification writing.

WORD PROCESSING SOFTWARE

Software programs for creating and editing text are numerous, powerful, and relatively easy to use—for simple documents like correspondence. These programs have evolved with capabilities to produce complex documents with graphics, automated links to numerical data, and publishing capabilities in printed and electronic forms. Virtually all architectural and engineering offices use PC-based word processing software. However, use for construction specifications and other sophisticated documents requires training that most firms do not provide. As a result, the power of current word processing programs is generally unused.

When put to use, word processing software such as the leading programs, Microsoft Word and Corel WordPerfect, in the hands of a technically savvy specifier trained in their use, enables one or a few persons to produce voluminous documents in several days. Briefly stated, commercially available text-editing software:

1. Provides call-up master text or previously written text, section by section, quickly and accurately.

2. Allows easy editing of characters, words, paragraphs, and larger blocks of text. Text can be deleted, modified, moved, duplicated, and added to with ease.

3. Provides for variable spacing, indenting, margin control, tabulating, replacing, and justifying.

4. Permits headings and footings to be added automatically, along with automatic page numbering within each section.

5. Stores edited text quickly and compactly for easy retrieval.

By using special formatting and text editing features, it is possible to:

1. Renumber paragraphs automatically

2. Cause notes to disappear

3. Change page formats (indentation, spacing, paragraph identifiers)

4. Generate lists of submittals, tests, standards, and the like

Text-editing programs for master guide specifications, published by ARCOM (*Masterworks* for *Masterspec*), CSRF (*Editspec™* for *SpecText™*), and *SpecsIntact* begin to harness the capabilities of PCs. These programs are discussed further below.

COMPUTERS AND PERIPHERAL EQUIPMENT

Except in large offices, mainframes and minicomputers are not used for word processing. PCs have taken over written document production. They have increasingly sophisticated performance characteristics, especially in architectural and engineering firms where graphics-capable PCs are the rule rather than the exception. Graphics-capable PCs easily handle word processing operations. PCs not only in commercial offices but in the microbusiness environment of home-based specifications consulting are networked. File sharing and printer sharing is affordable and the standard. Outmoded PCs are not trashed but continue in service as networked backups to new equipment.

Specific recommendations for computers and related peripheral equipment cannot be made. Technology advances too quickly. What can be safely recommended is the following:

1. *Central Processing Unit (CPU):* The CPU should be the fastest processor with the largest amount of usable memory and hard drive storage that can be afforded. Word processing software puts only modest demands on the CPU, but there are other considerations. Product information and professional development resources are becoming more video oriented, with heavy demands on the CPU and video processor if quick, steady video presentations are played.

2. *Graphics:* As just mentioned, video graphics demands are becoming heavy for everyone. Distance learning programs and downloaded product presentations require large screen monitors. Flat panel liquid crystal display (LCD) monitors are superseding cathode ray tube (CRT) displays. LCD monitors are flicker-free and have high color saturation. For size, go for the largest that the budget can bear to enable multiple windows to be open for drag-and-drop copying and editing. Video processors require less capability than for computer-based gaming.

3. *Storage:* Disk drives for reading and writing files include the infrequently used 3.5-inch floppy disk drive and a hard disk drive measured in hundreds of gigabytes. Networked computers require less storage capacity on individual workstations but huge capacity on the host. Recordable and rewritable digital video disks (DVDs) are taking over from compact disk (CD) drives. New storage technolo-

gies, including transportable random-access memory (RAM) storage devices, are being developed.

4. *Printers:* Architectural and engineering firms that produce only a few Project Manuals per month or only a few sections for Project Manuals assembled by others can use inkjet printers that output at least 12 pages per minute. Laser printers are the standard otherwise, and printers dedicated to specifications production should be as fast as the budget allows, with large paper capacities. The rush at the end of specifications production means that the printer cannot be too fast. Color inkjet and laser printers should be considered, but reproduction costs for multiple copies of color-enhanced Project Manuals also need to be considered.

5. *Telecommunications:* High-speed Internet connection at the specifier's workstation is essential. Since most architectural and engineering firms transmit large CAD files using high-speed connections, it should not be expensive to include this capability in the specifier's computer. Even home-based specifications consultants usually have high-speed cable modem or Digital Subscriber Line (DSL) connections.

WORD PROCESSING ENHANCEMENTS FOR SPECIFICATIONS

ARCOM for *Masterspec* and CSRF for *SpecText* have developed and market enhancements to Microsoft Word and Corel WordPerfect. For federal agency projects using *UFGS,* the NASA-developed *SpecsIntact* program does some similar functions. These programs use the computer to execute repetitive tasks, do global formatting, and generate reports on the content of the Project Manual. These should not be confused with knowledge-based specifications programs such as *Linx* by ARCOM and *SpecLink+™* by Building Systems Design, discussed later in this chapter.

Masterworks

Published by ARCOM for use with *Masterspec* master guide specifications, *Masterwork* is an add-in to the user's word processing program (Microsoft Word or Corel WordPerfect for Windows), purchased separately. *Masterworks* allows the specifier to do the following:

- Automatically select and edit paragraphs and specification options within a specifications section.
- Easily delete paragraphs and subordinate paragraphs within a specifications section.
- Change styles, fonts, margins, paragraph numbering, and other document formatting options from mouse-driven menus within a specifications section.
- Globally perform the above formatting tasks on all sections in the Project Manual, including custom page headers and footers.
- Apply different paragraph numbering schemes or locate and modify unique search strings in a list of files.
- Select between metric and English units of measure for all sections at once.
- Create basic and detailed tables of contents.
- Assemble all sections into a single document and wizards to produce an index and automatically generate project reports. It can choose from either instance reports (that pinpoint specifications containing a particular requirement) or detail reports (that list text from all files associated with a particular subject) for submittals, extra stock and materials, warranties, commissioning, field quality control, and maintenance service.
- From within *Masterworks* using drag-and-drop functionality, create new project folders and copy, move, rename, and delete files.
- *Masterworks,* in conjunction with Microsoft Word software, will batch process convert Word files to Adobe Acrobat Portable Document Format (PDF). This function is not available in *Masterworks* for Corel WordPerfect, but WordPerfect allows individual files to be published in (converted to) PDF.

For further information, contact:

ARCOM, Inc.
332 E. 500th Street
Salt Lake City, UT 84111
801/521-9162 or 800/424-5080
www.arcomnet.com

Editspec

Published by CSRF for use with *SpecText* master guide specifications, *Editspec* is a collection of section-editing tools and supplementary documents. *Editspec* is included with a subscription to

SpecText and works with the user's word processing program (Microsoft Word or Corel WordPerfect for Windows), purchased separately. *Editspec* offers the following functions:

- Create headers/footers in selected or all sections
- Insert and delete specifier's notes
- Insert project notes
- Insert [OR] statement
- Insert units of measure
- Accept inch-pound measurements
- Accept metric measurements
- Delete selected paragraphs
- Find and edit bracketed choices
- Delete brackets and fill-in-the-blanks
- Mass macro applicator

For further information, contact:

Construction Sciences Research Foundation, Inc.
The CSRF Support Center
P.O. Box 926
Bel Air, MD 21014-0926
410/838-7525 or 877/SPECTXT (773-2898)
www.spectext.com

KNOWLEDGE-BASED, COMPUTER-ASSISTED SPECIFICATIONS

As discussed above, there have been several attempts to harness the power of the computer to interact with a large database of knowledge about construction specifications and products. Bowne Information Systems' *Conspec,* Corbel's *Super-Spec,* McGraw-Hill/Sweet's *SweetSpec,* and ARCOM's *Masterspec Q&A* all attempted and failed to produce knowledge-based, computer-assisted specifications product programs. These programs were visionary, but their cost and the limitations of users' PCs and computer-to-computer communications doomed them.

In most knowledge-based specifying systems, the master text is not seen by the person answering the questions. The clerical aspects of text editing are not the concern, only the end product. As a result, the master is always resident in unaltered form, and it can be updated without affecting project-specific specifications if and when the specifier chooses.

The technical content of knowledge-based specifications programs is written by very knowledgeable specifiers. This knowledge base is an impressive resource but it is not a substitute for the specifier's knowledge. It is an aid that requires the specifier to be knowledgeable about construction specifications principles and practices and construction technology. A session with a demonstration version of a knowledge-based specifications program will make this apparent.

The greatest advantage of knowledge-based specifications programs is productivity when a knowledgeable user is combined with a comprehensive, well-prepared, up-to-date knowledge (data) base. One consideration in choosing a knowledge-based specifications program is its supplementary documentation of products and product selection. The more substantial these supplements are, the more fully the specifier will understand the editing options and the greater the satisfaction with the completed specifications.

Pressure to integrate production of drawings, specifications, cost estimates, and planchecking continues to build. Several organizations are pursuing the development of an all-encompassing computer program that will create drawings, specifications, and related documents based on parametric decisions of the architect or engineer. This search has been going on for over 30 years. Perhaps if the power of PCs becomes sufficient, if high-speed Internet telecommunications becomes adequate, if links between the very different software of CAD, specifications, and other information are created in usable form, and if architects, engineers, and specifiers are sufficiently trained to use the program, the long-promised, all-encompassing computer-assisted design (CADD) program will become a reality.

SpecLink+®

In 1994, Building Systems Design (BSD) of Atlanta, Georgia, developed *SpecLink,* a new specifications-intelligent linked system that relied on a fully integrated relational database rather than word processing files for its fundamental data structure. In addition to a master file containing hundreds of specification sections, the database included thousands of links between sections that multiply the productivity of the specifier. Instead of editing master specifications for a specific project by deleting unwanted or inappropriate text from the master, the user of *SpecLink* selected only text that was needed. As text selections were made, in any order, the links built into the database automatically included related text and excluded incompatible options, whether in the same section or in a related one. Only text that was selected either by the specifier or automatically by the program was assembled into the project specification.

SpecLink continued to develop, and the current program, *SpecLink+,* is enhanced to run under 32-bit PC-based operating systems rather than the 16-bit original version. Presently, the *SpecLink+* database contains over 700 master specification sections and over 110,000 data links. The best way to understand *SpecLink+* is to contact the publisher and obtain a demonstration version of the program.

A major criticism of *SpecLink+* is that a current subscription is required to access past projects. BSD's response is that *SpecLink+*-produced text may be exported as Microsoft Word-compatible Rich Text Format (RTF) documents. In fact, many users export files in this manner for final editing and formatting, although *SpecLink+* includes extensive capabilities for section paragraph numbering and page formatting.

For further information, contact:

Building Systems Design, Inc.
3520 Piedmont Road NE, Suite 415
Atlanta, GA 30305
888/BSD-SOFT (888/273-7638)
www.bsdsoftlink.com

Linx

Linx is an add-on to *Masterspec* published by ARCOM for the AIA. There is a separate subscription fee for *Linx. Linx* works with the full-length version of *Masterspec.*

ARCOM describes the use of *Linx* as the first major step in producing project-specific construction specifications. Basic product selections and basic editing decisions are made in *Linx.* When completed, the *Linx*-generated files are exported to the specifier's word processing software, which should include the *Masterworks* add-in software, for final editing and publication. This is different from *SpecLink+,* where the entire specifications production and publication process is performed.

Linx optimizes the edit-by-delete structure of *Masterspec* with a question-and-answer, criteria-based approach to editing. As the specifier answers questions, irrelevant specification text is eliminated. When a specification element is removed,

Linx automatically marks for deletion that element's subordinate text within the section. *Linx* also knows where structural and semantic links occur in any particular specification paragraph. It targets scattered requirements related to the choices made by the specifier and marks those for deletion as well. When the editing is completed, all text selected for deletion is marked as strikethrough text, and the specifier reviews the section prior to actual text deletion.

Linx-edited sections are saved and further edited by the specifier using the global and section text-specific tools of *Masterwork.* Since the resulting files are in Microsoft Word or Corel WordPerfect, there is no need to use a proprietary software editor, as required by *SpecLink+.* Access to archived files also is not dependent upon having a current subscription to the program since the files are in word processing file format.

For further information, contact:

ARCOM, Inc.
332 E. 500th Street
Salt Lake City, UT 84111
801/521-9162 or 800/424-5080
www.arcomnet.com

Precautions

A major problem associated with computer-assisted specifications is that novices are being required to use the programs to produce project-specific specifications with little training in the program and virtually no formal knowledge of the principles and procedures for construction specifications writing. Unlike online medical libraries and legal case law libraries, where only trained doctors and lawyers search the data to arrive at solutions and make decisions, computer-assisted specifications allow anyone to edit them.

Those who interact with the computer-assisted specifications program should be well versed in construction contracts, bidding requirements, building products, materials standards, products evaluation and selection, and the principles and procedures of construction specifications writing before being allowed to use the program. Attending training sessions conducted by the software publisher is also very important and, despite its substantial cost, could be offset by rapid attainment of productivity and avoidance of costly errors.

Chapter 20

Preliminary Project Description

UNIFORMAT®

CSI/CSC *MasterFormat*®—1995 edition organizes construction information according to construction requirements, products, and activities. This is the familiar 16-division format. *MasterFormat* is a uniform format for standard organization of construction specifications and other bidding and construction contract documents in the Project Manual. *MasterFormat* is also used for organizing detailed cost estimates, for filing of product information and other technical data, for identifying drawing objects, and for presenting construction market data. These uses occur during Design Development, Contract Documents production, bidding, and construction. Prior to Design Development, *MasterFormat* is cumbersome and overly detailed for schematic design and the early portion of design development.

Recognizing the shortcomings of *MasterFormat* and looking to models of construction information organization outside of North America, the AIA in the early 1970s developed for the General Services Administration (GSA) of the federal government a classification system for construction based on building and site elements. The resulting document was titled *UniFormat* and was primarily intended to standardize estimates and facilitate cost analysis and cost control. *UniFormat* has been used by federal agencies and by R.S. Means Co. in their *Means Assemblies Cost Data* for these purposes.

In 1988, ASTM Building Economics Subcommittee E06.81 formed a task group that included the GSA, CSI, R.S. Means, and the U.S. Department of Defense Tri-Services Committee. This task group had the responsibility to update *UniFormat* and have it formally approved as an ASTM standard.

The task group developed and updated *UniFormat,* and CSI published it as an "Interim" version for "trial use and comment." It was included in the 1992 edition of the CSI *Manual of Practice.* Meanwhile, the National Institute of Standards and Technology (NIST) developed an alternative version, titled *UniFormat II,* which was also published in the first half of 1992. The two formats were similar but not identical.

ASTM balloted and accepted *UniFormat II* as the basis for the formal standard. In late 1992, ASTM published ASTM E 1557, titled *Standard Classification for Building Elements and Related Sitework: UniFormat II.* This standard has subsequently been updated and, at the time of this writing, is designated ASTM E 1557-02. An abstract of this standard may be found online at *www.astm.org,* search for "ASTM E 1557."

UniFormat II added elements and expanded descriptions of many existing elements, making it suitable not only for cost analysis and control but also for project management and reporting at all stages of building life cycle-planning, programming, design, construction, operations, and disposal.

The working group, including representatives from CSI, the American Association of Cost Engineers, the American Society of Professional Estimators, GSA, the Naval Facilities Engineering Command, the U.S. Air Force, and the U.S. Army Corps of Engineers, continued development of *UniFormat II,* and ASTM published the result as ASTM E 1557-97. This standard has been revised and reissued as ASTM E 1557-02.

The *UniFormat II* working group determined the following:

- A need existed for a formally established classification system based on building systems and assemblies.

- The classification system should be based on *UniFormat.*

- The classification system should be expanded to encompass all types of construction rather than limited to building construction.

- The classification system should be coordinated with *MasterFormat.*

In 1999, a proposed update of *UniFormat II* was published by NIST as NISTIR 6389, which included expansion of the classifications to a fourth level. This document, dated 1999, is available for download and printing at *www.uniformat.com/6389.pdf*. Verify that it is current since the comparable ASTM E 1557 was updated in 2002.

In 1995, CSI and CSC began revising *UniFormat* to align it with ASTM E 1557 and to coordinate *UniFormat* with the 1995 edition of *MasterFormat*. In 1998, CSI/CSC formally published *UniFormat*, which had the following changes:

- Collected titles that were not physical building parts were relocated near the beginning of the document, with appropriate titles such as "Project Description."
- The format was revised to allow its use for classifying information necessary to solicit proposals and to contract for design-build projects.
- Category D "Services" and Category E "Equipment and Furnishings" were reorganized to reflect more of a systems approach, with generic functional categories.
- The title of Category F was changed from "Other Building Construction" to "Special Construction and Demolition."

The CSI version, *UniFormat*, and the ASTM E 1557 version, *UniFormat II*, have the same numbering scheme, and titles from Level One through Level Three under each element are identical. However, the CSI version added a Level Four for greater detail, added the Level One "Project Description" category, and added category Z - "General."

In 1999, NIST published a proposed revision to *UniFormat II* that expanded to Level Four and added category Z - General. Readers should obtain current copies of CSI/CSC *UniFormat*, ASTM E 1557, and the NIST document and study the commentaries for instructions for their use for production of project-specific documents.

Level One and Level Two categories of *UniFormat* are:

PROJECT DESCRIPTION

10 PROJECT DESCRIPTION
20 PROPOSAL, BIDDING AND CONTRACTING
30 COST SUMMARY

A SUBSTRUCTURE

A10 FOUNDATIONS
A20 BASEMENT CONSTRUCTION

B SHELL

B10 SUPERSTRUCTURE
B20 EXTERIOR ENCLOSURE
B30 ROOFING

C INTERIORS

C10 INTERIOR CONSTRUCTION
C20 STAIRS
C30 INTERIOR FINISHES

D SERVICES

D10 CONVEYING
D20 PLUMBING
D30 HEATING, VENTILATING AND AIR CONDITIONING (HVAC)
D40 FIRE PROTECTION
D50 ELECTRICAL

E EQUIPMENT AND FURNISHINGS

E10 EQUIPMENT
E20 FURNISHINGS

F SPECIAL CONSTRUCTION AND DEMOLITION

F10 SPECIAL CONSTRUCTION
F20 SELECTIVE DEMOLITION

G BUILDING SITEWORK

G10 SITE PREPARATION
G20 SITE IMPROVEMENTS
G30 SITE CIVIL/MECHANICAL UTILITIES
G40 SITE ELECTRICAL UTILITIES
G90 OTHER SITE CONSTRUCTION

Z GENERAL

Z10 GENERAL REQUIREMENTS
Z20 CONTINGENCIES

Obtain a copy of *UniFormat* from CSI that contains a "Master List of Numbers, Titles, Explanations and Related *MasterFormat* Numbers" for Level Three and Level Four headings, as well as an "Application Guide" and "Key Word Index." Expect an update after *MasterFormat*-2004 edition is published with its 49 division format. Contact *www.csinet.org.*

PRELIMINARY PROJECT DESCRIPTION

According to the CSI *Manual of Practice* (1996), the "Fundamentals and Formats" module FF/180, a Preliminary Project Description (PPD) is prepared to describe the scope and relationships of major project elements and is organized in terms of building systems and site components. PPDs are usually prepared during the initial design phases of a project. According to AIA Document B141, *Standard Form of Agreement between Owner and Architect,* the two phases of initial design are the "Schematic Design Phase" and the "Design Development Phase." According to EJCDC Document E-500, *Standard Form of Agreement Between Owner & Engineer for Professional Services,* these phases are the "Study and Report Phase" and the "Preliminary Design Phase." For consistency and brevity, the AIA terminology will be used in this discussion.

The purpose of the PPD, when first conceived, was preliminary cost estimating at the Schematic Design phase, when drawings were conceptual or did not exist. This purpose has grown. The PPD is now used to convey to the owner and architect or engineer knowledge of the various components and systems proposed for the project. PPDs accompany Requests for Proposal (RFPs) for design-build projects. *UniFormat* is used as the format for PPDs.

PPDs are particularly beneficial as a design management tool because they cause each design discipline to conceptualize baseline building systems early in the design process. Although the descriptions are provided in general or simple terms to suit the preliminary nature of the design, information can be communicated effectively to all involved in the project. PPDs enable more realistic scope descriptions and result in more adequate budgets for the project than mere square foot estimates based on mythical similar buildings in the general region. It is hoped that the result will be more accurate costing due to fewer unknowns, therefore reducing the need for large contingencies and allowances and the probability of cost overruns.

In preparing the PPD, the design team should think through the requirements of the project and document decisions and design criteria in broad terms. Refinements and changes can be made during development of the design as more information becomes available, design issues are resolved, and cost data is applied. A PPD may include both performance criteria and product descriptions, depending on the development of the design and the state of design decisions. It should provide information necessary for preparation of preliminary cost estimates, time schedules, and initial value engineering studies.

Following *UniFormat,* the PPD should be organized to describe groups of construction systems, assemblies, and components in a logical sequence from the ground up and from the outside in. The PPD should include proposed design solutions and decisions made or to be made, and it should serve as a master guide for the project from which subsequent project documentation will flow. Documentation through Schematic Design and most of the way through Design Development should:

- Reflect the building program in physical terms by providing preliminary descriptions of building products, assemblies, and systems, including prospective design solutions

- State the architect/engineer's understanding of the project delivery method (e.g., single prime contract, multiple prime contracts, construction management, design-build, design-bid-build, and phased construction), including considerations for the General Conditions of the Contract and construction contract administration

- Describe alternative products, assemblies, and systems that meet both the design criteria and budget constraints

- Document the basis for time schedules

- Document the basis for construction costs

- Aid the design process by documenting design decisions made and to be made, including responsibilities for decisions by project team members

- Provide a reference point on which subsequent decisions and phases of the project will be based, including the owner's formal acceptance and approval of the Schematic Design

There are two principal methods for defining building systems:

- System Descriptions: List and describe products, assemblies, and systems, including restrictions and possible alternatives that meet the design criteria

- Performance Criteria: List and describe the project as a system of components rather than actual products, with supplementary descriptions and including all pertinent criteria so that unsuitable systems can be readily identified and eliminated from consideration

Each of the PPD elements should be expanded as more information becomes available through the design development and value engineering processes. For example, a floor system under consideration in the B10 Superstructure element might be described as a 2-hour fire-rated, composite steel beam, steel deck and concrete slab system in 20-feet by 25-feet (6100 mm by 7600 mm) bay dimensions, capable of supporting a 75 psf (3.6 kPa) live load. With that information, a preliminary cost estimate for the proposed system and alternative systems could be developed for budget and value engineering purposes. The systems best suited to the design and to the project budget could then be identified and the PPD validated for the described system or the PPD could be modified to reflect an alternative system better suited for the project.

The floor system described above includes products from several different *MasterFormat* divisions that are combined under a PPD heading. This reflects the fact that the PPD describes *systems* rather than products typically categorized in the divisions of *MasterFormat*. The floor system example includes the following items from *MasterFormat*-1995 Edition:

- Reinforced portland cement concrete: Division 3
- Composite steel beams and steel decking: Division 5
- Sprayed fire-resistive material: Division 7

Many products will be identified under several PPD headings. Portland cement concrete for foundations, slabs-on-grade, paving, drainage structures, and light standard bases will be identified in the applicable PPD headings. Refer to the *UniFormat* publication by CSI for the "Application Guide" that describes this process further and the "Master List of Numbers, Titles, Explanations and Related *MasterFormat* Numbers" that cross-references *UniFormat* Level Three and Level Four headings with *MasterFormat* section numbers and titles.

A major benefit of performing an economic analysis based on an elemental framework instead of on a product-based classification is the reduction in time and costs for evaluating alternatives at an early stage in the design process. This should encourage more analyses of initial and life cycle costs and should optimize design and cost decisions for building elements. Refer to narratives in the NISTIR 6389, referenced above.

See *Appendix C—Sample Preliminary Project Description.*

COMPUTER-ASSISTED PPDS

Building Systems Design (BSD) has produced a computer-based program to produce PPDs and proposal documents related to design-build project delivery. The program, titled *PerSpective®,* uses the same software engine as BSD's computer-assisted specifications program, *SpecLink+.* This means that the program is PC-based and runs on top of a customized database program. It produces documents in *UniFormat.*

Three types of documents are produced by *PerSpective:*

- *Building Design Criteria:* An owner's RFP
- *Proposed Performance Specifications:* A designer-builder's proposal (response to RFP)
- *Instructions for Construction:* Short, descriptive specifications of proposed construction

PerSpective, according to BSD, includes the following features to expedite production of PPDs at the schematic design and design development phases of the design:

- Global switch to set the function of the document (Owner's RFP, Design-Builder's Proposal, or Instructions for Construction)
- Reuse of previous projects without danger of omission
- Global page formatting, one setting for all headers and footers, margins, etc.
- Automatic paragraph renumbering
- Edit-by-select rather than edit-by-delete—no data actually disappears
- Internal links to coordinate related text and prevent contradictions between elements
- Easy comparison of two project files
- Notes to editor with explanations of issues
- Seamless, automatic updating of reference standards without loss of user edits

For a detailed description of *PerSpective,* download a portion of the User Manual, titled "Chapter 2—Using *PerSpective®* Effectively," from the list of monographs associated with *PerSpective* on the BSD website: *www.bsdsoftlink.com.*

Chapter 21

Outline and Shortform Specifications

OUTLINE VERSUS SHORTFORM SPECIFICATIONS

PPDs, discussed in the previous chapter, and outline specifications have separate and distinct functions. The same is true of outline specifications and shortform specifications. Outline specifications are not suitable for construction contract documents and are intended to be used during development of the design, somewhat like the PPD but in a different format. Shortform specifications are suitable for construction contract documents and include more than design development information.

OUTLINE SPECIFICATIONS

Design Development (Outline) Specifications

After the project design has progressed from the Schematic Design phase to the Design Development phase, outline specifications become very useful and may often be required by the agreement between the owner and the architect or engineer. AIA Document *B141 Owner-Architect Agreement* requires "drawings and other documents to fix and describe the size and character of the project as to architectural, structural, mechanical and electrical systems, materials, and such other elements as may be appropriate," along with adjustments to the preliminary cost estimate. EJCDC Document *C-700* requires production of final design criteria, preliminary drawings, outline specifications, and written descriptions of the project, along with a revised estimate of the probable project cost. Outline specifications can meet the requirement for written descriptions.

Outline specifications aid in the design process and may be the basis for refined cost estimates, product and equipment schedules, and value engineering studies. They also may serve as a checklist for the project team when it selects products and methods during the Design Development phase. They are a means of communication among members of the project team and between the design professionals, the owner, and the construction manager.

Outline specifications help create a controlled decision-making process and encourage clarity in those decisions since they must be committed to concise written statements. Well-prepared outline specifications establish the understandings of the design team at the conclusion of Design Development.

PRELIMINARY PROJECT DESCRIPTIONS VERSUS OUTLINE SPECIFICATIONS

As stated above, PPDs and outline specifications have separate and distinct functions. The PPD, together with the schematic design drawings, define the components and systems proposed for the project. Both of these help the owner and the design team reach a more complete understanding of the project. The PPD serves as a basis for discussing building and site systems early in the life of the project but is based on less developed design criteria.

The PPD is organized in terms of building systems and site components. The outline specifications take information from the PPD and revise the format into the divisions of CSI *MasterFormat.* Information about building systems and site components is gathered from various PPD "elements" and expanded.

Outline specifications aid in the design process and help form the basis for revised cost estimates and schedules. As the design process continues, they become the basis for preparation of the construction specifications during the Construction Documents phase of the project. Outline specifications serve the project team as a checklist for choosing products and methods for later incorporation into the Project Manual. Properly developed outline specifications establish criteria for the final Contract Documents specifications. They also help to eliminate fragmented decision making, which can affect previous decisions and cause unnecessary changes and extra work.

Outline specifications briefly list materials, manufactured products, finishes, and methods to be used for the project. Their primary purpose is to provide information about the quality of products, unusual fabrication and installation requirements, and unusual quality assurance measures, such as mock-ups and field inspections and tests. This should provide descriptions of the design for use in preparing estimates of probable construction cost. Typically included are:

- Requirements for specified and acceptable manufacturers, materials, manufactured units, equipment, components, and accessories.

- Unusual requirements for material mixes, fabrications, and finishes.

- Unusual requirements for preparation, installation, erection, and application of products.

- Unusual general requirements for Part 1 - General of the section, such as submittals, samples, mock-ups, preconstruction conferences, special warranties, extra stock, and maintenance materials. Unusual requirements for qualifications for manufacturers, fabricators, or installers may also be included.

- Other requirements that may have a significant impact on quality, cost, or construction sequencing and duration.

Owners vary considerably in their familiarity with and understanding of building elements, products, and standards. Outline specifications help the owner overcome this difficulty. They also serve a variety of purposes for other entities, including lenders, estimators, construction managers, and code officials.

Outline specifications are more than a table of contents for the specifications to be produced during the Contract Documents phase. They are a record of decisions about specific materials, equipment, systems, methods, manufacturers, and special fabrication requirements. Outline specifications are not construction documents; their preliminary and incomplete nature makes them unsuitable for true construction specifications. Outline specifications should accomplish the following:

- Encourage and record product selection decisions early in the documentation process

- Assist accurate estimating of probable construction costs for budget management

- Assist in preparation and updating of preliminary construction schedules by clarifying the scope of products and special construction

requirements early in the documentation process

- Assist the owner in understanding what materials and systems are proposed

- Coordinate construction documentation of various design disciplines

- Provide an outline for preparation of the Project Manual, including construction specifications

General installation procedures, such as references to manufacturers' instructions, are not necessary in the outline specifications. However, identification of specific installation methods such as "gravel-surfaced 4-ply built-up asphalt roofing" or "thinset ceramic mosaic tile on anti-fracture membrane" are important because they have a direct relationship to quality levels and costs.

Information in the outline specifications should be sufficient for the reader of these specifications and the Design Development drawings to determine the types of products and their general locations on the project. Finish, door, and equipment schedules may be included on the drawings or in the outline specifications to indicate where various products will be used.

During the Design Development phase, the architect or engineer should request the owner's instructions for bidding, bonds, and insurance, General Conditions of the Contract, and construction contract administration requirements. If available, the outline specifications should include this requested information plus an abbreviated Division 1 - General Requirements. This will record the understanding of contract arrangements and provide an opportunity for the owner to review and confirm or request changes before production time is spent developing the final bidding and contract documents.

If it is intended to be used, Section 01100 - Summary should be edited in order to summarize the project and general scope of the work under the contract. It should also address the following items as applicable:

- Separate or multiple contracts
- Future work
- Coordination for fast-track and multiple prime contract requirements
- Phased construction
- Owner and contractor use of new and existing facilities

- Allowances, alternates bids, and unit prices
- Project scheduling
- Applicable codes
- Materials-testing responsibility
- Owner-furnished, contractor-installed items
- Code search reports, zoning requirements, fire protection reports, and other specific information that may be required for inclusion as appendices in outline specifications prepared for use by planning authorities or other public entities
- Access restrictions and security requirements
- Environmental restrictions and requirements

Outline Specifications Format

MasterFormat section numbers and titles are used for the outline specifications, typically using the section number and title from the level intended for the Contract Documents specifications. The three-part *SectionFormat* is not necessary for outline specifications, but listing the information in paragraphs following the same sequence and titles of the three-part contract documents specifications simplifies production during the contract documents phase.

See Appendix D for sample outline specifications.

Outline Specifications for Design Standards

A developing use for outline specifications is communication of design standards. Rather than produce and maintain full-length construction specifications, public agencies and corporations are producing outline specifications that incorporate basic information about products, installation, and quality assurance. These are furnished to architects and engineers performing design services for the owner.

SHORTFORM SPECIFICATIONS

Introduction to Shortform Specifications

"Less is more" is a familiar saying to architects. It is attributed to Ludwig Mies Van Der Rohe (1886–1969), an architect noted for his strong influence on American architecture, particularly the minimalistic "modern" style. To paraphrase his point in regard to construction specifications, reducing the specifications may yield more easily understood documents, with essential requirements uncluttered by superfluous text, resulting in better bids and lower construction costs.

Another American, Mark Twain, once wrote, "I didn't have time to write a short letter, so I wrote a long one instead." The point, applied to construction specifications, is that it takes greater effort and time to write concisely than the customary manner found in "longform" specifications. An informal survey of independent specifications consultants at a national meeting resulted in a consensus that "shortform" specifications take just about as long to produce as longform specifications. Whittling down a longform specification section and revising its language and specifications methods to transform it into a shortform specification actually takes more time than editing the longform specification for a typical project.

Shortform specifications attract great interest when they are presented, especially from those who do not appreciate highly detailed documents with very specific requirements for quality assurance and a well-brewed alphabet soup of ASTM, ANSI, AWI, ASHRAE, BHMA, SDI, and TCA reference standards. One of the hindrances to producing true construction specifications for light construction and simple projects is the burdensome nature of typical specifications. Project architects and engineers for these projects are unfamiliar and uncomfortable with "book specs" and prefer to get by with drawing notes and a sheet or two of "general notes" to satisfy plancheckers and supposedly fend off construction problems. The idea of more manageable specifications—shortform specifications—is very attractive for industrial shell buildings, commercial office interiors, residential, renovation, and other projects of limited scope and with less aggressive bidding and construction scenarios.

Credit should be given to the late Ben John Small, AIA, FCSI, for development of the concept of shortform specifications. He was an author, lecturer, and expert construction specifications writer in New York. His expertise is honored in the annual Ben John Small Memorial Award given each year by CSI to a professional member who has attained special proficiency and outstanding stature as a practicing specifications writer.

Another longtime advocate and expert in shortform specifications, and 1971 recipient of the Ben John Small Memorial Award, is Herman R. Hoyer, PE, Hon. CSI. In his 40+ years of practice as a construction specifications writer, Mr. Hoyer regularly used shortform specifications for construction contract specifications. He demonstrated conclusively that shortform specifications can be used not only for residential and light-commercial construction but for some heavier commercial and institutional construction as well.

Mr. Hoyer developed a successful set of shortform standard specifications for the San Francisco Bay Area Rapid Transit District that included not only architectural specifications but also an extensive list of site, rail transit, plumbing, mechanical, and electrical specifications. He proved that even public works construction can be contracted, under certain conditions, using shortform specifications.

In the simplest terms, shortform specifications are specifications reduced to the shortest length possible without reducing their effectiveness and without sacrificing any essential ingredients. They can even be shortened sufficiently to be reproduced in the construction contract drawings as "sheet specs." Unlike the outline specifications discussed previously in this chapter, shortform specifications are intended to be legally enforceable construction contract specifications.

The conclusion: Commercial and institutional construction, and simple residential construction, can likewise use shortform specifications—*if* the specifier knows how to write them and *if* the requirements of the project are suitable.

What is not appreciated generally—and this is where the quote from Mark Twain applies—is that it takes more than merely abridging or condensing a longform specification to create a shortform specification. Abridging a longform specification section invariably leads to omitting important elements or getting into dilemmas over what is important and unimportant.

In many ways, this is similar to the problems faced by NASA when the disasters of the space shuttles *Challenger* and *Columbia* occurred and when two Mars exploration missions failed. NASA had adopted the concept of "faster, better, cheaper" for its programs. A saying developed: "You can have only two. Which are you willing to give up?" Shortform specifications, to the uninitiated, hold similar promises but they also include a great risk of problems during construction, although certainly not of the same magnitude as NASA's.

Shortform Specifications Writing

Shortform specifications writing requires application of all the specification writing principles and procedures discussed in earlier chapters. Shortform specifications need to be written as such, and not merely as condensed longform or conventional specifications. This is why shortform specifications are discussed here, at the end of this book: *shortform specifications writing requires knowledge of how to write construction specifications.*

Shortform specifications writing needs to be approached with a single-minded purpose: make the specifications as concise as possible. This requires deviations from CSI formats and some recommended specifications writing procedures. However, it generally means choosing methods of specifying and applying other specifications writing procedures in optimum ways that reduce the amount of text.

Shortform specifications require fervent application of a basic specification writing principle: "Say it once and in the most appropriate location." Conciseness and clarity need to guide the writing process for shortform specifications.

Formats for Shortform Specifications

Contrary to recommended practices for typical master specifications, which advocate pre-editing specifications by using more narrowly focused specifications (narrow-scope specifications), shortform specifications should combine specifications information into what CSI *MasterFormat* calls Level Two or broad-scope sections. This is done to avoid repeating information. For example, rather than having separate narrow-scope sections such as Section 08211 - Flush Wood Veneer Doors, Section 08215 - Plastic Laminate-Faced Wood Doors, and Section 08218 - Stile and Rail Wood Doors, use broad-scope Section 08200 - Wood Doors.

In Division 15 - Mechanical and Division 16 - Electrical, make use of "basic materials and methods" sections to specify products that are used generally for plumbing, HVAC, and electrical systems. This eliminates repetition and is generally recommended for Division 15 and Division 16 specifications for even longform specifications. In Division 4 - Masonry, a basic materials and methods section can be used if there are several types of masonry that use similar setting mortar, grout, and accessories.

Perhaps the most radical change from longform specifications is to abandon the three-part format of CSI *SectionFormat*. The CSI *Manual of Practice* (1996) defines 10 articles for shortform specifications:

A. SYSTEM DESCRIPTION

B. SUBMITTALS

C. QUALITY ASSURANCE

D. WARRANTY

E. COMPONENTS

F. FABRICATION

G. PREPARATION

H. INSTALLATION

I. FIELD QUALITY CONTROL

J. SCHEDULES

With a more limited set of article headings, it is easier to focus on the topics to be specified. *Note:* There should be no obligation to address all the topics and use all the listed articles. Only use what is applicable and what is necessary.

Within a shortform specification section, the level of detail should be kept as broad as possible. If a high level of detail is required, then the use of shortform specifications should be questioned. Under article headings, shortform specifications should have only one or two paragraph levels.

Methods of Specifying for Shortform Specifications

Refer to Chapter 6 for a discussion of the four methods of specifying.

Some methods of specifying are more conducive to concise text than others. The reference standard specifying method is particularly useful, as is the proprietary method. The descriptive specifying method requires extensive text to adequately specify products and installation and is generally not helpful. The performance specifying method can be used for shortform specifications under some conditions.

The reference standard method is a concise way of specifying. For example, specifying wall tile installation as "Wall Tile Installation: TCA Handbook Method W243" incorporates by reference a huge amount of information regarding materials and procedures. Specifying "Wood Doors: AWI/AWMAC Architectural Woodwork Quality Standards, Section 1300, Custom Grade, Medium Density Overlay, particleboard core" is equivalent to a half-page detailed description. Specifying "Conductor Heads: SMACNA Architectural Sheet Metal Manual Figure 1-25F, 16 oz. (.55 mm) copper" not only incorporates by reference fabrication and installation information but can eliminate a detail on the drawings. The specifier must know

the standard and must
under the standard.

The proprietary specif
cise way of specifying
Asphalt Roofing: Man
incorporates a dozen p
manufacturer's catalog p
ASTM reference standards. "Surface Metal
ways: Wiremold 4000 System with factory-formed fittings and ivory polyester topcoat over ivory primer, capable of being field-painted" provides enough information for pricing (bidding) and purchasing required products.

The performance specifying method can result in very lengthy specifications for products and systems. It requires substantial text to specify required performance characteristics and the methods for validating that required performance is achieved. However, specifying performance in general terms can be concise. For example, temporary construction can be specified by the performance method: "Temporary Connections and Fees: Contractor shall arrange for services and pay all fees and service charges for temporary power, water, sewer, gas and other utility services necessary for the Work."

The descriptive method of specifying is generally unsuitable as the primary method of specifying. However, as a supplement to other methods, descriptions are appropriate if the information does not duplicate what is included in the referenced standard or the manufacturer's product data. The above examples for "wood doors" and "surface metallic raceway" include descriptive text in addition to the reference standard and the manufacturer's product name.

Language for Shortform Specifications

Refer to Chapter 14 for a discussion of specifications language.

The specifications should include, in a Series 0 introductory document to the Project Manual or in Division 1 in Section 01425 - Definitions and Interpretations, an explanation of language used in the Specifications (see Exhibit 21-1).

Similarly, terms and phrases used in the specifications should be defined in the Supplementary Conditions of the Contract to expand on the definitions in a typical document such as AIA *A201 - General Conditions of the Contract for Construction* or the terms and phrases should be specified in Division 1 in Section 01420 - References or Section 01425 - Definitions and Interpretations. Defined terms can substitute for several other words and

FICATIONS LANGUAGE

pecifications are written in a modified brief style consistent with clarity. In general, the words "the," "shall," "will," and "all" are not used. Where such words as "perform," "provide," "install," "erect," "furnish," "connect," "test," or words of similar meaning are used, it shall be understood that such words include the meaning of the phrase "The Contractor shall." The requirements indicated and specified apply to all Work of the same kind, class and type, even though the word "all" is not stated.

Frequently, Specifications are written using "streamlining" as recommended in the CSI *Manual of Practice*. Streamlining employs a colon (:) as a symbol for the words "shall be," "shall have," "shall conform with (or to)," or words of similar meaning appropriate to the context of the statement. For example, "Portland Cement: ASTM C 150, Type I" means the same as "Portland cement shall conform to ASTM C 150, Type I."

Exhibit 21-1. Sample explanation of specifications language.

clarify the intent of the Specifications. See Appendix B, Division 1, Section 01420 - References, for example text.

Although use of these terms seems initially to result in small reductions in text quantity, the savings add up and become substantial. Streamlined writing is the most effective way to reduce the quantity of text in construction specifications. In order for concise shortform specifications to be achieved, careful attention must be paid to the details of language and terminology.

Location of Information for Shortform Specifications

An obvious way to reduce the text in specifications is to put the information on the drawings. This does not mean substituting lengthy descriptions in the legends of keynotes or producing several sheets of General Notes to substitute for Division 1 - General Requirements. It means carefully using schedules and legends on the drawings that list information about products rather than sentences and paragraphs in the specifications that explain the information.

A finish materials schedule can identify the manufacturer, brand name, pattern, and color of products. Presented in tabular form, the information is very concise, especially when products come in multiple colors. This is very appropriate when the specifications are written using the proprietary method, where product options are very limited or prohibited. Similarly, door schedules, door hardware schedules, paint schedules, equipment schedules, and fixture schedules all may contain, within the schedule or in a related legend, sufficient information so that the specifications can simply refer to the manufacturer, catalog number, series, pattern, and color "as indicated on the Drawings."

Division 1 specifies general requirements applicable to all Division 2 through 16 sections of the specifications. Applying the principle "Say it once," Division 1 should include sufficient information so that what would normally be included in Part 1 - General of a longform specification section can be deleted except in extraordinary cases. By their very nature, projects using shortform specifications should not have administrative and quality assurance requirements. It is critical that Division 1 be well prepared and balance the need for brevity and completeness.

In Division 2 through 16 sections, cross references to Division 1 should be avoided. Understand that the specifications are written as from the owner to the contractor. Considerations for subcontractors, trades, and product suppliers should not be included, especially for shortform specifications. Since the contractor has read and should be familiar with the contents of Division 1, specifying submittals in a Division 2 through 16 section should not require the statement "Makes submittals as specified in Section 01330 - Submittals Procedures."

To reduce the size of Division 1, rely upon the General Conditions of the Contract and the Instructions to Bidders. Review AIA *A201 - General Conditions of the Contract for Construction* in Appendix E or review EJCDC *C-700 - Standard General Conditions of the Construction Contract* in Appendix F. Obtain copies of either AIA *A701 - Instructions to Bidders* or EJCDC *C-200 Guide to Preparation of Instructions to Bidders* and review these documents as well. It is apparent that many requirements typically specified in Division 1 are covered in these other contract documents. However, Division 1 amplifies and supplements what is included in these standard documents. The Supplementary Conditions of the Contact and the Supplementary Instructions to Bidders may be more appropriate locations for amplifying and supplementing information. Remember, shortform specifications are most appropriate for uncomplicated projects where extraordinary requirements are unnecessary.

For example, Paragraph 3.12 of AIA *A201 - General Conditions of the Contract for Construction* is titled "Shop Drawings, Product Data, and Samples," and it includes 10 subparagraphs containing considerable information on the submission and review of shop drawings, product data, and samples. Also in AIA *A201* is Subparagraph 4.2.7, a rather long paragraph dealing with the architect's review and approval of shop drawings, product data, and samples. So, unless the architect determines that requirements for sheet sizes of shop drawings, the number of copies to be submitted, and other procedural matters shall be specified, a Division 1 section on shop drawings, product data, and samples can be omitted.

Applications for Shortform Specifications

By their abbreviated nature, shortform specifications are appropriate for smaller, limited-scope, and less complex projects. The character of the project needs to be carefully considered.

A small project may have complex requirements for products, such as the reconstruction of a toilet room in a university classroom building. This project could require substantial dust and noise control provisions and could have relatively complicated temporary water, ventilation, power, and lighting requirements. In addition to requiring carefully controlled, selective demolition and carefully explained requirements for salvage and reuse of products, this small project could involve relatively small quantities of products such as light structural steel framing (overhead support for toilet partitions), concrete floor patching, waterproofing under ceramic tile flooring, tile, tile backing board for walls, and impact-resistant gypsum board for ceilings. Finishes on floors and walls would be included, as well as painting of gypsum board walls and ceilings. Countertops, toilet partitions, toilet room accessories, signage, custom wall mirrors, and joint sealers would need to be specified. Modifications and renewal of the plumbing systems and alterations to the fire sprinkler system could be required, as well as upgrades and replacements of electrical outlets and lighting. The quality requirements and the contracting method may mean that the specifications are just as substantial as those for a new building. In fact, due to the prob-

able lack of sophistication of the general contractor, the need for comprehensive specifications may be greater.

It is possible that a large warehouse shell building, without office interior construction, could be contracted with shortform specifications. The quantity of work may be immense, but its simplicity makes the project conducive to shortform specifications. However, if the quantities of some materials, such as roofing, are great, small factors can be subject to multiplying effects that have significant cost impacts. In such cases, blend shortform and longform specifications. Specify as appropriate to the project requirements.

Another project scenario where shortform specifications may work well is when construction is performed by the owner's own workforce. Modifications to and construction of new interior tenant space improvements may require minimal documentation. There may be "building standard products" that are stockpiled in a warehouse. The project manager and the construction superintendent may be very familiar with the products to be used. However, if portions of the work are subcontracted, documentation will be needed for the subcontracts, including specifications. In such cases, the directions of the project manager should be followed.

The persons in charge of a project may want construction specifications but do not want to be burdened by specific requirements. A project with a construction manager who wants the contract documents to only be sufficient to get a building permit is a candidate for shortform specifications. Code compliance needs to be demonstrated in the drawings and specifications. Beyond that, it becomes a matter of professional judgment for the architect or engineer and a business decision on whether to furnish greater or lesser detail in the contract documents.

Taking this further, design-build projects are candidates for shortform specifications. Remembering that shortform specifications are legally enforceable construction contract documents, while outline specifications are not, it can be argued that specifications presented in shortform format rather than in the abbreviated three-part format of outline specifications are more appropriate.

Shortform Specifications Masters

Contrary to the CSI *Manual of Practice*, it is not recommended to simply abridge a three-part master guide specification into a shortform specification version. That is, using a common master for shortform and longform specifications is not recommended. There are so many differences, as described above, between shortform and longform specifications that editing notes would not be adequate to describe all the necessary changes. Instead, office master shortform guide specifications should be developed if an office intends to produce shortform specifications.

Because longform specifications require writing methods that are not conducive to shortform specifications, such as the wordy descriptive method, condensing a longform specification requires virtual translation in many cases. The descriptive text would need to be winnowed out, leaving text based on reference standard and proprietary methods.

It would be a daunting task to convert office master guide specifications, which have been developed for competitively bid institutional projects, into shortform specifications for privately funded commercial projects where the general contractor is selected directly by the owner and not by competitive bidding. Creative architects and engineers who know word processing programs might consider developing "macro" programs to run inside the word processing program to eliminate unsuitable text. Alternatively, a specifier might develop a color scheme for text in the office master guide specification that would be automatically deleted for shortform specifications. These methods may work, but the work of developing and maintaining such master specifications would be prohibitive for all but large offices—and large offices tend not to do projects conducive to shortform specifications.

Shortform specifications should be based on shortform guide specifications. They should be developed using all of the recommendations above, including:

- Develop a concise but substantial set of Division 1 specifications that reflects the requirements of a construction contract suitable for shortform specifications. Four to six sections would suffice.

- Use industry-standard Instructions to Bidders (based on AIA *A701* or EJCDC *C-200*).

- Use industry-standard General Conditions of the Contract (based on AIA *A201* or EJCDC *C-700*) and develop office standard supplementary conditions of the contract conducive to concise specifications in Division 1 - General Requirements.

- Use a section format without parts, adapted from the 10 articles recommended by CSI and conducive to shortform specifications.

- Minimize descriptive text and write or rewrite using reference standard and proprietary methods of specifying. Utilize the performance method of specifying only when appropriate and suitable for shortform specifications.

- Include in Division 1 a section specifying the special phrases, terms, and punctuation used in shortform specifications so that the meaning and intent of the shortform text is clear by reference.

- Use schedules and tables on the drawings and in the specifications to list rather than describe products. Make it clear on the drawings where various products occur so that descriptions are not needed in the specifications. Coordinate this with keynotes and other notations on the drawings.

Commercially Available Shortform Master Guide Specifications

Shortform master guide specifications are available from several commercial sources. These are typically abridged versions of longform specifications and retain the three-part format of *SectionFormat*. If the above recommendations are followed, they need to be condensed, starting with elimination of the three-part format and use of the recommended article headings.

Commercially produced shortform master guide specifications should be scrutinized, both for what is included and what is not addressed. Shortform specifications need to include what is necessary to bid and construct the project, albeit in abbreviated form. If commercially produced specifications are written using descriptive text, the descriptions should be used as a guide to identify appropriate reference standards that cover the described attributes or to identify suitable proprietary products to specify.

Determine the level of detail represented in commercially produced shortform guide specifications. Edit the text to keep the level of detail simple. Consider the typical types of projects for which the shortform specifications will be used. Edit and adapt the text to be suitable.

Some resources for commercially produced shortform construction specifications are:

Master Shortform Specifications
Kalin Associates
154 Wells Avenue
Newton, MA 02459
617/964-5477 or 800/565-2546
www.kalinassociates.com

Easyspec: Construction Specifications, 2nd ed.
Bni Building News
www.buildersbook.com

Library of Specifications Sections, 6th ed.
Hans W. Meier (Builder's Book, Inc.)
www.buildersbook.com

Shortform specifications may be produced using BSDs *SpecLink+* by making limited selections from the available text for each section. Contact:

Building Systems Design, Inc. (BSD)
3520 Piedmont Road NE, Suite 415
Atlanta, GA 30305
404/365-8900 or 888/273-7638
www.bsdsoftlink.com

ARCOM offers a "Short Language Version" of their master guide specifications, *MasterSpec.* This should not be confused with shortform specifications. It uses CSI's three-part *SectionFormat* and addresses matters that are typically excluded from shortform specifications, as discussed above.

CSRF has published an abridged version of *SpecText,* called *SpecText II®. SpecText II's* broad-scope sections compare in length with *SpecText's* narrow-scope sections, yet the information in each paragraph is concise. Balance is maintained so that important subjects get greater attention than accessory materials and fine points of installation. No Part 1 text takes up more than 10% of any section. Division 1 is reduced to three concise broad-scope sections. *SpecText II* is a concise abridged library of master guide specifications but not true shortform specifications, as discussed herein.

Conclusion

Shortform specifications require knowledgeable, creative, and diligent use of the principles and practices of construction specifications writing, with the primary goal of minimizing the amount of text published in the Project Manual. Rather than being inferior construction specifications, shortform specifications can demonstrate the caliber of the specifier's writing ability. Shortform specifications embody the cardinal specifications writing principles that the text be "clear, correct, complete, and concise" and that the specifications "say it once and in the most appropriate location."

Appendix A

Project Manual Checklist

FOREWORD

The checklist following this introduction should be used as an example of an office standard Project Manual Checklist. It contains information for the project in general and identifies information needed specifically for bidding and construction. Also, it serves as a reminder of many items to be placed on the drawings but does not, and is not intended to, cover all particulars of a project. Rather, it should be used as a guide.

Each office or individual user must add to, delete from, or otherwise change the list as required by the project, including but not limited to applicable codes, ordinances, and standards of public authorities having jurisdiction, environmental factors of the project location, the contracting method, and practices of the architect or engineer.

Expand on information from the sample when preparing the office standard Project Manual Checklist. Extract information from typical sections in the specifications, especially Division 1 sections. For sections in Divisions 2 through 16, develop checklists for basic information for each section.

Although the checklist is intended for use during the Construction Documents (CD) phase of a proj-ect, many items should be identified and dissemi-tated during the Design Development (DD) phase. It is suggested that the document be used by the project architect or project engineer from the time DD drawings and outline specifications are started, with information added as the design progresses and design decisions are made.

USE OF THE PROJECT MANUAL CHECKLIST

Place a check mark at all items that apply to the project or underline items required for the project. Fill in blanks where information is required. Where no information is required, leave the spaces blank or the items unmarked.

At blank spaces, fill in additional information covering the particular item, such as the manufacturer's name, catalog page numbers, types, sizes, and other data.

List special materials, features, details, or other data that need to be noted in the specifications. List the names of equipment or other items required by the owner or as required to match existing items.

PROJECT MANUAL CHECKLIST

Project Title: _____

Project Identification No. (Owner) (Architect): _____

Project Location: _____

_____

Project Manual Due Dates: Preliminary Draft: _____

Plancheck Submission: _____

Plancheck Sign-Off: _____

Bid Sets Issued: _____

Construction Contract Set: _____

____ Attach copy of Project Directory to Project Manual Checklist, with names of contacts, telephone numbers, and e-mail addresses. Incorporate firm names, addresses and telephone numbers on Project Manual Title Page.

____ Review project scope with Architect and prepare preliminary Table of Contents of Project Manual. (Use Outline Specifications from DD phase, if available)

____ Bidding and Contract Requirements:

____ Issue to Owner, AIA *G612©*, "Owner's Instructions Regarding the Construction Contract, Insurance and Bonds, and Bidding Procedures"

____ Receive and incorporate information from completed AIA *G612©*

____ Regulatory Requirements

____ Applicable Codes

Building Code: _____

Plumbing Code: _____

Mechanical Code: _____

Electrical Code: _____

Fire Code: _____

____ Industrial Safety Regulations: _____

____ Accessibility Regulations: _____

____ Air Quality Regulations: _____

____ Civil (Site Construction) Specifications—Specifier's masters edited by Engineer.

____ Draft specifications to Civil Engineer for editing.

____ Incorporate spec editing by Civil Engineer.

____ Landscape Specifications—Specifier's masters edited by Landscape Architect.

____ Draft specifications to Landscape Architect for editing.

____ Incorporate spec editing by Landscape Architect.

____ Structural Engineering Specifications: Specifier's masters edited by Engineer.

    ____ Draft specifications to Structural Engineer for editing.

    ____ Incorporate spec editing by Structural Engineer.

____ Door Hardware Specifications: Specifier's master editing by Architectural Hardware Consultant; Door Hardware Schedule by consultant.

    ____ Draft specifications to Door Hardware Consultant.

    ____ Incorporate spec editing and Door Hardware Schedule by Door Hardware Consultant.

____ [_____] Specifications: By _____

    ____ Specifications received from consultant.

    ____ Make ready for reproduction.

____ [_____] Specifications: By _____

    ____ Specifications received from consultant.

    ____ Make ready for reproduction.

(add additional design disciplines as necessary)

____ Plumbing Specifications:

    ____ Specifications received from consultant.

    ____ Make ready for reproduction.

____ Fire Protection (Sprinklers) Specifications:

    ____ Specifications received from consultant.

    ____ Make ready for reproduction.

____ HVAC Specifications:

    ____ Specifications received from consultant.

    ____ Make ready for reproduction.

____ Electrical Power and Lighting Specifications:

    ____ Specifications received from consultant.

    ____ Make ready for reproduction.

____ Fire Detection and Alarm System Specifications:

    ____ Specifications received from consultant.

    ____ Make ready for reproduction.

____ Telecommunications Specifications:

    ____ Specifications received from consultant.

    ____ Make ready for reproduction.

____ Bidding Documents (if applicable): According to completed AIA *G612©*

    ____ Prepare (Invitation to Bid) (Advertisement for Bids).

    ____ Prepare Contractor's Qualifications procedures and forms (if applicable).

    ____ Prepare Instructions to Bidders (AIA *A701*) (Other) and, if applicable, Supplementary Instructions to Bidders.

____ Prepare Bid Form: Comply with instructions from (Owner) (Construction Manager).

____ Bid Information

Date: _____

Time: _____

Location: _____

____ Coordinate and incorporate Alternate Bids, Allowances, and Unit Prices.

____ Prepare and attach additional forms and certifications.

____ Prepare documents for "Information Available to (Bidders) (Contractor).

____ Geotechnical Data: Incorporate geotechnical data in Document 00320 - Geotechnical Data.

Prepared by: _____

Report title: _____

Dated: _____

Supplements: _____

Dated: _____

____ Site Survey: Incorporate geotechnical data in Document 00330 - Existing Conditions.

Prepared by: _____

Report title: _____

Dated: _____

____ Other information available to Bidders (identify, such has hazardous materials reports and project record drawings for existing construction).

____ Prepare Bid Form Supplements.

____ Schedule of Values.

____ Subcontractors List.

____ Contract Documents: According to completed AIA *G612*.

____ Obtain authorized copies of General Conditions of the Contract (AIA *A201*) (Custom General Conditions).

____ (If applicable) Prepare Supplementary Conditions of the Contract, according to instructions by Owner, Owner's legal counsel, Owner's insurance counsel and (if applicable) Construction Manager.

____ Obtain and incorporate applicable insurance, certification and other forms as required by (Owner) (Construction Manager).

____ Edit Division 1 - General Requirements.

____ Review with (Owner) (Construction Manager).

____ Coordinate Division 1 Specifications with Bidding Requirements.

____ Coordinate Division 1 Specifications with Conditions of the Contract.

____ Section 01100 - SUMMARY OF THE PROJECT: With Architect, edit master text and issue for information to Project Team.

____ Section 01210 - ALLOWANCES PROCEDURES: Edit with Architect and (Owner) (CM).

____ Sections 01230 - Contract Modification Procedures: Edit with Architect and (Owner) (CM).

____ Section 01250 - CONTRACT MODIFICATION PROCEDURES: Edit with Architect and (Owner) (CM).

____ Section 01290 - MEASUREMENT AND PAYMENT PROCEDURES: Edit with Architect and (Owner) (CM).

____ Section 01311 - PROJECT COORDINATION: Edit with Architect and (Owner) (CM).

____ Section 01312 - PROJECT MEETINGS: Edit with Architect and (Owner) (CM).

____ Section 01321 - CONSTRUCTION PROGRESS SCHEDULES: Edit with Architect and (Owner) (CM).

____ Section 01322 - CONSTRUCTION PROGRESS REPORTS: Edit with Architect and (Owner) (CM).

____ Section 01330 - SUBMITTALS PROCEDURES: Edit with Architect and (Owner) (CM).

____ Section 01410 - REGULATORY REQUIREMENTS: Edit with Architect and (Owner) (CM).

____ Section 01420 - REFERENCE STANDARDS AND ABBREVIATIONS: Edit with Architect and (Owner) (CM).

____ Section 01450 - QUALITY CONTROL: Edit with Architect and (Owner) (CM).

____ Section 01465 - CUTTING AND PATCHING: Edit with Architect and (Owner) (CM).

Section 01500 - TEMPORARY FACILITIES AND CONTROLS: Edit with Architect and (Owner) (CM).

____ Section 01580 - PROJECT IDENTIFICATION AND SIGNAGE: Edit with Architect and (Owner) (CM).

____ Section 01600 - PRODUCT REQUIREMENTS: Edit with Architect and (Owner) (CM).

____ Section 01722 - FIELD ENGINEERING: Edit with Architect and (Owner) (CM).

____ Section 01740 - CLEANING REQUIREMENTS: Edit with Architect and (Owner) (CM).

____ Section 01770 - CONTRACT CLOSEOUT PROCEDURES: Edit with Architect and (Owner) (CM).

____ Section 01783 - OPERATION AND MAINTENANCE DATA: Edit with Architect and (Owner) (CM).

____ Section 01785 - PRODUCT WARRANTIES AND BONDS: Edit with Architect and (Owner) (CM).

____ Section 01789 - PROJECT RECORD DOCUMENTS: Edit with Architect and (Owner) (CM).

____ Edit Division 2 - Site Construction: Develop Section checklist and edit master text. Incorporate Sections by consultants.

____ Edit Division 3 - Concrete: Develop Section checklist and edit master text. Incorporate Sections by consultants.

____ Edit Division 4 - Masonry: Develop Section checklist and edit master text. Incorporate Sections by consultants.

____ Edit Division 5 - Metals: Develop Section checklist and edit master text. Incorporate Sections by consultants.

____ Edit Division 6 - Wood and Plastics: Develop Section checklist and edit master text. Incorporate Sections by consultants.

____ Edit Division 7 - Thermal and Moisture Protection: Develop Section checklist and edit master text. Incorporate Sections by consultants.

____ Edit Division 8 - Doors and Windows: Develop Section checklist and edit master text. Incorporate Sections by consultants.

____ Edit Division 9 - Finishes: Develop Section checklist and edit master text. Coordinate with Finish Schedule on Drawings.

____ Edit Division 10 - Specialties: Develop Section checklist and edit master text.

____ Edit Division 11 - Equipment: Develop Section checklist and edit master text. Incorporate Sections by consultants.

____ Edit Division 12 - Furnishings: Develop Section checklist and edit master text.

____ Edit Division 13 - Special Construction: Develop Section checklist and edit master text. Incorporate Sections by consultants.

____ Edit Division 14 - Conveying Systems: Develop Section checklist and edit master text. Incorporate Sections by consultants.

____ Edit Division 15 - Mechanical: Develop Section checklist and edit master text. Incorporate Sections by consultants.

____ Edit Division 16 - Electrical: Develop Section checklist and edit master text. Incorporate Sections by consultants.

END OF CHECKLIST

Appendix B

Sample Division 1 - General Requirements

INTRODUCTION

Following this introduction are sample specifications sections for Division 1 - General Requirements. They are shortform specifications, typically, with more extensive content in some sections, consistent with the "say it once" principle of specifications writing and the recommendation for substantial Division 1 specifications, from the discussion of shortform specifications in Chapter 21.

These specifications should not be thoughtlessly copied. They must be reviewed and edited to suit the professional judgment of the responsible design professional. Adapt the text to the project's requirements, including bidding requirements and requirements of the Conditions of the Contract (General Conditions and Supplementary Conditions).

While some might find these shortform specifications more voluminous (substantial) than desired for brevity, it should be recalled that shortform specifications are suitable for contract documents, as opposed to a PPD (Appendix C) or outline specifications (Appendix D). Savings in the volume of Division 2 through 16 sections should be realized through proper use of substantial Division 1 specifications since repetition of requirements is eliminated. By comparison, there are often 25 Division 1 "longform" specifications for heavy commercial and institutional projects, with over 200 pages.

SECTION 01110 - SUMMARY OF THE PROJECT

A. THE PROJECT

1. Project Title: BRADBURY BUILDING, SUITE 230 IMPROVEMENTS (GHI Architects Project No. 04-121).

2. The Project: Tenant space improvements in existing office building for law firm.

3. Project Location: 2957 South Northbrook St., Sioux City, IA, as shown approximately on the Vicinity Map in the Drawings.

B. WORK INCLUDED IN THE CONTRACT

1. The Work: Construction and related services for Work indicated on Contract Drawings and in Contract Specifications. Refer to Drawings for building and site data and additional general information concerning the Work. The Work includes but is not limited to:
    a. Selective demolition of interior finishes, partitions, ceilings, doors and door frames.
    b. New interior non-load-bearing framing, finishes, custom casework, doors and door frames, glazing, specialties and equipment.
    c. Reconfiguration and extension of plumbing, HVAC and electrical power, lighting and signal systems.
    d. Residential food service equipment at office break room.
    e. Reconfiguration of wet-pipe fire sprinkler system, provided on design/build basis, to suit the requirements of reconfigured tenant space.
    f. Coordination of work being performed by others under separate contracts with the Owner, described in Article below titled "CONCURRENT WORK UNDER SEPARATE CONTRACTS."

2. Additional Information Available to Contractor: Project record drawings and specifications for building shell construction are available for review at the office of the Architect.

C. OWNER-FURNISHED PRODUCTS

1. Owner may furnish, for installation by Contractor, products which are identified on the Drawings and in the Specifications as OFCI (Owner-Furnished/Contractor-Installed).

2. Work under the Contract shall include all provisions necessary to fully incorporate such products into the Work, including, as necessary, fasteners, backing, supports, piping, conduit, conductors and other such provisions from point of service to point of connection, and field finishing.

3. Work Under Separate Contracts: Owner will award separate contracts for products and installation for the following work and other work as may be indicated on Drawings as NIC (Not in Contract). Work under the Contract shall include all provisions necessary to make such concurrent work under separate contracts complete in every respect and fully functional, including field finishing. Provide necessary backing, supports, piping, conduit, conductors and other such provisions from point of service to point of connection.
    a. Private branch exchange (PBX) telephone system.
    b. Local area network (LAN).
    c. Cable television antenna system.
    d. Office furniture system.
    e. Office furnishings, in addition to furniture system.
    f. Office equipment.

D. PERMITS, LICENSES AND FEES

1. Permits: For Work included in the Contract, Contractor will obtain all permits from authorities having jurisdiction and from serving utility companies and agencies.

2. Licenses: Contractor shall obtain and pay all licenses associated with construction activities, such as business licenses, contractors' licenses and vehicle and equipment licenses. All costs for licenses shall be included in the Contract Sum.

3. Assessments: Costs of assessments and connection fees shall not be included in the Contract Sum. Owner will pay all assessments and utility service connection fees.

4. Test and Inspection Fees: Contractor shall pay all fees charged by authorities having jurisdiction and from serving utility companies and agencies, for tests and inspections conducted by those authorities, companies and agencies. Owner will reimburse Contractor for actual amount of such fees, without mark-up.

E. **CONTRACTOR'S USE OF SITE AND PREMISES**

1. Contractor's Use of the Premises: During the construction period, Contractor shall have full use of the tenant space and designated site areas for storage and staging, as directed by building manager.
   a. Contractor's use of the tenant space shall be limited only by Owner's right to perform construction operations with its own forces or to employ separate contractors on portions of the Project, in accordance with the Conditions of the Contract.
   b. Contractor's use of surrounding areas shall be subject to approval and direction of the building manager, including maintenance of passageways, dust and debris control, and noise control.

2. Project Area: Future Suite 230, consisting of current Suites 230, 240 and 250. Use of other areas are subject to advanced approval of building manager.

3. Emergency Access: Provide pathways, drives, gates, directional signage and other provisions as required by authorities having jurisdiction, for emergency access to Project area(s).

4. Emergency Egress: Maintain all pathways, exitways, exit doors, drives, gates and other means of egress during construction, as required by authorities having jurisdiction.

5. Utility Outages and Shutdowns: Schedule utility outages and shutdowns to times and dates acceptable to other tenants, unless otherwise directed. Provide 48 hours notice of all utility outages and shutdowns. Provide minimum 48 hours notice to the Owner of all utility outages and shutdowns. Duration of outages and shutdowns shall not hinder normal activities on other tenants except as acceptable to building manager.

F. **OWNER'S USE OF SITE AND PREMISES**

1. Owner's Use of Site and Premises: Owner (tenant) reserves the right to occupy and to place and install equipment and furnishings in tenant space prior to Substantial Completion, provided that such occupancy does not interfere with completion of the Work within the Contract Time. Such partial occupancy by Owner shall not constitute acceptance of the total Work.

END OF SECTION 01110

SECTION 01210 - ALLOWANCE PROCEDURES

A. **ALLOWANCE PROCEDURES**

1. Allowance amounts below are for materials only. Include all other costs including installation in Contract Sum.

2. Coordinate allowances with requirements for related and adjacent Work.

3. Notify Owner and Architect of date when final decision on allowance items is required to avoid delays in the Work.

4. Furnish certification that quantities of products purchased are the actual quantities needed with reasonable allowance for cutting or installation losses, tolerances, mixing waste and similar margins.

5. Submit invoices or delivery slips to indicate actual quantities of materials delivered and costs. Indicate amounts of applicable trade discounts.

B. ALLOWANCES

1. Lump Sum Allowances:
   a. Custom hardwood casework in Law Library: $28,000.
   b. Custom conference table: $53,000.

2. Unit Cost Allowances:
   a. Carpet: $39 per square yard, material only. Installation by direct glue-down method shall be included in Contract Sum.

END OF SECTION 01210

SECTION 01230 - ALTERNATE BID PROCEDURES

A. ALTERNATE BID PROCEDURES

1. List price for each alternate in Bid Form. Include cost of modifications to other work to accommodate alternate. Include related costs such as overhead and profit.

2. Owner will determine which alternates are selected for inclusion in the Contract.

3. Alternates are described briefly in this Section. The Contract Documents define the requirements for alternates.

4. Coordinate alternates with related Work to ensure that Work affected by each selected alternate is properly accomplished.

B. ALTERNATE BID ITEMS

1. Alternate Bid No. 1 - Flush Wood Doors Veneer
   a. Base Bid condition: Provide interior flush wood veneer doors with plain-sliced white oak, book-matched and balanced, factory-finished with cherry stain and clear catalyzed varnish.
   b. Alternate Bid condition: Provide interior flush wood veneer doors with quarter-slice cherry, book-matched and balanced, factory-finished with clear catalyzed varnish.

2. Alternate Bid No. 2 - Music/Paging System
   a. Base Bid condition: Omit music/paging system but provide rough-in for future installation for system indicated on Electrical Drawings.
   b. Alternate Bid condition: Provide music/paging system as indicated on Electrical Drawings.

END OF SECTION 01230

SECTION 01250 - CONTRACT MODIFICATION PROCEDURES

A. CONTRACT MODIFICATION PROCEDURES

1. Minor Changes in the Work: Architect will issue Architect's Supplemental Instructions authorizing Minor Changes in the Work, not involving adjustment to the Contract Sum or the Contract Time, on AIA Document G710 - Architect's Supplemental Instructions.

2. Owner-Initiated Proposal Requests: Architect will issue a detailed description of proposed changes in the Work that may require adjustment to the Contract Sum or the Contract Time. If necessary, the description will include supplemental or revised Drawings and Specifications.
    a. Proposal Requests are for information only. Do not consider them instructions either to stop Work in progress or to execute the proposed change.
    b. Within time specified in Proposal Request after receipt of Proposal Request, submit a quotation estimating cost adjustments to the Contract Sum and the Contract Time necessary to execute the change.
        1) Include a list of quantities of products required or eliminated and unit costs, with total amount of purchases and credits to be made. If requested, furnish survey data to substantiate quantities.
        2) Indicate taxes, delivery charges, equipment rental, and amounts of trade discounts.
        3) Include an updated Contractor's Construction Schedule that indicates the effect of the change, including, but not limited to, changes in activity duration, start and finish times, and activity relationship. Use available total float before requesting an extension of the Contract Time.

3. Contractor-Initiated Proposals: If latent or unforeseen conditions require modifications to the Contract, Contractor may propose changes by submitting a request for a change.
    a. Include a statement outlining reasons for the change and the effect of the change on the Work. Provide a complete description of the proposed change. Indicate the effect of the proposed change on the Contract Sum and the Contract Time.
    b. Include a list of quantities of products required or eliminated and unit costs, with total amount of purchases and credits to be made. If requested, furnish survey data to substantiate quantities.
    c. Indicate taxes, delivery charges, equipment rental, and amounts of trade discounts.
    d. Include an updated Contractor's Construction Schedule that indicates the effect of the change, including, but not limited to, changes in activity duration, start and finish times, and activity relationship. Use available total float before requesting an extension of the Contract Time.
    e. Comply with requirements in Section 01600 - Product Requirements if the proposed change requires substitution of one product or system for product or system specified.

4. Proposal Request Form: Use AIA Document G709 for Proposal Requests.

5. Allowance Adjustment: Base each Change Order proposal on the difference between purchase amount and the allowance, multiplied by final measurement of work-in-place. Allow for cutting losses, tolerances, mixing wastes, normal product imperfections, and similar margins.
    a. Include installation costs only where indicated as part of the allowance.
    b. Prepare explanation and documentation to substantiate distribution of overhead costs and other margins claimed.
    c. Submit substantiation of a change in scope of work, if any, claimed in Change Orders related to unit-cost allowances. Owner reserves the right to establish the quantity of work-in-place by independent quantity survey, measure, or count.

6. Submit claims for increased costs because of a change in the allowance described in the Contract Documents, whether for the Purchase Order amount or Contractor's handling, labor, installation, overhead, and profit. Submit claims within 10 working days of receipt of the Change Order or Construction Change Directive authorizing Work to proceed. Owner may reject claims submitted later than 10 working days after such authorization.

7. Change Order Procedures: On Owner's approval of a Proposal Request, Architect will issue a Change Order for signatures of Owner and Contractor on AIA Document G701.

8. Construction Change Directive: Architect may issue a Construction Change Directive on AIA Document G714. Construction Change Directive instructs Contractor to proceed with a change in the Work, for subsequent inclusion in a Change Order.
   a. Construction Change Directive contains a complete description of change in the Work. It also designates method to be followed to determine change in the Contract Sum or the Contract Time.
   b. Documentation: Maintain detailed records on a time and material basis of work required by the Construction Change Directive.
      1) After completion of change, submit an itemized account and supporting data necessary to substantiate cost and time adjustments to the Contract.

END OF SECTION 01250

SECTION 01290 - PRICE AND PAYMENT PROCEDURES

A. SCHEDULE OF VALUES

1. Submit a printed schedule on AIA Form G703 - Application and Certificate for Payment Continuation Sheet. Contractor's standard form or electronic media printout will be considered.

2. Submit Schedule of Values in duplicate within 15 days after date of Owner-Contractor Agreement.

3. Format: Utilize the Table of Contents of this Project Manual. Identify each line item with number and title of the specification Section. Identify site mobilization.

4. Include in each line item the amount of Allowances specified in this section. For unit cost Allowances, identify quantities taken from Contract Documents multiplied by the unit cost to achieve the total for the item.

5. Include separately from each line item a direct proportional amount of Contractor's overhead and profit.

6. Revise schedule to list approved Change Orders, with each Application for Payment.

B. APPLICATIONS FOR PROGRESS PAYMENTS

1. Payment Period: Submit at intervals stipulated in the Agreement.

2. Present required information in typewritten form.

3. Form: AIA G702 - Application and Certificate for Payment and AIA G703 - Continuation Sheet including continuation sheets when required.

4. For each item, provide a column for listing each of the following:
   a. Item Number.
   b. Description of Work.
   c. Scheduled Values.
   d. Previous Applications.
   e. Work in Place and Stored Materials under this Application.
   f. Authorized Change Orders.

    g. Total Completed and Stored to Date of Application.
    h. Percentage of Complete.
    i. Remainder to be Accomplished.
    j. Retainage.

5. Execute certification by signature of authorized officer.

6. Use data from approved Schedule of Values. Provide dollar value in each column for each line item for portion of work performed and for stored Products.

7. List each authorized Change Order as a separate line item, listing Change Order number and dollar amount as for an original item of Work.

8. Submit three copies of each Application for Payment. Include the following with the application:
    a. Transmittal letter in format conforming to example furnished by Architect.
    b. Construction progress schedule, revised and current as specified in Section 01320.
    c. Partial release of liens covering Work on previous Application for Payment.

9. When Architect requires substantiating information, submit data justifying dollar amounts in question. Provide one copy of data with cover letter for each copy of submittal. Show application number and date, and line item by number and description.

C. APPLICATION FOR FINAL PAYMENT

1. Prepare Application for Final Payment as specified for progress payments, identifying total adjusted Contract Sum, previous payments, and sum remaining due.

2. Application for Final Payment will not be considered until the following have been accomplished:
    a. All closeout procedures specified in Section 01770 - Contract Closeout Procedures.
    b. Delivery of keys to Owner.
    c. Delivery of warranty and guaranty documentation.
    d. Completion of demonstration and training of Owner's personnel.
    e. Completion of all items on Correction List ("Punch List").

END OF SECTION 01290

SECTION 01310 - PROJECT MANAGEMENT AND COORDINATION

A. COORDINATION

1. Coordination: Coordinate construction operations included in various Sections of the Specifications to ensure efficient and orderly installation of each part of the Work. Coordinate construction operations, included in different Sections, that depend on each other for proper installation, connection, and operation.
    a. Schedule construction operations in sequence required to obtain the best results where installation of one part of the Work depends on installation of other components, before or after its own installation.
    b. Coordinate installation of different components with work by Owner under separate contracts to ensure maximum accessibility for required maintenance, service, and repair.
    c. Make adequate provisions to accommodate items scheduled for later installation under the Contract and for future work that may be noted.
    d. If necessary, prepare memoranda for distribution to each party involved, outlining special proce-

dures required for coordination. Include such items as required notices, reports, and list of attendees at meetings.

2. Administrative Procedures: Coordinate scheduling and timing of required administrative procedures with other construction activities and activities of other contractors to avoid conflicts and to ensure orderly progress of the Work. Such administrative activities include, but are not limited to, the following:
   a. Schedule of Values.
   b. Construction Progress Schedule.
   c. Requests for Interpretation (RFIs).
   d. Requests for substitution.
   e. Submittals: Shop drawings, product data, and samples.
   f. Test and inspection reports.
   g. Design-build drawings, product data, calculations, and permits.
   h. Manufacturer's instructions and field reports.
   i. Contract closeout activities.

3. Conservation: Coordinate construction activities to ensure that operations are carried out with consideration given to conservation of energy, water, and materials.

B. PROJECT MEETINGS

1. Project Meetings, General: Schedule and conduct meetings and conferences at Project site, unless otherwise indicated.
   a. Attendees: Inform participants and others involved, and individuals whose presence is required, of date and time of each meeting. Notify Owner and Architect of scheduled meeting dates and times.
   b. Agenda: Prepare the meeting agenda. Distribute the agenda to all invited attendees.
   c. Minutes: Record significant discussions and agreements achieved. Distribute the meeting minutes to everyone concerned, including Owner and Architect, within 2 working days of the meeting.

2. Preconstruction Conference: Schedule a preconstruction conference before starting construction, at a time convenient to Owner and Architect, but no later than 10 working days after execution of the Agreement. Hold the conference at Project site or another convenient location. Conduct the meeting to review responsibilities and personnel assignments.
   a. Attendees: Authorized representatives of Owner, Architect, and their consultants; Contractor and its superintendent; major subcontractors; manufacturers; suppliers; and other concerned parties shall attend the conference. All participants at the conference shall be familiar with Project and authorized to conclude matters relating to the Work.
   b. Agenda: Discuss items of significance that could affect progress, including the following:
      1) Preliminary construction schedule.
      2) Phasing.
      3) Critical sequencing of Work under the Contract and work under separate contracts by Owner.
      4) Designation of responsible personnel.
      5) Procedures for processing field decisions and Change Orders.
      6) Procedures for processing Applications for Payment.
      7) Distribution of the Contract Documents.
      8) Submittal procedures.
      9) Preparation of Record Documents.
      10) Use of the premises.
      11) Responsibility for temporary facilities and controls.
      12) Parking availability.
      13) Office, work, and storage areas.
      14) Security.
      15) Progress cleaning.
      16) Working hours.

3. Progress Meetings: Conduct progress meetings at weekly intervals. Coordinate dates of meetings with preparation of payment requests.
   a. Attendees: In addition to representatives of Owner and Architect, each contractor, subcontractor, supplier, and other entity concerned with current progress or involved in planning, coordination, or performance of future activities shall be represented at these meetings. All participants at the conference shall be familiar with Project and authorized to conclude matters relating to the Work.
   b. Agenda: Review and correct or approve minutes of previous progress meeting. Review other items of significance that could affect progress. Include topics for discussion as appropriate to status of Project.
      1) Contractor's Construction Schedule: Review progress since the last meeting. Determine whether each activity is on time, ahead of schedule, or behind schedule in relation to Contractor's Construction Schedule. Determine how construction behind schedule will be expedited; secure commitments from parties involved to do so. Discuss whether schedule revisions are required to ensure that current and subsequent activities will be completed within the Contract Time.
      2) Review present and future needs of each entity present, including the following:
         a) Interface requirements.
         b) Sequence of operations.
         c) Status of submittals.
         d) Deliveries.
         e) Off-site fabrication.
         f) Access.
         g) Site utilization.
         h) Temporary facilities and controls.
         i) Work hours.
         j) Progress cleaning.
         k) Quality and work standards.
         l) Change Orders.
         m) Documentation of information for payment requests.
   c. Reporting: Distribute minutes of the meeting to each party present and to parties who should have been present. Include a brief summary, in narrative form, of progress since the previous meeting and report.
   d. Schedule Updating: Revise Construction Progress Schedule after each progress meeting where revisions to the schedule have been made or recognized. Issue revised schedule concurrently with the report of each meeting.

END OF SECTION 01310

SECTION 01320 - CONSTRUCTION PROGRESS DOCUMENTATION

A. CONSTRUCTION PROGRESS SCHEDULE

1. Format: Bar chart. Include listings in chronological order according to the start date for each activity.
   a. Identify each activity with the applicable Specification Section Number.
   b. Show complete sequence of construction by activity, with dates for beginning and completion of each element of construction.
   c. Identify work of separate stages and other logically grouped activities.
   d. Include conferences and meetings in schedule.
   e. Show accumulated percentage of completion of each item, and total percentage of Work completed, as of the first day of each month.
   f. Provide separate schedule of submittal dates for shop drawings, product data, samples, and Owner-furnished products, products identified under Allowances and dates reviewed submittals will be required from Architect. Indicate decision dates for selection of finishes.

    g. Indicate delivery dates for Owner-furnished products.

    h. Coordinate content with Schedule of Values specified in Section 01290 - Price and Payment Procedures.

    i. Provide legend for symbols and abbreviations used.

2. Schedule Size: Maximum 22 × 17 inches (560 × 432 mm) or width required.

    a. Sheets shall fold neatly to multiple of 8½ × 11 inches (216 × 280 mm) for filing.

    b. Scale and spacing in content shall allow for notations and revisions.

B. REVIEW AND EVALUATION OF SCHEDULE

1. Review: Participate in joint review and evaluation of schedule with Architect at each submittal.

2. Evaluate project status to determine Work behind schedule and Work ahead of schedule.

3. After review, revise schedule as necessary as result of review, and resubmit within 5 working days.

C. UPDATING CONSTRUCTION PROGRESS SCHEDULE

1. Maintain schedules to record actual start and finish dates of completed activities.

2. Indicate progress of each activity to date of revision, with projected completion date of each activity.

3. Annotate diagrams to graphically depict current status of Work.

4. Identify activities modified since previous submittal, major changes in Work, and other identifiable changes.

5. Indicate changes required to maintain date of Substantial Completion.

6. Submit reports required to support recommended changes.

7. Provide narrative report to define problem areas, anticipated delays, and impact on the schedule. Report corrective action taken or proposed and its effect, including the effects of changes on schedules of separate contractors.

D. DISTRIBUTION OF SCHEDULE

1. Distribute copies of updated schedules to Contractor's project site file, to Subcontractors, suppliers, Architect, Owner, and other concerned parties.

2. Instruct recipients to promptly report, in writing, problems anticipated by projections shown in schedules.

END OF SECTION 01320

SECTION 01330 - SUBMITTAL PROCEDURES

A. SUBMITTALS FOR REVIEW BY ARCHITECT

1. Submit the following for review when specified in Specifications Sections:

    a. Product data.

    b. Shop drawings.

    c. Samples for selection.
    d. Samples for verification.

2. Make submittals to Architect for review for the limited purpose of checking for conformance with information given and the design concept expressed in the Contract Documents.

3. Samples will be reviewed only for aesthetic, color, or finish selection.

4. After review, provide copies and distribute submittals as specified below.

B.   SUBMITTALS FOR INFORMATION

1. When the following are specified in individual sections, submit them for information:
    a. Design data.
    b. Certificates.
    c. Test reports.
    d. Inspection reports.
    e. Manufacturer's instructions.
    f. Manufacturer's field reports.
    g. Other types specified.

2. Submittals for information shall be made for Architect's knowledge as Contract administrator or for Owner. No review action will be taken.

C.   SUBMITTALS FOR PROJECT CLOSEOUT

1. Submit the following at Contract Closeout when specified in Specifications Sections:
    a. Project record documents.
    b. Operation and maintenance data.
    c. Warranties.
    d. Bonds.
    e. Permits.
    f. Other types if specified.

D.   ADMINISTRATIVE REQUIREMENTS

1. Submission: Transmit each submittal with Letter of Transmittal, AIA Document G810, or other form containing substantially the same information, as acceptable to Architect.
    a. Deliver submittals to Architect at Architect's business address.
    b. Schedule submittals to expedite the Project and coordinate submission of related items.

2. Submittal Preparation: Sequentially number each submittal on the Transmittal Form. Provide space for Contractor and Architect review stamps. Revise submittals with original number and a sequential alphabetic suffix. Identify:
    a. Project.
    b. Contractor, Subcontractor and supplier, as applicable.
    c. Pertinent Drawing and detail number, and Specification Section and Title, as appropriate, on each copy.
    d. Variations from Contract Documents and Product or system limitations which may be detrimental to successful performance of the completed Work.
    e. When revised for resubmission, identify all changes made since previous submission.

3. Contractor's Review: Apply Contractor's stamp, signed or initialed, certifying that review, approval, verification of products required, field dimensions, adjacent construction Work, and coordination of information are in accordance with the requirements of the Work and Contract Documents.

4. Architect's Action: Architect will not review submittals that do not bear Contractor's approval stamp and will return them without action.
   a. Architect will review each submittal, make marks to indicate corrections or modifications required, and return it.
   b. Architect will stamp each submittal with an action stamp and will mark stamp appropriately to indicate action taken.
   c. For each submittal for review, allow 10 working days excluding delivery time to and from the Contractor.

5. Distribution: Architect will transmit reviewed submittals to Contractor for further action. Submittals with completed review actions shall be distributed by Contractor as appropriate.

6. Unsolicited Submittals: If not required by Contract Documents or requested by Architect, unsolicited submittals will not be recognized or processed, and will be returned to Contractor unreviewed.

7. Documents for Review:
   a. Small size sheets, not larger than 8½ × 11 inches (215 × 280 mm): Submit one copy; the Contractor shall make copies for construction as from original returned by the Architect.
   b. Larger sheets, not larger than 36 × 48 inches (910 × 1220 mm): Submit one reproducible transparency.

8. Documents for Information: Submit two copies.

9. Documents for Contract Closeout: Make one reproduction of submittal originally reviewed. Submit one extra of submittals for information.

10. Samples: Submit the number specified in individual Specification Sections, one of which will be retained by Architect.
    a. After review, produce duplicates.
    b. Retained samples will not be returned to Contractor unless specifically so specified.

END OF SECTION 01330

SECTION 01340 - REQUESTS FOR INTERPRETATION (RFI)

A. DEFINITIONS

1. Request for Interpretation: A document submitted by the Contractor requesting clarification of a portion of the Contract Documents, hereinafter referred to as an RFI.

B. CONTRACTOR'S REQUESTS FOR INTERPRETATION (RFIs)

1. Contractor's Requests for Interpretation (RFIs): Should Contractor be unable to determine from the Contract Documents the exact material, process, or system to be installed; or when the elements of construction are required to occupy the same space (interference); or when an item of Work is described differently at more than one place in the Contract Documents; the Contractor shall request that the Architect make an interpretation of the requirements of the Contract Documents to resolve such matters. Contractor shall comply with procedures specified herein to make Requests for Interpretation (RFIs).

2. Submission of RFIs: RFIs shall be prepared and submitted on a form provided by Architect.
   a. Forms shall be completely filled in, and if prepared by hand, shall be fully legible after copying by xerographic process.

    b. Each RFI shall be given a discrete, consecutive number.

    c. Each page of the RFI and each attachment to the RFI shall bear the University's project name, project number, date, RFI number, and a descriptive title.

    d. Contractor shall sign all RFIs attesting to good faith effort to determine from the Contract Documents the information requested for interpretation. Frivolous RFIs shall be subject to reimbursement from Contractor to Owner for fees charged by Architect, Architect's consultants and other design professionals engaged by the Owner.

3. Subcontractor-Initiated and Supplier-Initiated RFIs: RFIs from subcontractors and material suppliers shall be submitted through, be reviewed by and be attached to an RFI prepared, signed, and submitted by Contractor. RFIs submitted directly by subcontractors or material suppliers will be returned unanswered to the Contractor.

    a. Contractor shall review all subcontractor- and supplier-initiated RFIs and take actions to resolve issues of coordination, sequencing and layout of the Work.

    b. RFIs submitted to request clarification of issues related to means, methods, techniques and sequences of construction or for establishing trade jurisdictions and scopes of subcontracts will be returned without interpretation. Such issues are solely the Contractor's responsibility.

    c. Contractor shall be responsible for delays resulting from the necessity to resubmit an RFI due to insufficient or incorrect information presented in the RFI.

4. Requested Information: Contractor shall carefully study the Contract Documents to ensure that information sufficient for interpretation of requirements of the Contract Documents is not included. RFIs that request interpretation of requirements clearly indicated in the Contract Documents will be returned without interpretation.

    a. In all cases in which RFIs are issued to request clarification of issues related to means, methods, techniques and sequences of construction, for example, pipe and duct routing, clearances, specific locations of Work shown diagrammatically, apparent interferences and similar items, the Contractor shall furnish all information required for the Architect to analyze and/or understand the circumstances causing the RFI and prepare a clarification or direction as to how the Contractor shall proceed.

    b. If information included with this type of RFI by the Contractor is insufficient, the RFI will be returned unanswered.

5. Unacceptable Uses for RFIs: RFIs shall not be used for the following purposes:

    a. To request approval of submittals (use procedure specified in Section 01330 - Submittal Procedures).

    b. To request approval of substitutions (refer to Section 01630 - Product Substitution Procedures).

    c. To request changes that only involve change in Contract Time and Contract Sum (comply with provisions of the Contract General Conditions, as discussed in detail during pre-construction meeting).

    d. To request different methods of performing Work than those indicated in the Contract Drawings and Specifications (comply with provisions of the Contract General Conditions).

6. Disputed Requirements: In the event that the Contractor believes that a clarification by the Architect results in additional cost or time, Contractor shall not proceed with the Work indicated by the RFI until authorized to proceed by the Owner and Architect and claims, if any, are resolved in accordance with provisions in the General Conditions of the Contract.

7. RFI Log: Contractor shall prepare and maintain a log of RFIs, and at any time requested by the Architect, the Contractor shall furnish copies of the log showing all outstanding RFIs.

8. Review Time: Architect will return RFIs to Contractor within 10 working days of receipt. RFIs received after 12:00 noon shall be considered received on the next regular working day for the purpose of establishing the start of the 10-day response period.

END OF SECTION 01330

SECTION 01410 - REGULATORY REQUIREMENTS

A. AUTHORITY AND PRECEDENCE OF CODES, ORDINANCES AND STANDARDS

1. Authority: All codes, ordinances and standards referenced in the Drawings and Specifications shall have the full force and effect as though printed in their entirety in the Specifications.

2. Precedence:
   a. Where specified requirements differ from the requirements of applicable codes, ordinances and standards, the more stringent requirements shall take precedence.
   b. Where the Drawings or Specifications require or describe products or execution of better quality, higher standard or greater size than required by applicable codes, ordinances and standards, the Drawings and Specifications shall take precedence so long as such increase is legal.
   c. Where no requirements are identified in the Drawings or Specifications, comply with all requirements of applicable codes, ordinances and standards of authorities having jurisdiction.

B. APPLICABLE CODES, LAWS AND ORDINANCES

1. Applicable Codes, Laws and Ordinances:
   a. Performance of the Work shall be governed by all applicable laws, ordinances, rules and regulations of Federal, State and local governmental agencies and jurisdictions having authority over the Project.
   b. Performance of the Work shall meet or exceed the minimum requirements of the series of Codes published by the International Code Council (ICC) and the National Electrical Code (NEC), as adopted and interpreted by local authorities having jurisdiction.
   c. Performance of the Work shall be accomplished in conformance with all rules and regulations of public utilities, utility districts and other agencies serving the facility.
   d. Where such laws, ordinances, rules and regulations require more care or greater time to accomplish Work, or require better quality, higher standards or greater size of products, Work shall be accomplished in conformance to such requirements with no change to the Contract Time and Contract Sum, except where changes in laws, ordinances, rules and regulations occur subsequent to the execution date of the Agreement.

2. Date of Codes, Laws and Ordinances: The applicable edition of all codes shall be that adopted at the time of issuance of permits by the jurisdiction having authority and shall include all modifications and additions adopted by that jurisdiction. The applicable date of laws and ordinances shall be that of the date of performance of the Work.

END OF SECTION 01410

SECTION 01420 - REFERENCES

A. USE OF REFERENCES

1. References: The Drawings and Specifications contain references to various standards, standard specifications, codes, practices and requirements for products, execution, tests and inspections. These reference standards are published and issued by the agencies, associations, organizations and societies listed in this Section or identified in individual product specification Sections.
   a. Wherever term "Agency" occurs in Standard Specifications, it shall be understood to mean the term used for Owner for purposes of the Contract.

    b. Wherever term "Engineer" occurs in Standard Specifications, it shall be understood to mean Architect for purposes of the Contract.

    c. Standard Specifications shall be as amended and adopted by the jurisdiction in which the Project is located.

    d. Where reference is made to Standard Details, such reference shall be to the Standard Details accompanying the Standard Specifications, as amended and adopted by the jurisdiction in which the Project is located.

2. Relationship to Drawings and Specifications: Such references are incorporated into and made a part of the Drawings and Specifications to the extent applicable.

3. Referenced Grades Classes and Types: Where an alternative or optional grade, class or type of product or execution is included in a reference but is not identified on the Drawings or in the Specifications, provide the highest, best and greatest of the alternatives or options for the intended use and prevailing conditions.

4. ASTM and ANSI References: Specifications and Standards of the American Society for Testing and Materials (ASTM) and the American National Standards Institute (ANSI) are identified in the Drawings and Specifications by abbreviation and number only and may not be further identified by title, date, revision or amendment.

5. Copies of Reference Standards:

    a. Reference standards are not furnished with the Drawings and Specifications because it is presumed that the Contractor, subcontractors, manufacturers, suppliers, trades and crafts are familiar with these generally recognized standards of the construction industry.

    b. Copies of reference standards may be obtained from publishing sources.

6. Edition Date of References:

    a. When an edition or effective date of a reference is not given, it shall be understood to be the current edition or latest revision published as of the date of the permit issued by authorities having jurisdiction.

    b. All amendments, changes, errata and supplements as of the effective date shall be included.

7. Conflicting Requirements: Where compliance with two or more standards is specified and the standards establish different or conflicting requirements for minimum quantities or quality levels, comply with the most stringent requirement. Refer uncertainties and requirements that are different, but apparently equal, to Architect for a decision before proceeding.

    a. Minimum Quantity or Quality Levels: The quantity or quality level shown or specified shall be the minimum provided or performed. The actual installation may comply exactly with the minimum quantity or quality specified, or it may exceed the minimum within reasonable limits. To comply with these requirements, indicated numeric values are minimum or maximum, as appropriate, for the context of the requirements. Refer uncertainties to Architect for a decision before proceeding.

8. Jobsite Copies: Contractor shall obtain and maintain at the Project site copies of referenced codes and standards identified on the Drawings and in the Specifications in order to properly execute the Work, including:

    a. Local and State Building Codes: As referenced in Section 01410 - Regulatory Requirements.

    b. Safety Codes: Occupational Safety and Health Act (OSHA) regulations and State industrial safety laws and regulations, to extent applicable to the Work.

    c. General Standards:

        1) Building Code material, testing and installation standards.

        2) Underwriters Laboratories, Inc. (UL) Building Products Listing.

        3) Factory Mutual Research Organization (FM) Approval Guide.

        4) American Society for Testing and Materials (ASTM) Standards in Building Codes.

        5) American National Standards Institute (ANSI) standards.

d. Fire and Life Safety Standards: All referenced standards pertaining to fire rated construction and exiting.

e. Common Materials Standards: American Concrete Institute (ACI), American Institute of Steel Construction (AISC), American Welding Society (AWS), Gypsum Association (GA), National Fire Protection Association (NFPA), Tile Council of America (TCA) and Architectural Woodwork Institute (AWI) standards to the extent referenced within the Contract Specifications.

f. Research Reports: ICC Evaluation Service, Inc. (ICC-ES), Research Reports and National Evaluation Service, Inc. Reports (NER), for products not in conformance to prescribed requirements stated in applicable Building Code.

g. Product Listings: Approval documentation, indicating approval of authorities having jurisdiction for use of product within the applicable jurisdiction.

B. DEFINITIONS OF TERMS

1. Basic Contract Definitions: Words and terms governing the Work are defined in the Conditions of the Contract, as referenced in the Agreement.

2. Words and Terms Used on Drawings and in Specifications: Additional words and terms may be used in the Drawings and Specifications and are defined as follows:
   a. "And/or": If used, shall mean that either or both of the items so joined are required.
   b. "Applicable": As appropriate for the particular condition, circumstance or situation.
   c. "Approve(d)": Approval action shall be limited to the duties and responsibilities of the party giving approval, as stated in the [General] Conditions of the Contract. Approvals shall be valid only if obtained in writing and shall not apply to matters regarding the means, methods, techniques, sequences and procedures of construction. Approval shall not relieve the Contractor from responsibility to fulfill Contract requirements.
   d. "Directed": Limited to duties and responsibilities of the Owner or Architect as stated in the Conditions of the Contract, meaning as instructed by the Owner or Architect, in writing, regarding matters other than the means, methods, techniques, sequences and procedures of construction. Terms such as "directed," "requested," "authorized," "selected," "approved," "required," and "permitted" mean "directed by the Owner," "directed by the Architect," "requested by the Owner," and similar phrases. No implied meaning shall be interpreted to extend the responsibility of the Owner, Architect or other responsible design professional into the Contractor's supervision of construction.
   e. "Equal" or "Equivalent": As determined by Architect or other responsible design professional as being equivalent, considering such attributes as durability, finish, function, suitability, quality, utility, performance and aesthetic features.
   f. "Furnish": Means "supply and deliver, to the Project site, ready for unloading, unpacking, assembly, installation, and similar operations."
   g. "Indicated": The term indicated refers to graphic representations, notes, or schedules on the Drawings, or other Paragraphs or Schedules in the Specifications, and similar requirements in the Contract Documents. Terms such as "shown," "noted," "scheduled," and "specified" are used to help the reader locate the reference. There is no limitation on location.
   h. "Install": Describes operations at the Project site including the actual unloading, unpacking, assembly, erection, placing, anchoring, applying, working to dimension, finishing, curing, protecting, cleaning and similar operations.
   i. "Installer":
      1) "Installer" refers to the Contractor or an entity engaged by the Contractor, such as an employee, subcontractor or sub-subcontractor for performance of a particular construction activity, including installation, erection, application and similar operations. Installers are required to be experienced in the operations they are engaged to perform.
      2) "Experienced Installer": The term "experienced," when used with "installer," means having a minimum of 5 previous Projects similar in size to this Project, knowing the precautions necessary to perform the Work, and being familiar with requirements of authorities having jurisdiction over the Work.
   j. "Jobsite": Same as "Site."

k. "Necessary": With due considerations of the conditions of the Project and as determined in the professional judgment of the responsible design professional as being necessary for performance of the Work in conformance with the requirements of the Contract Documents, but excluding matters regarding the means, methods, techniques, sequences and procedures of construction.

l. "Noted": Same as "Indicated."

m. "Per": Same as "in accordance with," "according to" or "in compliance with."

n. "Products": Material, system or equipment.

o. "Project Site": Same as "Site."

p. "Proper": As determined by the Architect or other responsible design professional as being proper for the Work, excluding matters regarding the means, methods, techniques, sequences and procedures of construction, which are solely the Contractor's responsibility to determine.

q. "Provide": Means "furnish and install, complete and ready for the intended use."

r. "Regulation": Includes laws, ordinances, statutes and lawful orders issued by authorities having jurisdiction, as well as rules, conventions and agreements within the construction industry that control performance of the Work.

s. "Required": Necessary for performance of the Work in conformance with the requirements of the Contract Documents, excluding matters regarding the means, methods, techniques, sequences and procedures of construction, such as:
   1) Regulatory requirements of authorities having jurisdiction.
   2) Requirements of referenced standards.
   3) Requirements generally recognized as accepted construction practices of the locale.
   4) Notes, schedules and graphic representations on the Drawings.
   5) Requirements specified or referenced in the Specifications.
   6) Duties and responsibilities stated in the Bidding and Contract Requirements.

t. "Scheduled": Same as "Indicated."

u. "Selected": As selected by the Architect or other responsible design professional from the full selection of the manufacturer's products, unless specifically limited in the Contract Documents to a particular quality, color, texture or price range.

v. "Shown": Same as "Indicated."

w. "Site": Same as "Site of the Work" or "Project Site"; the area or areas or spaces occupied by the Project and including adjacent areas and other related areas occupied or used by the Contractor for construction activities, either exclusively or with others performing other construction on the Project. The extent of the Project Site is shown on the Drawings, and may or may not be identical with the description of the land upon which the Project is to be built.

x. "Supply": See "Furnish."

y. "Testing and Inspection Agency": An independent entity engaged to perform specific inspections or tests, at the Project Site or elsewhere, and to report on, and, if required, to interpret, results of those inspections or tests.

z. "Testing Laboratory" or "Testing Laboratories": Same as "Testing and Inspection Agency."

C. ABBREVIATIONS, ACRONYMS, NAMES AND TERMS, GENERAL

1. Abbreviations, Acronyms, Names and Terms: Where acronyms, abbreviations, names and terms are used in the Drawings, Specifications or other Contract Documents, they shall mean the recognized name of the trade association, standards-generating organization, authority having jurisdiction or other entity applicable.
   a. Refer to the Conditions of the Contract, referenced in Agreement and located elsewhere in the Contract Documents, for definitions of Contract terms.

2. Abbreviations, General: The following are commonly used abbreviations which may be found on the Drawings or in the Specifications:

AC or ac	Alternating current or air conditioning (depending upon context)
AMP or amp	Ampere
C	Celsius
CFM or cfm	Cubic feet per minute
CM or cm	Centimeter

CY or cy	Cubic yard
DC or dc	Direct current
DEG or deg	Degrees
F	Fahrenheit
FPM or fpm	Feet per minute
FPS or fps	Feet per second
FT or ft	Foot or feet
Gal or gal	Gallons
GPM or gpm	Gallons per minute
IN or in	Inch or inches
Kip or kip	Thousand pounds
KSF or ksf	Thousand pounds per square foot
KSI or ksi	Thousand pounds per square inch
KV or kv	Kilovolt
KVA or kva	Kilovolt amperes
KW or kw	Kilowatt
KWH or kwh	Kilowatt hour
LBF or lbf	Pounds force
LF or lf	Lineal foot
M or m	Meter
MM or mm	Millimeter
MPH or mph	Miles per hour
PCF or pcf	Pounds per cubic foot
PSF or psf	Pounds per square foot
PSI or psi	Pounds per square inch
PSY or psy	Per square yard
SF or sf	Square foot
SY or sy	Square yard
V or v	Volts

3. Abbreviations and Acronyms for Industry Organizations: Where abbreviations and acronyms are used in Specifications or other Contract Documents, they shall mean the recognized name of the entities indicated in the *Encyclopedia of Associations* or in the *National Trade and Professional Associations Directory of the United States,* identified below.

4. Undefined Abbreviations, Acronyms, Names and Terms: Words and terms not otherwise specifically defined in this Section, in the Instructions to Bidders, in the Conditions of the Contract, on the Drawings or elsewhere in the Specifications shall be as customarily defined by trade or industry practice, by reference standard and by specialty dictionaries such as the following:
   a. *Dictionary of Architecture and Construction, 3rd Edition* (Cyril M. Harris, McGraw-Hill Professional, 2000) ISBN: 0-0713517-8-7.
   b. *Encyclopedia of Associations,* online directory by Thomson Gale, accessible through many public libraries.
   c. *National Trade and Professional Associations of the United States,* Columbia Books, Inc. (Annapolis Junction, MD, 2004) ISBN: 0-9715487-8-1.

END OF SECTION 01420

SECTION 01450 - QUALITY CONTROL

A. REGULATORY REQUIREMENTS FOR TESTING AND INSPECTION

1. Building Code Requirements: Comply with requirements for testing and inspections of applicable Codes, including additional requirements for testing and inspection, as adopted and interpreted by local authorities having jurisdiction.

2. Requirements of Fire Regulations: Comply with testing and inspection requirements of State Fire Marshal and local Fire Marshal having jurisdiction.

B. QUALITY OF THE WORK

1. Quality of Products: Unless otherwise indicated or specified, all products shall be new, free of defects and fit for the intended use.

2. Quality of Installation: All Work shall be produced plumb, level, square and true, or true to indicated angle, and with proper alignment and relationship between the various elements.

3. Protection of Existing and Completed Work: Contractor shall take all measures necessary to preserve and protect existing and completed Work free from damage, deterioration, soiling and staining, until acceptance of the Work by Owner.

4. Standards and Code Compliance and Manufacturer's Instructions and Recommendations: Unless more stringent requirements are indicated or specified, comply with manufacturer's instructions and recommendations, reference standards and building code research report requirements in preparing, fabricating, erecting, installing, applying, connecting and finishing Work.

5. Deviations from Standards and Code Compliance and Manufacturer's Instructions and Recommendations: Document and explain all deviations from reference standards and building code research report requirements and manufacturer's product installation instructions and recommendations, including acknowledgment by the manufacturer that such deviations are acceptable and appropriate for the Project.

C. CONTRACTOR'S QUALITY CONTROL

1. Contractor's Quality Control: Contractor shall ensure that products, services, workmanship and site conditions comply with requirements of the Drawings and Specifications by coordinating, supervising, testing and inspecting the Work and by utilizing only suitably qualified personnel.

2. Quality Requirements: Work shall be accomplished in accordance with quality requirements of the Drawings and Specifications, including, by reference, all Codes, laws, rules, regulations and standards. When no quality basis is prescribed, the quality shall be in accordance with the best accepted practices of the construction industry for the locale of the Project for projects of this type.

3. Quality of Products: Unless otherwise indicated or specified, all products shall be new, free of defects and fit for the intended use.

4. Quality of Installation: All Work shall be produced plumb, level, square and true, or true to indicated angle, and with proper alignment and relationship between the various elements.

5. Protection of Existing and Completed Work: Take all measures necessary to preserve and protect existing and completed Work free from damage, deterioration, soiling and staining, until Acceptance by the Owner.

6. Standards and Code Compliance and Manufacturer's Instructions and Recommendations: Unless more stringent requirements are indicated or specified, comply with manufacturer's instructions and recommendations, reference standards and building code research report requirements in preparing, fabricating, erecting, installing, applying, connecting and finishing Work.

7. Deviations from Standards and Code Compliance and Manufacturer's Instructions and Recommendations: Document and explain all deviations from reference standards and building code research report

requirements and manufacturer's product installation instructions and recommendations, including acknowledgment by the manufacturer that such deviations are acceptable and appropriate for the Project.

8. Verification of Quality: Work shall be subject to verification of quality by Owner or Architect in accordance with provisions of the Conditions of the Contract.

D. INSPECTIONS AND TESTS BY AUTHORITIES HAVING JURISDICTION

1. Inspections and Tests by Authorities Having Jurisdiction: Contractor shall cause all tests and inspections required by authorities having jurisdiction to be made for Work under this Contract, including those by the Building Department, Public Works Department, Fire Department, Health Department and similar agencies.

E. INSPECTIONS AND TESTS BY SERVING UTILITIES

1. Inspections and Tests by Serving Utilities: Contractor shall cause all tests and inspections required by serving utilities to be made for Work under the Contract.

F. INSPECTIONS AND TESTS BY MANUFACTURER'S REPRESENTATIVES

1. Inspections and Tests by Manufacturer's Representatives: Contractor shall cause all specified tests and inspections to be conducted by materials or systems manufacturers. Additionally, all tests and inspections required by materials or systems manufacturers as conditions of warranty or certification of Work shall be made, the cost of which shall be included in the Contract Sum.

END OF SECTION 01450

SECTION 01500 - TEMPORARY FACILITIES AND CONTROLS

A. TEMPORARY UTILITIES

1. Temporary Utilities, General: Coordinate with building manager for points of connection, protection and payment of service charges.
    a. Provide temporary electrical power, lighting, water, heating and cooling, and ventilation as necessary for proper performance of the Work.
    b. Existing facilities may not be used unless approved by building manager.
    c. New permanent facilities may be used once permanent account is established with serving utility.

2. Temporary Electrical Power:
    a. Cost of Energy: Reimbursed by Contractor to building owner.
    b. Connection(s): Existing power service at location(s) as directed by building manager.
       1) Do not disrupt service to occupied tenant spaces and common areas except at dates and times approved by building manager.
       2) Exercise measures to conserve energy.
       3) Provide separate metering and reimburse Owner for cost of energy used.
    c. Provide temporary electric feeder from existing building electrical service.
    d. Provide power outlets for construction operations, with branch wiring and distribution boxes located at each floor. Provide flexible power cords as necessary.
    e. Provide over-current and ground-fault protection.
    f. Permanent convenience receptacles may be utilized during construction.

1. Temporary Lighting for Construction Purposes:
   a. Provide and maintain temporary lighting for construction operations to achieve a minimum lighting level of 2 watt/sq ft (21 watt/sq m).
   b. Provide and maintain 1 watt/sq ft (10.8 watt/sq m) lighting to exterior staging and storage areas after dark for security purposes.
   c. Provide and maintain 0.25 watt/sq ft (2.7 watt/sq m) H.I.D. lighting to interior work areas after dark for security purposes.
   d. Provide branch wiring from power source to distribution boxes with lighting conductors, pigtails, and lamps as required.
   e. Maintain lighting and provide routine repairs and lamp replacement.
   f. Permanent building lighting may be utilized during construction if cleaned and relamped at Contract Closeout.

3. Temporary Heating, Ventilating and Air Conditioning:
   a. Cost of Energy: Reimbursed by Contractor to building owner.
   b. Temporary Heating: Provide portable heaters as necessary to maintain suitable conditions for construction operations.
      1) Maintain minimum ambient temperature of 50 degrees F (10 degrees C) in all areas.
      2) Existing HVAC system may be used for heating if authorized by building manager.
         a) Exercise measures to conserve energy.
         b) Enclose building prior to activating temporary heat.
      3) Prior to operation of permanent HVAC system for temporary heating purposes, verify that installation is approved for operation, equipment is lubricated and filters are in place. Provide and pay for operation, maintenance, and regular replacement of filters and worn or consumed parts.

4. Temporary Cooling:
   a. Cost of Energy: Reimbursed by Contractor to building owner.
   b. Temporary Cooling: Provide cooling devices and cooling as needed to maintain suitable conditions for construction operations.
      1) Maintain maximum ambient temperature of 80 degrees F (26 degrees C) in areas where construction is in progress, unless indicated otherwise in Specifications.
      2) Existing HVAC system may be used for temporary cooling if authorized by building manager.
         a) Exercise measures to conserve energy.
         b) Enclose building prior to activating temporary heat.
      3) Prior to operation of permanent HVAC system for temporary cooling purposes, verify that installation is approved for operation, equipment is lubricated and filters are in place. Replace filters prior to Contract Closeout.

5. Temporary Ventilation:
   a. Cost of Energy: Reimbursed by Contractor to building owner.
   b. Temporary Ventilation: Provide portable fans. Use permanent HVAC system as authorized by building manager.
      1) Prior to operation of permanent HVAC system for temporary cooling purposes, verify that installation is approved for operation, equipment is lubricated and filters are in place. Replace filters prior to Contract Closeout.
   c. Permanent Ventilation: Extend and supplement HVAC system with temporary fan units as necessary to maintain clean air for construction operations and to expedite curing and drying.

6. Temporary Water Service:
   a. Cost of Water Used: Reimbursed by Contractor to building owner.
   b. Temporary Water: Provide and maintain suitable quality water service for construction operations at time of construction mobilization. Provide water as necessary at exterior and interior locations.
      1) Connect to existing water source.
      2) Extend branch piping with outlets located so that water is available by hoses with threaded connections.

   3) Exercise measures to conserve water.
   4) Provide separate metering and reimburse building owner for cost of water used.

7. Telephone Service
   a. Jobsite Telephone Service: Provide, maintain and pay for telephone service to Contractor's field staff, including cellular telephones to project manager and field superintendent.
   1) Facsimile service: Provide on-site fax machine.
   2) E-mail service: Provide on-site e-mail capability.

8. Temporary Sanitary Facilities:
   a. Temporary Toilets: Provide and maintain portable chemical toilets, located as approved by building manager. Provide at time of project mobilization. Remove temporary toilets prior to Contract Close-out.
   b. Existing Toilet Facilities: Use of existing toilet rooms is not permitted.

B. TEMPORARY BARRIERS, ENCLOSURES AND PASSAGEWAYS

1. Temporary Barriers, General: Provide temporary fencing, barriers and guardrails as necessary to provide for public safety, to prevent unauthorized entry to construction areas and to protect existing facilities and adjacent properties from damage from construction operations.
   a. Refer to temporary fencing and phasing plan in the Drawings. Comply with requirements indicated.
   b. Note requirements for continued occupancy and use of existing buildings and site areas during construction.
   c. Comply with requirements of applicable Building Code and authorities having jurisdiction, including industrial safety regulations.
   d. Maintain unobstructed access to fire extinguishers, fire hydrants, temporary fire-protection facilities, stairways and other access routes for firefighting.
   e. Paint temporary barriers and enclosures with appropriate colors, graphics and warning signs to inform personnel and public of possible hazard.
   f. Where appropriate and necessary, provide warning lighting, including flashing red or amber lights.

2. Temporary Chainlink Fencing:
   a. Portable Chain-Link Fencing: Minimum 2-inch (50-mm) 9-gage, galvanized steel, chain-link fabric fencing; minimum 6 feet (1.8 m) high with galvanized steel pipe posts; minimum 2⅜-inch- (60-mm-) OD line posts and 2⅞-inch- (73-mm-) OD corner and pull posts, with 1⅝-inch- (42-mm-) OD top and bottom rails.
   1) Provide concrete or galvanized steel bases for supporting posts.
   2) Provide protective barriers at bases to prevent tripping by pedestrians.
   b. Windscreen on Chain-Link Fencing: For screening of construction activities from view using closed mesh weave windscreen material.

3. Tarpaulins: Fire-resistive labeled with flame-spread rating of 15 or less.

4. Covered Passageways: Erect structurally adequate, protective, covered walkways for passage of persons along adjacent passageways.
   a. Coordinate installation details with Owner's requirements for continuing operations in adjoining facilities.
   b. Review design and details with Architect.
   c. Comply with applicable regulations of authorities having jurisdiction.
   d. Construct covered walkways using scaffold or shoring framing.
   e. Provide wood-plank overhead decking, protective plywood enclosure walls, handrails, barricades, warning signs, lights, safe and well-drained walkways, and similar provisions for protection and safe passage.
   f. Extend back wall beyond the structure to complete enclosure fence.
   g. Paint and maintain in a manner as directed by Architect.

5. Temporary Closures: Provide temporary closures for protection of construction, in progress and completed, from exposure, foul weather, other construction operations and similar activities. Provide temporary weathertight enclosure for building exterior.
   a. Where heating or cooling is needed and permanent enclosure is not complete, provide insulated temporary enclosures. Coordinate closures with ventilating and material drying or curing requirements to avoid dangerous conditions and effects such as mold.
   b. Vertical openings: Close openings of 25 sq ft (2.3 sq m) or less with plywood or similar materials.
   c. Horizontal openings: Close openings in floor or roof decks and horizontal surfaces with load-bearing, wood-framed construction.
   d. Install tarpaulins securely using wood framing and other suitable materials.
   e. Where temporary wood or plywood enclosure exceeds 100 sq ft (9.2 sq m) in area, use fire-retardant-treated material for framing and main sheathing.

6. Temporary Partitions: Erect and maintain temporary partitions and temporary closures to limit dust and dirt migration, including migration into existing facilities, to separate areas from fumes and noise and to maintain fire-rated separations.
   a. Dust barriers: Construct dustproof, floor-to-ceiling partitions of not less than nominal 4-inch (100-mm) studs, 2 layers of 3-mil (0.07-mm) polyethylene sheets, inside and outside temporary enclosure.
      1) Overlap and tape full length of joints.
      2) Include ⅝-inch thick gypsum board at temporary partitions serving as noise barrier.
      3) Insulate partitions to minimize noise transmission to adjacent occupied areas.
      4) Seal joints and perimeter of temporary partitions.
   b. Dust barrier passages: Where passage through dust barrier is necessary, provide gasketed doors or heavy plastic sheets that effectively prevent air passage.
      1) Construct a vestibule and airlock at each entrance to temporary enclosure with not less than 48 inches (1219 mm) between doors.
      2) Maintain water-dampened foot mats in vestibule where passage leads to existing occupied spaces.
      3) Equip doors with security locks.
   c. Fire-rated temporary partitions: Maintain fire-rated separations, including corridor walls and occupancy separations, by construction of stud partitions with gypsum board faces.
      1) Construction details shall comply with recognized time-rated fire-resistive construction. Typically, 1-hour rated partitions shall be 2 × 4 wood studs at 16 inches on center or 3½-inch metal studs at 16 inches on center, with ⅝-inch thick Type X gypsum board at both faces, with joints filled, taped and topped.
      2) Seal partition perimeters with acceptable fire stopping and smoke seal materials.
      3) Construct fire-rated temporary partitions whenever existing time-rate fire-resistive construction is removed for 12 hours or more.

7. HVAC Protection: Provide dust barriers at HVAC return grilles and air inlets to prevent spread of dust and clogging of filters.

8. Temporary Floor Protection: Protect existing floors from soiling and damage.
   a. Cover floor with 2 layers of 3-mil (0.07-mm) polyethylene sheets, extending sheets 18 inches (460 mm) up the side walls.
   b. Cover polyethylene sheets with ¾-inch (19-mm) fire-retardant plywood.
   c. Provide floor mats to clean dust from shoes.

9. Landscape Barriers: Provide barriers around trees and plants designated to remain.
   a. Locate barriers as directed outside of drip lines of trees and plants.
   b. Protect entire area under trees against vehicular traffic, stored materials, dumping, chemically injurious materials, and puddling or continuous running water.
   c. Contractor shall pay all costs to restore trees and plants within barriers that are damaged by construction activities. Restoration shall include replacement with plant materials of equal quality and size. Costs shall include all fines, if any, levied by authorities having jurisdiction.

10. Guardrails: Provide guardrails along tops of embankments and excavations. Along public walkways and areas accessible by the public, adjoining excavations, provide guardrails in addition to fencing.
    a. Guardrails shall be substantially and durably constructed of lumber, firmly anchored by posts embedded in concrete, and complying with Code requirements for temporary barriers.
    b. Guardrails shall comply with dimensional requirements and accommodate loads as prescribed by applicable Building Code for permanent guardrails.

11. Security Closures: Provide temporary closures of openings in exterior surfaces to prevent entry of unauthorized persons. Provide doors with self-closing hardware and locks.

12. Weather Closures: Provide temporary weather-tight closures at exterior openings to prevent intrusion of water, to create acceptable working conditions, to protect completed Work and to maintain temporary heating, cooling and ventilation. Provide access doors with self-closing hardware and locks.

13. Temporary Access, Passage and Exit Ways: Construct temporary stairs, ramps and covered walkways, with related doors, gates, closures, guardrails, handrails, lighting and protective devices, to maintain access and exit ways to existing facilities to remain operational.
    a. Design and location of temporary construction shall be by Contractor, subject to review by Architect and authorities having jurisdiction.
    b. Provide temporary lighting, illuminated interior exit signage, nonilluminated directional and instructional signage, and temporary security alarms for temporary exits and exit passageways.
    c. Temporary measures shall suit and connect to existing building systems.

C. SECURITY

1. Security, General: Provide security and facilities to protect Work, existing facilities and Owner's operations from unauthorized entry, vandalism and theft.
    a. Protect Work, existing premises and Owner's operations from theft, vandalism and unauthorized entry.
    b. Initiate program in coordination with Owner's existing security system at project mobilization.
    c. Maintain program throughout construction period until Owner acceptance precludes the need for Contractor security.

2. Security Program:
    a. Protect Work, existing premises and Owner's operations from theft, vandalism and unauthorized entry.
    b. Initiate program in coordination with Owner's existing security system at project mobilization.
    c. Maintain program throughout construction period until Owner acceptance precludes the need for Contractor security.

3. Entry Control:
    a. Restrict entrance of persons and vehicles into Project site and existing facilities.
    b. Allow entrance only to authorized persons with proper identification.
    c. Maintain log of workers and visitors, make available to Owner on request.
    d. Owner will control entrance of persons and vehicles related to Owner's operations.
    e. Contractor shall control entrance of persons and vehicles related to Owner's operations.
    f. Coordinate access of Owner's personnel to site in coordination with Owner's security forces.

4. Personnel Identification: Provide identification badge to each person authorized to enter premises. Include on badge person's photograph, name, assigned number, expiration date and employer. Require return of badges at expiration of person's employment on the Work.

5. Coordination: Coordinate with security programs of Owner and building manager.

D. VEHICULAR ACCESS AND PARKING

1. Coordination:
    a. Coordinate haul routes with authorities having jurisdiction.
    b. Provide and maintain access to fire hydrants, free of obstructions.
    c. Coordinate designated construction parking areas with building manager. Restrict on-site parking to designated area. Parking on adjacent public thoroughfares shall be subject to approval of public safety authorities having jurisdiction.
    d. Parking adjacent to building entrances will be designated for construction use prior to 8 a.m. and after 6 p.m. for unloading and loading of construction products.

2. Construction Parking:
    a. Use designated areas of existing parking lot only, as approved by building manager.
    b. Do not allow heavy vehicles or construction equipment in parking areas except drive lanes constructed for load fire apparatus and garbage collection vehicles.
    c. Arrange for temporary parking areas to accommodate use of construction personnel.
    d. When site space is not adequate, use off-site parking.

3. Construction Parking Control:
    a. Control vehicular parking to prevent interference with traffic and parking by building tenants and access by emergency vehicles.
    b. Monitor parking of construction personnel's vehicles in existing facilities. Maintain vehicular access to and through parking areas.
    c. Provide parking authorization signage for windshields of construction vehicles and worker's personal vehicles parked on site.

4. Restoration of Parking Lot: At completion of construction, clean and restore paving to equal or better condition to that at start of construction.
    a. Apply pavement seal coat if necessary to restore asphaltic concrete paving.
    b. Repaint parking and traffic control markings damaged by construction traffic and other construction-related activities.

E. WASTE MANAGEMENT

1. Waste Removal: Provide waste removal facilities and services as required to maintain the site and existing facilities in clean and orderly condition.
    a. Provide containers with lids. Dispose of waste off-site periodically.
    b. Open free-fall chutes are not permitted. Terminate closed chutes into appropriate containers with lids.

2. Waste Management: Separate and dispose of construction waste in compliance with waste management regulations of authorities having jurisdiction.

F. FIELD OFFICES

1. Contractor's Field Office: Weathertight, with lighting, electrical outlets, heating, cooling equipment, and equipped with sturdy furniture, drawing rack and drawing display table.
    a. Provide space for Project meetings, with table and chairs to accommodate 6 persons.
    b. Locate offices minimum distance of 30 feet (10 m) from existing building, at location in existing parking lot assigned by building manager.

G. PROTECTION OF INSTALLED WORK

1. Protection of Installed Work, General: Provide temporary protection for installed products. Control traffic in immediate area to minimize damage.

2. Protective Coverings: Provide protective coverings at walls, projections, jambs, sills and soffits of openings as necessary to prevent damage from construction activities, such as coatings applications, and as necessary to prevent other than normal atmospheric soiling.

3. Interior Traffic Protection:
   a. Protect finished floors, stairs and other surfaces from traffic, soiling, wear and marring.
   b. Provide temporary covers of plywood, reinforced kraft paper or temporary rugs and mats, as necessary. Temporary covers shall not slip or tear under normal use.
   c. Prohibit traffic and storage on waterproofed and roofed surfaces and on landscaped areas.
   d. Protect newly fine graded, seeded and planted areas with barriers and flags to designate such areas as closed to pedestrian and vehicular traffic.

H. REMOVAL OF TEMPORARY FACILITIES AND CONTROLS

1. Removal of Temporary Facilities and Controls:
   a. Remove temporary utilities, equipment, facilities, and materials prior to Substantial Completion review.
   b. Remove underground installations to a minimum depth of 2 ft (600 mm). Grade site as indicated.
   c. Clean and repair damage caused by installation or use of temporary facilities and controls.
   d. Restore existing and permanent facilities used during construction to original condition. Restore permanent facilities used during construction to condition equal to or better than at commencement of construction.

END OF SECTION 01500

SECTION 01600 - PRODUCT REQUIREMENTS

A. GENERAL PRODUCT REQUIREMENTS

1. Products, General: Items purchased for incorporation in the Work, whether purchased for the Project or taken from previously purchased stock, including materials, equipment, assemblies, fabrications and systems.
   a. Named Products: Items identified by manufacturer's product name, including make or model designations indicated in the manufacturer's published product data.
   b. Materials: Products that are shaped, cut, worked, mixed, finished, refined or otherwise fabricated, processed or installed to form a part of the Work.
   c. Equipment: A product with operating parts, whether motorized or manually operated, that requires connections such as wiring or piping.

2. Specific Product Requirements: Refer to requirements of Section 01450 - Quality Control and individual Specifications Sections in Divisions 2 through 16 for specific requirements for products.

3. Minimum Requirements: Specified requirements for products are minimum requirements. Refer to general requirements for quality of the Work specified in Section 01450 - Quality Control and elsewhere herein.

4. Product Selection: Provide products that fully comply with the Contract Documents, and are undamaged and unused at installation. Comply with additional requirements specified herein in Article titled "PRODUCT OPTIONS."

5. Standard Products: Where specific products are not specified, provide standard products of types and kinds that are suitable for the intended purposes and that are usually and customarily used on similar

projects under similar conditions. Products shall be as selected by Contractor and subject to review and acceptance by Architect.

6. Product Completeness: Provide products complete with all accessories, trim, finish, safety guards and other devices and details needed for a complete installation and for the intended use and effect. Comply with additional requirements specified herein in Article titled "SYSTEM COMPLETENESS."

7. Code Compliance: All products, other than commodity products prescribed by Code, shall be governed by current ICC Evaluation Service, Inc. (ICC ES) Research Report or National Evaluation Report (NER) as interpreted and required by authority having jurisdiction. Refer to additional requirements specified in Section 01410 - Regulatory Requirements.

8. Interchangeability: To the fullest extent possible, provide products of the same kind from a single source. Products required to be supplied in quantity shall be the same product and interchangeable throughout the Work. When options are specified for the selection of any of two or more products, the product selected shall be compatible with products previously selected.

9. Product Nameplates and Instructions:
   a. Except for required Code-compliance labels and operating and safety instructions, locate nameplates on inconspicuous, accessible surfaces. Do not attach manufacturer's identifying nameplates or trademarks on surfaces exposed to view in occupied spaces or to the exterior.
   b. Provide a permanent nameplate on each item of service-connected or power-operated equipment. Nameplates shall contain identifying information and essential operating data such as the following example:
      1) Name of manufacturer
      2) Name of product
      3) Model and serial number
      4) Capacity
      5) Operating and Power Characteristics
      6) Labels of Tested Compliance with Codes and Standards
   c. Refer to additional requirements specified in Division 15 - Mechanical and Division 16 - Electrical.
   d. For each item of service-connected or power-operated equipment, provide operating and safety instructions, permanently affixed and of durable construction, with legible machine lettering. Comply with all applicable requirements of authorities having jurisdiction and listing agencies.

B. PRODUCT OPTIONS

1. Products Specified by Description: Where Specifications describe a product, listing characteristics required, with or without use of a brand name, provide a product that has the specified attributes and otherwise complies with specified requirements.

2. Products Specified by Performance Requirements: Where Specifications require compliance with performance requirements, provide product(s) that comply and are recommended by the manufacturer for the intended application. Verification of manufacturer's recommendations may be by product literature or by certification of performance from manufacturer.

3. Products Specified by Reference to Standards: Where Specifications require compliance with a standard, provided product shall fully comply with the standard specified. Refer to general requirements specified in Section 01420 - References regarding compliance with referenced standards, standard specifications, codes, practices and requirements for products.

4. Products Specified by Identification of Manufacturer and Product Name or Number:
   a. "Specified Manufacturer": Provide the specified product(s) of the specified manufacturer.
      1) If only one manufacturer is specified, without "acceptable manufacturers" being identified, provide only the specified product(s) of the specified manufacturer.

2) If the phrase "or equal" is stated or reference is made to the "or equal provision," products of other manufacturers may be provided if such products are equivalent to the specified product(s) of the specified manufacturer. Equivalence shall be demonstrated by submission of information in compliance with requirements specified herein under the Article titled "SUBSTITUTIONS."

b. "Acceptable Manufacturers": Product(s) of the named manufacturers, if equivalent to the specified product(s) of the specified manufacturer, will be acceptable in accordance with the requirements specified herein in the Article titled "SUBSTITUTIONS," except considerations regarding changes in Contract Time and Contract Sum will be waived if no increase in Contract Time or Contract Sum results from use of such equivalent products.

c. Unnamed manufacturers: Products of unnamed manufacturers will be acceptable only as follows:

1) Unless specifically stated that substitutions will not be accepted or considered, the phrase "or equal" shall be assumed to be included in the description of specified product(s). Equivalent products of unnamed manufacturers will be accepted in accordance with the "or equal" provision specified herein, below.

2) If provided, products of unnamed manufacturers shall be subject to the requirements specified herein in the Article titled "SUBSTITUTIONS."

d. Quality basis: Specified product(s) of the specified manufacturer shall serve as the basis by which products of named acceptable manufacturers and products of unnamed manufacturers will be evaluated. Where characteristics of the specified product are described, where performance characteristics are identified or where reference is made to industry standards, such characteristics are specified to facilitate evaluation of products by identifying the most significant attributes of the specified product(s).

5. Products Specified by Combination of Methods: Where products are specified by a combination of attributes, including manufacturer's name, product brand name, product catalog or identification number, industry reference standard, or description of product characteristics, provide products conforming to all specified attributes.

6. "Or Equal" Provision: Where the phrase "or equal" or the phrase "or approved equal" is included, product(s) of unnamed manufacturer(s) may be provided as specified above in subparagraph titled "Unnamed manufacturers."

a. The requirements specified herein in the Article titled "SUBSTITUTIONS" shall apply to products provided under the "or equal" provision except, if the proposed product(s) are determined to be equivalent to the specified product(s) of the specified manufacturer, the requirement specified for substitutions to result in a net reduction in Contract Time or Contract Sum will be waived.

b. Use of product(s) under the "or equal" provision shall not result in any delay in completion of the Work, including completion of portions of the Work for use by Owner or for work under separate contract by Owner.

c. Use of product(s) under the "or equal" provision shall not result in change in Contract Sum and Contract Time. Should additional costs be incurred, including costs for re-design and for fees for plancheck review and permit, costs shall be paid by Contractor with no change in Contract Sum and Contract Time.

d. Use of product(s) under the "or equal" provision shall not require substantial change in the intent of the design, in the opinion of the Architect. The intent of the design shall include functional performance and aesthetic qualities.

1) Should changes in dimensions, configurations, locations and interfaces between products be necessary due to use of other than the specified products of the specified manufacturer, such changes shall be made by the Contractor, subject to review by the Architect, at no change in Contract Sum and Contract Time.

e. The determination of equivalence will be made by the Architect and such determination shall be final.

7. Visual Matching: Where Specifications require matching a sample, the decision by the Architect on whether a proposed product matches shall be final. Where no product visually matches but the product complies with other requirements, comply with provisions for substitutions for selection of a matching product in another category.

8. Selection of Products: Where requirements include the phrase "as selected from manufacturer's standard colors, patterns and textures" or a similar phrase, selections of products will be made by indicated party or, if not indicated, by the Architect. The Architect will select color, pattern and texture from the product line of submitted manufacturer if all other specified provisions are met.

C. SUBSTITUTIONS

1. Substitutions: Requests by Contractor to deviate from specified requirements for products, materials, equipment and methods, or to provide products other than those specified, shall be considered requests for substitutions except under the following conditions:
   a. Substitutions are requested during the bidding period and accepted prior to execution of the Contract. Acceptance shall be in the form of written Addendum to the Bidding documents or revision to the Drawings or Specifications for use as Construction Contract Documents.
   b. Changes in products, materials, equipment and methods of construction are directed by the Owner or Architect.
   c. Contractor options for provision of products and construction methods are specifically stated in the Contract Documents.
   d. Change in products, materials, equipment and methods of construction is required for compliance with Codes, ordinances, regulations, orders and standards of authorities having jurisdiction.

2. Substitution Provisions: Refer to substitution provisions of the Bidding and Contract Requirements, in addition to the requirements specified herein. Provisions for consideration and acceptance of substitutions shall be as follows:
   a. Documentation: Substitutions will not be considered if they are indicated or implied on shop drawings, product data or sample submittals. All requests for substitution shall be by separate written request from Contractor. Contractor shall utilize Substitution Request form provided by Owner.
   b. Cost and Time Considerations: Substitutions will not be considered unless a net reduction in Contract Sum or Contract Time results to the Owner's benefit, including redesign costs, life cycle costs, plancheck and permit fees, changes in related Work and overall performance of building systems.
   c. Design Revision: Substitutions will not be considered if acceptance will require substantial revision of the Contract Documents or will substantially change the intent of the design, in the opinion of the Architect. The intent of the design shall include functional performance and aesthetic qualities.
   d. Data: It shall be the responsibility of the Contractor to provide adequate data demonstrating the merits of the proposed substitution, including cost data and information regarding changes in related Work.
   e. Determination by Architect: Architect will determine the acceptability of proposed substitutions and will notify Contractor, in writing within a reasonable time, of acceptance or rejection. The determination by the Architect regarding functional performance and aesthetic quality shall be final.
   f. Non-Acceptance: If a proposed substitution is not accepted, Contractor shall immediately provide the specified product.
   g. Substitution Limitation: Only one request for substitution will be considered for each product.

3. Request for Substitution Process:
   a. Contractor shall prepare a request for substitution and submit the request to the Architect for review and acceptance. Submit a minimum of 4 copies. Form and other administrative requirements shall be as directed by the Architect.
   b. Substitution requests shall included complete product data, including drawings and descriptions of products, fabrication details and installation procedures. Include samples where applicable or requested.
   c. Substitution requests shall include appropriate product data for the specified product(s) of the specified manufacturer, suitable for use in comparison of characteristics of products.
      1) Include a written, point-by-point comparison of characteristics of the proposed substitute product with those of the specified product.
      2) Include a detailed description, in written or graphic form as appropriate, indicating all changes or modifications needed to other elements of the Work and to construction to be performed by the

Owner and by others under separate contracts with Owner that will be necessary if the proposed substitution is accepted.

    d. Substitution requests shall include a statement indicating the substitution's effect on the Construction Schedule. Indicate the effect of the proposed substitution on overall Contract Time and, as applicable, on completion of portions of the Work for use by Owner or for work under separate contracts by Owner.

    e. Except as otherwise specified, substitution requests shall include detailed cost data, including a proposal for the net change, if any, in the Contract Sum.

    f. Substitution requests shall include signed certification that the Contractor has reviewed the proposed substitution and has determined that the substitution is equivalent or superior in every respect to product requirements indicated or specified in the Contract Documents, and that the substitution is suited for and can perform the purpose or application of the specified product indicated or specified in the Contract Documents.

    g. Substitution requests shall include a signed waiver by the Contractor for change in the Contract Time or Contract Sum because of the following:

        1) Substitution fails to perform adequately.

        2) Substitution requires changes in other elements of the Work.

        3) Substitution causes problems in interfacing with other elements of the Work.

        4) Substitution is determined to be unacceptable by authorities having jurisdiction.

    h. If, in the opinion of the Architect, the substitution request is incomplete or has insufficient data to enable a full and thorough review of the intended substitution, the substitution may be summarily refused and determined to be unacceptable.

4. Contract Document Revisions:

    a. Should a Contractor-proposed substitution or alternative sequence or method of construction require revision of the Contract Drawings or Specifications, including revisions for the purposes of determining feasibility, scope or cost, or revisions for the purpose of obtaining review and approval by authorities having jurisdiction, revisions will be made by Architect or other consultant of Owner who is the responsible design professional, as approved in advance by Owner.

    b. Services of Architect or other responsible design professional for researching and reporting on proposed substitutions or alternative sequence and method of construction shall be paid by Contractor when such activities are considered additional services to the design services contracts of the Owner with Architect or other responsible design professional.

    c. Costs of services by Architect or other responsible design professional of the Owner shall be paid on a time and materials basis, based on current hourly fee schedules, with reproduction, long distance telephone and shipping costs reimbursable at cost plus usual and customary mark-up for handling and billing.

    d. Such fees shall be paid whether or not the proposed substitution or alternative sequence or method of construction is ultimately accepted by Owner and a Change Order is executed.

    e. Such fees shall be paid from Contractor's portion of savings if a net reduction in Contract Sum results. If fees exceed Contractor's portion of net reduction, Contractor shall pay all remaining fees unless otherwise agreed in advance by Owner.

    f. Such fees owed shall be deducted from the amount owed Contractor on the Application for Payment next made following completion of revised Contract Drawings and Specifications or completion of research and other services. Owner will then pay Architect or other consultant of the Owner.

D. SYSTEM COMPLETENESS

1. System Completeness:

    a. The Contract Drawings and Specifications are not intended to be comprehensive directions on how to produce the Work. Rather, the Drawings and Specifications describe the design intent for the completed Work.

    b. It is intended that all equipment, systems and assemblies be complete and fully functional even though not fully described. Provide all products and operations necessary to achieve the design intent described in the Contract Documents.

    c. Refer to related general requirements specified in Section 01410 - Regulatory Requirements regarding compliance with minimum requirements of applicable codes, ordinances and standards.

2. Omissions and Misdescriptions: Contractor shall report to Architect immediately when elements essential to proper execution of the Work are discovered to be missing or misdescribed in the Drawings and Specifications or if the design intent is unclear.
    a. Should an essential element be discovered as missing or misdescribed prior to receipt of Bids, an Addendum will be issued so that all costs may be accounted for in the Contract Sum.
    b. Should an obvious omission or misdescription of a necessary element be discovered and reported after execution of the Agreement, Contractor shall provide the element as though fully and correctly described, and a no-cost Change Order shall be executed.
    c. Refer to related general requirements specified in Section 01310 - Project Management and Coordination regarding construction interfacing and coordination.

E. TRANSPORTATION, DELIVERY AND HANDLING

1. Transportation, Delivery and Handling, General: Comply with manufacturer's instructions and recommendations for transportation, delivery and handling, in addition to the following.

2. Transportation: Transport products by methods to avoid product damage.

3. Delivery:
    a. Schedule delivery to minimize long-term storage and prevent overcrowding construction spaces. Coordinate with installation to ensure minimum holding time for items that are flammable, hazardous, easily damaged or sensitive to deterioration, theft and other losses.
    b. Deliver products in undamaged condition in manufacturer's original sealed container or packaging system, complete with labels and instructions for handling, storing, unpacking, protecting and installing.

4. Handling:
    a. Provide equipment and personnel to handle products by methods to prevent soiling, marring or other damage.
    b. Promptly inspect products on delivery to ensure that products comply with contract documents, quantities are correct, and to ensure that products are undamaged and properly protected.

F. STORAGE AND PROTECTION

1. Storage and Protection, General: Store and protect products in accordance with manufacturer's instructions, with seals and labels intact and legible.
    a. Periodically inspect to ensure that products are undamaged and are maintained under required conditions.
    b. Products damaged by improper storage or protection shall be removed and replaced with new products at no change in Contract Sum or Contract Time.
    c. Store sensitive products in weathertight enclosures.

2. Inspection Provisions: Arrange storage to provide access for inspection and measurement of quantity or counting of units.

3. Structural Considerations: Store heavy materials away from the structure in a manner that will not endanger supporting construction.

4. Weather-Resistant Storage:
    a. Store moisture-sensitive products above ground, under cover in a weathertight enclosure or covered with an impervious sheet covering. Provide adequate ventilation to avoid condensation.
    b. Maintain storage within temperature and humidity ranges required by manufacturer's instructions.

   c. For exterior storage of fabricated products, place products on raised blocks, pallets or other supports, above ground and in a manner to not create ponding or misdirection of runoff. Place products on sloped supports above ground.
   d. Store loose granular materials on solid surfaces in a well-drained area. Prevent mixing with foreign matter.

5. Protection of Completed Work:
   a. Provide barriers, substantial coverings and notices to protect installed Work from traffic and subsequent construction operations.
   b. Remove protective measures when no longer required and prior to Substantial Completion review of the Work.
   c. Comply with additional requirements specified in Section 01500 - Temporary Facilities and Controls.

G. INSTALLATION OF PRODUCTS

1. Installation of Products:
   a. Comply with manufacturer's instructions and recommendations for installation of products, except where more stringent requirements are specified, are necessary due to Project conditions or are required by authorities having jurisdiction.
   b. Anchor each product securely in place, accurately located and aligned with other Work.
   c. Clean exposed surfaces and provide protection to ensure freedom from damage and deterioration at time of Substantial Completion review. Refer to additional requirements specified in Section 01740 - Cleaning Requirements.

END OF SECTION 01600

SECTION 01740 - CLEANING REQUIREMENTS

A. SUBMITTALS

1. Product List: Submit complete list of all cleaning agents and materials for Owner's review and approval.

2. Cleaning Procedures: Submit description of cleaning processes, agents and materials to be used for final cleaning of the Work. Processes and degree of cleanliness shall be as directed by Architect. All cleaning processes, agents and materials shall be subject to Architect's review and approval.

B. QUALITY ASSURANCE

1. Cleaning and Disposal Requirements, General: Conduct cleaning and disposal operations in compliance with all applicable codes, ordinances and regulations, including waste management and environmental protection laws, rules and practices.

C. CLEANING MATERIALS

1. Cleaning Agents and Materials: Use only those cleaning agents and materials which will not create hazards to health or property and which will not damage or degrade surfaces.
   a. Use only those cleaning agents, materials and methods recommended by manufacturer of the material to be cleaned.
   b. Use cleaning materials only on surfaces recommended by cleaning agent manufacturer.

D. CLEANING DURING CONSTRUCTION

1. Waste Management: Control accumulation of debris, waste materials and rubbish; periodically dispose of debris, waste and rubbish off-site in a legal manner.

2. Cleaning, General: Clean sidewalks, driveways and streets frequently to maintain public thoroughfares free of dust, debris and other contaminants.

3. Cleaning of Existing Facilities: Clean surfaces in existing buildings where alteration and renovation Work is being performed or where other construction activities have caused soiling and accumulation of dust and debris.
   a. Clean dust and soiling from floor surfaces.
   b. Clean dust from horizontal and vertical surfaces, including lighting fixtures.
   c. Clean HVAC filters.

4. Parking Area Cleaning: Keep parking areas clear of construction debris, especially debris hazardous to vehicle tires.

5. Thoroughfare Clearing and Cleaning: Keep site accessways, parking areas and building access and exit facilities clear of mud, soiling and debris.
   a. Remove mud, soil and debris and dispose of in a manner which will not be injurious to persons, property, plant materials and site.
   b. Comply with runoff control requirements stated above and as required by authorities having jurisdiction.

6. Cleaning Frequency: At a minimum, clean Work areas daily.

7. Failure to Clean: Should cleaning by Contractor not be sufficient or acceptable to Architect, especially regarding paths of travel, Owner may engage cleaning service to perform cleaning and deduct costs for such cleaning from sums owed to Contractor.

E. CONTRACT COMPLETION REVIEW CLEANING, GENERAL

1. Contract Completion Review Cleaning, General: Execute a thorough cleaning prior to Contract Completion review by Architect. Complete final cleaning before submitting final Application for Payment. Employ professional building cleaners to thoroughly clean building.

F. INTERIOR CLEANING

1. Interior Cleaning:
   a. Clean each surface or unit to the condition expected in a normal commercial building cleaning and maintenance program.
   b. Remove grease, mastic, adhesives, dust, dirt, stains, fingerprints, labels and other foreign materials from all visible interior and exterior surfaces.
   c. Remove dust from all horizontal surfaces not exposed to view, including light fixtures, ledges and plumbing fixtures.
   d. Clean all horizontal surfaces to dust-free condition, including tops of door and window frames, tops of doors and interiors of cabinets and casework.
   e. Owner will perform sanitary cleaning of food service, toilet and shower spaces, fixtures and equipment.
   f. Remove waste and surplus materials, rubbish and temporary construction facilities, utilities and controls.

2. Accessories and Fixtures Cleaning: Clean building accessories, including toilet partitions, fire extinguisher cabinets, lockers and toilet accessories, all plumbing fixtures and all lighting fixture lenses and trim.

3. Glass and Mirror Cleaning: Clean and polish all glass and mirrors as specified in Section 08810 - Glass and Glazing.

4. Metalwork: Clean and buff all metalwork to be free of soiling and fingerprints. Mirror finished metalwork shall be buffed to high luster.

5. Floor Cleaning:
   a. Exposed concrete floors: Thoroughly sweep and wet mop floors in enclosed spaces. At parking areas and ramps, sweep and hose off floor surface.
   b. Ceramic tile flooring: Thoroughly sweep and mop tile flooring. Comply with specific requirements of tile and installation materials manufacturers for cleaning materials.
   c. Resilient flooring: Thoroughly sweep all resilient flooring. Damp wash and wax (as appropriate) all resilient flooring. Comply with specific requirements in applicable resilient flooring Sections and notes of the Drawings.
   d. Carpet cleaning: Comply with accepted industry practices for cleaning commercial carpet, subject to review and acceptance by Architect. Vacuum, spot clean and generally clean carpet using commercial carpet cleaning solution, scrubbers and solution extraction-type vacuuming equipment.

6. Ventilation System Cleaning: Replace filters and clean heating and ventilating equipment used for temporary heating, cooling and ventilation.

G. EXTERIOR CLEANING

1. Building Exterior Cleaning: Clean exterior of adjacent facilities where construction activities have caused soiling and accumulation of dust and debris.
   a. Wash down exterior surfaces to remove dust.
   b. Clean exterior surfaces of mud and other soiling.
   c. Clean exterior side of windows, including window framing.

2. Site Cleaning: Broom clean exterior paved surfaces. Rake clean other surfaces of the grounds.
   a. Wash down and scrub where necessary all paving soiled as a result of construction activities. Thoroughly remove mortar droppings, paint splatters, stains and adhered soil.
   b. Remove from the site all construction waste, unused materials, excess soil and other debris resulting from the Work. Legally dispose of waste.

H. CLEANING INSPECTION

1. Cleaning Inspection: Prior to Final Payment or acceptance by Owner for partial occupancy or beneficial use of the premises, Contractor and Architect shall jointly conduct an inspection of interior and exterior surfaces to verify that entire Work is acceptably clean.

2. Inadequate Cleaning: Should final cleaning be inadequate, as determined by Architect, and Contractor fails to correct conditions, Owner may engage cleaning service under separate contract and deduct cost from Final Payment.

END OF SECTION 01740

SECTION 01770 - CONTRACT CLOSEOUT PROCEDURES

A. COMPLETION REVIEWS

1. Correction (Punch) List: Contractor shall prepare and distribute a typewritten, comprehensive list of items to be completed and corrected (punch list) to make the Work ready for acceptance by the Owner.

2. Substantial Completion Review: On a date mutually agreed by the Owner and Contractor, a meeting shall be conducted at the Project site to determine whether the Work is satisfactory and complete for filing a Notice of Completion (Substantial Completion).
   a. Contractor shall provide 3 working days' notice to Owner for requested date of Substantial Completion meeting.
   b. In addition to conducting a walk-through of the facility and reviewing the punch list, the purpose of the meeting shall include submission of warranties, guarantees and bonds to the Owner, submission of operation and maintenance data (manuals), provision of specified extra materials to the Owner, and submission of other Contract closeout documents and materials as required and if not already submitted.
   c. Contractor shall correct the punch list and record additional items as may be identified during the walk-through, including notations of corrective actions to be taken.
   d. Contractor shall retype the punch list and distribute it within 3 working days to those attending the meeting.

3. Final Completion Inspection: Submit a written request for final inspection for Acceptance of the Work and Contract closeout.
   a. Upon receipt of request, Architect will either proceed with inspection or notify Contractor of unfulfilled requirements.
   b. Architect will prepare a final Certificate for Payment after satisfactory inspection or will notify Contractor of construction that requires completion or correction before Certificate will be issued.

B. CONTRACT CLOSEOUT

1. Closeout Actions: Before requesting Substantial Completion review and filing of Notice of Completion, complete the following.

2. Keying: Make final changeover of permanent locks and deliver keys to Owner. Advise Owner's personnel of changeover in security provisions.

3. System Starting, Testing and Adjusting:
   a. Complete startup testing and inspection.
   b. Perform adjustments.
   c. Balance HVAC system and submit balance report.

4. Temporary Facilities and Controls: Terminate and remove temporary facilities from Project site, along with mockups, construction tools and similar elements.
   a. Advise Owner of changeover to permanent utilities.
   b. Coordinate with Owner the establishment of permanent accounts and metering of utility services.

C. FINAL COMPLETION SUBMITTALS

1. Final Completion Submittals, General: Prior to final Application for Payment, complete and submit the following.

2. Agency Document Submittals: Submit to Owner all documents required by authorities having jurisdiction, including serving utilities and other agencies. Submit original versions of all permit cards, with final sign-off by inspectors. Submit all certifications of inspections and tests.

3. Project Record Drawings: Maintain and submit one set of blue- or black-line white prints of Contract Drawings and Shop Drawings.
   a. Mark Record Prints to show the actual installation where installation varies from that shown originally. Require individual or entity who obtained record data, whether individual or entity is Installer, subcontractor, or similar entity, to prepare the marked-up Record Prints.

    b. Give particular attention to information on concealed elements that cannot be readily identified and recorded later.

    c. Mark record sets with erasable, red-colored pencil. Use other colors to distinguish between changes for different categories of the Work at the same location.

    d. Identify and date each Record Drawing; include the designation "PROJECT RECORD DRAWING" in a prominent location. Organize into manageable sets; bind each set with durable paper cover sheets. Include identification on cover sheets.

4. Project Record Specifications: Submit one copy of Project Manual, including Addenda and Contract Modifications. Mark copy to indicate actual products installed where installation varies from that indicated in Specifications, Addenda, and Contract Modifications.

    a. Give particular attention to information on concealed products and installations that cannot be readily identified and recorded later.

    b. Mark copy with the proprietary name and model number of products, materials and equipment furnished, including substitutions and product options selected.

5. Operating and Maintenance Data Submittals: Assemble a complete set of operation and maintenance data indicating the operation and maintenance of each system, subsystem and piece of equipment not part of a system. Include operation and maintenance data required in individual Specification Sections and as follows:

    a. Operation Data: Include emergency instructions and procedures, system and equipment descriptions, operating procedures and sequence of operations.

    b. Maintenance Data: Include manufacturer's information, list of spare parts, maintenance procedures, maintenance and service schedules for preventive and routine maintenance, and copies of warranties and bonds.

    c. Organize operation and maintenance manuals into suitable sets of manageable size. Bind and index data in heavy-duty, 3-ring, vinyl-covered, loose-leaf binders, in thickness necessary to accommodate contents, with pocket inside the covers to receive folded oversized sheets. Identify each binder on front and spine with the printed title "OPERATION AND MAINTENANCE MANUAL," Project name and subject matter of contents.

6. Guaranty and Warranty Submittals: Submit written guaranties and warranties prior to Final Completion Inspection.

    a. Organize warranty documents into an orderly sequence based on the Table of Contents of the Project Manual.

    b. Bind warranties and bonds in heavy-duty, 3-ring, vinyl-covered, loose-leaf binders, thickness as necessary to accommodate contents, and sized to receive 8½ by 11 inch paper.

7. Products Submittals: Submit to Owner all documents and products required by Specifications to be submitted, including the following:

    a. Keys and keying schedule.

    b. Tools, spare parts, extra materials and similar items, delivered to location designated by Owner. Label products with manufacturer's name and model number where applicable.

    c. Test reports and certificates of compliance.

8. Certificates of Compliance and Test Report Submittals: Submit to Owner certificates and reports as specified and as required by authorities having jurisdiction, including the following:

    a. Sterilization of water systems.

    b. Sanitary sewer system tests.

    c. Gas system tests.

    d. Lighting, power and signal system tests.

    e. Ventilation equipment and air balance tests.

    f. Fire sprinkler system tests, if applicable.

    g. Roofing inspections and tests, if applicable.

9. Lien and Bonding Company Releases: Submit to Owner evidence of satisfaction of encumbrances on Project by completion and submission of The American Institute of Architects Forms G706 - Contractor's Affidavit of Payment of Debts and Claims, G706A - Contractor's Affidavit of Release of Liens, and (if applicable) G707 - Consent of Surety. Comply also with other requirements of Owner, as directed. Signatures shall be notarized.

10. Warranty Documents: Prepare and submit to Owner all warranties and bonds as specified in Section 01785 - Warranties and Bonds.

11. Insurance: Advise Owner of pending insurance changeover requirements.

12. Final Payment: Make final Application for Payment as specified in Section 01290 - Price and Payment Procedures.

END OF SECTION 01770

END OF DIVISION 1 SPECIFICATIONS

Appendix C

Sample Preliminary Project Description

INTRODUCTION

Following is a sample Preliminary Project Description (PPD) for a mythical project to construct the shell and core of a privately funded medical office building. This document represents project information at the end of the Schematic Design phase or early portion of the Design Development phase of the project.

The sample PPD does not include cross references to the construction specifications that *UniFormat*® (1998 edition) includes. Including additional specifications information, with appropriate *MasterFormat*® cross references, requires design information beyond what is available at the Schematic Design phase. The more comprehensive PPD, with *MasterFormat* references, is valuable for detailed estimates of probable construction costs and value engineering of various design options. However, it exceeds what is known about the design at the end of Schematic Design, when the drawings are sketchy and requirements are still to be determined.

Note: The technical content of the following PPD should not be used for actual projects. Select and verify products in compliance with actual project requirements, including applicable codes, regulations, and environmental and budgetary criteria.

PRELIMINARY PROJECT DESCRIPTION (PPD)

PROJECT DESCRIPTION

10 PROJECT DESCRIPTION

1010 Project Summary
1. Construct new two-story medical office building and interior core, approximately 15,300 sf, IBC Type V construction, 10'-8" floor to floor height, 4' high parapets, interior ceiling height 9'-0", plus site development of vehicle parking, pedestrian paving, landscaping, site lighting and site amenities.

1020 Project Program
1. The Project shall comply with applicable Codes, ordinances, regulations and standards of the City of Columbus, OH, including the series of Codes published by the International Code Council (ICC).
   a. Design and products shall comply with accessibility regulations adopted by State of Ohio.
   b. Design and products shall comply with State of Ohio energy conservation laws and regulations.
   c. Construction waste management shall comply with City of Columbus ordinances.
2. Construction product selections and execution shall be comparable to those of medical office buildings in suburban setting, with emphasis on low maintenance and an image of professionalism.

1040 Owner's Work
1. Under separate, concurrent contracts, Owner will construct tenant space improvements for dental, medical and laboratory occupants.
2. Under separate, concurrent contracts, Owner will contract for signage, graphics and artwork in common areas.

20 PROPOSAL, BIDDING AND CONTRACTING

2010 Delivery Method
1. Building shell and interior core construction will be constructed under a single prime contract. Contractor will be selected by competitive bidding by invited, prequalified General Contractors.
2. A Construction Manager has been engaged by the Owner and will act as the Owner's representative during pre-construction and construction activities.
3. The Construction Manager will administer the Contract for construction, with the Architect and the Architect's consultants acting in advisory roles.

2020 Qualifications Requirements
1. Prospective General Contractors will be invited by the Owner, through the Construction Manager, to submit Statements of Qualifications.
2. Prospective General Contractors will be prequalified by the Construction Manager and the Owner.

2040 Bid Requirements
1. Prequalified General Contractors will be invited by Owner, through the Construction Manager, to submit competitive bids on a date and at a time determined by the Construction Manager.
2. Instructions to Bidders will be prepared and issued by the Construction Manager.
3. Information available to Bidders includes:
   a. Site survey.
   b. Geotechnical investigation.

2050 Contracting Requirements
1. Agreement Form: AIA A101 - Standard Form of Agreement Between Owner and Contractor - Stipulated Sum 1997 edition.

2. General Conditions of the Contract: AIA A201 - General Conditions of the Contract for Construction 1997 edition.
3. Supplementary Conditions of the Contract: As developed by Construction Manager and Owner's legal and insurance counsels.

30 COST SUMMARY

3010 Elemental Cost Estimate
1. Project Budget: Initial budget is $1,350,000, as of August 2004.

3020 Assumptions and Qualifications
1. Construction Schedule:
   a. Assumed Notice to Proceed: February 10, 2005.
   b. Assumed start of construction: March 10, 2005.
   c. Assumed Contract closeout (shell and core construction): November 10, 2005.
2. Cost Escalation: Assumed to be 24% per year through February 2005 and 36% per year from March 2005 through March 2006.
3. Qualifications: Delay of Notice to Proceed beyond March 30, 2005 will require re-estimate of probable construction cost.

3030 Allowances
1. Door hardware allowance: $43,200.
2. Lobby floor tile: $37,000.
3. Elevator cab interior: $18,000.

3040 Alternates
1. Roofing: Built-up asphalt roofing system vs. reinforced thermoplastic olefin (TPO) single-ply roofing system.
2. HVAC Units: Products by Trane only.
3. Music/Paging System: Add for common areas.

A SUBSTRUCTURE

A10 FOUNDATIONS

A1010 Standard Foundations
1. Perimeter Wall Foundations: Continuous spread footings of steel reinforced portland cement concrete, extending 36" below frost line.
2. Column Foundations: Spread footings of steel reinforced portland cement concrete, extending 24" below finished floor.

A1030 Slabs on Grade
1. Aggregate Base: Crushed stone, 4" deep.
2. Vapor Retarder: Coextruded PVC sheet, complying with ASTM E 1745, Class A permeance and Class B puncture resistance, and 0.012 perms when tested according to ASTM E 96.
3. Concrete Floor Slab: Portland cement concrete, 4" thick with minimum #4 steel reinforcing bars at 16" on center each way, cured by sheet method only, with Floor flatness (F_F) and floor levelness (F_L) according to ACI 117, as follows:
   a. Interior concrete slab on grade floors, to be covered by direct application of ceramic tile: Smooth trowel finish with light broom texture; F_F35-F_L25 SOV (Specified Overall Value)/F_F24-F_L16 MLV (Minimum Local Value).
   b. Interior concrete slab on grade floors to receive adhesively applied or loosely laid floor covering: Smooth trowel finish; F_F35-F_L25 SOV/F_F24-F_L16 MLV.

      c. Interior floors to remain exposed, in service areas and equipment rooms: Smooth trowel finish; F_F35-F_L25 SOV/F_F24-F_L16 MLV.

B SHELL

B10 SUPERSTRUCTURE

B1010 Floor Construction
    1. Floor Structural Frame: Braced structural steel frame of steel columns and beams.
      a. Typical wide flange members: ASTM A 572, Grade 50.
      b. Wide flange members used in braced frame: ASTM A 992.
      c. Other steel shapes, bars and plates: ASTM A 572, Grade 50.
    2. Floor Deck: Composite concrete and steel decking, concrete mix design with $F'_c = 3000$ psi, regular weight aggregate, with shear stud connectors and steel reinforcing of #4 bars at maximum 24″ on centers, finishes as for concrete slabs-on-grade.

B1020 Roof Construction
    1. Roof Structural Frame: Steel joists and joist-girders, complying with SJI Specifications, with columns of varying height to achieve primary roof drainage:
      a. Typical steel roof joists: SJI - Open Web Steel Joists, K-Series.
      b. Long-span steel roof joists: SJI - Longspan Steel Joists, LH-Series.
      c. Joist girders: SJI - Joist Girders, G-Series.
    2. Light Steel Framing: ASTM A 572, Grade 50 or ASTM A 36.
    3. Roof Deck: Steel decking, hot-dipped galvanized.

B20 EXTERIOR CLOSURE

B2010 Exterior Walls
    1. Exterior Wall Framing: Cold-formed, light gage steel studs, C-shape, galvanized finish, 6″ wide, metal thickness as designed by manufacturer according to American Iron and Steel Institute (AISI) - Specification for the Design of Cold Formed Steel Structural Members, for L/240 deflection.
    2. Exterior Wall Sheathing: Glass-mat gypsum sheathing board complying with ASTM C 79, ⅝″ thick, non-rated, DensGlass® Gold exterior gypsum sheathing, or equal.
    3. Weather-Resistant Barriers:
      a. Primary barrier: Typar® HouseWrap as manufactured by Reemay, Inc., complying with ASTM E 1677, Type I air retarder.
      b. Secondary barrier: Rubberized sheet membrane flashing, suitable for above-grade temperatures, installed at parapet, wall openings and sloped and horizontal projections.
    4. Sheet Metal Flashing and Trim: Hot-dipped galvanized, minimum.
    5. Exterior Finish: Smooth finish plaster.
      a. Base coats: Polymer-modified portland cement plaster.
      b. Finish coat: Acrylic basecoat with fiberglass reinforcing mesh, sandable non-aggregate smooth portland cement coating and patching compound, and three-coat acrylic paint (epoxy primer + two coats finish).
    6. Insulation: Fiberglass batt, friction-fit, R-19, Flexible, resilient, noncombustible blankets of mineral or glass fiber, with vapor-retarding kraft facing at concealed conditions and fire-resistant, foil-reinforced-kraft (FRK) facing exposed conditions.
    7. Decorative Shapes: Proprietary plaster-coated expanded polystyrene foam shapes, adhered to plaster base coat with acrylic basecoat.
    8. Plaster Control Joints, reveals and trim: Hot-dipped galvanized steel, minimum 0.0207″ thick.
    9. Wall Louvers: Extruded aluminum, drainable blades, nominal 6″ deep, finish to match aluminum storefront framing, with insect screens.

B2020 Exterior Windows
1. Design Criteria: According to IBC for project location, framing members sized for net deflection not to exceed L/175 times span or maximum ¾″, and thermal expansion and contraction of 180 degrees F.
2. Windows: Aluminum storefront sections with operable sash, matching storefront.
3. Storefronts and Entrance Framing: Aluminum extrusions with thermal break, nominal 2″ wide by 4½″ deep, for 1″ insulating glass.
4. Finish: Polyvinylidine fluoride (PVDF) coating, PPG Duranar XL metallic finish to be selected, with clear top coat at doors.
    a. Primer: Minimum 03. mil (01. mil) dry film thickness.
    b. Color coat: 1.0 mil (–0 mil, +01. mil) dry film thickness.
    c. Top coat: 08. mil (–0 mil, +01. mil) dry film thickness.
5. Glass: Insulating glass units consisting of outboard light of gray tinted glass and interior light of low-E clear glass, 1″ thick and using tempered safety glass where required by IBC.
6. Joint Sealers:
    a. Exterior: Single-component non-sag urethane sealant, ASTM C 834 Type S, Grade NS, Class 25.
    b. Interior: Single-component, non-sag, mildew-resistant, acrylic-emulsion sealant, paintable.

B2030 Exterior Doors
1. Exterior Flush Doors: Comply with ANSI A250/SDI-100.
    a. Steel door frames: Formed steel (hollow metal), minimum 16 gage, galvanized with shop primer finish.
    b. Steel doors: Flush steel, Level 2 and Physical Performance Level B (Heavy Duty), Model 1 Full Flush, 16 gage (0.053″/1.3 mm) (0.042″/1.0 mm) sheet steel faces. Provide louvered doors at Electrical and Mechanical rooms.
    c. Steel doors and frames finish: Field-applied, two-component aliphatic urethane coating.
2. Aluminum Entrance Doors: Narrow stile design, with finish and glazing to match aluminum storefront and entrance framing.
    a. Hinges: Continuous aluminum.
    b. Closers: Concealed overhead.
    c. Exit devices: Von Duprin 99 Series.
    d. Pulls: Offset 1″ diameter stainless steel rod.
    e. Locksets: Adams-Rite, with lock cylinder to match building keying.
    f. Finish: Match aluminum storefront framing, including clear topcoat.

B30 ROOFING

B3010 Roof Construction
1. Roof Deck Insulation: Composite perlite and polyurethane foam insulation of varying thickness to achieve secondary roof drainage and cricketing, minimum insulation thickness to achieve R-30 rating.
2. Roofing: single ply flexible membrane system of reinforced thermoplastic olefin (TPO), fully-adhered, light color, 60 mil thickness and reinforced (no primary liquefied plasticizers modifiers permitted).
    a. Listing: UL Class A, Wind Uplift: I-90 minimum.
    b. Total System Warranty: 15 year manufacturer's warranty, no dollar limit.
3. Parapet Copings: Factory-manufactured, formed aluminum with galvanized steel pans and anchoring system, aluminum finish to match aluminum-framed storefronts and windows.
4. Counterflashing: Recessed galvanized steel reglets and stainless steel counterflashing.

B3020 Roof Openings
1. Unit Skylights: Formed acrylic sheet on metal curb assembly, double dome, with outboard sheet of gray tinted acrylic and interior sheet of clear acrylic, sizes as indicated on the Drawings.
2. Roof Hatch: 36″ square, thermal insulated, with integral curb, aluminum hatch and safety railings.

C INTERIORS

C10   INTERIOR CONSTRUCTION

C1010 Partitions
1. Non-Load Bearing Metal Framing: Minimum 22 gage galvanized metal studs, 16″ on center, 3⅝″ wide unless greater width required according to manufacturer's published tables for L/240 maximum deflection.
   a. Provide deep leg bottom track.
   b. Provide flexible head track where studs extend from floor to underside of building framing or floor or roof decking.
   c. Provide 18 gage studs or doubled 22 gage studs at door openings.
   d. Provide proprietary backing plates for attachment of casework and accessories to partitions.
2. Gypsum Board: ⅝″ Type X fire-resistive typically.
3. Acoustical Insulation: Acoustical batts, typical except where partition has cased opening.
4. Firestopping and Smoke Seals: Empty openings and openings containing cables, pipes, ducts, conduits and penetrating items through fire resistance rated walls and partitions, including voids around:
   a. Structural members.
   b. Pipes, conduits, cable trays, cables and wires.
   c. HVAC ductwork.

C1020 Interior Doors
1. Flush Wood Swinging Doors: Plain sliced white oak veneer, bookmatched and balanced, with cherry stain and clear varnish, factory-finished, reinforced for door hardware, edges to match face veneer, 1¾″ thick.
   a. Non-rated: Particleboard core except laminated veneer lumber (LVL) where glazed opening exceeds 14″ wide and 24″ high.
   b. 20-minute fire-rated: Particleboard core.
   c. Fire-rated 45-minutes and greater: Mineral fiber core.
   d. Door hardware: Comply with applicable wheelchair accessibility and handicapped persons' use.
      1) Hinges: 3-knuckle design, plain bearing at non-rated doors and ballbearing at doors with closers.
      2) Closers: Surface overhead, adjustable closing force, LCN 4040 series, parallel arm.
      3) Exit devices: Von Duprin 99 Series.
      4) Locksets: Schlage L9000 Series, backset 2¾″, lever handle 17A design.
      5) Keying: Schlage Primus patented keyway, with interchangeable cores, masterkeyed.
      6) Metal finish: BHMA 613 oil-rubbed bronze, except BHMA 625 polished chrome in toilet rooms.
2. Access Doors: Rated and non-rated, galvanized steel typically except stainless steel in toilet rooms.
   a. Frames: 16 gage.
   b. Doors and panels: 14 gage.
3. Glazing:
   a. Non-rated doors: Clear float glass, tempered.
   b. Fire-rated doors: Proprietary fire-resistive glass, clear.

C1030 Fittings
1. Toilet Partitions: Solid plastic (HDPE polymer) panels, 10 percent recycled materials, marble finish, color to be selected, commercial-grade hardware, except hardware at accessible stalls shall comply with applicable wheelchair accessibility and handicapped persons' use requirements.
   a. Toilet compartments: Floor-to-ceiling pilasters.
   b. Urinal screens: Stainless steel post, floor to ceiling, continuous wall channel at wall anchorage.
2. Wall Protection: At all corridors.
   a. Corner guards: Resilient plastic cover over aluminum retainer, from top of base to ceiling, at all corners. Provide custom covers for corners less than 90 deg.
   b. Wall bumper/handrail assembly: Resilient plastic cover over aluminum support, handrail grip in compliance with applicable handicapped use requirements.

    c. Wall protection: Resilient plastic, impact- and abrasion-resistant, from top of base to underside of wall bumper/handrail assembly.

C20 STAIRS

C2010 Stair Construction
1. Service Stairs: Pre-engineered, factory-manufactured steel stairs with concrete-filled metal pan treads and landings, with tubular steel handrails and guardrails.
2. Lobby Stairs: Ornamental steel stairs, custom fabricated, with grout-filled metal pan treads and landing, with glass-supported bronze railing.

C2020 Stair Finishes
1. Service Stair Finish: Factory-primed and field-painted semi-gloss finish on steel components; exposed concrete treads and landings with hardener/sealer.
2. Lobby Stair Finish: Treads and landing with ceramic tile finish to match Lobby floor, railing with powder-coated metallic paint finish, exposed stringers and landing framing clad with pre-formed aluminum panels with powder-coated paint finish to match railing.

C30 INTERIOR FINISHES

C3010 Wall Finishes
1. Lobby Walls:
    a. Feature wall: Wood paneling, plain sliced white oak, bookmatched, cherry stain and clear varnish finish.
    b. Other walls: Vinyl wallcovering to match typical corridors.
2. Corridor Walls: Vinyl wallcovering over prepared gypsum board, heavy weight, pattern and color to be selected.
3. Toilet Rooms: Glazed ceramic tile over portland cement tile backer board, floor to ceiling, tile size nominal 8", with 2" accent trim, thinset.
4. Mailroom and Service Stairs: Gypsum board with GA 214 Level 4 finish, one coat PVA sealer and two coats semi-gloss acrylic enamel paint.
5. Leasing Office: Gypsum board with GA 214 Level 5 finish, one coat PVA sealer and two coats eggshell low-sheen acrylic enamel paint.

C3020 Floor Finishes
1. Lobby Floor: Porcelain tile, matte finish nominal 18" square field tile with 3" × 3" and 2" × 6" polished accent tile, "medium" bed latex portland cement mortar set over anti-fraction sheet membrane on portland cement concrete slab on grade.
2. Corridors: Carpet, multi-level loop, manufacturer, series, pattern and color to be selected, heavy commercial grade, one color throughout, glue-down installation on prepared concrete slab, with resilient rubber base.
3. Leasing Office: Building tenant space standard carpet, cut pile, commercial grade, direct glue-down on prepared concrete slab, with resilient rubber base.
4. Mailroom: Vinyl composition tile, 12" square, white/gray heavy texture, with resilient rubber base and rubber reducer strips.

C3030 Ceiling Finishes
1. Lobby Ceiling: Linear metal acoustical ceiling, polished, tinted bronze finish, with matching light fixture trim and linear HVAC diffusers, nominal 4" wide metal strips with 1" gap and black acoustical insulation.
2. Corridors: Suspended T-bar grid ceiling, 9/16" bottom flange with ¼" reveal, with 24" × 24" fine, non-directional acoustical panels with regressed edges for flush bottom with ceiling grid, white paint finish on grid and panels.

3. Leasing Office and Mailroom: Tenant space standard suspended T-bar grid ceiling, ⁹⁄₁₆″ bottom flange, with 24″ × 24″ fine, non-directional acoustical panels with regressed edges for flush bottom with ceiling grid, white paint finish on grid and panels.
4. Toilet Rooms: Suspended T-bar grid for direct attachment of gypsum board, with ½″ proprietary Fire-code C gypsum board, GA 214 Level 4 finish, one coat PVA sealer and two coats semi-gloss acrylic enamel paint.

D SERVICES

D10 CONVEYING SYSTEMS

D1010 Elevators and Lifts
1. Hydraulic Passenger Elevator: Pre-engineered telescopic holeless design.
   a. Performance:
      1) Capacity: 4,500 pounds.
      2) Landings served: 2.
      3) Number of openings: 2, in line.
      4) Travel distance: As indicated on the Drawings.
      5) Speed: 100 feet per minute, full load, up.
      6) Operation: Single Car Automatic.
   b. Car design: Lightweight design, sheet steel walls with applied panels covered with custom hardwood veneer paneling and custom bronze handrails at sides and rear.
      1) Car size: 5′8″ wide by 7′9½″ deep, clear inside. Clear inside dimensions of car shall comply with all applicable wheelchair accessibility and medical emergency regulations.
      2) Car inside height: 8′0″; clear from floor to underside of metal ceiling.
      3) Car doors: Net opening 4′0″ by 7′0″, two-speed single slide opening, handing as indicated on the Drawings.
      4) Car front: Satin bronze.
      5) Ceiling: Metal panels, satin bronze.
      6) Lighting: Custom recessed HID fixtures, mounted in metal ceiling.
   c. Corridor hoistway frame and doors: Satin bronze.
   d. Corridor fixtures: Premium grade, satin bronze.
   e. Acoustical isolation and insulation: Mufflers and insulation on operating components, flexible connections on hydraulic and electrical piping, acoustical isolators at piping penetrations, acoustical insulation and resilient furring at gypsum board finishes.

D20 PLUMBING SYSTEMS

D2010 Plumbing Fixtures
1. Fixture Types, Shell and Core Construction: Lavatories, water closets and urinals, Kohler as quality basis, white color typically, accessible design.
   a. Water Closets: Vitreous china, siphon jet, wall-hung, water conserving 1.5 gallon type, with infrared sensor flush valve and open-front solid plastic seat.
   b. Urinals: Vitreous china, wall-hung, siphon jet, water conserving 1 gallon, with infrared sensor flush valve.
   c. Lavatories: Undercounter-mount, with infrared sensor 0.5 gpm flow-restricted faucet, hot and cold water.
   d. Service sink: Floor sink with wall-mounted faucet with hose and nozzle.
   e. Drinking fountains: Integral chiller units.
   f. Hose bibbs: At building exterior, freeze-proof, and in mechanical room.

D2020 Domestic Water Distribution
1. System shall include water piping, fittings, valves, specialties and insulation.

      a. Domestic water service: Metered service to building and fire system, extending from water main to 5′ outside building.
        1) Service piping: As required by serving utility.
        2) Meter: By serving utility, paid by Contractor and reimbursed at no mark-up by Owner.
      b. Building domestic water distribution: Through mains, risers and branches to plumbing fixtures and equipment, using copper piping throughout.
  2. Circulation Pump: On domestic hot water system.
  3. Hot Water Heater: Commercial-grade hot water heater with 200-gallon storage capacity, with heat exchanger for heating mediums from hot water boiler and rooftop solar panels.

D2030 Sanitary Waste
  1. System includes waste and vent piping, fittings and connections.
      a. Plumbing fixture drains by gravity through soil, waste and vent stacks, to 5′ outside building.
      b. Piping: Cast-iron no-hub on drain lines and ABS on vent lines, if permitted by Code.

D2040 Rain Water Drainage
  1. Roof Drains: Cast iron, with sump and integral overflow, suitable for roofing type.
  2. Roof Drainage Piping: No-hub cast iron, from roof to connection to site underground storm drain line. Overflow lines shall exit at planter or paving adjacent to building entrances for visibility.

D2060 Miscellaneous Piping Systems
  1. Natural Gas System:
      a. Service from site service connection 5′ from building to mechanical rooms and rooftop HVAC units.
      b. Tenant space service: Routed from mechanical room through both floors, for connection of branches to serve tenant spaces as necessary.

D30   HEATING, VENTILATING, AND AIR CONDITIONING SYSTEMS

D3010 Energy Supply
  1. Flat Plate Solar Collectors: Rooftop mounted, for domestic hot water boosting, with heat exchanger tank, pumps and controls.

D3020 Heat Generation
  1. Low Pressure Heating Hot Water Boiler: Capacity to suit domestic hot water and HVAC uses, natural gas-fired.
      a. Domestic hot water heating by heat exchanger in hot water tank.
      b. Boiler controls integrated with flat plate solar collectors and domestic hot water service.
  2. Boiler flue: Commercial grade, triple-insulated, stainless steel with isolators, connectors and flue cap.

D3030 Refrigeration
  1. Chilled Water System:
      a. Packaged water-to-air heat pumps, rooftop mounted with vibration isolation.
      b. Storage tank.
      c. Chilled water system controls, to integrate with energy management and control system.

D3040 HVAC Distribution
  1. Air Distribution System:
      a. Air handling units: One serving each floor, drawing in and filtering outside air with electronic air cleaning device.
      b. Sheet metal ductwork: Constructed to SMACNA standards, with thermal and acoustical insulation.
        1) Provide fire dampers where ductwork passes through fire-rated construction.
        2) Interconnect fire dampers with fire alarm system, to close dampers.
      c. Air inlets: Return air grilles in T-bar acoustical panel ceilings and gypsum board ceilings.

   d. Air outlets: Linear metal diffusers for Lobby ceiling and painted metal diffusers suitable for specified T-bar acoustical panel ceiling system.
2. Hydronic Distribution System: 2-pipe system hot water distribution system from hot water boiler to air terminal units.
3. Exhaust Fans: Serving toilet rooms, ducted to rooftop fans.

D3050 Terminal and Packaged Units
1. Terminal Heat Transfer Units: Fancoil units in Tenant spaces, with refrigerant coil for cooling and hydronic heating coil for space heating.

D3060 HVAC Controls
1. Pre-engineered control system for control of ventilation, heating and cooling and integrated with access control and lighting control.

D3080 HVAC Testing, Adjusting, and Balancing
1. Testing Agency: Independent testing agency engaged by Contractor and paid by Owner.
2. HVAC Controls: Checking and adjusting for proper operation.
3. Air Balancing: For Lobby, corridors and Leasing Office.

D40   FIRE PROTECTION SYSTEMS

D4010 Sprinkler System
1. Wet-Pipe Fire Sprinkler System: Design-build system complying with NFPA 13 and requirements of Fire authority having jurisdiction.
   a. Locate riser and check station in Mechanical Room.
   b. Provide recessed white sprinkler heads typically.
   c. Provide exposed brass sprinkler heads in service areas and elevator hoistway.
   d. Interconnect flow sensors with fire detection and alarm system.

D4020 Standpipes
1. Fire Protection Standpipe System: Located where required by Fire authority having jurisdiction, with capacity and connections according to standards of Fire Department.

D4030 Fire Protection Specialties
1. Fire Extinguisher Cabinets: Recessed cabinets with fire-rated tub, located where required by Fire authority having jurisdiction.
   a. Cabinet door design: Solid face with "FIRE EXTINGUISHER" notation.
   b. Fire extinguishers: Dry-chemical type typically and carbon dioxide type in equipment rooms; capacity and installation height according to requirements of Fire authority.

D50   ELECTRICAL SYSTEMS

D5010 Electrical Service and Distribution
1. Electric Power: Comply with NFPA 70 (National Electrical Code) and requirements of serving electric utility.
   a. New underground electric service from on-site transformer on concrete pad at site perimeter, by serving electric utility. Serving utility shall be responsible for medium voltage trenching, conduit and conductors.
   b. New utility-owned meter and customer-owned switchgear in electric utility room.
   c. Customer-owned main switchgear.
2. Distribution: Distribution panels, transformers, and subpanels within building. Line voltage to be determined.

3. Electric Branch Circuit Panelboards: In utility closets on each floor, to serve tenant spaces, with separate metering capability for each tenant space. Provide branch circuit panelboards for common areas, elevator, HVAC systems and exterior lighting.
4. Motor Control Centers: For HVAC equipment.

D5020 Lighting and Branch Wiring
1. Electric Branch Wiring: Comply with NFPA 70 (NEC), no aluminum conductors.
   a. Provide commercial-grade wiring devices.
   b. Provide isolated ground to each tenant space.
   c. Provide electrical identification for all circuits and outlets.
2. Interior Lighting, Common Areas:
   a. Lobby lighting: High-intensity discharge (HID), recessed in linear metal ceiling system plus HID display lighting for artwork.
   b. Corridors: Recessed fluorescent luminaires in T-bar ceiling grid, 24″ × 24″, gold parabolic reflectors.
   c. Mailroom and Leasing Office (building tenant space standard): 24″ × 48″ recessed fluorescent in T-bar ceiling grid, four-lamp, by-level switching in Leasing Office with occupancy sensors.
   d. Emergency exit lighting: On battery system, for corridors and stairways, as required by Code, integrated in general lighting luminaires. Batteries shall be located above ceiling in corridors.

D5030 Communication and Security System
1. Fire Detection and Alarm System: Addressable system, integrated with building automation system and HVAC system, as required by Fire authority having jurisdiction.
   a. Smoke detectors shall be as required by authority having jurisdiction, including smoke detectors in HVAC ductwork.
   b. Connections to fire dampers and ductwork smoke detectors shall be included.
   c. Comply with visual and audible alarm requirements of authority having jurisdiction.
2. Telephone System:
   a. Service by serving telephone company, from street to terminal board in electric room, including high-speed data communications.
   b. Telephone service to tenant spaces from electrical room to telephone terminal board at each floor; conduit with conductors routed to tenant spaces.
   c. Telephone equipment not in contract (NIC).
3. High-Speed Data Communications: Provide separate open conduit with pull cord, from telecommunications terminal board.
4. Cable Television Service: From street to telecommunications terminal board, by serving utility. From terminal board to terminal board at each floor by Contractor, with coaxial cable to each tenant space for connection and equipment by tenant.
5. Public address and music system: Scope and performance characteristics to be determined.

E EQUIPMENT AND FURNISHINGS

E10   EQUIPMENT

E1090 Residential Equipment
1. Residential Kitchen Equipment: For office break areas (Leasing Office only):
   a. Undercabinet microwave oven: General Electric, white-on-white, Profile series.
   b. Refrigerator: General Electric, white-on-white, Profile series, with icemaker and chilled water dispenser, nominal 27 cubic feet. Provide water filters under sink and serving sink faucet (cold water), hot water dispenser and refrigerator ice maker/chilled water storage.
   c. Hot water dispenser: Mounted in sink.
   d. Garbage disposer: In-Sink-Erator.

E20   FURNISHINGS

E2010 Fixed Furnishings

1. Custom Plastic-Laminate Faced Casework: For office break areas (Leasing Office only), AWI Custom Grade.
   a. Cabinets: Flush overlay design, solid color matte finish laminate as selected by Architect, polished chrome plated 4″ wire pull, base cabinet height to suit accessibility regulations.
   b. Countertops: Solid surfacing countertop, 1-inch front edge, 4″ back and side splashes, Avonite brand, pattern and color as selected by Architect.
2. Window Treatment: Vertical blinds, all exterior windows, PVC vanes, perforated, color as selected by Architect, heavy commercial grade.

F SPECIAL CONSTRUCTION AND DEMOLITION

F10   SPECIAL CONSTRUCTION

F1040 Special Facilities

1. Koi Pond: Interior pool with natural stone surround, waterfall, filtration and circulation system and underwater lighting; custom-design to be determined. Plants and fish by Owner.

G BUILDING SITEWORK

G10   SITE PREPARATION

G1010 Site Clearing

1. Tree Relocation: Existing oak tree, transplanted to location indicated on Drawings, including temporary removal and storage on site during building and underground utility construction.
2. Clearing and grubbing of site to 24″ below new finish grades.

G1020 Site Demolition and Relocations

1. Building Demolition: Existing storage buildings, including garage adjacent to existing residence, including removal of foundations, footings and concrete slabs on grade; removal of basement walls and concrete slabs on grade.
2. Structure Moving: Existing residence, under separate contract by Owner.
3. Paving: All existing asphaltic concrete paving, gravel surfacing, portland cement concrete sidewalks, curbs and gutters within property lines.
4. Utility Demolition: Removal of underground water, natural gas, sanitary sewer and storm drainage lines; removal of existing septic tank and leach field; removal of overhead electric power and telephone lines, including removal of all utility poles.
5. Earthwork: Overexcavate and recompact soil at building pad, drive lanes and parking areas, according to geotechnical report furnished by Owner's geotechnical engineering consultant.
   a. Under pavement subject to vehicular traffic: 95% relative compaction.
   b. Under pavement subject to pedestrian traffic: 90% relative compaction.
   c. Under building footings, foundations and slabs on grade: 95% relative compaction.
   d. Landscape areas: 90% relative compaction except top 12″ scarified for planting.
6. Storm Water Pollution Protection: Develop and implement Storm Water Pollution Protection Plan as acceptable to authority having jurisdiction.

G20   SITE IMPROVEMENTS

G2020 Parking Areas

1. Parking Lot Base Course: Granular base course, depth and material according to geotechnical report furnished by Owner's geotechnical engineering consultant.

2. Flexible Parking Lot Pavement: Asphaltic concrete paving, medium size aggregate with seal coat, complying with State Department of Transportation standards, with thickness of paving in drive lanes and at trash enclosure to suit loads of fire apparatus and garbage collection vehicle.
3. Decorative Concrete Paving: Imprinted ("stamped") integral color concrete at driveway entrance to parking lot.
4. Pavement Markings: Comply with State Department of Transportation standards for traffic control markings and City standard designs for parking stalls; provide wheelchair accessible stall markings according to requirements of authority having jurisdiction.
5. Parking and Traffic Control Signage: As required by authority having jurisdiction, complying with State Department of Transportation standards.

G2030 Pedestrian Paving
1. Walkway Paving, from Parking Areas to Building Entrances: Integral color concrete with medium abrasive blast finish and water repellent.
2. Walkway Paving, from Building Emergency Exits and Pedestrian Paving: Natural color concrete with medium broom texture and water repellent.

G2040 Site Development
1. Site Perimeter Fencing: Omit at street frontage, 8' high precast concrete planks with brick pilasters.
2. Monument Sign: Non-illuminated, brick base and ends, precast concrete sign panel, design to be determined; sign illumination by ground-mounted HID fixtures, both sides.

G2050 Landscaping
1. Underground Irrigation System: For trees, plants and ground covers, commercial grade, with automatic controller.
2. Planting Area Preparation: Re-use stockpiled topsoil and add additional topsoil as necessary; fertilize planting soil according to agronomic report commissioning and paid by Contractor; provide anodized aluminum edging at planting area perimeter.
3. Specimen Tree: Transplant existing oak tree.
4. Trees, Shrubs and Ground Cover: New trees, shrubs and ground covers, species and sizes to be developed, with commercial-grade supports and accessories.
5. Landscape Maintenance: For 1 year from Contract closeout, including fertilizing, insect control, pruning and irrigation system maintenance and adjustments.

G30   SITE CIVIL AND MECHANICAL UTILITIES

G3010 Water Supply
1. Domestic Water Supply: By serving utility to water meter inside property line at street side of property. Service line from meter to building by Contractor. Owner will pay for utility charges, including connection fees.
2. Fire Protection Water Supply: By serving utility to point of connection of building fire sprinkler system. Owner will pay for utility charges, including connection fees.

G3020 Sanitary Sewer System
1. Sewer Main: From point of connection in street to building point of connection 5' outside of building foundation, by serving sanitary sewer agency. Owner will pay for utility charges, including connection fees. Include cleanouts in sanitary sewer lines.

G3030 Storm Sewer System
1. Storm Drain Main: From point of connection in street to point of connection to site storm drain point of connection 5' inside of property line, by serving storm sewer agency. Owner will pay for utility charges, including connection fees.
2. Onsite Storm Drainage: Underground RCP storm drain lines, precast concrete catch basins with galvanized steel gratings. Include connections for roof rainwater leader connections.

G40   SITE ELECTRIC UTILITIES

G4010 Electrical Distribution
1. Transformer and Disconnect: By serving electric utility, on concrete pad at site perimeter.
2. Underground Electric Service Line: From transformer to building electrical room and meter.

G4020 Site Lighting System
1. Parking Lot Illumination: Pole mounted, cut-off high-pressure sodium, providing illumination according to City maximum allowable.
2. Exterior Building Illumination: For security, all sides of building, wall-mounted except ground-mounted on street side, HID luminaires.
3. Walkway Illumination: Precast concrete bollards with HID lamps.
4. Landscape Illumination: In-ground recessed HID luminaires, uplighting for trees.

Z GENERAL

Z10   GENERAL REQUIREMENTS

Z1010 Administration
1. Project Management, Payment Procedures: To be determined and issued by Construction Manager.
   a. Retainage: 10%.
   b. Lien releases: Each month, covering previous month's application for payment.
2. Project Coordination:
   a. Project Meetings:
      1) Weekly construction progress meetings.
      2) Monthly construction progress/payment application review meeting.
      3) Meeting administration by Construction Manager; requirements to be determined.
3. Construction Progress Schedule: CPM schedule produced by Contractor, with degree of detail and administration according to directions of Construction Manager.
4. Design Submittals: For deferred approval and design-build elements, as directed by Construction Manager and according to requirements of authority having jurisdiction.
5. Contract Modifications: Procedures as directed by Construction Manager, in accordance to the Conditions of the Contract for:
   a. Requests for Interpretation (RFIs).
   b. Change Order Requests.
   c. Requests for Proposal.
   d. Change Orders.
   e. Construction Change Directives.
   f. Architect's Supplemental Instructions.
6. Construction Submittals: Requirements specified by Architect, consistent with General Conditions of the Contract.
   a. Product data.
   b. Shop drawings.
   c. Calculations.
   d. Certifications.
   e. Installation Instructions.
   f. Operation and maintenance data.
   g. Extra materials and spare parts.
   h. Warranties and guaranties.

Z1020 Quality Requirements
1. Quality Assurance and Control:
   a. Independent testing and inspection agency: Selected and paid by Owner.

    b. Regulatory requirements: Contractor's responsibilities for compliance with applicable Codes, ordinances and standards.

    c. Contractor's quality control: According to Conditions of the Contract.

    d. Field Mock-Ups: Exterior wall with window/storefront.

Z1030 Temporary Facilities and Controls

1. Temporary Utilities: By Contractor, with utility service charges included in Contract Sum.
   a. Temporary water.
   b. Temporary heating and ventilating.
   c. Temporary power and lighting.
2. Field Offices: Separate mobile offices for Contractor and for Construction Manager/Testing Laboratory.
   a. Contractor's office shall include conference space for 12 persons.
   b. Construction Manager/Testing Laboratory office shall include furnishings, office equipment, consumables and cleaning service.
3. Temporary Barriers and Enclosures:
   a. Make building weathertight as soon as practicable and maintain closures.
   b. Secure site with temporary chainlink fencing with windscreen and gates.
4. Project Identification: Project sign according to design provided by Architect.

Z1040 Project Closeout

1. Requirements for Substantial Completion Review: "Punch List," according to instructions to be specified by Architect and Construction Manager.
2. Start-Up and Adjusting:
   a. Testing and inspection of building services:
      1) Domestic water, including sterilization.
      2) HVAC systems, including refrigerant, hot water and air systems.
      3) Electrical power: As required by authority having jurisdiction.
   b. Elevator inspection: As required by authority having jurisdiction; ride shall be adjusted to satisfaction of Architect.
3. Final Completion Inspection:
   a. Inspections completed according to requirements of authority having jurisdiction.
   b. "Punch List" (correction list) completion review.
4. Completion Submittals:
   a. Project record drawings.
   b. Site survey.
   c. Operation and maintenance instructions.
   d. Warranties and guaranties.
   e. Final keying of door hardware.
   f. Spare parts and extra materials.
5. Final Payment: Requirements as specified by Architect and administered by Construction Manager.
   a. Procedures for final application for payment.
   b. Filing of Notice of Completion with County recorder.
   c. Requirements for lien releases.
   d. Final payment and commencement of one-year Correction Period.

Z1050 Permits, Insurance and Bonds

1. Permits and Approvals:
   a. Zoning, planning and design reviews and permits: By Owner and Architect.
   b. Building permits: Obtained and paid by Contractor; reimbursed by Owner without mark-up.
2. Construction Insurance: Types and coverages to be provided by Owner through Construction Manager.
3. Bonds:
   a. Faithful Performance Bond: 100% of Contract Sum.
   b. Labor and Material Payment Bond: 100% of Contract Sum.

END OF DOCUMENT

Appendix D

Sample Outline Specifications

INTRODUCTION

Following is a sample set of Outline Specifications for a mythical school modernization project. These specifications have been prepared to demonstrate various levels of detail in Outline Specifications. Some sections have well-developed content, while others are merely generalized statements about what will be specified. Inconsistencies in writing style are intentionally included for the reader to identify and consider how improvement may be made.

Outline specifications are produced during the Design Development phase of a project. Several versions should be produced, if there is time, with information and greater detail added as the design develops. The more information is provided, the more accurate the estimates of probable construction cost should be. Also, with design decisions more clearly expressed prior to beginning the Contract Documents phase, the more efficient and consistent will be the production of the Contract Drawings and Contract Specifications.

Note: The technical content of these outline specifications should not be used for actual projects. Select and verify products in compliance with actual project requirements, including applicable codes, regulations, and environmental and budgetary criteria.

FRED ROGERS ELEMENTARY SCHOOL MODERNIZATION

Owner: Smallville Unified School District
Architect: PDQ Associates, LLC, Architecture - Planning - Interior Design - Kitchenware
Civil Engineer: Doze, Digg and Phil, Inc., Civil Engineers
Landscape Architect: Green Side Up, Landscape Architecture
Structural Engineer: William F. Shaky & Associates, Inc., Structural Engineers
Mechanical Engineer: Pypes and Dux, Consulting Engineers
Electrical Engineer: Sparks & Terror, Consulting Engineers
July 17, 2004

OUTLINE SPECIFICATIONS

SERIES 0 - BIDDING AND CONTRACT REQUIREMENTS

District-produced documents:

ADVERTISEMENT FOR BIDS
INSTRUCTIONS TO BIDDERS
BID FORM
AGREEMENT FORM
GENERAL CONDITIONS OF THE CONTRACT

DIVISION 1 - GENERAL REQUIREMENTS

01100 - Summary of Work

A. Project Description:
    1. Construction type.
    2. Number of stories/buildings.
    3. Floor area.

B. Summary of Work under the Contract

C. Summary of concurrent Work under separate contracts

D. Owner-Furnished/Contractor-Installed (OFCI) products

E. Obtaining and paying for permits, licenses and fees

F. Construction sequence and schedule

G. Use of Work area
    1. Contractor's use of Work area
    2. Owner's continued occupancy and use of existing facilities

01230 - Alternate Bid Procedures

A. Alternate Bid items to be determined. Preliminary list of Alternate Bids:
1. Roofing:
   a. Base Bid: Built-up asphalt roofing system; see Section 07511.
   b. Alternate Bid: Cold-applied built-up roofing system; see Section 07520.

01250 - Contract Modification Procedures

A. Requirements for Requests for Interpretation (RFIs).

B. Requests for Proposal (RFPs).

C. Change Orders.

D. Construction Change Directives.

E. Architect's Supplemental Instructions.

F. Reconciliation of Change Orders.

01290 - Measurement and Payment Procedures

A. Preparation and submission of Applications for Payment.

B. Requirements for substantiating data.

C. Final application for payment and Contract closeout.

01310 - Project Management and Coordination

A. Coordination responsibilities of Contractor.

B. Requirements and procedures for meetings.

01320 - Construction Progress Documentation

A. Construction Progress Schedule: CPM schedule using Primavera software, version as directed by Construction Manager.

B. Submittals Schedule: Coordinated with construction progress schedule.

01330 - Submittals Procedures

A. Product Submittals: Requirements for preparation, submission, review and filing for:
1. Product data.
2. Shop drawings.

   3. Samples.
   4. Certifications.

B. Test and Inspection Reports: Requirements during construction and at Contract Closeout.

01410 - Regulatory Requirements

A. Authority and precedence of Codes, ordinances and standards.

B. Applicable Codes, laws, ordinances and regulations applicable to performance of Work under the Contract.

01420 - Reference Standards, Abbreviations and Terminology

A. Resources and application of reference standards.

B. Typical abbreviations.

C. Terminology and definitions.

01450 - Quality Control

A. Regulatory requirements for tests and inspections.

B. Independent testing and inspection agency (Testing Laboratory), selection, payment and limitations on authority.

C. Contractor's responsibilities for quality control.

D. Observation of construction by Architect and other responsible design professionals.

E. Summary of required inspections and tests.

01500 - Temporary Facilities and Controls

A. Temporary Utilities and Services: Requirements for:
   1. Heating and cooling during construction.
   2. Ventilation during construction.
   3. Temporary water service.
   4. Temporary sanitary facilities.
   5. Temporary power and lighting.
   6. Temporary telephone service.

B. Temporary construction barriers, enclosures and passageways.

C. Runoff control, including Storm Water Pollution Protection Plan (SWPPP).

D. Field offices and storage sheds, including separate mobile offices for Contractor and for Construction Manager/Project Inspector.

E. Parking and traffic controls.

F. Work area security.

G. Removal of construction facilities and temporary controls, including restoration.

01600 - Product Requirements

A. General requirements for quality, completeness and integration of products.

B. Requirements applicable to product options.

C. Requirements for product substitutions.

D. General requirements for delivery, storage and handling.

01700 - Execution Requirements

A. Execution Requirements: Requirements for application, erection and installation of products, including:
    1. Examination.
    2. Preparation.
    3. Execution.
    4. Cleaning.
    5. Starting and adjusting.
    6. Protection of completed Work.

B. Progress cleaning.

C. Contract closeout procedures.

D. Operation and maintenance data.

E. Product warranties and guaranties.

F. Project record documents.

01810 - Commissioning

A. Requirements for preparation of Commissioning Plan.

B. System performance evaluations.

C. General requirements for testing, adjusting and balancing.

DIVISION 2 - SITE CONSTRUCTION

02222 - Selective Demolition

A. Selective Demolition: Removal of portions of existing building, as indicated on Drawings, Including:
    1. Removal of building utility services, such as power and signal circuits and including capping and identification.

   2. Removal of designated building equipment and fixtures.
   3. Removal of designated walls, partitions and components, including cutting of new openings in existing construction for new doors, plumbing HVAC and electrical components.
   4. Removal and protection of existing fixtures, materials and equipment items indicated as "salvage."

B. Handling and disposal of removed materials, including recycling requirements.

02225 - Removals and Relocations

A. Removal and salvage of designated existing products.

B. Relocation and reinstallation of salvaged products.

02230 - Site Clearing

A. Clearing of plant life and grass, surface rocks and debris.

B. Removal of minor existing construction within Project area.

C. Grubbing of root systems of trees and shrubs, abandoned utility lines and structures, and other below grade obstructions.

D. Protection of trees, landscaping, site improvements, utilities, and other items not scheduled for clearing or that might be damaged by construction activities.

E. Handling and disposal of debris, including compliance with waste management regulations.

02300 - Earthwork

A. Site Preparation: Excavation, filling, backfilling and grading of site, including:
   1. Stockpiling of soil for later use under the Contract.
   2. Shoring and underpinning.
   3. Foundation excavations.
   4. Dewatering.

B. Trenching and backfilling for underground utilities.

C. Capillary break stone fill under slabs-on-grade: Washed, evenly graded mixture of crushed stone, or crushed or uncrushed gravel, ASTM D 448, coarse aggregate grading size 57, with 100 percent passing a 1½ inch (38 mm) sieve and not more than 5 percent passing a No. 8 (2.36 mm) sieve to provide, when compacted, a smooth and even surface below slabs on grade.

02530 - Sanitary Sewerage System

A. New sanitary sewer lines from new toilet rooms to existing sanitary sewer line in street.

02551 - Site Gas Distribution

A. New natural gas service line to new building addition from on-site gas main.

02741 - Asphaltic Concrete Paving

A. Patching of existing asphaltic concrete (a.c.) paving where cut for new underground utilities.

B. New a.c. paving over Class 2 aggregate base at playground expansion, including fine aggregate top course of a.c.

C. Repair of existing a.c. paving at locations designated on Drawings.

02751 - Portland Cement Concrete Paving

A. New natural-color portland cement concrete paving at locations designated on Drawings, with medium texture broom finish.

B. New integral-color, exposed aggregate portland cement concrete at reconstructed Courtyard.

02765 - Pavement Marking

A. Parking and traffic control markings on asphaltic concrete paving and portland cement concrete curbs.

B. Playground markings on asphaltic concrete paving.

02810 - Landscape Irrigation Systems

A. Modifications to existing landscape irrigation system to suit reconfigured and expanding planting areas for trees, shrubs and groundcovers.

B. New irrigation system, including controller, for reconstructed turf playfields.

02821 - Chain Link Fencing

A. Modifications to existing chainlink fencing, including accessible gate hardware and additional fencing to enclose expanded turf playfields; fencing shall be green PVC coated to match existing.

02910 - Planting Preparation

A. Removal and disposal of existing plant materials in existing planting areas.

B. Removal of existing turf and sterilization of existing soil.

C. Soil amendments and fertilizers in planting areas, including turf areas.

D. Weed abatement.

E. Black anodized extruded aluminum edgings at perimeter of planting areas where no adjoining paving.

02920 - Lawns and Grasses

A. Hydroseeding of turf areas.

02950 - Landscape Planting

A. New trees at parking lot adjacent to Administration Building.

B. Shrubs and ground cover at reconfigured planting areas adjacent to new building addition.

DIVISION 3 - CONCRETE

03300 - Cast-in-Place Concrete

A. Formwork: With shoring, bracing and anchorage, designed by Contractor, comply with ACI 318.
   1. Lumber: Douglas fir or douglas fir-larch, grade appropriate for intended use, sound and undamaged straight edges, solid knots.

B. Forming Materials: For form surfaces in contact with concrete at exposed conditions, comply with ACI 301.
   1. Forms for exposed finish concrete: Plywood, metal, metal-framed plywood faced, or other acceptable panel-type materials to provide continuous, straight, smooth, exposed surfaces. Furnish in largest practicable sizes to minimize number of joints and to conform to joint system shown on Drawings.
   2. Forms for unexposed finish concrete: Plywood, lumber, metal or another acceptable material. Provide lumber dressed on at least two edges and one side for tight fit. When unexposed concrete is intended to receive waterproofing, provide forming materials as for exposed finish concrete.
   3. Fillets for chamfered corners: Wood molding at plywood or lumber forms; rigid plastic at steel, fiberglass and plastic forms.

C. Reinforcing:
   1. Reinforcing steel bars: ASTM 615, type and grade as noted on Structural Drawings.
   2. Welded steel wire fabric: Not used. Provide reinforcing steel bars as concrete slabs-on-grade.

D. Concrete Materials:
   1. Portland cement: ASTM C 150, Type I or Type II, gray color.
   2. Aggregates: ASTM C 33
   3. Admixtures: Comply with ASTM C 494.
      a. Water-reducing admixture: ASTM C 494, Type A.
      b. Accelerating or retarding admixtures: ASTM C 494 for Type C or Type B.
      c. Plasticizer: Conform to ASTM C 494, Type F.
   4. Bonding compound: Polyvinyl acetate, acrylic or styrene butadiene base. Provide polyvinyl acetate compound at interior locations only.

E. Concrete Mixes:
   1. Cast-in-place concrete footings and foundations, $F_c = 3000$ psi at 28 days.

F. Cast-in-place retaining wall, $F_c = 3000$ psi at 28 days.
   1. Concrete slab-on-grade floors at new building addition, $F_c = 4000$ psi at 28 days, ACI 117 tolerances:

G. Floor Slab Finishes:
   1. Floated finishes: Depressions between high spots shall not exceed ¼-inch under a 10-foot straightedge.
   2. Troweled finishes: Achieve level surface plane so that depressions between high spots do not exceed ⅛-inch, using a 10-foot straightedge.

03650 - Cementitious Underlayment

A. Cementitious Underlayment:
1. Self-leveling underlayment for floor slab surface restoration and leveling, ASTM C 349, 4100 psi in 28 days.
2. Surface patching and filling underlayment: Suitable from feather-edge to 1-inch thick, fast-setting.
3. Slab crack filler.

B. Manufacturers:
1. Specified manufacturer: Ardex, Inc., Coraopolis, PA (412/264-4240).
2. Acceptable manufacturers:
   a. Sonneborn Building Products, Minneapolis, MN (612/835-3434 or 800/243-6739).
   b. Dayton Superior Corp., Miamisburg, OH (513/866-0711 or 800/745-3700).
   c. Symons Corp., Des Plaines, IL (708/298-3200 or 800/800-7966).

DIVISION 4 - MASONRY

04822 - Reinforced Concrete Unit Masonry

A. Hollow Load Bearing Units: ASTM C 90, Grade N, Type I, two core type, modular sized to 6 × 8 × 16, unless otherwise indicated on Drawings.
1. Weight Classification: Medium weight.
2. Face designs and colors:
   a. Type 1: Smooth, Precision masonry units, gray color, for locations to receive applied finish.
   b. Type 2: Split Face (S1S and SVSC), tan color to be selected.
   c. Type 3: Smooth, Precision masonry unite, black color with honed surface one side and exposed ends.
   d. Type 4: Smooth, Precision masonry units, color to match split-face units, with medium abrasive blast finish.

B. Reinforcing: ASTM A 615, yield grade as indicated on Drawings; deformed billet steel bars; plain finish. Comply with requirements specified in Section 03200 - Reinforcing Steel.

C. Veneer Anchors: Fleming Masonry Anchoring System manufactured by Meadow Burke Products, Converse, TX (210/658-6883), consisting of formed steel anchor channel attached to structure and t-shaped steel anchors designed to engage a continuous wire embedded in veneer mortar joint. Anchoring system shall allow vertical or horizontal adjustment but resist tension and compression forces perpendicular to plane of wall, for attachment over sheathing to wall studs.

D. Mortar and Grout Materials:
1. Portland cement: ASTM C 150, Type as indicated on (Structural) Drawings.
   a. Mortar, concealed conditions: Gray color.
   b. Mortar, exposed conditions: Tinted color to match concrete masonry units.
2. Mortar materials:
   a. Mortar aggregate: ASTM C 144, standard masonry type, clean, dry, protected against dampness, freezing, and foreign matter.
   b. Hydrated lime: ASTM C 207, Type as indicated on Structural Drawings.
   c. Premix mortar: ASTM C 387, using gray cement, Normal strength.
   d. Quicklime: ASTM C 5, non-hydraulic type.
3. Grout materials:
   a. Grout aggregate: ASTM C404.
   b. Grout course aggregate: Maximum ⅜-inch size; 200 percent by volume.
   c. Grout fine aggregate: Washed river sand; 225 percent by volume.

E. Mortar and Grout Mixes:
   1. Combined compressive strength, masonry unit and mortar assembly: F'_m as indicated on (Structural) Drawings. Strength shall be verified by masonry prism tests.
   2. Mortar compressive strength: As indicated on (Structural) Drawings.
   3. Grout compressive strength: As indicated on (Structural) Drawings.

F. Adjustable Masonry-Veneer Anchors: Fleming Masonry Anchoring System manufactured by Meadow Burke Products, consisting of formed steel anchor channel attached to structure and t-shaped steel anchors designed to engage a continuous wire embedded in veneer mortar joint.

DIVISION 5 - METALS

05120 - Structural Steel

A. Steel Shapes:
   1. Typical wide flange members: ASTM A 572 for Grade 50.
   2. Wide flange members used in braced frame: ASTM A 992.
   3. Other steel shapes, bars and plates: ASTM A 36.

B. Structural Steel Pipe: ASTM A53, Grade B, Type E or S.

C. Structural Steel Tubing: ASTM A500, Grade B, $F_y = 46$ ksi.

D. Anchors and Fasteners:
   1. Anchor bolts: ASTM A 307, Grade C.
   2. Standard bolts: ASTM A 307, Grade A.
   3. Plain washers: ASTM F 844 plain (flat) unhardened steel washers.
   4. Nuts: ASTM A 563, Heavy Hex, Grade B, plain (non-zinc coated).
   5. High-strength threaded fasteners: Heavy hex structural bolts, ASTM A 325, Type 1, Supplementary Requirements S.1, with threads included in shear plane and marked "A 325 T," unless otherwise noted on Contract Drawings.
      a. Washers:
         1) Hardened type: ASTM F 436, Type 1, style as required.
         2) Direct tension load indicating type: ASTM F 959 Type 325 or Type 490.
      a. Nuts: ASTM A 563, Heavy Hex, Grade C, plain (non-zinc coated).

E. Welding Materials: AWS D1.1, type as required for materials being welded. Provide electrodes as indicated on Structural Drawings.

F. Shop Primer: According to SSPC-PS Guide 7.00, Guide for Selecting One-Coat Shop Painting Systems, gray color at exposed members.

05300 - Metal Deck

A. Materials:
   1. Galvanized sheet steel: Zinc-coated (galvanized) steel sheet, ASTM A 653/A 653M, Structural Steel (SS), Grade 33 (230), G60 (Z180) zinc coating; structural quality.
   2. Welding rods: AWS D1.1 and AWS D1.3, type as indicated on the Drawings.
   3. Galvanizing repair: Where galvanized surfaces are damaged, prepare surfaces and repair in accordance with procedures specified in ASTM A 780. Use cold galvanizing compound field touch-up, ZRC Zinc Rich Coating or equivalent.

B. Steel Roof Decking: Corrugated steel deck, fabricated from galvanized sheet steel, product, gage, depth and width as indicated on Structural Drawings. Comply with applicable ICC Evaluation Service, Inc. (ICC ES) Evaluation Report, current edition.
1. End laps: Flush, unless otherwise indicated on the Structural Drawings.
2. Side laps: Interlocking, unless otherwise indicated.

C. Minimum Structural Properties: As indicated on Drawings.

05500 - Metal Fabrications

A. Light structural steel framing members and structural steel support members, with required bracing, welding and fasteners.

B. Hot-Dipped Galvanizing: General requirements.

C. Miscellaneous metal fabrications, including:
1. Loose bearing and leveling plates.
2. Steel angle nosings and thresholds.
3. Rough hardware.
4. Sleeves for penetrations through structural members and stud partitions.
5. Fixed metal ladders (roof access ladders).
6. Trash enclosure gates.

D. General requirements for anchors and fasteners:
1. Expansion anchors
2. Powder-actuated driven fasteners.
3. Grouting compounds.

DIVISION 6 - WOOD AND PLASTICS

06100 - Rough Carpentry

A. Lumber for Framing: Softwood lumber, manufactured in compliance with PS 20 - American Softwood Lumber Standard and according to WCLIB or WWPA grading standards as applicable, nominal dimensions indicated.
1. Sill plates: Pressure preservative treated, douglas fir, No. 1 grade or better.
2. Studs: No. 1 grade, douglas fir or douglas fir-larch.
3. Posts:
   a. 4-inches thick, 4-inches wide: No. 1 grade, douglas fir or douglas fir-larch.
   b. 4-inches thick, 6-inches and wider: Select Structural douglas fir.
4. Ceiling and roof joists:
   a. 2- to 4-inches thick, 4-inches wide: No. 2 or better douglas fir or douglas fir-larch.
   b. 2- to 4-inches thick, 6-inches and wider: No. 1 douglas fir or douglas fir-larch.
5. Roof beams: 6-inches and thicker, 6-inches and wider: Select Structural douglas fir.
6. Blocking and bridging: No. 2 and better, douglas fir or douglas fir-larch.
7. Surfacing: S4S, unless otherwise indicated.
8. Moisture Content: All lumber shall be kiln-dried to percent specified below. Air season in place, protected from rain and high humidity conditions, no less than 15 days before applying finish materials.
   a. Concealed lumber: 19 percent maximum moisture content at time of dressing and shipment, unless otherwise indicated.
   b. Exposed lumber and timber: 15 percent at time of delivery, unless otherwise indicated.

B. Construction Panels: APA Performance-Rated Panels, Group 1 Series, PS 1.
   1. Plywood panels for roof sheathing: Douglas fir, Structural I, APA RATED SHEATHING.
      a. Exposure Durability Classification: EXTERIOR.
      b. Thickness: As indicated on the Drawings.
      c. Edge detail, low slope roofs: Square if all edges supported on framing or tongue-and-groove (T&G) if edges are unsupported. Plyclips will not be acceptable.
      d. Edge detail, high slope roofs: Square.
   2. Plywood panels for wall sheathing: Douglas fir, Structural I, APA Rated Sheathing.
      a. Exposure Durability Classification: Exterior.
      b. Thickness: As indicated on the Drawings.
      c. Edge detail: Square.

C. Framing Anchors and Connectors: Simpson Strong-Tie Co., Pleasanton, CA (510/460-9912) or equal.

06181 - Glue Laminated Structural Units

A. Wood Species for Glue Laminated Wood Structural Units:
   1. Lumber for members not exposed to weather: Douglas fir-larch lumber conforming to WCLIB or WWPA grading standards.
   2. Lumber for members exposed to weather: Alaskan Cedar (Pacific Coast Yellow), also known as Alaskan Yellow Cedar (Chamaecyparis Nootkatensis), heartwood.

B. Glue Laminated Wood Structural Units Fabrication: ANSI/AITC A190.1 and AITC 117, using laminating stock and adhesives of type to suit specified Service Grade and exposure.
   1. Members not exposed to view and not exposed to weather: Douglas fir, Industrial appearance grade, Service Grade 24F-V8. Simple span members may be 24F-V4.
   2. Members exposed to view but protected from weather by roof: Douglas fir, Architectural appearance grade, Service Grade 24F-V8.
   3. Members exposed to view and not protected by roof: Alaskan Yellow Cedar, Architectural appearance grade, Service Grade 20F-V12.

C. Connectors, Anchors, and Accessories: Provide stock- and custom-fabricated connectors of structural steel (ASTM A 36) shapes, plates and bars, welded into assemblies of types and sizes indicated, with steel bolts (ASTM A 307), lag bolts, and other fasteners, as indicated and as necessary.

06200 - Finish Carpentry

A. Softwood Lumber and Moldings: For opaque, painted finish, AWI Custom Grade, WWPA Grading Rules, C-Select.

B. Hardwood Lumber and Moldings: For transparent finish, AWI Custom Grade, white oak.

C. Construction Panels: For telecommunication backboards, minimum APA C-D PLUGGED.
   1. Exposure Durability Classification: EXPOSURE 1.
   2. Thickness: As indicated on the Drawings, or, if not otherwise indicated, not less than $^{15}/_{32}$-inch thick.
   3. Fire-retardant treatment: If required by authorities having jurisdiction or serving utility.

06410 - Custom Casework

A. Casework, Classrooms: Plastic-laminate faced casework, AWI Custom Grade, Flush Overlay Construction.
   1. Casework panels: ANSI A208.2, Grade MD, medium density fiberboard (MDF), formaldehyde-free.

2. Plastic laminate, exposed surfaces: NEMA LD 3 and ANSI A161.2., solid color, matte finish, as selected by Architect from manufacturer's full selection of standard and limited production colors.
   a. Vertical applications: NEMA Type GP28 (0.028-inch nominal thickness) or NEMA Type PF42 (0.039-inch nominal thickness).
   b. Horizontal applications, other than countertops, and mill option for vertical applications: NEMA Type PF42 (0.039-inch nominal thickness).
3. Plastic laminate backing: NEMA LD-3, BK 20, high pressure paper base laminate without a decorative finish; Style ND, Type IV, 0.020-inch thick, smooth surface finish, for backing at countertops and other concealed locations.
4. Cabinet liner: High-pressure cabinet liner laminate (HPL), 0.020-inch thick. Provide white color, unless otherwise directed.
5. Panel edging: PVC edge band, to match exposed panel face.

B. Casework, Reception Desk in Administration: Hardwood veneer, AWI Premium Grade, Flush Overlay construction.
   1. Casework panels: Hardwood plywood complying with HPMA standard, veneer (plywood) core.
   2. Hardwood lumber: White oak.
   3. Hardwood veneer: White oak, rotary sliced, book matched.
   4. Finish: Cherry stain and catalyzed polyurethane varnish.

DIVISION 7 - THERMAL AND MOISTURE PROTECTION

07210 - Building Insulation

A. Thermal Batt Insulation: Flexible, resilient, noncombustible blankets of mineral or glass fiber, complying with ASTM C665, type, class and facing material as indicated following.
   1. Concealed conditions: ASTM C 665, Type III, Class C, Category 1, vapor-retarding kraft facing.
   2. Exposed conditions: ASTM C 655, Type III, Class A, Category 1, fire-resistant, foil-reinforced-kraft (FRK) facing.

07511 - Built-Up Asphalt Roofing

A. (Base Bid) Roofing: Hot-applied built-up asphalt roofing system consisting of base ply plus two interplies of asphalt impregnated and coated glass fiber roofing felts, with granule-surfaced glass fiber cap sheet and modified bitumen (SBS) flashing materials, manufacturer's 10 year No Dollar Limit warranty.

07520 - Cold Process Built-Up Roofing

A. (Alternate Bid) Roofing: Cold-applied built-up roofing consisting of rosin paper underlayment, base ply of mineral-surfaced asphalt-coated fiberglass roofing sheet, three plies of coated fiberglass roofing sheet, adhered with cold-applied asphalt based adhesive and surfaced with fiberglass-reinforced asphalt emulsion, and coated with aluminum emulsion reflective surfacing, manufacturer's 15 year No Dollar Limit warranty.

07620 - Sheet Metal Flashing and Trim

A. Extruded aluminum fascia and gravel stop, painted.

B. Flashings at roof projections to be stainless steel.

07710 - Manufactured Roof Specialties

A. Factory-manufactured reglets and counterflashing; stainless steel sheet metal.

B. Factory-manufactured, formed aluminum parapet copings, prefinished, custom color.

07720 - Roof Accessories

A. Roof hatch with integral sheet metal curb, aluminum cover, with safety railing system.

07840 - Firestopping and Smoke Seals

A. Through-Penetration Firestopping at Fire-Rated Construction: Provide firestopping materials and assemblies to seal all penetrations at all fire barriers. Firestopping assemblies shall be listed in the UL Fire Resistance Directory under categories XHCR and XHEZ, providing that such assemblies conform to the construction type, penetration type, annual space requirements and fire-rating requirement for each distinct condition, and that the system shall be symmetrical for wall applications.

07920 - Joint Sealants

A. Elastomeric Sealant Standard: Provide manufacturer's standard chemically curing, elastomeric sealant of base polymer indicated which complies with ASTM C 920 requirements, including those referenced for Type, Grade, Class, and Uses.
   1. Exterior elastomeric joint sealant: One-Part Neutral-Curing Silicone Sealant, Type S, Grade NS, Class 50; suitable for Uses NT, M, G, A and, as applicable to joint substrates indicated, O.
   2. Interior sanitary joint sealant: One-Part Mildew-Resistant Silicone Sealant (Sealant Type 2), Type S, Grade NS, Class 25; suitable for Uses NT, G, A and, as applicable to non-porous joint substrates indicated, O; formulated with fungicide.
   3. Interior floor joint sealant: One-Part Moisture-Cured Polyurethane (Sealant Type 3), Type M, Grade NS, Class 25

B. Interior Calk (Painter's Calk): Non-movement joints not intended to prevent passage of moisture or water, one-part, non-sag, mildew-resistant, acrylic-emulsion sealant complying with ASTM C834, formulated to be paintable.

C. Joint Backing: Preformed, compressible, resilient, non-waxing, non-extruding strips of flexible, non-gassing plastic foam, non-absorbent to water and gas.

DIVISION 8 - DOORS AND WINDOWS

08110 - Steel Doors and Frames

A. Steel Door Frames: Full-formed sheet steel frames for doors, transoms, sidelights, borrowed lights, fixed windows and other openings, of types and styles as shown on Drawings and schedules, with concealed fastenings, welded construction, complying with ANSI A250.
   1. Exterior door frames: 14 gage minimum gage steel, galvanized steel, shop primer finish.
   2. Interior door frames: 16 gage minimum steel, primer painted steel finish.

B. Steel Doors: Full flush panel steel doors, complying with requirements indicated below by reference to ANSI 250.8 for level and model and ANSI A250.4 for physical-endurance level.

1. Exterior steel doors: Level 3 and Physical Performance Level A (Extra Heavy Duty), Model 2 (Seamless), 0.053-inch (1.3 mm) thick faces.
2. Interior steel doors: Not used.

08210 - Wood Doors

A. Flush Wood Doors: Comply with AWI's "Architectural Woodwork Quality Standards Illustrated."
1. Classroom interior doors: Custom (Grade A faces), Natural Birch, rotary cut, for opaque (painted) finish.
2. Administration Building doors: Custom (Grade A faces), White Oak, plain sliced, bookmatched, for cherry color stain and clear polyurethane varnish finish.

B. Door Construction:
1. Non-fire-rated doors: Particleboard core.
2. 20-minute fire-rated doors: Particleboard core.
3. 45-minute and greater fire-rated doors: Mineral core.
4. Door edges: Hardwood to match door face.

C. Warranty: Lifetime of installation.

08310 - Access Doors and Panels

A. Specified Manufacturer: Karp Associates, Inc. (www.karpinc.com)

B. Acceptable Manufacturers:
1. Larsen's Manufacturing Co. (www.larsenmfg.com)
2. Nystrom Access Doors & Hatches (www.nystrom.com)
3. Or equal.

C. Access Panels in Walls: Sizes as indicated on Drawings.
1. Gypsum board walls, dry locations, non-rated: Karp DSC-214M, plain steel.
2. Gypsum board walls, damp locations, non-rated: Karp DSC-214M, stainless steel.
3. Gypsum board walls, ceramic tile wall finish, non-rated: Karp DSB-214SM, flush-mounted with face of tile, stainless steel finish.
4. Gypsum board walls, dry locations, fire-rated: Karp KRP-150FR, UL-listed B-Label assembly, 20 gage steel door in 16 gage steel frame, filled with 2-inch thick fire-rated insulation, with automatic closer, self-latching bolt-type latch.

D. Access Panels in Ceilings: Sizes as indicated on Drawings.
1. Gypsum board ceilings, dry locations, non-rated.
2. Gypsum board ceilings, dry locations, fire-rated.
3. Gypsum board ceilings, damp locations, fire-rated.

E. Keying: Lock cylinder keyed to building lock system; coordinate with Section 08710 - Door Hardware.

08710 - Door Hardware

A. Door Hardware: Match District Standards and existing campus hardware.

B. Keying: Integrate with existing grand masterkey system.

08800 - Glazing

A. Interior Glass: Clear float glass.

B. Exterior Glass: Gray float, PPG Solargray, ¼-inch thick, to match existing glass.

C. Safety Glass: Tempered where required by US Consumer Product Safety Commission Standard 16CFR1201 CI and CII.

DIVISION 9 - FINISHES

09110 - Non-Load Bearing Metal Framing

A. Interior Light Gage Metal Framing: ASTM C 645, minimum yield strength 33 ksi, size as indicated on Drawings, galvanized, minimum 22 gage, actual gage according to manufacturer's product data to limit deflection to L/360.

B. Flexible Head Track: Proprietary deflection track, steel sheet top runner manufactured to prevent cracking of gypsum board applied to interior partitions resulting from deflection of structure above; in thickness indicated for studs and in width to accommodate depth of studs.

09210 - Gypsum Plaster

A. Work Included: Patching of existing gypsum plaster. Match existing materials and finishes. Substitute materials will be considered in accordance with provisions specified in Section 01600 - Product Requirements.

09250 - Gypsum Board

A. Gypsum Board:
  1. Typical: ASTM C36, Type X (special fire-resistant), typically 48-inches wide and ⅝-inch thick.
  2. Impact-resistant gypsum board: For corridors and other locations indicated, USG Fiberock Brand VHI Panels, or equal, consisting of face layers of fiberglass scrim embedded in a high-density layer of gypsum and cellulose fibers, with perlite core, producing panels with high resistance to abrasion, indentation and penetration.

B. Finishing Materials and Accessories:
  1. Joint treatment materials, general: ASTM C 475.
  2. Finishing or topping compound: Factory-mixed compound, specifically formulated and manufactured for use as filling and finishing compound.
  3. Cornerbead: USG No. 800 or equal.
  4. Edge Trim: USG No. 200-B or equal.
  5. Casing Bead: USG No. 66, square edge, or equal.

C. Installation and Finishing: Comply with GA-201 and GA-216.
  1. Concealed Locations: GA-214, Level 1. Provide this level of finish at locations such as plenum areas above ceilings, in attics, in areas where assembly would generally be concealed.
  2. Semi-Exposed Locations: GA-214, Level 2. Provide this level of finish at locations such as under paneling and in service spaces not exposed to public view.
  3. Locations with Texture Topping and Paint Finish: GA-214, Level 3. Provide this level of finish with orange peel spray-texture topping compound covering no less than 80 percent of surface. Otherwise, provide Level 4 finish and not less than 60 percent of surface with spray-texture topping compound.

4. Locations to Receive Flat and Eggshell Paint on Smooth Surface: GA-214, Level 4. Provide this level of finish at locations with flat and eggshell paint finishes. Gypsum board shall be covered with spray-applied topping with not less than 60 percent of surface covered with topping compound.

09310 - Ceramic Tile

A. Floor Tile: Unglazed ceramic mosaic floor, thinset over crack isolation membrane on cured portland cement concrete slab on grade, similar to TCA Handbook Method F122, using latex-modified portland cement setting mortar.

B. Wall Tile: Glazed wall tile installation, thinset over portland cement tile backer board, using latex-portland cement setting mortar, TCA Handbook Method W244.

09510 - Acoustical Panel Ceilings

A. Acoustical Panel Ceilings:
   1. Classrooms: Armstrong Silhouette XL bolt-slot ceiling grid with ⁹⁄₁₆-inch bottom flange and white painted finish, with Armstrong Fine Fissured panels, catalog no. 1821.
   2. Offices: Armstrong Prelude XL ceiling grid with ⁹⁄₁₆-inch bottom flange and white painted finish, with Armstrong Fine Fissured panels, catalog no. 1821.

B. Seismic Bracing: Comply with details indicated on the Drawings, requirements of International Building Code (IBC).

09650 - Resilient Flooring

A. Resilient Tile Flooring: ASTM F 1066, Composition 1, Class 2 (nonasbestos formulated), homogeneous through thickness, Tarkett series as indicated on the Drawings.

B. Resilient Base and Accessories: ASTM F 1861, Type TS, Group 1, Style A & B, surface applied, smooth finish, solid colors, with molded inside and outside corners and end stops.

C. Finish: Coordinate with District maintenance program. Field-applied finish shall yield minimum static coefficient of friction of 0.5, when tested according to ASTM D 2047.

09900 - Painting

A. Air Quality Regulations: Comply with State and regional air quality regulations.

B. Waste Management: Comply with local regulations for recycling of unused paints and coatings.

C. Painting: Typically primer plus minimum two finish coats at unpainted surfaces and minimum one coat. Finishes:
   1. Exterior steel: As specified in Section 09970 - Coatings for Exterior Steel.
   2. Exterior plaster and concrete: 100 percent acrylic, low sheen.
   3. Interior gypsum board and gypsum plaster: 100 percent acrylic, eggshell (low-sheen) finish except toilet rooms semi-gloss finish.
   4. Interior wood, opaque finish: 100 percent acrylic, semi-gloss finish.
   5. Interior wood, stain and varnish finish: Wood stain plus two coats of clear satin polyurethane varnish.
   6. Interior steel: 100 percent acrylic, semi-gloss finish.

D. Colors: As indicated on the Drawings. Colors by Sherwin Williams Company are indicated.

09970 - Coatings for Exterior Steel

A. Coatings on Exterior Steel Framing, Decking, Door Frames, Doors, Flashing and Sheet Metal: Products by Tnemec Company, Inc., North Kansas City, MO (816/474-3400) or equal.
   1. Primer, plain steel: Tnemec Series 90-97 Tnemec-Zinc, two-component catalyzed epoxy coating (2.5–3.5 mils DFT).
   2. Primer, galvanized steel: Tnemec Series 66 Epoxoline, two-component catalyzed epoxy coating (2–3 mils DFT).
   3. Finish: Tnemec Series 75 Endura-Shield, pigmented, aliphatic, polyurethane coating, semi-gloss sheen (3 mils DFT minimum).

DIVISION 10 - SPECIALTIES

10110 - Visual Display Boards

A. Porcelain Enamel Markerboards:
   1. Fixed markerboards.
   2. 3-panel vertical sliding markerboards, wall-mounted.
   3. 3-panel horizontal sliding markerboards, casework-mounted.

10170 - Solid Plastic Toilet Partitions

A. Toilet Room Partitions: Solid plastic (HDPE polymer) panels, 10 percent recycled materials, Comtec Industries, Inc., Moosic, PA (570/348-0997) to match District standards, complying with applicable wheelchair accessibility and dependent persons use requirements.
   1. Toilet compartments: Floor-to-ceiling design, with pilasters anchored to floor and ceiling, and side panels and doors standard 55-inches high.
   2. Urinal screens: Floor-to-ceiling end pilaster design, with 42-inch high by 24-inch deep panel.

10506 - Metal Lockers Refurbishment

A. Preparation and repainting of existing steel lockers.

B. Replacement of worn and damaged hardware, including accessible door latches, shelves and hooks.

10810 - Toilet Accessories

A. New satin finish stainless steel toilet roof accessories, including recessed paper towel dispensers, soap dispensers, toilet paper dispensers, seat cover dispensers, grab bars and metal-framed mirrors.

DIVISION 11 - EQUIPMENT

11130 - Projection Screens

A. Front Projection Screens: Manual pull-down, wall-mounted, Da-Lite Model C Roller-Type Screen, with CSR (Controlled Screen Return) feature, standard format, 70-inches high by 70-inches wide, glass bead-coated, optical-quality flexible fabric, with standard black masking borders, with black extra drop fabric, fire retardant and mildew resistant.

DIVISION 12 - FURNISHINGS

12490 - Window Blinds

A. Horizontal Slat Louver Blinds: Nominal 1-inch aluminum blinds, two colors as selected by Architect, at Administration Building offices.

DIVISION 13 - SPECIAL CONSTRUCTION

13130 - Fabric Structures

A. Sunshade Fabric Structure: Fabric structure, including framing, cables, connectors and fabric. Fabric structures shall be designed, engineered and fabricated by manufacturer to configurations indicated on the Contract Drawings.

DIVISION 14 - CONVEYING SYSTEM

14420 - Wheelchair Lift

A. Wheelchair Lift: Electric-powered, interior platform (wheelchair) lift for access to Stage in existing Multipurpose Room, complying with applicable wheelchair accessibility and dependent persons use regulations.

DIVISION 15 - MECHANICAL

Section 15050 - Basic Mechanical Materials and Methods

A. Section Includes: Requirements for mechanical work.

B. Coordination:
   1. Contract Documents: Diagrammatic, showing physical relationships within mechanical work and its interface with other work.
   2. Coordination Drawings: Where clearances are limited and where elements of mechanical work (or combinations of mechanical and other work) are located with precision to fit into available space, prepare Coordination Drawings (Shop Drawings) at suitable scale showing required dimensions and submit for approval.
   3. Utility Connections:
      a. Coordinate the connection of mechanical system with utilities and services.
      b. Comply with regulations of utility suppliers.
      c. The Contract Documents indicate available information on existing utilities and services, and for new services (if any) to be provided.
         1) Notify Architect if discrepancies are found.
      d. Coordinate mechanical utility interruptions with the Owner and the Utility Company.
         1) Plan work to minimize duration of interruptions.

C. Mechanical and Electrical Coordination:
   1. Responsibility: Unless otherwise indicated, provide motors and controls for Division 15 equipment, set in place, and wire in accordance with specifications.
   2. Control Wiring:
      a. Consists of wiring in pilot circuits of contactors, starters, relays, and wiring for valve and damper operators that is not a part of mechanical work.

b. For single phase devices where power current passes through controller, wiring between controller and device shall be considered control wiring; wiring to device from electric panel is considered power wiring.

D. Coordination of Mechanical Openings: Coordinate mechanical openings with related or adjacent work.

E. Coordination with Other Work: Coordinate mechanical work with related or adjacent work.
   1. Cutting and Patching: Refer to Section 01465.
   2. Chases, Inserts and Openings:
      a. Provide measurements, drawings and layouts so that openings, inserts and chases in new construction can be built in as construction progresses.
      b. Check sizes and locations of openings provided.
      c. Any cutting and patching made necessary by failure to provide measurements, drawings and layouts at the proper time shall be done at no additional cost to the Owner.
   3. Support Dimensions: Provide dimensions and drawings so that concrete bases and other equipment supports to be provided under other Sections of the Specifications can be built at the proper time.

F. Regulatory Requirements:
   1. Comply with governing Codes and requirements of authorities having jurisdiction.
   2. Inspections and Tests: Arrange for all required inspections and tests.

G. Execution: Provide accessory materials and components required for mechanical work.
   1. Excavation.
   2. Access panels.
   3. Equipment bases and supports.
   4. Drip pans.
   5. Seismic bracing.

Section 15060 - Hangers and Supports

A. Section Includes: Mechanical supports and anchors.

B. Regulatory Requirements:
   1. Comply with MSS Standard Practice SP-69, published by Manufacturer's Standardization Society of Valve and Fitting Industry for type and size.

C. Materials:
   1. Pipe hangers:
      a. Use adjustable pipe hangers on suspended pipe.
      b. Chain, wire or perforated strap hangers will not be permitted.
      c. Isolate hangers coming in contact with bare copper pipe with dielectric hanger liners.
      d. Provide supports between piping and building structure where necessary to prevent swaying.
      e. Provide dielectric isolation between bare copper pipe and steel studs in all walls.
   2. Pipe anchors.

Section 15071 - Mechanical Vibration and Seismic Controls

A. Section Includes:
   1. Vibration isolators.
   2. Air mounting systems.
   3. Roof curbs.
   4. Seismic restraints.
   5. Vibration isolation bases.

B. Quality Assurance
   1. Provide seismic restraint devices with horizontal and vertical load testing and analysis acceptable to authorities having jurisdiction, showing maximum seismic restraint ratings.
      a. Ratings based on independent testing are preferred to ratings based on calculations.
      b. Calculations (including combining shear and tensile loads) to support seismic restraint designs must be signed and sealed by a qualified professional engineer. Testing and calculations must include both shear and tensile loads and 1 test or analysis at 45 degrees to the weakest mode.
   2. Welding: Qualify procedures and personnel according to AWS D1.1, Structural Welding Code—Steel.

C. Materials and Components:
   1. Elastomeric isolator pads.
   2. Elastomeric mounts.
   3. Restrained elastomeric mounts.
   4. Restrained spring isolators.
   5. Housed spring mounts.
   6. Elastomeric hangers.
   7. Spring hangers with vertical limit stop.
   8. Thrust limits.
   9. Pipe riser resilient support.
   10. Resilient pipe guides.
   11. Air mounts.
   12. Roof curb rails.
   13. Support assemblies.
   14. Spring isolators.
   15. Elastomeric isolator pads.
   16. Snubber bushings.
   17. Water seal.
   18. Seismic snubbers.
   19. Restraining cables.
   20. Anchor bolts.
   21. Isolation equipment base and brackets.

Section 15075 - Mechanical Identification

A. Section Includes: Mechanical identification materials.

B. Quality Assurance:
   1. Regulatory Requirements:
      a. ASME Compliance: Comply with ASME A13.1, Scheme for the Identification of Piping Systems, for letter size, length of color field, colors and viewing angles of identification devices for piping.

C. Materials:
   1. Equipment nameplates.
   2. Pipe markers.
   3. Valve tags.
   4. Warning tags.

D. Execution:
   1. Ductwork identification.
   2. Piping system identification.
   3. Valve identification.
   4. Mechanical equipment identification.
   5. Nonpotable water identification.

Section 15080 - Mechanical Insulation

A. Section Includes: Insulation for mechanical materials, equipment and components.

B. Quality Assurance:
1. Regulatory Requirements:
    a. Fire Test Response Characteristics: As determined by testing materials identical to those specified in this Section according to ASTM E 84, by a testing and inspecting agency acceptable to authorities having jurisdiction. Factory label insulation and jacket materials and sealer and cement material containers with appropriate markings of applicable testing and inspecting agency.
        1) Insulation Installed Indoors: Flame spread rating of 25 or less, and smoke developed rating of 50 or less.
        2) Insulation Installed Outdoors: Flame spread rating of 75 or less, and smoke developed rating of 150 or less.

C. Materials:
1. Semirigid and flexible duct, plenum and breeching insulation.
2. Insulating cements.
3. Preformed, rigid and flexible pipe insulation.
4. Field applied jackets.
5. Accessories and attachments.
6. Sealing compounds.
7. Pipe and fitting covers.
8. Metallic pipe and fitting covers.
9. Duct insulation.
10. Outdoor duct insulation.
11. Equipment insulation.
12. Breeching and stack insulation.
13. Heat tracing for grease piping.
14. Heat tracing for exterior piping.

Section 15110 - Valves

A. Section Includes: Mechanical valves.

B. Materials and Components:
1. Valve Types and Sizes:
    a. Where type or body material is not indicated, provide valve with pressure class selected from MSS or ANSI standards, based on the maximum pressure and temperature in the piping system.
    b. Except as otherwise indicated, provide valve of same size as connecting pipe size.
    c. Ball valves or butterfly valves may be used in water lines in lieu of gate valves when pressure and temperature ratings are adequate.
2. Valves:
    a. Gate valves.
    b. Globe valves.
    c. Angle valves.
    d. Ball valves.
    e. Butterfly valves.
    f. Check valves.
    g. Plug valves.
    h. Combination balancing and shutoff valves.

Section 15140 - Domestic Water Piping

A. Section Includes:
   1. Pressure reducing valves.
   2. Backflow preventers.
   3. Water hammer arresters.
   4. Disinfection of water lines.

B. Materials:
   1. Domestic water piping and condensate drain piping.
   2. Compressed air piping.

Section 15150 - Sanitary Waste and Vent Piping

A. Section Includes: Sanitary sewer and vent piping.

B. Soil and Vent Piping:
   1. Above Ground:
      a. Cast iron No-Hub, neoprene gasket and stainless steel sleeve.
      b. DWV copper with DWV fittings, solder joint.
   2. Underground:
      a. Cast iron No-Hub, neoprene gasket and stainless steel sleeve.
      b. MG cast iron coupling and neoprene gasket.

Section 15410 - Plumbing Fixtures

A. Section Includes: Plumbing fixtures.

B. Materials: Provide factory fabricated fixtures with trim, carriers, valves and accessories as required for complete installation.
   1. Water closets.
   2. Urinals.
   3. Lavatories.
   4. Stainless steel sinks.
   5. Service sinks.
   6. Circulating pumps.
   7. Booster pumps.

Section 15430 - Plumbing Specialties

A. Section Includes:
   1. Backflow preventers.
   2. Water regulators.
   3. Balancing valves.
   4. Strainers.
   5. Outlet boxes.
   6. Hose stations.
   7. Trap seal primer valves.
   8. Drain valves.
   9. Valve tags and charts.

    10. Dielectric fittings.
    11. Prefabricated concrete pits.
    12. Water reels.
    13. Miscellaneous piping specialties.
    14. Sleeve penetration systems.
    15. Flashing materials.
    16. Cleanouts.
    17. Floor drains.
    18. Trench drains.
    19. Roof drains.

B. Performance Requirements:
   1. Provide components and installation capable of producing piping systems with following minimum working pressure ratings, unless otherwise indicated:
     a. Domestic Water Piping: 125 psig (860 kPa).
     b. Sanitary Waste and Vent Piping: 10 foot head of water (30 kPa).
     c. Storm Drainage Piping: 10 foot head of water (30 kPa).
     d. Force Main Piping: 100 psig (690 kPa).

C. Quality Assurance:
   1. Regulatory Requirements:
     a. Comply with ASME B31.9, Building Services Piping, for piping materials and installation.
     b. NSF Compliance:
       1) Comply with NSF 14, Plastics Piping Components and Related Materials, for plastic domestic water piping components. Include marking NSF-pw on plastic potable water piping and NSF-dwv on plastic drain, waste and vent piping.
       2) Comply with NSF 61, Drinking Water System Components—Health Effects, Sections 1 through 9, for potable domestic water plumbing specialties.
     c. Provide specialties bearing label, stamp or other markings of specified testing agency.
     d. Listed and labeled as defined in NFPA 70, Article 100, by a testing agency acceptable to authorities having jurisdiction and marked for intended use.

Section 15485 - Electric Domestic Water Heaters

A. Section Includes: Electric water heater.

B. Quality Assurance:
   1. Regulatory Requirements:
     a. Electrical Components, Devices, and Accessories: Listed and labeled as defined in NFPA 70, Article 100, by testing agency acceptable to authorities having jurisdiction, and marked for intended use.
     b. Fabricate and label water heater, hot water storage tanks to comply with ASME Boiler and Pressure Vessel Code: Section VIII, Pressure Vessels, Division 1.
     c. Comply with performance efficiencies prescribed for ASHRAE 90.1, Energy Efficient Design of New Buildings except Low-Rise Residential Buildings for commercial water heaters.

C. Equipment:
   1. Household, electric water heaters.
   2. Tankless, electric water heaters.
   3. Commercial, electric water heaters.
   4. Compression tanks.
   5. Accessories.

Section 15720 - Air Handling Units

A. Section Includes: Air handling units.

B. Performance Ratings:
   1. Capacity Ratings:
      a. Fans: Certified according to ARI 430.
      b. Coils: Certified according to ARI 410.
   2. Horsepower: Do not increase or decrease fan motor horsepower without written approval from Architect.

C. Quality Assurance:
   1. Fire Ratings: Provide thermal and/or acoustical insulation which conforms to the following ratings when tested in accordance with ASTM E 84 (NFPA 255):
      a. Flame Spread: 25 or less.
      b. Fuel Contributed: 50 or less.
      c. Smoke Developed: 50 or less.

D. Materials:
   1. Air handling units.
   2. Make up air unit (gas fired).
   3. Cooling coils.
   4. Filters.
   5. Dampers.
   6. Horizontal fan coil units.
   7. Console fan coil units.
   8. Vertical stacked fan coil units.

Section 15738 - Air Conditioning Units

A. Section Includes: Air conditioning units.

B. Quality Assurance:
   1. Regulatory Requirements:
      a. Applicable provisions of NFPA 70 - National Electric Code.
      b. Provide air handling unit thermal insulation with flame spread index of 25 or less, fuel contributed index of 50 or less, and smoke developed index of 50 or less.
      c. Air Movement and Control Association (AMCA) standards as applicable to testing and rating fans, and testing louvers, dampers and shutters.
      d. Sheet Metal and Air Conditioning Contractors National Association (SMACNA) ductwork construction standards as applicable.
      e. Provide refrigerant coils complying with construction and testing standards of ANSI/ASHRAE 15 Safety Code for Mechanical Refrigeration.
      f. ASHRAE recommendations applicable to packaged air handling units.
      g. Provide central station packaged air handling units which comply with Air Conditioning and Refrigeration Institute (ARI) Standard 430 and display ARI's certification symbols.
      h. Provide electric components for air handling units which have been listed and labeled by Underwriters Laboratories International (ULI).
      i. Applicable portions of the International Building Code (IBC) as adopted by authorities having jurisdiction.

C. Equipment:
   1. Air conditioning unit including mixing box, fans and motors, compressors, filters, coils, drip pan, thermal insulation, outdoor casing, roof curb, internal isolators of moving equipment and complete factory assembly of all components.

Section 15810 - Metal Ducts

A. Section Includes: Ductwork and installation products.

B. Quality Assurance:
1. Regulatory Requirements: Comply with Sheet Metal and Air Conditioning Contractors' National Association (SMACNA) recommendations for fabrication, construction and details, and installation procedures, except as otherwise indicated.

C. Materials:
1. Ductwork materials.
2. Round duct.
3. Miscellaneous accessories.

Section 15820 - Duct Accessories

A. Quality Assurance:
1. Regulatory Requirements:
   a. Comply with SMACNA recommendations for fabrication, construction and details, and installation procedures, except as otherwise indicated.

B. Materials:
1. Flexible duct, low pressure.
2. Flexible duct, high pressure.
3. Louvers.
4. Fire dampers (standard width).
5. Fire dampers (thin line).
6. Combination fire/smoke dampers.
7. Ceiling radiation dampers.
8. Miscellaneous ductwork accessories.

Section 15835 - Fans

A. Section Includes: Fans for building mechanical systems, including:
1. Centrifugal Fans:
   a. Centrifugal Fans for Indoor Installations: Belt-driven with housing, wheel, fan shaft, bearings, motor and disconnect switch, drive assembly, support structure.
   b. Tubular Centrifugal Fans: Tubular inline, belt driven with housing, wheel, outlet guide vanes, fan shaft, bearings, drive assembly, motor, mounting brackets, accessories.
   c. Inline Centrifugal Fans: Inline, belt driven with housing, wheel, outlet guide vanes, fan shaft, bearings, drive assembly, motor and disconnect switch, mounting brackets, accessories.
2. Axial Fans:
   a. Propeller Axial Fans: Belt driven or direct drive propeller fan with fan blades, hub, housing, orifice ring, motor, drive, accessories.
   b. Vaneaxial Axial Fans: Belt driven or direct drive, variable pitch or adjustable pitch, vaneaxial fan with fan wheel and housing, straightening vane section, factory mounted motor, inlet cone section, accessories.

Section 15850 - Air Outlets and Inlets

A. Section Includes: Grilles, registers and diffusers.

B. Regulatory Requirements:
1. Listed and labeled as defined in NFPA 70, Article 100, by a testing agency acceptable to authorities having jurisdiction and marked for intended use.
2. Install air terminal units according to NFPA 90A, Standard for the Installation of Air Conditioning and Ventilating Systems.
3. Provide products tested in accordance with Air Diffusion Council test codes and bearing ADC seal.

C. Materials: Compatible with ceiling construction.
1. Grilles, registers and diffusers.
2. Linear slot diffusers.
3. Bypass single duct air terminal units
4. Dual duct air terminal units.
5. Fan powered air terminal units.
6. Induction air terminal units.
7. Shutoff single duct air terminal units.
8. Integral diffuser air terminal units.

Section 15890 - Air Filters

A. Air Filters: ASHRAE 52, ARI 850, NFPA 90A, 90B.
1. Replaceable (Throwaway) Panel Filters: Flat panels, interlaced glass fiber media, 20 gauge (.0329-inch) (.8 mm) galvanized steel frame, 20 gauge (0.0329-inch) (.8 mm) galvanized steel duct holding frame.
2. Cleanable (Washable) Panel Filters: Flat panels, zinc electroplated steel screening media, 18 gage (0.0358-inch) (.9 mm) galvanized steel duct holding frames.
3. Extended Surface Disposable Panels Filters: Fibrous material media in deep pleats, galvanized steel frame, holding frames.
4. Extended Surface Nonsupported Media Filters: Fibrous material with flexible internal supports, galvanized steel frame, duct holding frames.
5. Automatic (Self Renewing) Roll Filters: Automatic, motor driven type, with fibrous glass material media, galvanized steel holding frame, [manual] [automatic] control.
6. Activated Carbon Filters: Carbon trays in deep V arrangement with replaceable panel prefilter, activated carbon media, 14 gage (0.0625-inch) (1.6 mm) epoxy coated steel frame, 16 gage (.0598-inch) (1.5 mm) galvanized steel duct holding frames.
7. High Efficiency Particulate Air (HEPA) Filters: UL 586 fibrous glass media with vinyl coated aluminum separators, galvanized steel frame, suitable media to frame side bond and face gasket, duct holding frames.
8. Electronic Air Cleaners: Galvanized steel assembly with electronic agglomerator and prefilters, ionizing wire media, self-contained power pack, safety accessories.
9. Front and Rear Access Filter Frames: Aluminum framing members, prefilters, sealers.
10. Side Service Housings: 16 gauge (0.0598-inch) (1.5 mm) galvanized steel side service housings, prefilters, access doors, sealers.
11. Filter Gauges: Diaphragm type with suitable filter gauge range, manometer type filter gauge.

Section 15900 - HVAC Instrumentation and Controls

A. Section Includes:
1. Control equipment for HVAC systems and components, including control components for terminal heating and cooling units not supplied with factory wired controls.
2. Instruments, devices, wiring and conduits, and accessories required for a complete installation.

B. Equipment:
1. Space Thermostats: Heavy duty type room thermostats suitable for industrial environment with adjustable sensitivity and calibrated dial.

2. Smoke detectors
3. Fire alarm supervisory control circuits.
4. Automatic dampers and smoke dampers.
5. Damper operator.
6. Interlocks.

Section 15940 - Sequence of Operations

A. Section Includes:
   1. Control equipment for HVAC systems and components, including control components for terminal heating and cooling units not supplied with factory wired controls.
   2. Instruments, devices, wiring and conduits, and accessories required for a complete installation.

B. Sequence of Operation and Points List: To be developed.

Section 15950 - Testing, Adjusting, Balancing

A. Section Includes: Testing requirements.

B. Quality Assurance:
   1. Qualification: Work shall be done by a firm certified by the National Environmental Balancing Bureau (NEBB) or the Associated Air Balance Council (AABC).
   2. Industry Standards: Comply with one or all of the following:
      a. HVAC Systems Testing, Adjusting, Balancing published by SMACNA.
      b. Procedural Standards for Testing, Adjusting, Balancing of Environmental Systems published by NEBB.
      c. Procedural Standards for Certified Testing of Clean Rooms published by NEBB.

C. Testing, Adjusting and Balancing:
   1. Air systems.
   2. Hydronic systems.
   3. Detailed requirements.
   4. Report.

DIVISION 16 ELECTRICAL

Section 16060 - Grounding and Bonding

A. Description:
   1. Primary grounding.
   2. Substation grounding.
   3. Power system grounding.
      a. Communication system grounding.
      b. Electrical equipment and raceway grounding and bonding.

B. Materials:
   1. Grounding Accessories: Ground rods and connectors.
   2. Secondary Grounding:
      a. Electrical grounding and bonding systems indicated; with assembly of materials, including, but not limited to, cables/wires, connectors, solderless lug terminals, grounding electrodes and plate

electrodes, bonding jumper braid, surge arresters and additional accessories needed for a complete installation.
  b. Where more than one type component product meets indicated requirements, selection is Installer's option. Where materials or components are not indicated, provide products which comply with NEC, UL and IEEE requirements and with established industry standards for those applications indicated.
  c. Conductors: Unless otherwise indicated, provide electrical grounding conductors for grounding system connections that match power supply wiring materials and are sized according to NEC.

Section 16070 - Hangers and Supports

A. Description:
  1. Conduit and equipment supports.
  2. Fastening hardware.

B. Materials:
  1. Support Channel: Galvanized or painted steel.
  2. Hardware: Corrosion resistant.

Section 16075 - Electrical Identification

A. Description:
  1. Nameplates and tape labels.
  2. Wire and cable markers.

B. Materials:
  1. Nameplates: Engraved three layer laminated plastic, white letters on a black background (normal power), white letters on a red background (emergency power).
  2. Wire and Cable Markers: Cloth markers, split sleeve or tubing type.
  3. Conduit Color Coding Identification:
     a. Paint or tape the following systems with the appropriate color on conduits every 30 feet and every pullbox.
     b. Fire Alarm System: Yellow.
     c. Security System: Purple.
     d. Emergency Power: Red.

Section 16080 - Electrical Testing

A. Description:
  1. Electrical testing.
  2. Accessories required for complete testing.

B. Quality Assurance:
  1. IES LM-50 Photometric Measurement of Roadway Lighting Installations
  2. IEEE 81 IEEE Guide for Measuring Earth Resistivity, Ground Impedance and Earth Surface Potentials of a Ground System
  3. National Electrical Testing Association (NETA) Standards for Testing.

C. Testing:
  1. Testing Equipment: Provide Megger tester for megohmmeter and earth tester.
  2. Tools: Provide tools required for opening test facilities, making wire connections, splicing, and other purposes.

3. Ground Mat Test.
4. 600 Volt Wire and Cable Test.

Section 16120 - Conductors and Cable

A. Description:
  1. Building wire.
  2. Cable.
  3. Wiring connections and terminations.

B. Materials:
  1. Building Wire:
     a. Thermoplastic insulated Building Wire: NEMA WC 5.
     b. Feeders and Branch Circuits Larger Than 6 AWG: Copper, stranded conductor, 600 volt insulation, THW, THHN/THWN, or XHHW. Aluminum alloy (AA-8000) wire in sizes 1/0 and larger may be substituted for copper for feeders if ampacity is equal to or greater than copper ampacity and if voltage drop is equal to or less than copper voltage drop.
     c. Feeders and Branch Circuits 6 AWG and Smaller: Copper conductor, 600 volt insulation, THW, THHN/THWN, or XHHW, 6 and 8 AWG, stranded conductor; smaller than 8 AWG, solid conductor.
     d. Control Circuits: Copper, stranded conductor 600 volt insulation, THW or THHN/THWN.
  2. Remote Control and Signal Cable:
     a. Control Cable for Class 1 Remote Control and Signal Circuits: Copper conductor, 600 volt insulation, rated 75 degrees C, individual conductors twisted together and covered with a PVC jacket.
     b. Control Cable for Class 2 or Class 3 Remote Control and Signal Circuits: Copper conductor, 300 volt insulation, rated 60 degrees C, individual conductors twisted together, and covered with a PVC jacket; UL listed.

Section 16130 - Raceways and Boxes

A. Description:
  1. Raceways:
     a. Surface metal raceways.
     b. Multioutlet assemblies.
     c. Auxiliary gutters (wireways).
  2. Conduit:
     a. Rigid metal conduit and fittings.
     b. Intermediate metal conduit and fittings.
     c. Electrical metallic tubing and fittings.
     d. Flexible metal conduit and fittings.
     e. Liquidtight flexible metal conduit and fittings.
     f. Non-metallic conduit and fittings.
     g. Liquidtight flexible non-metallic conduit.
  3. Boxes:
     a. Wall and ceiling outlet boxes.
     b. Pull and junction boxes.
  4. Cabinets and Enclosures:
     a. Hinged cover enclosures.
     b. Cabinets.
     c. Terminal blocks and accessories.

B. Materials:
  1. Raceways:
     a. Surface metal raceway.

   b. Multiple outlet assembly.
   c. Auxiliary gutters.
2. Conduit:
   a. Rigid metal conduit and fittings.
   b. Intermediate metal conduit (IMC) and fittings.
   c. Electrical metallic tubing and fittings.
   d. Flexible metal conduit and fittings.
   e. Liquidtight flexible conduit and fittings.
   f. Plastic conduit and fittings.
   g. Conduit Support: Conduit Clamps, Straps, and Supports: Steel or malleable iron.
   h. Nonmetallic conduit and ducts: Liquidtight flexible nonmetallic conduit.
3. Boxes:
   a. Outlet boxes.
   b. Pull and junction boxes.
4. Cabinets and Enclosures:
   a. Hinged cover enclosures.
   b. Cabinets.
   c. Terminal blocks and accessories.

Section 16140 - Wiring Devices

A. Description:
   1. Wall switches.
   2. Wall dimmers.
   3. Receptacles.
   4. Device plates and box covers.

B. Materials:
   1. Wall Switches:
      a. Wall Switches for Lighting Circuits: NEMA WD 1; FS-W S-896; AC general use snap switch with toggle handle, rated 20 amperes and 120-277 volts AC.
      b. Color selected by Architect.
   2. Receptacles:
      a. Convenience and Straight-blade Receptacles: NEMA WD 1; FS W-C-596.
      b. Locking Blade Receptacles: NEMA WD 5.
      c. Convenience Receptacle Configuration: NEMA WD 1; Type 5-20 R.
      d. Specific-Use Receptacle Configuration: NEMA WD 1 or WD 5; type as indicated on Drawings.
      e. GFCI Receptacles: Duplex convenience receptacle with integral ground fault current interrupter.
      f. Color selected by Architect.
   3. Wall Plates:
      a. Decorative Cover Plate: Smooth high abuse nylon or lexan in dry locations, stainless steel in damp or wet locations (kitchens, toilet rooms, etc.).
      b. Weatherproof Cover Plate: Gasketed thermoplastic.
      c. Color as selected.

Section 16150 - Wiring Connections

A. Description: Electrical connections to equipment under other work or furnished by Owner.

B. Materials:
   1. Cords and Caps:
      a. Straightblade Attachment Plug: NEMA WD 1.; FS W-C-596.
      b. Locking Blade Attachment Plug: NEMA WD 5.

    c. Attachment Plug Configuration: Match receptacle configuration at outlet provided for equipment.

    d. Cord Construction: Oil resistant thermoset insulated Type SO multiconductor flexible cord with identified equipment grounding conductor, suitable for extra hard usage.

    e. Cord Size: Suitable for connected load of equipment and rating of branch circuit overcurrent protection.

Section 16410 - Enclosed Switches and Circuit Breakers

A. Description:

    1. Individually mounted enclosed switches and circuit breakers, rated 600 V and less, used for disconnecting and protection functions.

B. Quality Assurance:

    1. Regulatory Requirements:

        a. Electrical Components, Devices, and Accessories: Listed and labeled as defined in NFPA 70, Article 100, by a testing agency acceptable to authorities having jurisdiction, and marked for intended use.

        b. Comply with NFPA 70.

C. Materials:

    1. Enclosed, Nonfusible Switch: NEMA KS 1, Type HD, with lockable handle, interlocked with cover.

    2. Enclosed, Fusible Switch, 800 A and Smaller: NEMA KS 1, Type HD, with clips to accommodate specified fuses, and lockable handle, interlocked with cover.

    3. Molded Case Circuit Breaker: NEMA AB 1, with interrupting capacity to meet available fault currents.

        a. Thermal Magnetic Circuit Breakers: Inverse time-current element for low-level overloads, and instantaneous magnetic trip element for short circuits. Adjustable magnetic trip setting for circuit breaker frame sizes 250 A and larger.

        b. Adjustable Instantaneous Trip Circuit Breakers: Magnetic trip element with front-mounted, field-adjustable trip setting.

        c. Current Limiting Circuit Breakers: Frame sizes 400 A and smaller; let-through ratings less than NEMA FU 1, RK-5.

        d. GFCI Circuit Breakers: Single- and two-pole configurations with 30 mA trip sensitivity.

    4. Molded Case Circuit-Breaker Features and Accessories: Standard frame sizes, trip ratings, and number of poles.

        a. Lugs: Suitable for number, size, trip ratings, and material of conductors.

        b. Application Listing: Appropriate for application; Type SWD for switching fluorescent lighting loads; Type HACR for heating, air conditioning, and refrigerating equipment.

        c. Ground Fault Protection: Integrally mounted relay and trip unit with adjustable pickup and time delay settings, push to test feature, and ground-fault indicator.

        d. Shunt Trip: 120 volt trip coil energized from separate circuit, set to trip at 55 percent of rated voltage.

    5. Listed for environmental conditions of installed locations, including:

        a. Outdoor Locations: NEMA 250, Type 3R.

        b. Other Wet or Damp Indoor Locations: NEMA 250, Type 4.

Section 16441 - Switchboards

A. Description: Main and distribution switchboards.

B. Materials:

    1. Switchboard Construction and Ratings: Factory assembled, dead front, metal enclosed, front accessible and self supporting switchboard assembly conforming to NEMA PB2, and complete from incoming line terminals to load side terminations.

        a. Rating: Indicated on drawings.

2. Switching and Overcurrent Protective Devices:
   a. Fusible switch assemblies, 600 amperes and less.
3. Fuses:
   a. Fuses 600 Amperes and Less: Dual element, current limiting, time delay, UL Class J, RK 1 or RK 5 as scheduled on Drawings.
   b. Fuses 601 Amperes and Larger: Current limiting, time delay one time fuse, 600 volts, UL Class L.
   c. Interrupting Rating: 200,000 rms amperes.
4. Instruments and Sensors:
   a. Metering.
   b. Current Transformers: ANSI C57.13; 5 ampere secondary, bar or window type, with secondary winding and secondary shorting device, primary/secondary ratio as required.
   c. Ground Fault Sensor: Zero sequence type.
   d. Ground Fault Relay: Adjustable ground fault sensitivity from 200 to 1200 amperes, time delay adjustable from 0 to 15 seconds.

Section 16442 - Panelboards

A. Description: Lighting and appliance branch circuit panelboards.

B. Materials:
1. Branch Circuit Panelboards
   a. Lighting and Appliance Branch Circuit Panelboards: NEMA PB1; circuit breaker type. FS W-P-115; Type I, Class 1.
   b. Enclosure: NEMA PB 1; Type 1 or Type 3R. as required.
   c. Cabinet Size: 5¾-inches (146 mm) deep; 20-inches (508 mm) wide.
   d. Provide flush or surface cabinet front as indicated on the Drawings with concealed trim clamps, concealed hinge and flush lock all keyed alike. Finish in manufacturer's standard gray enamel.
   e. Provide with copper bus, minimum AIC rating of 10 K and ratings as scheduled on Drawings. Provide copper ground bus in all panelboards.
   f. Molded Case Circuit Breakers: NEMA AB 1; FS W-C-375; bolt-on type thermal magnetic trip circuit breakers, with common trip handle for all poles. Provide circuit breakers UL listed as Type SWD for circuit breaker switched lighting circuits. Provide UL Class A ground fault interrupter circuit breakers where scheduled on Drawings.
   g. Series rate system with upstream overcurrent device.

Section 16461 - Dry Type Transformers (600 Volts or Less)

A. Description: Dry type two winding transformers.

B. Materials:
1. Dry Type Two Winding Transformers:
   a. Dry Type Transformers: ANSI/NEMA ST 20; factory assembled, air cooled dry type transformers; ratings as shown on the Drawings.
   b. Insulation system and average winding temperature rise for rated KVA.

Section 16510 - Lighting

A. Description:
1. Interior luminaires and accessories.
2. Exterior luminaires and accessories.
3. Lamps.
4. Ballasts.

B. Materials:
  1. Interior Luminaires and Accessories:
     a. Fluorescent Luminaires: FS W-F-414; provide hinged frames with spring latches, and virgin acrylic lenses unless scheduled otherwise on Drawings.
     b. Recessed Fluorescent Luminaires: Provide trim type and accessories required for installation in ceiling system installed.
     c. Exit Signs: Stencil face, 6-inch (150-mm) high red letters on white background, directional arrows as indicated, universal mounting type as scheduled.
     d. HID Luminaires: Pre-wired, with integral ballast.
     e. Recessed fixtures shall be thermally protected and UL listed.
  2. Exterior Luminaries and Accessories:
     a. Enclosures: Complete with gaskets to form weatherproof assembly.
     b. Provide low temperature ballasts, with reliable starting to 0 degrees F (−17 degrees C).
  3. Lamps:
     a. General Use Incandescent Lamps: Inside frosted type, rated 130 volts.
     b. Incandescent Reflector Lamps: Shape as scheduled, rated 130 volts.
     c. Fluorescent Lamps: Energy saving type compatible with ballasts provided as scheduled on the drawings.
     d. Metal Halide HID Lamps: As scheduled on Drawings.
     e. High Pressure Sodium Lamps: As scheduled on Drawings.
  4. Fluorescent Ballasts, Magnetic: ANSI C82.1; high power factor type; nominal 430 ma Lamp Ballasts: Low energy type.
  5. Fluorescent Ballasts, Electronic: Frequency of operation of 20 kHz or greater, and operate without visible flicker; power factor of 95 percent or above; sound rated A.

END OF OUTLINE SPECIFICATIONS

Appendix E

Sample AIA Documents

INTRODUCTION

Following are AIA documents referenced in Chapters 1 through 21 and noted as being included in this appendix. These sample documents have been reproduced with the permission of the AIA for educational purposes only.

Clean copies of these documents, for use in Project Manuals, should be obtained from the AIA. Most local chapter offices stock these documents, or they may be obtained by contacting:

The American Institute of Architects (AIA)
1735 New York Avenue, N.W.
Washington, DC 20006
202/626-7300 or 800/365-2724
www.aia.org/documents

1997 EDITION

AIA DOCUMENT | A701-1997

Instructions to Bidders

This document has important legal consequences. Consultation with an attorney is encouraged with respect to its completion or modification.

TABLE OF ARTICLES

© 1997 AIA®
AIA DOCUMENT A701-1997
INSTRUCTIONS TO BIDDERS

The American Institute
of Architects
1735 New York Avenue, N.W.
Washington, D.C. 20006-5292

1

ARTICLE 1 DEFINITIONS

1.1 Bidding Documents include the Bidding Requirements and the proposed Contract Documents. The Bidding Requirements consist of the Advertisement or Invitation to Bid, Instructions to Bidders, Supplementary Instructions to Bidders, the bid form, and other sample bidding and contract forms. The proposed Contract Documents consist of the form of Agreement between the Owner and Contractor, Conditions of the Contract (General, Supplementary and other Conditions), Drawings, Specifications and all Addenda issued prior to execution of the Contract.

1.2 Definitions set forth in the General Conditions of the Contract for Construction, AIA Document A201, or in other Contract Documents are applicable to the Bidding Documents.

1.3 Addenda are written or graphic instruments issued by the Architect prior to the execution of the Contract which modify or interpret the Bidding Documents by additions, deletions, clarifications or corrections.

1.4 A Bid is a complete and properly executed proposal to do the Work for the sums stipulated therein, submitted in accordance with the Bidding Documents.

1.5 The Base Bid is the sum stated in the Bid for which the Bidder offers to perform the Work described in the Bidding Documents as the base, to which Work may be added or from which Work may be deleted for sums stated in Alternate Bids.

1.6 An Alternate Bid (or Alternate) is an amount stated in the Bid to be added to or deducted from the amount of the Base Bid if the corresponding change in the Work, as described in the Bidding Documents, is accepted.

1.7 A Unit Price is an amount stated in the Bid as a price per unit of measurement for materials, equipment or services or a portion of the Work as described in the Bidding Documents.

1.8 A Bidder is a person or entity who submits a Bid and who meets the requirements set forth in the Bidding Documents.

1.9 A Sub-bidder is a person or entity who submits a bid to a Bidder for materials, equipment or labor for a portion of the Work.

ARTICLE 2 BIDDER'S REPRESENTATIONS

2.1 The Bidder by making a Bid represents that:

2.1.1 The Bidder has read and understands the Bidding Documents or Contract Documents, to the extent that such documentation relates to the Work for which the Bid is submitted, and for other portions of the Project, if any, being bid concurrently or presently under construction.

2.1.2 The Bid is made in compliance with the Bidding Documents.

2.1.3 The Bidder has visited the site, become familiar with local conditions under which the Work is to be performed and has correlated the Bidder's personal observations with the requirements of the proposed Contract Documents.

2.1.4 The Bid is based upon the materials, equipment and systems required by the Bidding Documents without exception.

© 1997 A I A®
AIA DOCUMENT A701-1997
INSTRUCTIONS TO BIDDERS

The American Institute
of Architects
1735 New York Avenue, N.W.
Washington, D.C. 20006-5292

2

ARTICLE 3 BIDDING DOCUMENTS

3.1 COPIES

3.1.1 Bidders may obtain complete sets of the Bidding Documents from the issuing office designated in the Advertisement or Invitation to Bid in the number and for the deposit sum, if any, stated therein. The deposit will be refunded to Bidders who submit a bona fide Bid and return the Bidding Documents in good condition within ten days after receipt of Bids. The cost of replacement of missing or damaged documents will be deducted from the deposit. A Bidder receiving a Contract award may retain the Bidding Documents and the Bidder's deposit will be refunded.

3.1.2 Bidding Documents will not be issued directly to Sub-bidders unless specifically offered in the Advertisement or Invitation to Bid, or in supplementary instructions to bidders.

3.1.3 Bidders shall use complete sets of Bidding Documents in preparing Bids; neither the Owner nor Architect assumes responsibility for errors or misinterpretations resulting from the use of incomplete sets of Bidding Documents.

3.1.4 The Owner and Architect may make copies of the Bidding Documents available on the above terms for the purpose of obtaining Bids on the Work. No license or grant of use is conferred by issuance of copies of the Bidding Documents.

3.2 INTERPRETATION OR CORRECTION OF BIDDING DOCUMENTS

3.2.1 The Bidder shall carefully study and compare the Bidding Documents with each other, and with other work being bid concurrently or presently under construction to the extent that it relates to the Work for which the Bid is submitted, shall examine the site and local conditions, and shall at once report to the Architect errors, inconsistencies or ambiguities discovered.

3.2.2 Bidders and Sub-bidders requiring clarification or interpretation of the Bidding Documents shall make a written request which shall reach the Architect at least seven days prior to the date for receipt of Bids.

3.2.3 Interpretations, corrections and changes of the Bidding Documents will be made by Addendum. Interpretations, corrections and changes of the Bidding Documents made in any other manner will not be binding, and Bidders shall not rely upon them.

3.3 SUBSTITUTIONS

3.3.1 The materials, products and equipment described in the Bidding Documents establish a standard of required function, dimension, appearance and quality to be met by any proposed substitution.

3.3.2 No substitution will be considered prior to receipt of Bids unless written request for approval has been received by the Architect at least ten days prior to the date for receipt of Bids. Such requests shall include the name of the material or equipment for which it is to be substituted and a complete description of the proposed substitution including drawings, performance and test data, and other information necessary for an evaluation. A statement setting forth changes in other materials, equipment or other portions of the Work, including changes in the work of other contracts that incorporation of the proposed substitution would require, shall be included. The burden of proof of the merit of the proposed substitution is upon the proposer. The Architect's decision of approval or disapproval of a proposed substitution shall be final.

3.3.3 If the Architect approves a proposed substitution prior to receipt of Bids, such approval will be set forth in an Addendum. Bidders shall not rely upon approvals made in any other manner.

© 1997 AIA®
AIA DOCUMENT A701-1997
INSTRUCTIONS TO BIDDERS

The American Institute
of Architects
1735 New York Avenue, N.W.
Washington, D.C. 20006-5292

3

3.3.4 No substitutions will be considered after the Contract award unless specifically provided for in the Contract Documents.

3.4 ADDENDA

3.4.1 Addenda will be transmitted to all who are known by the issuing office to have received a complete set of Bidding Documents.

3.4.2 Copies of Addenda will be made available for inspection wherever Bidding Documents are on file for that purpose.

3.4.3 Addenda will be issued no later than four days prior to the date for receipt of Bids except an Addendum withdrawing the request for Bids or one which includes postponement of the date for receipt of Bids.

3.4.4 Each Bidder shall ascertain prior to submitting a Bid that the Bidder has received all Addenda issued, and the Bidder shall acknowledge their receipt in the Bid.

ARTICLE 4 BIDDING PROCEDURES

4.1 PREPARATION OF BIDS

4.1.1 Bids shall be submitted on the forms included with the Bidding Documents.

4.1.2 All blanks on the bid form shall be legibly executed in a non-erasable medium.

4.1.3 Sums shall be expressed in both words and figures. In case of discrepancy, the amount written in words shall govern.

4.1.4 Interlineations, alterations and erasures must be initialed by the signer of the Bid.

4.1.5 All requested Alternates shall be bid. If no change in the Base Bid is required, enter "No Change."

4.1.6 Where two or more Bids for designated portions of the Work have been requested, the Bidder may, without forfeiture of the bid security, state the Bidder's refusal to accept award of less than the combination of Bids stipulated by the Bidder. The Bidder shall make no additional stipulations on the bid form nor qualify the Bid in any other manner.

4.1.7 Each copy of the Bid shall state the legal name of the Bidder and the nature of legal form of the Bidder. The Bidder shall provide evidence of legal authority to perform within the jurisdiction of the Work. Each copy shall be signed by the person or persons legally authorized to bind the Bidder to a contract. A Bid by a corporation shall further give the state of incorporation and have the corporate seal affixed. A Bid submitted by an agent shall have a current power of attorney attached certifying the agent's authority to bind the Bidder.

4.2 BID SECURITY

4.2.1 Each Bid shall be accompanied by a bid security in the form and amount required if so stipulated in the Instructions to Bidders. The Bidder pledges to enter into a Contract with the Owner on the terms stated in the Bid and will, if required, furnish bonds covering the faithful performance of the Contract and payment of all obligations arising thereunder. Should the Bidder refuse to enter into such Contract or fail to furnish such bonds if required, the amount of the bid security shall be forfeited to the Owner as liquidated damages, not as a penalty. The amount of the bid security shall not be forfeited to the Owner in the event the Owner fails to comply with Paragraph 6.2.

© 1 9 9 7   A I A ®
AIA DOCUMENT A701-1997
INSTRUCTIONS TO BIDDERS

The American Institute
of Architects
1735 New York Avenue, N.W.
Washington, D.C. 20006-5292

4

4.2.2 If a surety bond is required, it shall be written on AIA Document A310, Bid Bond, unless otherwise provided in the Bidding Documents, and the attorney-in-fact who executes the bond on behalf of the surety shall affix to the bond a certified and current copy of the power of attorney.

4.2.3 The Owner will have the right to retain the bid security of Bidders to whom an award is being considered until either (a) the Contract has been executed and bonds, if required, have been furnished, or (b) the specified time has elapsed so that Bids may be withdrawn or (c) all Bids have been rejected.

4.3   SUBMISSION OF BIDS

4.3.1 All copies of the Bid, the bid security, if any, and any other documents required to be submitted with the Bid shall be enclosed in a sealed opaque envelope. The envelope shall be addressed to the party receiving the Bids and shall be identified with the Project name, the Bidder's name and address and, if applicable, the designated portion of the Work for which the Bid is submitted. If the Bid is sent by mail, the sealed envelope shall be enclosed in a separate mailing envelope with the notation "SEALED BID ENCLOSED" on the face thereof.

4.3.2 Bids shall be deposited at the designated location prior to the time and date for receipt of Bids. Bids received after the time and date for receipt of Bids will be returned unopened.

4.3.3 The Bidder shall assume full responsibility for timely delivery at the location designated for receipt of Bids.

4.3.4 Oral, telephonic, telegraphic, facsimile or other electronically transmitted bids will not be considered.

4.4   MODIFICATION OR WITHDRAWAL OF BID

4.4.1 A Bid may not be modified, withdrawn or canceled by the Bidder during the stipulated time period following the time and date designated for the receipt of Bids, and each Bidder so agrees in submitting a Bid.

4.4.2 Prior to the time and date designated for receipt of Bids, a Bid submitted may be modified or withdrawn by notice to the party receiving Bids at the place designated for receipt of Bids. Such notice shall be in writing over the signature of the Bidder. Written confirmation over the signature of the Bidder shall be received, and date- and time-stamped by the receiving party on or before the date and time set for receipt of Bids. A change shall be so worded as not to reveal the amount of the original Bid.

4.4.3 Withdrawn Bids may be resubmitted up to the date and time designated for the receipt of Bids provided that they are then fully in conformance with these Instructions to Bidders.

4.4.4 Bid security, if required, shall be in an amount sufficient for the Bid as resubmitted.

ARTICLE 5   CONSIDERATION OF BIDS

5.1   OPENING OF BIDS

At the discretion of the Owner, if stipulated in the Advertisement or Invitation to Bid, the properly identified Bids received on time will be publicly opened and will be read aloud. An abstract of the Bids may be made available to Bidders.

5.2   REJECTION OF BIDS

The Owner shall have the right to reject any or all Bids. A Bid not accompanied by a required bid security or by other data required by the Bidding Documents, or a Bid which is in any way incomplete or irregular is subject to rejection.

© 1997   AIA®
AIA DOCUMENT A701-1997
INSTRUCTIONS TO BIDDERS

The American Institute
of Architects
1735 New York Avenue, N.W.
Washington, D.C. 20006-5292

5

5.3 ACCEPTANCE OF BID (AWARD)

5.3.1 It is the intent of the Owner to award a Contract to the lowest qualified Bidder provided the Bid has been submitted in accordance with the requirements of the Bidding Documents and does not exceed the funds available. The Owner shall have the right to waive informalities and irregularities in a Bid received and to accept the Bid which, in the Owner's judgment, is in the Owner's own best interests.

5.3.2 The Owner shall have the right to accept Alternates in any order or combination, unless otherwise specifically provided in the Bidding Documents, and to determine the low Bidder on the basis of the sum of the Base Bid and Alternates accepted.

ARTICLE 6 POST-BID INFORMATION

6.1 CONTRACTOR'S QUALIFICATION STATEMENT

Bidders to whom award of a Contract is under consideration shall submit to the Architect, upon request, a properly executed AIA Document A305, Contractor's Qualification Statement, unless such a Statement has been previously required and submitted as a prerequisite to the issuance of Bidding Documents.

6.2 OWNER'S FINANCIAL CAPABILITY

The Owner shall, at the request of the Bidder to whom award of a Contract is under consideration and no later than seven days prior to the expiration of the time for withdrawal of Bids, furnish to the Bidder reasonable evidence that financial arrangements have been made to fulfill the Owner's obligations under the Contract. Unless such reasonable evidence is furnished, the Bidder will not be required to execute the Agreement between the Owner and Contractor.

6.3 SUBMITTALS

6.3.1 The Bidder shall, as soon as practicable or as stipulated in the Bidding Documents, after notification of selection for the award of a Contract, furnish to the Owner through the Architect in writing:

.1 a designation of the Work to be performed with the Bidder's own forces;

.2 names of the manufacturers, products, and the suppliers of principal items or systems of materials and equipment proposed for the Work; and

.3 names of persons or entities (including those who are to furnish materials or equipment fabricated to a special design) proposed for the principal portions of the Work.

6.3.2 The Bidder will be required to establish to the satisfaction of the Architect and Owner the reliability and responsibility of the persons or entities proposed to furnish and perform the Work described in the Bidding Documents.

6.3.3 Prior to the execution of the Contract, the Architect will notify the Bidder in writing if either the Owner or Architect, after due investigation, has reasonable objection to a person or entity proposed by the Bidder. If the Owner or Architect has reasonable objection to a proposed person or entity, the Bidder may, at the Bidder's option, (1) withdraw the Bid or (2) submit an acceptable substitute person or entity with an adjustment in the Base Bid or Alternate Bid to cover the difference in cost occasioned by such substitution. The Owner may accept the adjusted bid price or disqualify the Bidder. In the event of either withdrawal or disqualification, bid security will not be forfeited.

6.3.4 Persons and entities proposed by the Bidder and to whom the Owner and Architect have made no reasonable objection must be used on the Work for which they were proposed and shall not be changed except with the written consent of the Owner and Architect.

6

ARTICLE 7 PERFORMANCE BOND AND PAYMENT BOND

7.1 BOND REQUIREMENTS

7.1.1 If stipulated in the Bidding Documents, the Bidder shall furnish bonds covering the faithful performance of the Contract and payment of all obligations arising thereunder. Bonds may be secured through the Bidder's usual sources.

7.1.2 If the furnishing of such bonds is stipulated in the Bidding Documents, the cost shall be included in the Bid. If the furnishing of such bonds is required after receipt of bids and before execution of the Contract, the cost of such bonds shall be added to the Bid in determining the Contract Sum.

7.1.3 If the Owner requires that bonds be secured from other than the Bidder's usual sources, changes in cost will be adjusted as provided in the Contract Documents.

7.2 TIME OF DELIVERY AND FORM OF BONDS

7.2.1 The Bidder shall deliver the required bonds to the Owner not later than three days following the date of execution of the Contract. If the Work is to be commenced prior thereto in response to a letter of intent, the Bidder shall, prior to commencement of the Work, submit evidence satisfactory to the Owner that such bonds will be furnished and delivered in accordance with this Subparagraph 7.2.1.

7.2.2 Unless otherwise provided, the bonds shall be written on AIA Document A312, Performance Bond and Payment Bond. Both bonds shall be written in the amount of the Contract Sum.

7.2.3 The bonds shall be dated on or after the date of the Contract.

7.2.4 The Bidder shall require the attorney-in-fact who executes the required bonds on behalf of the surety to affix thereto a certified and current copy of the power of attorney.

ARTICLE 8 FORM OF AGREEMENT BETWEEN OWNER AND CONTRACTOR

Unless otherwise required in the Bidding Documents, the Agreement for the Work will be written on AIA Document A101 Standard Form of Agreement Between Owner and Contractor Where the Basis of Payment Is a Stipulated Sum.

© 1997 AIA®
AIA DOCUMENT A701-1997
INSTRUCTIONS TO BIDDERS

The American Institute
of Architects
1735 New York Avenue, N.W.
Washington, D.C. 20006-5292

7

© 1997 AIA®
AIA DOCUMENT A701-1997
INSTRUCTIONS TO BIDDERS

The American Institute
of Architects
1735 New York Avenue, N.W.
Washington, D.C. 20006-5292

8

9/97

Contract Administration
G612 Owner's Instructions to the Architect Regarding the Construction Contract

GENERAL INFORMATION

PURPOSE. AIA Document G612 is directed toward the most common method of arriving at a construction contract, the bidding process. Though it is recommended that all three parts of G612 be given to the Owner simultaneously, they may be staggered.

USE OF CURRENT AIA DOCUMENTS. Prior to using any AIA Contract Document, users should consult the AIA Contract Documents Web site, *www.aia.org/documents,* or visit a local component chapter to be certain they are using current document editions.

COPYRIGHT. AIA Document G612 is a copyrighted work and may not be excerpted from in substantial part or re-posted via any print or electronic means without the express written permission of the AIA.

To report copyright violations of AIA Contract Documents, email The American Institute of Architects' legal counsel, *copyright@aia.org.*

COMPLETING AIA DOCUMENT G612

AIA Document G612 is divided into a chronological sequence based on the sequence of information requirements on a construction project. Part A of the document relates to contracts, Part B relates to insurance and bonds, and Part C deals with bidding procedures.

Questions relating to the contract may alert an Owner to construction methods and procedures that may affect drawings and specifications and may be overlooked until such documents are almost ready to be issued for bid purposes (e.g., construction to be performed by Owner's own forces; fast-track; phased occupancy). Because many items relating to the contract will have some bearing on the development of construction documents, it is important to place Part A in the Owner's hands at the earliest possible phase of the Project, allowing sufficient time to incorporate relevant instructions into those documents.

The use of General Conditions and Supplementary Conditions is often overlooked for cost-plus or phased (fast-track) construction contracts when the Owner-Contractor Agreement is entered into prior to completion of Drawings and Specifications and assembly of other material for the Project Manual. The General Conditions are as important for those forms of agreement as they are for stipulated-sum contracts based on bidding and should be part of those agreements when signed. When used with B141-1997, the project summary completed in B141-1997 should be checked for consistency with the completed G612.

Responses relating to insurance and bonds cannot appropriately be given until specific General Conditions are chosen, as the General Conditions document normally contains any basic insurance requirements. For that reason, Part B should be given to the Owner along with Part A.

The early date by which instructions in Part A and Part B will be required may be too early for the determination of bidding procedures requested in Part C. It would be appropriate, if timing is questioned, to suggest that Owner replies to Part C may be delayed for a time but should be returned to the Architect prior to completion of the Construction Documents. It is imperative that comments are returned before the completion of the Construction Documents so proper arrangements can be made for issuing sets of bidding documents in anticipation of receipt of bids.

The Architect should acquaint the Owner with the G612 document and its purpose, and should be prepared to explain the need for the requested information and provide necessary assistance, particularly by furnishing any related documents which may contribute to obtaining sufficient information for purposes of the construction agreement and bidding procedures. When discussing contractual documents and insurance, it is essential that the Architect keep in mind that the Architect's advice is for guidance purposes only and that relevant decisions in these matters are for the Owner's lawyers and insurance advisors to make. A letter similar to the following example may be used to transmit the questionnaire and related forms to the Owner.

Dear Owner:

Although we are just starting the construction documents for your project, it is important that we have, at your earliest convenience, information regarding the construction contract, insurance, surety bonds and bidding instructions. Your reply can have considerable bearing on many decisions we will have to make during preparation of drawings and specifications.

Information regarding bidding procedures could be delayed until a later date if these questions cannot be responded to at the same time the other material would be forthcoming.

For your convenience, enclosed in duplicate, is a tri-part questionnaire seeking answers to our major concerns involving the construction process. Along with the questionnaire we have enclosed a standard form of construction agreement, related General Conditions and a standard form of Instructions to Bidders; those are for your information, review and approval or other recommendations. We have also enclosed a recent sample Supplementary Conditions, which illustrates the kind of modifications needed to tailor the standard document to a specific project.

Both you and the Contractor will be required to maintain certain insurance coverages during the construction period. These are set forth in the General Conditions, Article 11, along with the bonding option, and suggestions for typical modifications related to coverages, limits and bonds are found in the accompanying suggested Supplementary Conditions.

We urge that you review all of the enclosures with legal and insurance counsel; we will, of course, be available for explanation, interpretation and advice as may be needed. Your insurance advisor should review the suggested requirements in careful detail and evaluate the protection you and the Contractor should be advised to carry.

We would appreciate an early response to these inquiries. Only one copy of the questionnaire and documents need to be returned to us.

Yours truly,
Q. A. Architect

CHANGES FROM AIA DOCUMENTS G610 AND G611

1. Format Changes

Information requests relating to the construction contract have been grouped together in Part A, Owner's Instructions Regarding the Construction Contract.

2. Changes in Content

AIA Document G612 requests that the Owner provide a detailed description of the building-site property, state whether the Contractor or Contractors will be selected by bidding or negotiation, and indicate which Instructions to Bidders will be used. If multiple contracts are to be utilized, the Owner is asked to indicate how coordination of separate Contractors will be accomplished. In addition, minor changes have been made in the requests for information regarding insurance and bonds to take account of changes in the available policy forms.

COMPLETING AIA DOCUMENT G612

1. The information requests contained in AIA Document G612 are fairly straightforward and require little explanation. However, the information supplied in response to certain requests may have ramifications that are not immediately apparent. These situations are explained below.

2. Part A—Item 6

The single-contract arrangements listed in this request may be effected using the forms of agreement listed in Item 7 and the General Conditions listed in Item 8. Portions of construction by the Owner's own forces and phased construction (fast-track) may be effected by using these documents with suitable modifications. Multiple contracts may also be executed using these documents.

3. Part A—Item 6

If the Contractor(s) are to be selected by negotiation, Part C need not be completed.

4.   Part A—Items 12-16

The information requested herein relates specifically to information that may be incorporated in the Agreement and the Conditions of the Contract Between Owner and Contractor as well as division 1 of the Specifications.

5.   Parts A and B

Contract language embodying the requirements listed herein may be found in AIA Document A511, *Supplementary Conditions to the Contract for Construction.*

EXECUTION OF AIA DOCUMENT G612

The completed AIA Document G612 should be signed and dated by the Owner or a person authorized to act on the Owner's behalf. The Owner should retain at least one copy and forward the original to the Architect.

AIA Contract Administration

G612 Owner's Instructions to the Architect—*Part A, Page 1 of 5*

Owner (*Name and address*):

Date:
Project Title:
Project Number:

Architect (*Name and address*):

NOTATION TO OWNER—In consultation with your attorney and other appropriate advisors, complete this form which will provide your instructions regarding requirements for Contract Documents for this Project. Please return the completed form to your Architect. After reviewing your instructions, the Architect will proceed with the preparation of construction-related documents. Respond carefully to every question.

1. What is the Project title to be used in the Contract Documents?
    _____
    _____

2. What is the legal name and address of the Owner as you wish it to be stated in the Contract Documents?
    _____
    _____

3. How will the Owner be designated?

    ☐ Corporation
    ☐ Partnership
    ☐ Individual
    ☐ Other *(Specify)*_____

    If a corporation, what is the state of incorporation? _____

    Is it qualified to do business at the Project location? ☐ Yes ☐ No

    Is the Owner, identified in the Contract Documents, the owner of the Project site? ☐ Yes ☐ No

4. What is the name and title of the Owner's Representative?
    _____

    Supplementary Conditions of the Contract and General Requirements ☐ may or ☐ may not be discussed and reviewed directly with Owner's attorney, _____,
    whose address is _____.
    Telephone (   ) _____
    Fax (   ) _____
    Email _____

    Supplementary Conditions of the Contract and General Requirements ☐ may or ☐ may not be discussed and reviewed directly with Owner's insurance advisor, _____,
    whose address is _____.
    Telephone (   ) _____
    Fax (   ) _____
    Email _____

Contract Administration
G612 Owner's Instructions to the Architect—*Part A, Page 2 of 5*

During the construction of the Project, will the Owner's employees be responsible for on-site field representation?
☐ Yes ☐ No

> 5. What is the legal description of the Project site, including the legal name and address of the property owner if different from the Owner identified above?
>
> _____
> _____
> _____
> _____
>
> How will the Owner provide a site survey to the Contractor? ☐ Directly from Owner ☐ Directly from the surveyor
> If from the surveyor, list surveyor's name, address and date of survey.
>
> _____
> _____
>
> How will the Owner provide a subsurface investigation report? ☐ Directly from Owner ☐ Directly from the geotechnical engineer
> If from the geotechnical engineer, list the geotechnical engineer's name, address, date of report and report number.
>
> _____
> _____
>
> Are special surveys required? ☐ Yes ☐ No
> If so, describe.
>
> _____
> _____
>
> Will the Contractor be required to make a videotaped survey of existing conditions prior to starting selective demolition or construction? ☐ Yes ☐ No
> If so, specify which areas of the building or areas surrounding the site require a survey.
>
> _____
> _____

6. How will the Project be contracted?
   ☐ Single contract, stipulated sum
   ☑ Single contract, Cost of the Work plus a fee
   ☐ Multiple contracts, stipulated sum
   ☐ Multiple contracts, Cost of the Work plus a fee
   ☐ Portions of construction by Owner's own forces

   Describe phasing of construction or fast-track project delivery requirements, if any.

   _____
   _____

   (Note: A contract may be phased without being fast-tracked. Please describe any specific criteria for project delivery requirements and attach any available information.)

7. When contracting for the Project, what will be the form(s) of agreement between Owner and Contractor(s)?

   _____

   (Note: Refer to the Instructions for a complete list of related AIA Owner-Contractor Agreements.)

8. What will be the form of the general conditions of the contract for construction?
    ☐ AIA Document A201, *General Conditions of the Contract for Construction*
    ☐ AIA Document _____

    Do the administrative responsibilities of the Architect during construction, as defined in the Owner-Architect agreement, differ from those specified in the form of general conditions of the contract for construction being used?
    ☐ Yes ☐ No
    If so, how do you wish to describe the Architect's different responsibilities to the Contractor?
    _____
    _____

9. Are any portions of the bidding requirements to be included in the Contract Documents (advertisement or invitation to bid, instructions to bidders, sample forms, the Contractor's bid or portions of Addenda relating to bidding requirements)? ☐ Yes ☐ No
    If so, specify which portions. _____

    How many copies of Drawings and Project Manuals will be furnished to the Contractor at the Owner's cost?
    _____

10. *(Note: The Owner has a right under AIA general conditions to perform construction and to award separate contracts.)*
    In addition to the general contract for construction, will there be any separate construction contracts? ☐ Yes ☐ No
    If so, summarize scope of such separate contracts. _____
    _____

    Are there works of separate contracts to be performed concurrently for the Owner on the Project? ☐ Yes ☐ No
    If multiple, separate, contracts as described above are to be used, how will the Owner coordinate the activities of the Contractors?
    ☐ Through the Owner's own forces
    ☐ Through another service contract
    If so, who is providing the service and what is the scope? _____
    _____

    If other service contracts exist, please attach a copy of the construction phase duties.

    Are there any items to be furnished or installed by the Owner's own forces? ☐ Yes ☐ No
    If so, describe. _____
    _____

    Do any of these Owner-furnished items require coordination (such as special scheduling, sequencing or inclusion on the Drawings or Specifications) of Work? ☐ Yes ☐ No
    If so, describe. _____
    _____

11. *(Note: According to the AIA General Conditions, within seven days of receipt the Architect will issue or withhold certificate for payment to the Owner.)*
    What day of the month will the Architect receive the Application for Payment?
    ☐ No later than the_____day of the_____month.

    Unless otherwise provided, will the form of Application for Payment and Certificate for Payment be AIA Document G702 and AIA Document G703 and the Continuation Sheet for G702? ☐ Yes ☐ No
    If no, please attach sample document.

Contract Administration
G612 Owner's Instructions to the Architect—*Part A, Page 4 of 5*

Will payment be made for completed portions of Construction Change Directives prior to execution of a Change Order? ☐ Yes ☐ No

Should Applications for Payment be accompanied by the Contractor's partial waiver of liens for Work? ☐ Yes ☐ No

Is a preliminary copy of the Application for Payment required for review by the Owner and Architect prior to submittal of each application? ☐ Yes ☐ No

12. When will the Contractor's applications for payment be paid?
    ☐ By the_____day of each month
    ☐ Other *(Specify)*_____

13. Retainage:
    (a) What percentage of retainage of each progress payment to the Contractor will be retained?
    ☐ _____% of each payment.
    ☐ _____% of each payment until the Work is 50% complete, after which remaining partial payments shall be paid in full without reduction of previous retainage.
    ☐ _____% of each payment (calculated separately for each Work category) until the Work is 50% complete, after which remaining partial payments shall be paid in full without reduction of previous retainage.
    ☐ Other *(Specify)*_____

    (b) Upon Substantial Completion, what will the retainage be reduced to?
    ☐ _____% of completed work
    ☐ $ _____
    ☐ Architect's estimate of incomplete or defective Work remaining in contract (lump sum)
    Are there any other inspections or approvals after Substantial Completion that will be required prior to release of retainage or final payment, which are critical to your use and occupancy of the Project? ☐ Yes ☐ No
    If so, describe in detail. _____
    _____

    (c) Will retained amounts be paid into an escrow account in a financial institution chosen by the Contractor and approved by the Owner, the interest earnings from which accrue to the benefit of the Contractor? ☐ Yes ☐ No

    (d) On a cost-plus-fee contract, will retainage be held on the Contractor's fee?
    ☐ Yes ☐ No
    If so, indicate percentage:_____%

    Will retainage on the Contractor's fee be released upon Substantial Completion? ☐ Yes ☐ No
    If no, specify circumstances. _____
    _____

    Will there be any early releases of retainage to some subcontractors? ☐ Yes ☐ No
    If so, specify which subcontractors. _____
    _____

    (e) Will retainage information be published in the Supplementary Conditions? ☐ Yes ☐ No
    (Note: The primary location of retainage information is in the Agreement. Reference may be made in the Supplementary Conditions if subcontractors or other are to be made aware of these requirements.)

14. Are liquidated damages required? ☐ Yes ☐ No
If required, liquidated damages shall be assessed in the amount of $_____ per day for each calendar day necessary to achieve Substantial Completion beyond the date set forth in the AIA Owner-Contractor Agreement.
(NOTE: When liquidated damages are stipulated, it is important that Subcontractors be made aware of this provision of the Contract. Therefore, in addition to inclusion in the AIA Owner-Contractor Agreement, it is recommended that language be included in the Instructions to Bidders and Supplementary Conditions.)

15. Do you know about, or have you been informed of, any hazardous materials or pollutants at the existing site or in the building? ☐ Yes ☐ No
If so, specify. _____
How will these conditions be abated or mitigated? _____
_____

Will this be done prior to start of demolition or construction? ☐ Yes ☐ No
Will such abatement be done under separate contracts with the Owner? ☐ Yes ☐ No
(Note: If the Owner or Contractor is abating hazardous materials or pollutants concurrently with other construction activities, disclosure in the Supplementary Conditions and instructions to bidders is advisable.)

16. Miscellaneous Provisions
Will any of the following conditions require addition or modification to the Contract Documents? ☐ Yes ☐ No
If so, please attach any available information describing each condition checked.
☐ Equal opportunity requirements
☐ Tax exemptions
☐ Extensions of time criteria
☐ Submission of payment applications to additional parties
☐ Monthly affidavits or release of liens
☐ Lender requirements
☐ Cost savings provisions
☐ Reuse of salvaged materials permitted or required
☐ Substitution and product change procedures
☐ Special time periods during which the Contractor cannot perform construction
☐ Progress schedule format to which the Contractor must conform
☐ Wage or labor standards to which the Contractor must conform
☐ Utility fee or easement criteria
☐ Environmental impact fee criteria
☐ Land use criteria
☐ Building permit criteria
☐ Tenant allowances
☐ Inspection, sampling and testing allowances
☐ Owner's contingency allowances
☐ Portions of buildings and site to remain occupied and used during renovation
☐ Coordination drawings
☐ Tax requirements for historic structures
☐ Construction photographs
☐ Project record document criteria

Owner's Representative _____ By _____ Date _____

▉AIA Contract Administration
G612 Owner's Instructions to the Architect—*Part B, Page 1 of 5*

Owner (*Name and address*):

Date:
Project Title:
Project Number:

Architect (*Name and address*):

NOTATION TO OWNER—In consultation with your attorney and other appropriate advisors, complete this form, which will provide your instructions regarding requirements for Contract Documents for this Project. Please return the completed form to your Architect. After reviewing your instructions, the Architect will proceed with the preparation of construction-related documents. Please respond to every question.

1. **Certificates and Forms**
    Will certificates of insurance, per Subparagraph 11.1.3 of AIA Document A201-1997, be on ACORD Form 25-S, supplemented by AIA Document G715, *Supplemental Attachment*? ☐ Yes ☐ No
    If no, attach a sample of the required form(s).

2. **Contractor's liability insurance**
    Specify the minimum limits of insurance described in Subparagraph 11.1.2 of AIA Document A201-1997.

    a. Workers' compensation insurance
    (Note: Workers' compensation is generally required by statute in most states, with several important exceptions. Exceptions depend upon the occupation or the minimum number of workers employed by that business. The Owner can mandate workers' compensation insurance even for those businesses that are exempt by requiring Voluntary workers' compensation as noted below. In addition to each state having applicable workers' compensation laws, federal and foreign laws may apply to the Contractor's or Subcontractor's employees. Where the Work includes construction involving the following categories, specific coverage may be required for maritime work, longshoremen, harbor work, work at or outside U.S. boundaries, and benefits required by labor union contracts. Please note such requirements below or by separate attachment.)

    Are limits in excess of those required by statute to be provided? ☐ Yes ☐ No
    If so, limits for such insurance shall be as follows: $_____
    $_____   Each accident
    $_____   Disease, policy limit
    $_____   Disease, each employee
    Voluntary workers' compensation (by any exempt entities):
    Will private entities exempt from coverage on account of the number of employees or occupation maintain voluntary compensation coverage at the same limits specified for mandatory coverage? ☐ Yes ☐ No
    If so, indicate dollar limits of coverage below:
    $_____   Each accident
    $_____   Disease, policy limit
    $_____   Disease, each employee

    b. Commercial general liability insurance
    Will commercial general liability insurance, including coverage for premises operations, independent contractors' protective, products completed operations, contractual liability, personal injury and property damage (including coverage for explosion, collapse and underground hazards) be required of the Contractor? ☐ Yes ☐ No
    If so, indicate dollar limits of coverage below:
    $_____   Each occurrence
    $_____   General aggregate
    $_____   Personal and advertising injury

$ _____   Products completed operations aggregate

Will the policy be endorsed to have the general aggregate per Project? ☐ Yes ☐ No

If so, state the general aggregate amount. $ _____
Will the Contractual liability insurance include coverage sufficient to meet obligations equivalent to those stipulated under Paragraph 3.18 AIA Document A201, 1997 edition? ☐ Yes ☐ No
If no, specify the coverage desired. _____
_____

Will products and completed operations insurance shall be maintained for a minimum period of at least _____
( ) year(s) after either 90 days following Substantial Completion or final payment, whichever is earlier?
☐ Yes ☐ No
If no, specify. _____
_____

How much automobile liability insurance (owned, non-owned and hired vehicles) for bodily injury and property damage is required?
$_____   Each occurrence

c. What will be the umbrella or excess liability limit?
$_____
$_____   Over primary insurance
$_____   Retention for self-insured hazards, each occurrence

d. What will be the aircraft liability (owned and non-owned), when applicable? *(Select one)*
☐ With limits proposed by the Contractor for the Owner's approval
☐ With the following limits:
(1) Bodily injury:
$_____   Each person

$_____   Each occurrence
(2) Property damage:
$_____   Each occurrence

e. What will be the watercraft liability (owned and non-owned), when applicable? *(Select one)*
☐ With limits proposed by the Contractor for the Owner's approval
☐ With the following limits:
(1) Bodily injury:
$_____   Each person
$_____   Each occurrence
(2) Property damage:
$_____   Each occurrence

f. Will there be other contractor's liability insurance? ☐ Yes ☐ No
If so, describe. _____
_____

Are any revisions required with regard to hazardous substances or other items, or the Architect's role with regard to the same items? ☐ Yes ☐ No
If so, provide exact written language for insertion into Supplementary Conditions.

3.   Owner's liability insurance
Per Paragraph 11.2 of AIA Document A201, 1997 edition, will the Owner maintain its usual liability insurance?
☐ Yes ☐ No
If no, please specify scope of the Owner's liability insurance as you wish to see it described in the conditions of the Contract._____
_____

4.   Project management protective liability insurance
Will the Contractor be required to provide project management protective liability insurance? ☐ Yes ☐ No
If so, it shall have the following limits:
(a) Bodily injury: $_____   Each occurrence

(b) Property damage: $_____   Each occurrence

(c) Aggregate limit, bodily injury and property damage: $ _____

5.   Property insurance
(a) Will the Owner purchase builder's risk coverage with special causes of loss (including coverage for all material and equipment to be incorporated or used in the Project when stored off-site or in transit)? ☐ Yes ☐ No
(Note: If you answered no to the above question, see question 5i.)
If so, identify the type of form used for the policy:
☐ Completed Value
☐ Reporting
☐ Other *(Specify).* _____

(b) What will be the monetary limits of insurance?
☐ Contract Sum, including future amendments
☐ Other amount *(Specify).* _____

(c) Will any of the following named perils be required, either by specific endorsement or separate policies?
☐ Yes ☐ No
If so, identify below:
☐ Government ordered demolition ☐ Earthquake ☐ Flood

(d) If the Owner provides property insurance, will it be written with a deductible? ☐ Yes ☐ No
If so, identify below:
☐ a deductible of not more than $_____ (aggregate) or
☐ a deductible of not more than $_____ per occurrence.
Will there be an aggregate deductible applicable to the entire Project? ☐ Yes ☐ No
If no, provide description of portions of Project subject to an aggregate deductible.
_____
_____

(NOTE: If coverage for alterations and additions to existing structures is to be included under the Owner's existing coverage, specific instructions should be included under Item 6 below.)

(e) Should the property insurance required by Paragraph 11.4 of AIA Document A201, 1997 edition, cover machinery, tools or equipment owned or rented by the Contractor that are utilized in the performance of the Work, but not incorporated into the permanent improvements? ☐ Yes ☐ No

(f) Will the Owner provide boiler and machinery insurance? ☐ Yes ☐ No
If so, specify the limits and objects to be insured: _____
_____

(g) Will the Owner provide loss of use insurance? ☐ Yes ☐ No
    The Contractor shall provide this insurance with limits of $_____.

(h) List any additions/modifications to the specified coverages:

(i) If you answered no to question 5a, will the Contractor be required to carry builder's risk with special causes of loss form property insurance? ☐ Yes ☐ No

Will the limits of such insurance be the Contract Sum, including future amendments? ☐ Yes ☐ No
If so, will the limits of such insurance also include the value of separate contracts and Owner-furnished items?
☐ Yes ☐ No
Will there be any dollar limits of insurance for Contractor provided property insurance? ☐ Yes ☐ No
If so, state how much. $_____
Will the Owner provide partial property insurance? ☐ Yes ☐ No
If so, specify scope limits:_____
Can the Contractor, at the Contractor's own expense, provide insurance coverage for materials stored off the site after written approval of the Owner at the value established in the approval, and also for portions of the Work in transit until such materials are permanently attached to the Work? ☐ Yes ☑ No
If no, specify how you wish insurance on materials off the site to be handled._____
_____

If the Owner is damaged by the failure of the Contractor to purchase and maintain property insurance without so notifying the Owner in writing, will the Contractor be required to bear all reasonable costs attributable thereto?
☐ Yes ☐ No
Will the Contractor be responsible for deductibles? ☐ Yes ☐ No
Shall Contractor provided property insurance be written with a specified maximum deductible per occurrence?
☐ Yes ☐ No
If so, specify the maximum deductible. $_____
Specify special instructions for Contractor provided property insurance.
_____
_____

6.   Other instructions related to bonds or insurance
    (If none, please indicate.)
    Are any special coverages required with regard to alterations or additions to existing structures? ☐ Yes ☐ No
    Are any revisions required with regard to hazardous substances or other items, or the Owner's, Contractor's or Architect's role with regard to the same items? ☐ Yes ☐ No
    If so, provide exact written language for insertion into Supplementary Conditions.

7.   Bonds
    Are performance bonds and payment bonds required? ☐ Yes ☐ No
    (a) If so, the required bonds shall be in the amount of *(Select one option for each bond)*:

Performance	☐ 100% of Contract Sum	☐____% of Contract Sum	☐ $_____
Payment	☐ 100% of Contract Sum	☐____% of Contract Sum	☐ $_____

Contract Administration

G612 Owner's Instructions to the Architect—*Part B, Page 5 of 5*

(b) If so, the form of bonds shall be:

☐ AIA Document A312

☐ Other

(If other, describe and furnish sample copy if available)

(c) Special instructions:

_____

Owner's Representative

_____   _____

By   Date

AIA Contract Administration
G612 Owner's Instructions to the Architect—*Part C, Page 1 of 3*

Owner (*Name and address*):

Date:
Project Title:
Project Number:

Architect (*Name and address*):

NOTATION TO OWNER—This form provides your instructions regarding requirements for bidding procedures for this Project. Please return the completed form to your Architect. After reviewing your instructions, the Architect will proceed with the preparation of necessary bidding documentation. Please respond to every question.

1. What method will be used for selecting the Contractor(s)?
   (a) ☐ Bidding
      ☐ Open and competitive
      ☐ By invitation only
      ☐ Other *(Specify)*
   (b) ☐ Negotiating
      ☐ A single Contractor
      ☐ Multiple Contractors
      ☐ Other *(Specify)*

   Are there any special eligibility requirements for contractor selection? ☐ Yes ☐ No
   If so, describe requirements._____

   Will the list of invited bidders be included in the bidding requirements? ☐ Yes ☐ No
   Will bidders be required to list subcontractors in their proposals? ☐ Yes ☐ No
   Will the Bidder be basing the Bid Sum or Time of Completion on partial drawings and specifications?
   ☐ Yes ☐ No
   Will any bidders or sub-bidders be pre-qualified? ☐ Yes ☐ No
   If so, please list work categories requiring pre-qualification. _____

   _____

   Are there any additional instructions on the method of selection or qualification of bidders? ☐ Yes ☐ No
   If so, describe_____

   _____

2. Who will prepare the instructions to bidders, ☐ Owner or ☐ Architect?
   If the Architect will prepare the instructions, please attach any standard forms or specific language for special bidding provisions you wish to see incorporated into the instructions to bidders.

3. Who prepares the proposal form?
      ☐ Owner
      ☐ Architect

4. How will bids be solicited?
      ☐ Public advertisement arranged by ☐ Owner or ☐ Architect
      ☐ Private invitation issued by ☐ Owner or ☐ Architect

Contract Administration

G612 Owner's Instructions to the Architect—*Part C, Page 2 of 3*

Will Bid opening be ☐ public or ☐ private?

Will there be a pre-bid conference? ☐ Yes ☐ No

If so, specify when and where._____

5. Will the cost of the performance and payment bonds be included in the ☐ base bid or ☐ alternate?

Will property insurance by the Contractor be bid as an alternate? ☐ Yes ☐ No

Is an itemized breakdown of the bid price required? ☐ Yes ☐ No

If so, identify those items of Work._____

_____

Will the Contractor be required to obtain more than one subcontractor bid on any item of work? ☐ Yes ☒ No

If so, list items._____

_____

6. Will bid security be required? ☐ Yes ☐ No

If so, in the amount of

☐ $_____, or

☐ _____% of the total bid in the form of:

☐ A bid bond (AIA Document A310 in conjunction with AIA Document A312), or

☐ Other_____

(Describe other acceptable types of security or bond forms, if any.)

7. Where will copies of the Bidding Documents be made available for reference of the bidders?

☐ Plan rooms designated by the Owner

☐ Plan rooms selected by the Architect

☐ Owner's office

☐ Architect's office

☐ Other *(Specify)*_____

Who can provide copies of Bidding Documents?

☐ Owner ☐ Architect ☐ Printer

Are there any limits on the number of sets of Bidding Documents to be issued to each bidder? ☐ Yes ☐ No

If so, describe the limitations._____

8. What date and time is required for receipt of bids?

☐ *(Specify)*_____

☐ Will be determined later by the Owner

☐ Will be determined by the Architect

9. Where shall bids be received?

☐ At the Architect's office

☐ Other *(Specify name and address of recipient)*_____

_____

10. Who will prepare the bid tabulation forms?

☐ Owner

☐ Architect

11. Will bids be publicly opened and read aloud? ☐ Yes ☐ No

If opened in private, will bid tabulation be furnished to Bidders? ☐ Yes ☐ No

Contract Administration
G612 Owner's Instructions to the Architect—*Part C, Page 3 of 3*

12.   Are copies, in addition to the usual original signed bid, required? ☐ Yes ☐ No
        If so, specify how many copies._____

13.   How many calendar days after receipt of bids must a bid remain open for acceptance?_____

14.   If a Contract is awarded, when will construction at the site commence?
            ☐ Upon execution of the Agreement
            ☐ Upon, but not before, receipt of a notice to proceed
            ☐ Not earlier than_____days after award of the Contract
            ☐ Other *(Specify)*_____

15.   Will the required time of Substantial Completion be stipulated in the Bidding Documents? ☑ Yes ☐ No
        If so, Work shall be substantially complete:
            ☐ _____calendar days after the Date of Commencement
            ☐ By, _____ *(Date)*
            ☐ In the number of calendar days stipulated by the bidder in the bid form

        Will a preliminary schedule be required to accompany the bid? ☐ Yes ☐ No
        Will bidders be required to incorporate any milestone dates into the preliminary schedule? ☐ Yes ☐ No
        If so, describe._____

        Will designated portions of the Work require Substantial Completion in advance of the rest of the Project?
        ☐ Yes ☐ No
        Will these portions be identified on Drawings? ☐ Yes ☐ No
        If no, describe the scope and anticipated date(s) of completion for such designated portion(s) of the Work as you
        wish to see them incorporated in the Drawings. _____
        _____
        _____

16.   Special instructions:
        (Note: Please describe checked items by separate attachment.)
            ☐ Unit price proposal language
            ☐ Substitution criteria
            ☐ Fee proposal language
            ☐ Overhead or profit limits
            ☑ Waivers
            ☐ Non-Collusion affidavit
            ☐ Qualification statement

_____
Owner's Representative

_____   _____   _____
By   Who represents *(Name of Owner)*   Date

1997 EDITION

AIA DOCUMENT | A201-1997

General Conditions of the Contract for Construction

This document has important legal consequences. Consultation with an attorney is encouraged with respect to its completion or modification.

This document has been approved and endorsed by The Associated General Contractors of America.

TABLE OF ARTICLES

© 1997 AIA®
AIA DOCUMENT A201-1997
GENERAL CONDITIONS
OF THE CONTRACT FOR
CONSTRUCTION

The American Institute
of Architects
1735 New York Avenue, N.W.
Washington, D.C. 20006-5292

CAUTION: You should use an original AIA document with the AIA logo printed in red. An original assures that changes will not be obscured as may occur when documents are reproduced.

1

INDEX

© 1997 A I A®
AIA DOCUMENT A201-1997
GENERAL CONDITIONS
OF THE CONTRACT FOR
CONSTRUCTION

The American Institute
of Architects
1735 New York Avenue, N.W.
Washington, D.C. 20006-5292

3

© 1997 AIA®
AIA DOCUMENT A201-1997
GENERAL CONDITIONS
OF THE CONTRACT FOR
CONSTRUCTION

The American Institute
of Architects
1735 New York Avenue, N.W.
Washington, D.C. 20006-5292

4

The American Institute of Architects is pleased to provide this sample copy of an AIA Contract Document for education purposes. Created with the consensus of contractors, attorneys, architects and engineers, the AIA Contract Documents represent over 110 years of legal precedent.

Reproduced with permission of The American Institute of Architects, 1735 New York Avenue NW, Washington D.C. 20006. For more information or to purchase AIA Contract Documents, visit www.aia.org.

© 1997 AIA®
AIA DOCUMENT A201-1997
GENERAL CONDITIONS
OF THE CONTRACT FOR
CONSTRUCTION

The American Institute
of Architects
1735 New York Avenue, N.W.
Washington, D.C. 20006-5292

7

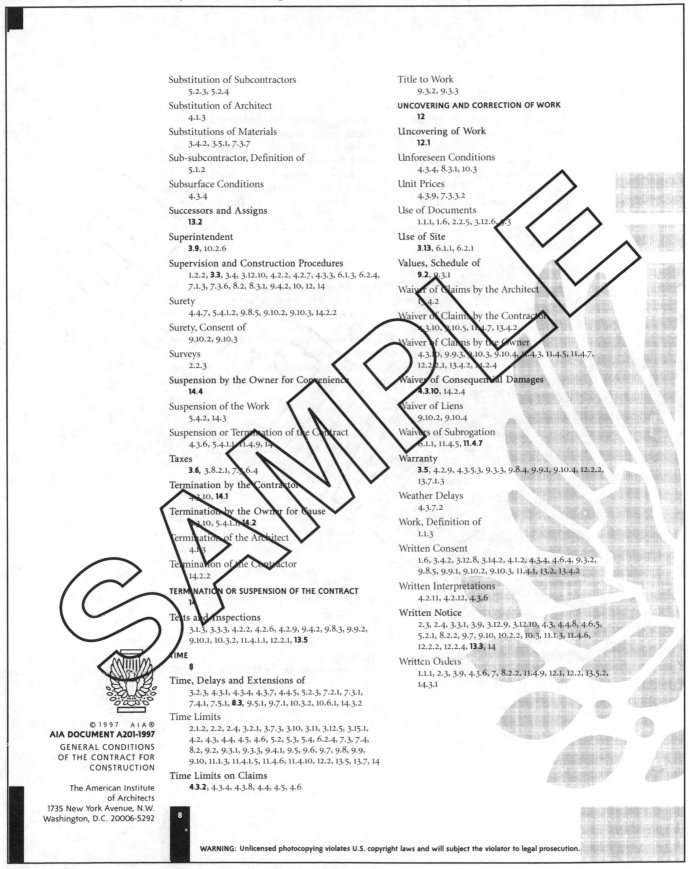

© 1997 AIA®
AIA DOCUMENT A201-1997
GENERAL CONDITIONS
OF THE CONTRACT FOR
CONSTRUCTION

The American Institute
of Architects
1735 New York Avenue, N.W.
Washington, D.C. 20006-5292

8

ARTICLE 1   GENERAL PROVISIONS

1.1   BASIC DEFINITIONS

1.1.1   THE CONTRACT DOCUMENTS

The Contract Documents consist of the Agreement between Owner and Contractor (hereinafter the Agreement), Conditions of the Contract (General, Supplementary and other Conditions), Drawings, Specifications, Addenda issued prior to execution of the Contract, other documents listed in the Agreement and Modifications issued after execution of the Contract. A Modification is (1) a written amendment to the Contract signed by both parties, (2) a Change Order, (3) a Construction Change Directive or (4) a written order for a minor change in the Work issued by the Architect. Unless specifically enumerated in the Agreement, the Contract Documents do not include other documents such as bidding requirements (advertisement or invitation to bid, Instructions to Bidders, sample forms, the Contractor's bid or portions of Addenda relating to bidding requirements).

1.1.2   THE CONTRACT

The Contract Documents form the Contract for Construction. The Contract represents the entire and integrated agreement between the parties hereto and supersedes prior negotiations, representations or agreements, either written or oral. The Contract may be amended or modified only by a Modification. The Contract Documents shall not be construed to create a contractual relationship of any kind (1) between the Architect and Contractor, (2) between the Owner and a Subcontractor or Sub-subcontractor, (3) between the Owner and Architect or (4) between any persons or entities other than the Owner and Contractor. The Architect shall, however, be entitled to performance and enforcement of obligations under the Contract intended to facilitate performance of the Architect's duties.

1.1.3   THE WORK

The term "Work" means the construction and services required by the Contract Documents, whether completed or partially completed, and includes all other labor, materials, equipment and services provided or to be provided by the Contractor to fulfill the Contractor's obligations. The Work may constitute the whole or a part of the Project.

1.1.4   THE PROJECT

The Project is the total construction of which the Work performed under the Contract Documents may be the whole or a part and which may include construction by the Owner or by separate contractors.

1.1.5   THE DRAWINGS

The Drawings are the graphic and pictorial portions of the Contract Documents showing the design, location and dimensions of the Work, generally including plans, elevations, sections, details, schedules and diagrams.

1.1.6   THE SPECIFICATIONS

The Specifications are that portion of the Contract Documents consisting of the written requirements for materials, equipment, systems, standards and workmanship for the Work, and performance of related services.

1.1.7   THE PROJECT MANUAL

The Project Manual is a volume assembled for the Work which may include the bidding requirements, sample forms, Conditions of the Contract and Specifications.

1.2   CORRELATION AND INTENT OF THE CONTRACT DOCUMENTS

1.2.1   The intent of the Contract Documents is to include all items necessary for the proper execution and completion of the Work by the Contractor. The Contract Documents are

© 1997   A I A ®
AIA DOCUMENT A201-1997
GENERAL CONDITIONS
OF THE CONTRACT FOR
CONSTRUCTION

The American Institute
of Architects
1735 New York Avenue, N.W.
Washington, D.C. 20006-5292

9

complementary, and what is required by one shall be as binding as if required by all; performance by the Contractor shall be required only to the extent consistent with the Contract Documents and reasonably inferable from them as being necessary to produce the indicated results.

1.2.2 Organization of the Specifications into divisions, sections and articles, and arrangement of Drawings shall not control the Contractor in dividing the Work among Subcontractors or in establishing the extent of Work to be performed by any trade.

1.2.3 Unless otherwise stated in the Contract Documents, words which have well-known technical or construction industry meanings are used in the Contract Documents in accordance with such recognized meanings.

1.3 CAPITALIZATION
1.3.1 Terms capitalized in these General Conditions include those which are (1) specifically defined, (2) the titles of numbered articles and identified references to Paragraphs, Subparagraphs and Clauses in the document or (3) the titles of other documents published by the American Institute of Architects.

1.4 INTERPRETATION
1.4.1 In the interest of brevity the Contract Documents frequently omit modifying words such as "all" and "any" and articles such as "the" and "an" but the fact that a modifier or an article is absent from one statement and appears in another is not intended to affect the interpretation of either statement.

1.5 EXECUTION OF CONTRACT DOCUMENTS
1.5.1 The Contract Documents shall be signed by the Owner and Contractor. If either the Owner or Contractor or both do not sign all the Contract Documents, the Architect shall identify such unsigned Documents upon request.

1.5.2 Execution of the Contract by the Contractor is a representation that the Contractor has visited the site, become generally familiar with local conditions under which the Work is to be performed and correlated personal observations with requirements of the Contract Documents.

1.6 OWNERSHIP AND USE OF DRAWINGS, SPECIFICATIONS AND OTHER INSTRUMENTS OF SERVICE
1.6.1 The Drawings, Specifications and other documents, including those in electronic form, prepared by the Architect and the Architect's consultants are Instruments of Service through which the Work to be executed by the Contractor is described. The Contractor may retain one record set. Neither the Contractor nor any Subcontractor, Sub-subcontractor or material or equipment supplier shall own or claim a copyright in the Drawings, Specifications and other documents prepared by the Architect or the Architect's consultants, and unless otherwise indicated the Architect and the Architect's consultants shall be deemed the authors of them and will retain all common law, statutory and other reserved rights, in addition to the copyrights. All copies of Instruments of Service, except the Contractor's record set, shall be returned or suitably accounted for to the Architect, on request, upon completion of the Work. The Drawings, Specifications and other documents prepared by the Architect and the Architect's consultants, and copies thereof furnished to the Contractor, are for use solely with respect to this Project. They are not to be used by the Contractor or any Subcontractor, Sub-subcontractor or material or equipment supplier on other projects or for additions to this Project outside the scope of the Work without the specific written consent of the Owner, Architect and the Architect's consultants. The Contractor, Subcontractors, Sub-subcontractors and material or equipment suppliers are authorized to use and reproduce applicable portions of the Drawings, Specifications and other documents prepared by the Architect and the Architect's consultants appropriate to and for use in

© 1997 AIA®
AIA DOCUMENT A201-1997
GENERAL CONDITIONS
OF THE CONTRACT FOR
CONSTRUCTION

The American Institute
of Architects
1735 New York Avenue, N.W.
Washington, D.C. 20006-5292

10

the execution of their Work under the Contract Documents. All copies made under this authorization shall bear the statutory copyright notice, if any, shown on the Drawings, Specifications and other documents prepared by the Architect and the Architect's consultants. Submittal or distribution to meet official regulatory requirements or for other purposes in connection with this Project is not to be construed as publication in derogation of the Architect's or Architect's consultants' copyrights or other reserved rights.

ARTICLE 2   OWNER

2.1   GENERAL

2.1.1   The Owner is the person or entity identified as such in the Agreement and is referred to throughout the Contract Documents as if singular in number. The Owner shall designate in writing a representative who shall have express authority to bind the Owner with respect to all matters requiring the Owner's approval or authorization. Except as otherwise provided in Subparagraph 4.2.1, the Architect does not have such authority. The term "Owner" means the Owner or the Owner's authorized representative.

2.1.2   The Owner shall furnish to the Contractor within fifteen days after receipt of a written request, information necessary and relevant for the Contractor to evaluate, give notice of or enforce mechanic's lien rights. Such information shall include a correct statement of the record legal title to the property on which the Project is located, usually referred to as the site, and the Owner's interest therein.

2.2   INFORMATION AND SERVICES REQUIRED OF THE OWNER

2.2.1   The Owner shall, at the written request of the Contractor, prior to commencement of the Work and thereafter, furnish to the Contractor reasonable evidence that financial arrangements have been made to fulfill the Owner's obligations under the Contract. Furnishing of such evidence shall be a condition precedent to commencement or continuation of the Work. After such evidence has been furnished, the Owner shall not materially vary such financial arrangements without prior notice to the Contractor.

2.2.2   Except for permits and fees, including those required under Subparagraph 3.7.1, which are the responsibility of the Contractor under the Contract Documents, the Owner shall secure and pay for necessary approvals, easements, assessments and charges required for construction, use or occupancy of permanent structures or for permanent changes in existing facilities.

2.2.3   The Owner shall furnish surveys describing physical characteristics, legal limitations and utility locations for the site of the Project, and a legal description of the site. The Contractor shall be entitled to rely on the accuracy of information furnished by the Owner but shall exercise proper precautions relating to the safe performance of the Work.

2.2.4   Information or services required of the Owner by the Contract Documents shall be furnished by the Owner with reasonable promptness. Any other information or services relevant to the Contractor's performance of the Work under the Owner's control shall be furnished by the Owner after receipt from the Contractor of a written request for such information or services.

2.2.5   Unless otherwise provided in the Contract Documents, the Contractor will be furnished, free of charge, such copies of Drawings and Project Manuals as are reasonably necessary for execution of the Work.

2.3   OWNER'S RIGHT TO STOP THE WORK

2.3.1   If the Contractor fails to correct Work which is not in accordance with the requirements of the Contract Documents as required by Paragraph 12.2 or persistently fails to carry out Work in

AIA DOCUMENT A201-1997
GENERAL CONDITIONS
OF THE CONTRACT FOR
CONSTRUCTION

The American Institute
of Architects
1735 New York Avenue, N.W.
Washington, D.C. 20006-5292

11

accordance with the Contract Documents, the Owner may issue a written order to the Contractor to stop the Work, or any portion thereof, until the cause for such order has been eliminated; however, the right of the Owner to stop the Work shall not give rise to a duty on the part of the Owner to exercise this right for the benefit of the Contractor or any other person or entity, except to the extent required by Subparagraph 6.1.3.

2.4   OWNER'S RIGHT TO CARRY OUT THE WORK

2.4.1   If the Contractor defaults or neglects to carry out the Work in accordance with the Contract Documents and fails within a seven-day period after receipt of written notice from the Owner to commence and continue correction of such default or neglect with diligence and promptness, the Owner may after such seven-day period give the Contractor a second written notice to correct such deficiencies within a three-day period. If the Contractor within such three-day period after receipt of such second notice fails to commence and continue to correct any deficiencies, the Owner may, without prejudice to other remedies the Owner may have, correct such deficiencies. In such case an appropriate Change Order shall be issued deducting from payments then or thereafter due the Contractor the reasonable cost of correcting such deficiencies, including Owner's expenses and compensation for the Architect's additional services made necessary by such default, neglect or failure. Such action by the Owner and amounts charged to the Contractor are both subject to prior approval of the Architect. If payments then or thereafter due the Contractor are not sufficient to cover such amounts, the Contractor shall pay the difference to the Owner.

ARTICLE 3   CONTRACTOR

3.1   GENERAL

3.1.1   The Contractor is the person or entity identified as such in the Agreement and is referred to throughout the Contract Documents as if singular in number. The term "Contractor" means the Contractor or the Contractor's authorized representative.

3.1.2   The Contractor shall perform the Work in accordance with the Contract Documents.

3.1.3   The Contractor shall not be relieved of obligations to perform the Work in accordance with the Contract Documents either by activities or duties of the Architect in the Architect's administration of the Contract, or by tests, inspections or approvals required or performed by persons other than the Contractor.

3.2   REVIEW OF CONTRACT DOCUMENTS AND FIELD CONDITIONS BY CONTRACTOR

3.2.1   Since the Contract Documents are complementary, before starting each portion of the Work, the Contractor shall carefully study and compare the various Drawings and other Contract Documents relative to that portion of the Work, as well as the information furnished by the Owner pursuant to Subparagraph 2.2.3, shall take field measurements of any existing conditions related to that portion of the Work and shall observe any conditions at the site affecting it. These obligations are for the purpose of facilitating construction by the Contractor and are not for the purpose of discovering errors, omissions, or inconsistencies in the Contract Documents; however, any errors, inconsistencies or omissions discovered by the Contractor shall be reported promptly to the Architect as a request for information in such form as the Architect may require.

3.2.2   Any design errors or omissions noted by the Contractor during this review shall be reported promptly to the Architect, but it is recognized that the Contractor's review is made in the Contractor's capacity as a contractor and not as a licensed design professional unless otherwise specifically provided in the Contract Documents. The Contractor is not required to ascertain that the Contract Documents are in accordance with applicable laws, statutes, ordinances, building codes, and rules and regulations, but any nonconformity discovered by or made known to the Contractor shall be reported promptly to the Architect.

3.2.3 If the Contractor believes that additional cost or time is involved because of clarifications or instructions issued by the Architect in response to the Contractor's notices or requests for information pursuant to Subparagraphs 3.2.1 and 3.2.2, the Contractor shall make Claims as provided in Subparagraphs 4.3.6 and 4.3.7. If the Contractor fails to perform the obligations of Subparagraphs 3.2.1 and 3.2.2, the Contractor shall pay such costs and damages to the Owner as would have been avoided if the Contractor had performed such obligations. The Contractor shall not be liable to the Owner or Architect for damages resulting from errors, inconsistencies or omissions in the Contract Documents or for differences between field measurements or conditions and the Contract Documents unless the Contractor recognized such error, inconsistency, omission or difference and knowingly failed to report it to the Architect.

3.3   SUPERVISION AND CONSTRUCTION PROCEDURES

3.3.1 The Contractor shall supervise and direct the Work, using the Contractor's best skill and attention. The Contractor shall be solely responsible for and have control over construction means, methods, techniques, sequences and procedures and for coordinating all portions of the Work under the Contract, unless the Contract Documents give other specific instructions concerning these matters. If the Contract Documents give specific instructions concerning construction means, methods, techniques, sequences or procedures, the Contractor shall evaluate the jobsite safety thereof and, except as stated below, shall be fully and solely responsible for the jobsite safety of such means, methods, techniques, sequences or procedures. If the Contractor determines that such means, methods, techniques, sequences or procedures may not be safe, the Contractor shall give timely written notice to the Owner and Architect and shall not proceed with that portion of the Work without further written instructions from the Architect. If the Contractor is then instructed to proceed with the required means, methods, techniques, sequences or procedures without acceptance of changes proposed by the Contractor, the Owner shall be solely responsible for any resulting loss or damage.

3.3.2 The Contractor shall be responsible to the Owner for acts and omissions of the Contractor's employees, Subcontractors and their agents and employees, and other persons or entities performing portions of the Work for or on behalf of the Contractor or any of its Subcontractors.

3.3.3 The Contractor shall be responsible for inspection of portions of Work already performed to determine that such portions are in proper condition to receive subsequent Work.

3.4   LABOR AND MATERIALS

3.4.1 Unless otherwise provided in the Contract Documents, the Contractor shall provide and pay for labor, materials, equipment, tools, construction equipment and machinery, water, heat, utilities, transportation, and other facilities and services necessary for proper execution and completion of the Work, whether temporary or permanent and whether or not incorporated or to be incorporated in the Work.

3.4.2 The Contractor may make substitutions only with the consent of the Owner, after evaluation by the Architect and in accordance with a Change Order.

3.4.3 The Contractor shall enforce strict discipline and good order among the Contractor's employees and other persons carrying out the Contract. The Contractor shall not permit employment of unfit persons or persons not skilled in tasks assigned to them.

3.5   WARRANTY

3.5.1 The Contractor warrants to the Owner and Architect that materials and equipment furnished under the Contract will be of good quality and new unless otherwise required or permitted by the Contract Documents, that the Work will be free from defects not inherent in the quality required or permitted, and that the Work will conform to the requirements of the Contract

© 1997   AIA®
AIA DOCUMENT A201-1997
GENERAL CONDITIONS
OF THE CONTRACT FOR
CONSTRUCTION

The American Institute
of Architects
1735 New York Avenue, N.W.
Washington, D.C. 20006-5292

13

Documents. Work not conforming to these requirements, including substitutions not properly approved and authorized, may be considered defective. The Contractor's warranty excludes remedy for damage or defect caused by abuse, modifications not executed by the Contractor, improper or insufficient maintenance, improper operation, or normal wear and tear and normal usage. If required by the Architect, the Contractor shall furnish satisfactory evidence as to the kind and quality of materials and equipment.

3.6   TAXES

3.6.1   The Contractor shall pay sales, consumer, use and similar taxes for the Work provided by the Contractor which are legally enacted when bids are received or negotiations concluded, whether or not yet effective or merely scheduled to go into effect.

3.7   PERMITS, FEES AND NOTICES

3.7.1   Unless otherwise provided in the Contract Documents, the Contractor shall secure and pay for the building permit and other permits and governmental fees, licenses and inspections necessary for proper execution and completion of the Work which are customarily secured after execution of the Contract and which are legally required when bids are received or negotiations concluded.

3.7.2   The Contractor shall comply with and give notices required by laws, ordinances, rules, regulations and lawful orders of public authorities applicable to performance of the Work.

3.7.3   It is not the Contractor's responsibility to ascertain that the Contract Documents are in accordance with applicable laws, statutes, ordinances, building codes, and rules and regulations. However, if the Contractor observes that portions of the Contract Documents are at variance therewith, the Contractor shall promptly notify the Architect and Owner in writing, and necessary changes shall be accomplished by appropriate Modification.

3.7.4   If the Contractor performs Work knowing it to be contrary to laws, statutes, ordinances, building codes, and rules and regulations without such notice to the Architect and Owner, the Contractor shall assume appropriate responsibility for such Work and shall bear the costs attributable to correction.

3.8   ALLOWANCES

3.8.1   The Contractor shall include in the Contract Sum all allowances stated in the Contract Documents. Items covered by allowances shall be supplied for such amounts and by such persons or entities as the Owner may direct, but the Contractor shall not be required to employ persons or entities to whom the Contractor has reasonable objection.

3.8.2   Unless otherwise provided in the Contract Documents:
   .1 allowances shall cover the cost to the Contractor of materials and equipment delivered at the site and all required taxes, less applicable trade discounts;
   .2 Contractor's costs for unloading and handling at the site, labor, installation costs, overhead, profit and other expenses contemplated for stated allowance amounts shall be included in the Contract Sum but not in the allowances;
   .3 whenever costs are more than or less than allowances, the Contract Sum shall be adjusted accordingly by Change Order. The amount of the Change Order shall reflect (1) the difference between actual costs and the allowances under Clause 3.8.2.1 and (2) changes in Contractor's costs under Clause 3.8.2.2.

3.8.3   Materials and equipment under an allowance shall be selected by the Owner in sufficient time to avoid delay in the Work.

3.9 SUPERINTENDENT

3.9.1 The Contractor shall employ a competent superintendent and necessary assistants who shall be in attendance at the Project site during performance of the Work. The superintendent shall represent the Contractor, and communications given to the superintendent shall be as binding as if given to the Contractor. Important communications shall be confirmed in writing. Other communications shall be similarly confirmed on written request in each case.

3.10 CONTRACTOR'S CONSTRUCTION SCHEDULES

3.10.1 The Contractor, promptly after being awarded the Contract, shall prepare and submit for the Owner's and Architect's information a Contractor's construction schedule for the Work. The schedule shall not exceed time limits current under the Contract Documents, shall be revised at appropriate intervals as required by the conditions of the Work and Project, shall be related to the entire Project to the extent required by the Contract Documents, and shall provide for expeditious and practicable execution of the Work.

3.10.2 The Contractor shall prepare and keep current, for the Architect's approval, a schedule of submittals which is coordinated with the Contractor's construction schedule and allows the Architect reasonable time to review submittals.

3.10.3 The Contractor shall perform the Work in general accordance with the most recent schedules submitted to the Owner and Architect.

3.11 DOCUMENTS AND SAMPLES AT THE SITE

3.11.1 The Contractor shall maintain at the site for the Owner one record copy of the Drawings, Specifications, Addenda, Change Orders and other Modifications, in good order and marked currently to record field changes and selections made during construction, and one record copy of approved Shop Drawings, Product Data, Samples and similar required submittals. These shall be available to the Architect and shall be delivered to the Architect for submittal to the Owner upon completion of the Work.

3.12 SHOP DRAWINGS, PRODUCT DATA AND SAMPLES

3.12.1 Shop Drawings are drawings, diagrams, schedules and other data specially prepared for the Work by the Contractor or a Subcontractor, Sub-subcontractor, manufacturer, supplier or distributor to illustrate some portion of the Work.

3.12.2 Product Data are illustrations, standard schedules, performance charts, instructions, brochures, diagrams and other information furnished by the Contractor to illustrate materials or equipment for some portion of the Work.

3.12.3 Samples are physical examples which illustrate materials, equipment or workmanship and establish standards by which the Work will be judged.

3.12.4 Shop Drawings, Product Data, Samples and similar submittals are not Contract Documents. The purpose of their submittal is to demonstrate for those portions of the Work for which submittals are required by the Contract Documents the way by which the Contractor proposes to conform to the information given and the design concept expressed in the Contract Documents. Review by the Architect is subject to the limitations of Subparagraph 4.2.7. Informational submittals upon which the Architect is not expected to take responsive action may be so identified in the Contract Documents. Submittals which are not required by the Contract Documents may be returned by the Architect without action.

3.12.5 The Contractor shall review for compliance with the Contract Documents, approve and submit to the Architect Shop Drawings, Product Data, Samples and similar submittals required by

15

the Contract Documents with reasonable promptness and in such sequence as to cause no delay in the Work or in the activities of the Owner or of separate contractors. Submittals which are not marked as reviewed for compliance with the Contract Documents and approved by the Contractor may be returned by the Architect without action.

3.12.6 By approving and submitting Shop Drawings, Product Data, Samples and similar submittals, the Contractor represents that the Contractor has determined and verified materials, field measurements and field construction criteria related thereto, or will do so, and has checked and coordinated the information contained within such submittals with the requirements of the Work and of the Contract Documents.

3.12.7 The Contractor shall perform no portion of the Work for which the Contract Documents require submittal and review of Shop Drawings, Product Data, Samples or similar submittals until the respective submittal has been approved by the Architect.

3.12.8 The Work shall be in accordance with approved submittals except that the Contractor shall not be relieved of responsibility for deviations from requirements of the Contract Documents by the Architect's approval of Shop Drawings, Product Data, Samples or similar submittals unless the Contractor has specifically informed the Architect in writing of such deviation at the time of submittal and (1) the Architect has given written approval to the specific deviation as a minor change in the Work, or (2) a Change Order or Construction Change Directive has been issued authorizing the deviation. The Contractor shall not be relieved of responsibility for errors or omissions in Shop Drawings, Product Data, Samples or similar submittals by the Architect's approval thereof.

3.12.9 The Contractor shall direct specific attention, in writing or on resubmitted Shop Drawings, Product Data, Samples or similar submittals, to revisions other than those requested by the Architect on previous submittals. In the absence of such written notice the Architect's approval of a resubmission shall not apply to such revisions.

3.12.10 The Contractor shall not be required to provide professional services which constitute the practice of architecture or engineering unless such services are specifically required by the Contract Documents for a portion of the Work or unless the Contractor needs to provide such services in order to carry out the Contractor's responsibilities for construction means, methods, techniques, sequences and procedures. The Contractor shall not be required to provide professional services in violation of applicable law. If professional design services or certifications by a design professional related to systems, materials or equipment are specifically required of the Contractor by the Contract Documents, the Owner and the Architect will specify all performance and design criteria that such services must satisfy. The Contractor shall cause such services or certifications to be provided by a properly licensed design professional, whose signature and seal shall appear on all drawings, calculations, specifications, certifications, Shop Drawings and other submittals prepared by such professional. Shop Drawings and other submittals related to the Work designed or certified by such professional, if prepared by others, shall bear such professional's written approval when submitted to the Architect. The Owner and the Architect shall be entitled to rely upon the adequacy, accuracy and completeness of the services, certifications or approvals performed by such design professionals, provided the Owner and Architect have specified to the Contractor all performance and design criteria that such services must satisfy. Pursuant to this Subparagraph 3.12.10, the Architect will review, approve or take other appropriate action on submittals only for the limited purpose of checking for conformance with information given and the design concept expressed in the Contract Documents. The Contractor shall not be responsible for the adequacy of the performance or design criteria required by the Contract Documents.

© 1997 AIA®
AIA DOCUMENT A201-1997
GENERAL CONDITIONS
OF THE CONTRACT FOR
CONSTRUCTION

The American Institute
of Architects
1735 New York Avenue, N.W.
Washington, D.C. 20006-5292

16

7.2   CHANGE ORDERS

7.2.1   A Change Order is a written instrument prepared by the Architect and signed by the Owner, Contractor and Architect, stating their agreement upon all of the following:

  .1   change in the Work;

  .2   the amount of the adjustment, if any, in the Contract Sum; and

  .3   the extent of the adjustment, if any, in the Contract Time.

7.2.2   Methods used in determining adjustments to the Contract Sum may include those listed in Subparagraph 7.3.3.

7.3   CONSTRUCTION CHANGE DIRECTIVES

7.3.1   A Construction Change Directive is a written order prepared by the Architect and signed by the Owner and Architect, directing a change in the Work prior to agreement on adjustment, if any, in the Contract Sum or Contract Time, or both. The Owner may by Construction Change Directive, without invalidating the Contract, order changes in the Work within the general scope of the Contract consisting of additions, deletions or other revisions, the Contract Sum and Contract Time being adjusted accordingly.

7.3.2   A Construction Change Directive shall be used in the absence of total agreement on the terms of a Change Order.

7.3.3   If the Construction Change Directive provides for an adjustment to the Contract Sum, the adjustment shall be based on one of the following methods:

  .1   mutual acceptance of a lump sum properly itemized and supported by sufficient substantiating data to permit evaluation;

  .2   unit prices stated in the Contract Documents or subsequently agreed upon;

  .3   cost to be determined in a manner agreed upon by the parties and a mutually acceptable fixed or percentage fee; or

  .4   as provided in Subparagraph 7.3.6.

7.3.4   Upon receipt of a Construction Change Directive, the Contractor shall promptly proceed with the change in the Work involved and advise the Architect of the Contractor's agreement or disagreement with the method, if any, provided in the Construction Change Directive for determining the proposed adjustment in the Contract Sum or Contract Time.

7.3.5   A Construction Change Directive signed by the Contractor indicates the agreement of the Contractor therewith, including adjustment in Contract Sum and Contract Time or the method for determining them. Such agreement shall be effective immediately and shall be recorded as a Change Order.

7.3.6   If the Contractor does not respond promptly or disagrees with the method for adjustment in the Contract Sum, the method and the adjustment shall be determined by the Architect on the basis of reasonable expenditures and savings of those performing the Work attributable to the change, including, in case of an increase in the Contract Sum, a reasonable allowance for overhead and profit. In such case, and also under Clause 7.3.3.3, the Contractor shall keep and present, in such form as the Architect may prescribe, an itemized accounting together with appropriate supporting data. Unless otherwise provided in the Contract Documents, costs for the purposes of this Subparagraph 7.3.6 shall be limited to the following:

  .1   costs of labor, including social security, old age and unemployment insurance, fringe benefits required by agreement or custom, and workers' compensation insurance;

  .2   costs of materials, supplies and equipment, including cost of transportation, whether incorporated or consumed;

  .3   rental costs of machinery and equipment, exclusive of hand tools, whether rented from the Contractor or others;

27

construed to negate, abridge, or reduce other rights or obligations of indemnity which would otherwise exist as to a party or person described in this Paragraph 3.18.

3.18.2 In claims against any person or entity indemnified under this Paragraph 3.18 by an employee of the Contractor, a Subcontractor, anyone directly or indirectly employed by them or anyone for whose acts they may be liable, the indemnification obligation under Subparagraph 3.18.1 shall not be limited by a limitation on amount or type of damages, compensation or benefits payable by or for the Contractor or a Subcontractor under workers' compensation acts, disability benefit acts or other employee benefit acts.

ARTICLE 4   ADMINISTRATION OF THE CONTRACT
4.1   ARCHITECT
4.1.1   The Architect is the person lawfully licensed to practice architecture or an entity lawfully practicing architecture identified as such in the Agreement and is referred to throughout the Contract Documents as if singular in number. The term "Architect" means the Architect or the Architect's authorized representative.

4.1.2   Duties, responsibilities and limitations of authority of the Architect as set forth in the Contract Documents shall not be restricted, modified or extended without written consent of the Owner, Contractor and Architect. Consent shall not be unreasonably withheld.

4.1.3   If the employment of the Architect is terminated, the Owner shall employ a new Architect against whom the Contractor has no reasonable objection and whose status under the Contract Documents shall be that of the former Architect.

4.2   ARCHITECT'S ADMINISTRATION OF THE CONTRACT
4.2.1   The Architect will provide administration of the Contract as described in the Contract Documents, and will be an Owner's representative (1) during construction, (2) until final payment is due and (3) with the Owner's concurrence, from time to time during the one-year period for correction of Work described in Paragraph 12.2. The Architect will have authority to act on behalf of the Owner only to the extent provided in the Contract Documents, unless otherwise modified in writing in accordance with other provisions of the Contract.

4.2.2   The Architect, as a representative of the Owner, will visit the site at intervals appropriate to the stage of the Contractor's operations (1) to become generally familiar with and to keep the Owner informed about the progress and quality of the portion of the Work completed, (2) to endeavor to guard the Owner against defects and deficiencies in the Work, and (3) to determine in general if the Work is being performed in a manner indicating that the Work, when fully completed, will be in accordance with the Contract Documents. However, the Architect will not be required to make exhaustive or continuous on-site inspections to check the quality or quantity of the Work. The Architect will neither have control over or charge of, nor be responsible for, the construction means, methods, techniques, sequences or procedures, or for the safety precautions and programs in connection with the Work, since these are solely the Contractor's rights and responsibilities under the Contract Documents, except as provided in Subparagraph 3.3.1.

4.2.3   The Architect will not be responsible for the Contractor's failure to perform the Work in accordance with the requirements of the Contract Documents. The Architect will not have control over or charge of and will not be responsible for acts or omissions of the Contractor, Subcontractors, or their agents or employees, or any other persons or entities performing portions of the Work.

© 1 9 9 7   A I A ®
AIA DOCUMENT A201-1997
GENERAL CONDITIONS
OF THE CONTRACT FOR
CONSTRUCTION

The American Institute
of Architects
1735 New York Avenue, N.W.
Washington, D.C. 20006-5292

18

LIVERPOOL
JOHN MOORES UNIVERSITY
AVRIL ROBARTS LRC
TEL 0151 231 4022

4.2.4 Communications Facilitating Contract Administration. Except as otherwise provided in the Contract Documents or when direct communications have been specially authorized, the Owner and Contractor shall endeavor to communicate with each other through the Architect about matters arising out of or relating to the Contract. Communications by and with the Architect's consultants shall be through the Architect. Communications by and with Subcontractors and material suppliers shall be through the Contractor. Communications by and with separate contractors shall be through the Owner.

4.2.5 Based on the Architect's evaluations of the Contractor's Applications for Payment, the Architect will review and certify the amounts due the Contractor and will issue Certificates for Payment in such amounts.

4.2.6 The Architect will have authority to reject Work that does not conform to the Contract Documents. Whenever the Architect considers it necessary or advisable, the Architect will have authority to require inspection or testing of the Work in accordance with Subparagraphs 13.5.2 and 13.5.3, whether or not such Work is fabricated, installed or completed. However, neither this authority of the Architect nor a decision made in good faith either to exercise or not to exercise such authority shall give rise to a duty or responsibility of the Architect to the Contractor, Subcontractors, material and equipment suppliers, their agents or employees, or other persons or entities performing portions of the Work.

4.2.7 The Architect will review and approve or take other appropriate action upon the Contractor's submittals such as Shop Drawings, Product Data and Samples, but only for the limited purpose of checking for conformance with information given and the design concept expressed in the Contract Documents. The Architect's action will be taken with such reasonable promptness as to cause no delay in the Work or in the activities of the Owner, Contractor or separate contractors, while allowing sufficient time in the Architect's professional judgment to permit adequate review. Review of such submittals is not conducted for the purpose of determining the accuracy and completeness of other details such as dimensions and quantities, or for substantiating instructions for installation or performance of equipment or systems, all of which remain the responsibility of the Contractor as required by the Contract Documents. The Architect's review of the Contractor's submittals shall not relieve the Contractor of the obligations under Paragraphs 3.3, 3.5 and 3.12. The Architect's review shall not constitute approval of safety precautions or, unless otherwise specifically stated by the Architect, of any construction means, methods, techniques, sequences or procedures. The Architect's approval of a specific item shall not indicate approval of an assembly of which the item is a component.

4.2.8 The Architect will prepare Change Orders and Construction Change Directives, and may authorize minor changes in the Work as provided in Paragraph 7.4.

4.2.9 The Architect will conduct inspections to determine the date or dates of Substantial Completion and the date of final completion, will receive and forward to the Owner, for the Owner's review and records, written warranties and related documents required by the Contract and assembled by the Contractor, and will issue a final Certificate for Payment upon compliance with the requirements of the Contract Documents.

4.2.10 If the Owner and Architect agree, the Architect will provide one or more project representatives to assist in carrying out the Architect's responsibilities at the site. The duties, responsibilities and limitations of authority of such project representatives shall be as set forth in an exhibit to be incorporated in the Contract Documents.

4.2.11 The Architect will interpret and decide matters concerning performance under, and requirements of, the Contract Documents on written request of either the Owner or Contractor.

© 1997 A I A ®
AIA DOCUMENT A201-1997
GENERAL CONDITIONS
OF THE CONTRACT FOR
CONSTRUCTION

The American Institute
of Architects
1735 New York Avenue, N.W.
Washington, D.C. 20006-5292

19

The Architect's response to such requests will be made in writing within any time limits agreed upon or otherwise with reasonable promptness. If no agreement is made concerning the time within which interpretations required of the Architect shall be furnished in compliance with this Paragraph 4.2, then delay shall not be recognized on account of failure by the Architect to furnish such interpretations until 15 days after written request is made for them.

4.2.12 Interpretations and decisions of the Architect will be consistent with the intent of and reasonably inferable from the Contract Documents and will be in writing or in the form of drawings. When making such interpretations and initial decisions, the Architect will endeavor to secure faithful performance by both Owner and Contractor, will not show partiality to either and will not be liable for results of interpretations or decisions so rendered in good faith.

4.2.13 The Architect's decisions on matters relating to aesthetic effect will be final if consistent with the intent expressed in the Contract Documents.

4.3 CLAIMS AND DISPUTES

4.3.1 Definition. A Claim is a demand or assertion by one of the parties seeking, as a matter of right, adjustment or interpretation of Contract terms, payment of money, extension of time or other relief with respect to the terms of the Contract. The term "Claim" also includes other disputes and matters in question between the Owner and Contractor arising out of or relating to the Contract. Claims must be initiated by written notice. The responsibility to substantiate Claims shall rest with the party making the Claim.

4.3.2 Time Limits on Claims. Claims by either party must be initiated within 21 days after occurrence of the event giving rise to such Claim or within 21 days after the claimant first recognizes the condition giving rise to the Claim, whichever is later. Claims must be initiated by written notice to the Architect and the other party.

4.3.3 Continuing Contract Performance. Pending final resolution of a Claim except as otherwise agreed in writing or as provided in Subparagraph 9.7.1 and Article 14, the Contractor shall proceed diligently with performance of the Contract and the Owner shall continue to make payments in accordance with the Contract Documents.

4.3.4 Claims for Concealed or Unknown Conditions. If conditions are encountered at the site which are (1) subsurface or otherwise concealed physical conditions which differ materially from those indicated in the Contract Documents or (2) unknown physical conditions of an unusual nature, which differ materially from those ordinarily found to exist and generally recognized as inherent in construction activities of the character provided for in the Contract Documents, then notice by the observing party shall be given to the other party promptly before conditions are disturbed and in no event later than 21 days after first observance of the conditions. The Architect will promptly investigate such conditions and, if they differ materially and cause an increase or decrease in the Contractor's cost of, or time required for, performance of any part of the Work, will recommend an equitable adjustment in the Contract Sum or Contract Time, or both. If the Architect determines that the conditions at the site are not materially different from those indicated in the Contract Documents and that no change in the terms of the Contract is justified, the Architect shall so notify the Owner and Contractor in writing, stating the reasons. Claims by either party in opposition to such determination must be made within 21 days after the Architect has given notice of the decision. If the conditions encountered are materially different, the Contract Sum and Contract Time shall be equitably adjusted, but if the Owner and Contractor cannot agree on an adjustment in the Contract Sum or Contract Time, the adjustment shall be referred to the Architect for initial determination, subject to further proceedings pursuant to Paragraph 4.4.

© 1997 AIA®
AIA DOCUMENT A201-1997
GENERAL CONDITIONS
OF THE CONTRACT FOR
CONSTRUCTION

The American Institute
of Architects
1735 New York Avenue, N.W.
Washington, D.C. 20006-5292

20

4.3.5   Claims for Additional Cost. If the Contractor wishes to make Claim for an increase in the Contract Sum, written notice as provided herein shall be given before proceeding to execute the Work. Prior notice is not required for Claims relating to an emergency endangering life or property arising under Paragraph 10.6.

4.3.6   If the Contractor believes additional cost is involved for reasons including but not limited to (1) a written interpretation from the Architect, (2) an order by the Owner to stop the Work where the Contractor was not at fault, (3) a written order for a minor change in the Work issued by the Architect, (4) failure of payment by the Owner, (5) termination of the Contract by the Owner, (6) Owner's suspension or (7) other reasonable grounds, Claim shall be filed in accordance with this Paragraph 4.3.

4.3.7   CLAIMS FOR ADDITIONAL TIME

4.3.7.1 If the Contractor wishes to make Claim for an increase in the Contract Time, written notice as provided herein shall be given. The Contractor's Claim shall include an estimate of cost and of probable effect of delay on progress of the Work. In the case of a continuing delay only one Claim is necessary.

4.3.7.2 If adverse weather conditions are the basis for a Claim for additional time, such Claim shall be documented by data substantiating that weather conditions were abnormal for the period of time, could not have been reasonably anticipated and had an adverse effect on the scheduled construction.

4.3.8   Injury or Damage to Person or Property. If either party to the Contract suffers injury or damage to person or property because of an act or omission of the other party, or of others for whose acts such party is legally responsible, written notice of such injury or damage, whether or not insured, shall be given to the other party within a reasonable time not exceeding 21 days after discovery. The notice shall provide sufficient detail to enable the other party to investigate the matter.

4.3.9   If unit prices are stated in the Contract Documents or subsequently agreed upon, and if quantities originally contemplated are materially changed in a proposed Change Order or Construction Change Directive so that application of such unit prices to quantities of Work proposed will cause substantial inequity to the Owner or Contractor, the applicable unit prices shall be equitably adjusted.

4.3.10 Claims for Consequential Damages. The Contractor and Owner waive Claims against each other for consequential damages arising out of or relating to this Contract. This mutual waiver includes:

    .1 damages incurred by the Owner for rental expenses, for losses of use, income, profit, financing, business and reputation, and for loss of management or employee productivity or of the services of such persons; and

    .2 damages incurred by the Contractor for principal office expenses including the compensation of personnel stationed there, for losses of financing, business and reputation, and for loss of profit except anticipated profit arising directly from the Work.

This mutual waiver is applicable, without limitation, to all consequential damages due to either party's termination in accordance with Article 14. Nothing contained in this Subparagraph 4.3.10 shall be deemed to preclude an award of liquidated direct damages, when applicable, in accordance with the requirements of the Contract Documents.

4.4   RESOLUTION OF CLAIMS AND DISPUTES

4.4.1   Decision of Architect. Claims, including those alleging an error or omission by the Architect but excluding those arising under Paragraphs 10.3 through 10.5, shall be referred initially to the Architect for decision. An initial decision by the Architect shall be required as a

© 1997 AIA®
AIA DOCUMENT A201-1997
GENERAL CONDITIONS
OF THE CONTRACT FOR
CONSTRUCTION

The American Institute
of Architects
1735 New York Avenue, N.W.
Washington, D.C. 20006-5292

21

condition precedent to mediation, arbitration or litigation of all Claims between the Contractor and Owner arising prior to the date final payment is due, unless 30 days have passed after the Claim has been referred to the Architect with no decision having been rendered by the Architect. The Architect will not decide disputes between the Contractor and persons or entities other than the Owner.

4.4.2 The Architect will review Claims and within ten days of the receipt of the Claim take one or more of the following actions: (1) request additional supporting data from the claimant or a response with supporting data from the other party, (2) reject the Claim in whole or in part, (3) approve the Claim, (4) suggest a compromise, or (5) advise the parties that the Architect is unable to resolve the Claim if the Architect lacks sufficient information to evaluate the merits of the Claim or if the Architect concludes that, in the Architect's sole discretion, it would be inappropriate for the Architect to resolve the Claim.

4.4.3 In evaluating Claims, the Architect may, but shall not be obligated to, consult with or seek information from either party or from persons with special knowledge or expertise who may assist the Architect in rendering a decision. The Architect may request the Owner to authorize retention of such persons at the Owner's expense.

4.4.4 If the Architect requests a party to provide a response to a Claim or to furnish additional supporting data, such party shall respond, within ten days after receipt of such request, and shall either provide a response on the requested supporting data, advise the Architect when the response or supporting data will be furnished or advise the Architect that no supporting data will be furnished. Upon receipt of the response or supporting data, if any, the Architect will either reject or approve the Claim in whole or in part.

4.4.5 The Architect will approve or reject Claims by written decision, which shall state the reasons therefor and which shall notify the parties of any change in the Contract Sum or Contract Time or both. The approval or rejection of a Claim by the Architect shall be final and binding on the parties but subject to mediation and arbitration.

4.4.6 When a written decision of the Architect states that (1) the decision is final but subject to mediation and arbitration and (2) a demand for arbitration of a Claim covered by such decision must be made within 30 days after the date on which the party making the demand receives the final written decision, then failure to demand arbitration within said 30 days' period shall result in the Architect's decision becoming final and binding upon the Owner and Contractor. If the Architect renders a decision after arbitration proceedings have been initiated, such decision may be entered as evidence, but shall not supersede arbitration proceedings unless the decision is acceptable to all parties concerned.

4.4.7 Upon receipt of a Claim against the Contractor or at any time thereafter, the Architect or the Owner may, but is not obligated to, notify the surety, if any, of the nature and amount of the Claim. If the Claim relates to a possibility of a Contractor's default, the Architect or the Owner may, but is not obligated to, notify the surety and request the surety's assistance in resolving the controversy.

4.4.8 If a Claim relates to or is the subject of a mechanic's lien, the party asserting such Claim may proceed in accordance with applicable law to comply with the lien notice or filing deadlines prior to resolution of the Claim by the Architect, by mediation or by arbitration.

4.5   MEDIATION

4.5.1 Any Claim arising out of or related to the Contract, except Claims relating to aesthetic effect and except those waived as provided for in Subparagraphs 4.3.10, 9.10.4 and 9.10.5 shall, after initial decision by the Architect or 30 days after submission of the Claim to the Architect, be subject to mediation as a condition precedent to arbitration or the institution of legal or equitable proceedings by either party.

4.5.2 The parties shall endeavor to resolve their Claims by mediation which, unless the parties mutually agree otherwise, shall be in accordance with the Construction Industry Mediation Rules of the American Arbitration Association currently in effect. Request for mediation shall be filed in writing with the other party to the Contract and with the American Arbitration Association. The request may be made concurrently with the filing of a demand for arbitration but, in such event, mediation shall proceed in advance of arbitration or legal or equitable proceedings, which shall be stayed pending mediation for a period of 60 days from the date of filing, unless stayed for a longer period by agreement of the parties or court order.

4.5.3 The parties shall share the mediator's fee and any filing fees equally. The mediation shall be held in the place where the Project is located, unless another location is mutually agreed upon. Agreements reached in mediation shall be enforceable as settlement agreements in any court having jurisdiction thereof.

4.6 ARBITRATION

4.6.1 Any Claim arising out of or related to the Contract, except Claims relating to aesthetic effect and except those waived as provided for in Subparagraphs 4.3.10, 9.10.4 and 9.10.5, shall, after decision by the Architect or 30 days after submission of the Claim to the Architect, be subject to arbitration. Prior to arbitration, the parties shall endeavor to resolve disputes by mediation in accordance with the provisions of Paragraph 4.5.

4.6.2 Claims not resolved by mediation shall be decided by arbitration which, unless the parties mutually agree otherwise, shall be in accordance with the Construction Industry Arbitration Rules of the American Arbitration Association currently in effect. The demand for arbitration shall be filed in writing with the other party to the Contract and with the American Arbitration Association, and a copy shall be filed with the Architect.

4.6.3 A demand for arbitration shall be made within the time limits specified in Subparagraphs 4.4.6 and 4.6.1 as applicable, and in other cases within a reasonable time after the Claim has arisen, and in no event shall it be made after the date when institution of legal or equitable proceedings based on such Claim would be barred by the applicable statute of limitations as determined pursuant to Paragraph 13.7.

4.6.4 Limitation on Consolidation or Joinder. No arbitration arising out of or relating to the Contract shall include, by consolidation or joinder or in any other manner, the Architect, the Architect's employees or consultants, except by written consent containing specific reference to the Agreement and signed by the Architect, Owner, Contractor and any other person or entity sought to be joined. No arbitration shall include, by consolidation or joinder or in any other manner, parties other than the Owner, Contractor, a separate contractor as described in Article 6 and other persons substantially involved in a common question of fact or law whose presence is required if complete relief is to be accorded in arbitration. No person or entity other than the Owner, Contractor or a separate contractor as described in Article 6 shall be included as an original third party or additional third party to an arbitration whose interest or responsibility is insubstantial. Consent to arbitration involving an additional person or entity shall not constitute consent to arbitration of a Claim not described therein or with a person or entity not named or described therein. The foregoing agreement to arbitrate and other agreements to arbitrate with an additional person or entity duly consented to by parties to the Agreement shall be specifically enforceable under applicable law in any court having jurisdiction thereof.

© 1 9 9 7   A I A ®
AIA DOCUMENT A201-1997
GENERAL CONDITIONS
OF THE CONTRACT FOR
CONSTRUCTION

The American Institute
of Architects
1735 New York Avenue, N.W.
Washington, D.C. 20006-5292

23

4.6.5   Claims and Timely Assertion of Claims. The party filing a notice of demand for arbitration must assert in the demand all Claims then known to that party on which arbitration is permitted to be demanded.

4.6.6   Judgment on Final Award. The award rendered by the arbitrator or arbitrators shall be final, and judgment may be entered upon it in accordance with applicable law in any court having jurisdiction thereof.

ARTICLE 5   SUBCONTRACTORS
5.1   DEFINITIONS
5.1.1 A Subcontractor is a person or entity who has a direct contract with the Contractor to perform a portion of the Work at the site. The term "Subcontractor" is referred to throughout the Contract Documents as if singular in number and means a Subcontractor or an authorized representative of the Subcontractor. The term "Subcontractor" does not include a separate contractor or subcontractors of a separate contractor.

5.1.2 A Sub-subcontractor is a person or entity who has a direct or indirect contract with a Subcontractor to perform a portion of the Work at the site. The term "Sub-subcontractor" is referred to throughout the Contract Documents as if singular in number and means a Sub-subcontractor or an authorized representative of the Sub-subcontractor.

5.2   AWARD OF SUBCONTRACTS AND OTHER CONTRACTS FOR PORTIONS OF THE WORK
5.2.1 Unless otherwise stated in the Contract Documents or the bidding requirements, the Contractor, as soon as practicable after award of the Contract, shall furnish in writing to the Owner through the Architect the names of persons or entities (including those who are to furnish materials or equipment fabricated to a special design) proposed for each principal portion of the Work. The Architect will promptly reply to the Contractor in writing stating whether or not the Owner or the Architect, after due investigation, has reasonable objection to any such proposed person or entity. Failure of the Owner or Architect to reply promptly shall constitute notice of no reasonable objection.

5.2.2 The Contractor shall not contract with a proposed person or entity to whom the Owner or Architect has made reasonable and timely objection. The Contractor shall not be required to contract with anyone to whom the Contractor has made reasonable objection.

5.2.3 If the Owner or Architect has reasonable objection to a person or entity proposed by the Contractor, the Contractor shall propose another to whom the Owner or Architect has no reasonable objection. If the proposed but rejected Subcontractor was reasonably capable of performing the Work, the Contract Sum and Contract Time shall be increased or decreased by the difference, if any, occasioned by such change, and an appropriate Change Order shall be issued before commencement of the substitute Subcontractor's Work. However, no increase in the Contract Sum or Contract Time shall be allowed for such change unless the Contractor has acted promptly and responsively in submitting names as required.

5.2.4 The Contractor shall not change a Subcontractor, person or entity previously selected if the Owner or Architect makes reasonable objection to such substitute.

5.3   SUBCONTRACTUAL RELATIONS
5.3.1 By appropriate agreement, written where legally required for validity, the Contractor shall require each Subcontractor, to the extent of the Work to be performed by the Subcontractor, to be bound to the Contractor by terms of the Contract Documents, and to assume toward the Contractor all the obligations and responsibilities, including the responsibility for safety of the

©1997   AIA®
AIA DOCUMENT A201-1997
GENERAL CONDITIONS
OF THE CONTRACT FOR
CONSTRUCTION

The American Institute
of Architects
1735 New York Avenue, N.W.
Washington, D.C. 20006-5292

24

Subcontractor's Work, which the Contractor, by these Documents, assumes toward the Owner and Architect. Each subcontract agreement shall preserve and protect the rights of the Owner and Architect under the Contract Documents with respect to the Work to be performed by the Subcontractor so that subcontracting thereof will not prejudice such rights, and shall allow to the Subcontractor, unless specifically provided otherwise in the subcontract agreement, the benefit of all rights, remedies and redress against the Contractor that the Contractor, by the Contract Documents, has against the Owner. Where appropriate, the Contractor shall require each Subcontractor to enter into similar agreements with Sub-subcontractors. The Contractor shall make available to each proposed Subcontractor, prior to the execution of the subcontract agreement, copies of the Contract Documents to which the Subcontractor will be bound, and, upon written request of the Subcontractor, identify to the Subcontractor terms and conditions of the proposed subcontract agreement which may be at variance with the Contract Documents. Subcontractors will similarly make copies of applicable portions of such documents available to their respective proposed Sub-subcontractors.

5.4 CONTINGENT ASSIGNMENT OF SUBCONTRACTS
5.4.1 Each subcontract agreement for a portion of the Work is assigned by the Contractor to the Owner provided that:

    .1 assignment is effective only after termination of the Contract by the Owner for cause pursuant to Paragraph 14.2 and only for those subcontract agreements which the Owner accepts by notifying the Subcontractor and Contractor in writing; and

    .2 assignment is subject to the prior rights of the surety, if any, obligated under bond relating to the Contract.

5.4.2 Upon such assignment, if the Work has been suspended for more than 30 days, the Subcontractor's compensation shall be equitably adjusted for increases in cost resulting from the suspension.

ARTICLE 6 CONSTRUCTION BY OWNER OR BY SEPARATE CONTRACTORS
6.1 OWNER'S RIGHT TO PERFORM CONSTRUCTION AND TO AWARD SEPARATE CONTRACTS
6.1.1 The Owner reserves the right to perform construction or operations related to the Project with the Owner's own forces, and to award separate contracts in connection with other portions of the Project or other construction or operations on the site under Conditions of the Contract identical or substantially similar to these including those portions related to insurance and waiver of subrogation. If the Contractor claims that delay or additional cost is involved because of such action by the Owner, the Contractor shall make such Claim as provided in Paragraph 4.3.

6.1.2 When separate contracts are awarded for different portions of the Project or other construction or operations on the site, the term "Contractor" in the Contract Documents in each case shall mean the Contractor who executes each separate Owner-Contractor Agreement.

6.1.3 The Owner shall provide for coordination of the activities of the Owner's own forces and of each separate contractor with the Work of the Contractor, who shall cooperate with them. The Contractor shall participate with other separate contractors and the Owner in reviewing their construction schedules when directed to do so. The Contractor shall make any revisions to the construction schedule deemed necessary after a joint review and mutual agreement. The construction schedules shall then constitute the schedules to be used by the Contractor, separate contractors and the Owner until subsequently revised.

6.1.4 Unless otherwise provided in the Contract Documents, when the Owner performs construction or operations related to the Project with the Owner's own forces, the Owner shall be deemed to be subject to the same obligations and to have the same rights which apply to the

© 1 9 9 7   A I A ®
AIA DOCUMENT A201-1997
GENERAL CONDITIONS
OF THE CONTRACT FOR
CONSTRUCTION

The American Institute
of Architects
1735 New York Avenue, N.W.
Washington, D.C. 20006-5292

25

Contractor under the Conditions of the Contract, including, without excluding others, those stated in Article 3, this Article 6 and Articles 10, 11 and 12.

6.2   MUTUAL RESPONSIBILITY

6.2.1   The Contractor shall afford the Owner and separate contractors reasonable opportunity for introduction and storage of their materials and equipment and performance of their activities, and shall connect and coordinate the Contractor's construction and operations with theirs as required by the Contract Documents.

6.2.2   If part of the Contractor's Work depends for proper execution or results upon construction or operations by the Owner or a separate contractor, the Contractor shall, prior to proceeding with that portion of the Work, promptly report to the Architect apparent discrepancies or defects in such other construction that would render it unsuitable for such proper execution and results. Failure of the Contractor so to report shall constitute an acknowledgment that the Owner's or separate contractor's completed or partially completed construction is fit and proper to receive the Contractor's Work, except as to defects not then reasonably discoverable.

6.2.3   The Owner shall be reimbursed by the Contractor for costs incurred by the Owner which are payable to a separate contractor because of delays, improperly timed activities or defective construction of the Contractor. The Owner shall be responsible to the Contractor for costs incurred by the Contractor because of delays, improperly timed activities, damage to the Work or defective construction of a separate contractor.

6.2.4   The Contractor shall promptly remedy damage wrongfully caused by the Contractor to completed or partially completed construction or to property of the Owner or separate contractors as provided in Subparagraph 10.2.5.

6.2.5   The Owner and each separate contractor shall have the same responsibilities for cutting and patching as are described for the Contractor in Subparagraph 3.14.

6.3   OWNER'S RIGHT TO CLEAN UP

6.3.1   If a dispute arises among the Contractor, separate contractors and the Owner as to the responsibility under their respective contracts for maintaining the premises and surrounding area free from waste materials and rubbish, the Owner may clean up and the Architect will allocate the cost among those responsible.

ARTICLE 7   CHANGES IN THE WORK

7.1   GENERAL

7.1.1   Changes in the Work may be accomplished after execution of the Contract, and without invalidating the Contract, by Change Order, Construction Change Directive or order for a minor change in the Work, subject to the limitations stated in this Article 7 and elsewhere in the Contract Documents.

7.1.2   A Change Order shall be based upon agreement among the Owner, Contractor and Architect; a Construction Change Directive requires agreement by the Owner and Architect and may or may not be agreed to by the Contractor; an order for a minor change in the Work may be issued by the Architect alone.

7.1.3   Changes in the Work shall be performed under applicable provisions of the Contract Documents, and the Contractor shall proceed promptly, unless otherwise provided in the Change Order, Construction Change Directive or order for a minor change in the Work.

7.2   CHANGE ORDERS

7.2.1   A Change Order is a written instrument prepared by the Architect and signed by the Owner, Contractor and Architect, stating their agreement upon all of the following:

    .1   change in the Work;

    .2   the amount of the adjustment, if any, in the Contract Sum; and

    .3   the extent of the adjustment, if any, in the Contract Time.

7.2.2   Methods used in determining adjustments to the Contract Sum may include those listed in Subparagraph 7.3.3.

7.3   CONSTRUCTION CHANGE DIRECTIVES

7.3.1   A Construction Change Directive is a written order prepared by the Architect and signed by the Owner and Architect, directing a change in the Work prior to agreement on adjustment, if any, in the Contract Sum or Contract Time, or both. The Owner may by Construction Change Directive, without invalidating the Contract, order changes in the Work within the general scope of the Contract consisting of additions, deletions or other revisions, the Contract Sum and Contract Time being adjusted accordingly.

7.3.2   A Construction Change Directive shall be used in the absence of total agreement on the terms of a Change Order.

7.3.3   If the Construction Change Directive provides for an adjustment to the Contract Sum, the adjustment shall be based on one of the following methods:

    .1   mutual acceptance of a lump sum properly itemized and supported by sufficient substantiating data to permit evaluation;

    .2   unit prices stated in the Contract Documents or subsequently agreed upon;

    .3   cost to be determined in a manner agreed upon by the parties and a mutually acceptable fixed or percentage fee; or

    .4   as provided in Subparagraph 7.3.6.

7.3.4   Upon receipt of a Construction Change Directive, the Contractor shall promptly proceed with the change in the Work involved and advise the Architect of the Contractor's agreement or disagreement with the method, if any, provided in the Construction Change Directive for determining the proposed adjustment in the Contract Sum or Contract Time.

7.3.5   A Construction Change Directive signed by the Contractor indicates the agreement of the Contractor therewith, including adjustment in Contract Sum and Contract Time or the method for determining them. Such agreement shall be effective immediately and shall be recorded as a Change Order.

7.3.6   If the Contractor does not respond promptly or disagrees with the method for adjustment in the Contract Sum, the method and the adjustment shall be determined by the Architect on the basis of reasonable expenditures and savings of those performing the Work attributable to the change, including, in case of an increase in the Contract Sum, a reasonable allowance for overhead and profit. In such case, and also under Clause 7.3.3.3, the Contractor shall keep and present, in such form as the Architect may prescribe, an itemized accounting together with appropriate supporting data. Unless otherwise provided in the Contract Documents, costs for the purposes of this Subparagraph 7.3.6 shall be limited to the following:

    .1   costs of labor, including social security, old age and unemployment insurance, fringe benefits required by agreement or custom, and workers' compensation insurance;

    .2   costs of materials, supplies and equipment, including cost of transportation, whether incorporated or consumed;

    .3   rental costs of machinery and equipment, exclusive of hand tools, whether rented from the Contractor or others;

© 1997   A I A ®
AIA DOCUMENT A201-1997
GENERAL CONDITIONS
OF THE CONTRACT FOR
CONSTRUCTION

The American Institute
of Architects
1735 New York Avenue, N.W.
Washington, D.C. 20006-5292

27

construed to negate, abridge, or reduce other rights or obligations of indemnity which would otherwise exist as to a party or person described in this Paragraph 3.18.

3.18.2 In claims against any person or entity indemnified under this Paragraph 3.18 by an employee of the Contractor, a Subcontractor, anyone directly or indirectly employed by them or anyone for whose acts they may be liable, the indemnification obligation under Subparagraph 3.18.1 shall not be limited by a limitation on amount or type of damages, compensation or benefits payable by or for the Contractor or a Subcontractor under workers' compensation acts, disability benefit acts or other employee benefit acts.

ARTICLE 4   ADMINISTRATION OF THE CONTRACT
4.1   ARCHITECT
4.1.1   The Architect is the person lawfully licensed to practice architecture or an entity lawfully practicing architecture identified as such in the Agreement and is referred to throughout the Contract Documents as if singular in number. The term "Architect" means the Architect or the Architect's authorized representative.

4.1.2   Duties, responsibilities and limitations of authority of the Architect as set forth in the Contract Documents shall not be restricted, modified or extended without written consent of the Owner, Contractor and Architect. Consent shall not be unreasonably withheld.

4.1.3   If the employment of the Architect is terminated, the Owner shall employ a new Architect against whom the Contractor has no reasonable objection and whose status under the Contract Documents shall be that of the former Architect.

4.2   ARCHITECT'S ADMINISTRATION OF THE CONTRACT
4.2.1   The Architect will provide administration of the Contract as described in the Contract Documents, and will be an Owner's representative (1) during construction, (2) until final payment is due and (3) with the Owner's concurrence, from time to time during the one-year period for correction of Work described in Paragraph 12.2. The Architect will have authority to act on behalf of the Owner only to the extent provided in the Contract Documents, unless otherwise modified in writing in accordance with other provisions of the Contract.

4.2.2   The Architect, as a representative of the Owner, will visit the site at intervals appropriate to the stage of the Contractor's operations (1) to become generally familiar with and to keep the Owner informed about the progress and quality of the portion of the Work completed, (2) to endeavor to guard the Owner against defects and deficiencies in the Work, and (3) to determine in general if the Work is being performed in a manner indicating that the Work, when fully completed, will be in accordance with the Contract Documents. However, the Architect will not be required to make exhaustive or continuous on-site inspections to check the quality or quantity of the Work. The Architect will neither have control over or charge of, nor be responsible for, the construction means, methods, techniques, sequences or procedures, or for the safety precautions and programs in connection with the Work, since these are solely the Contractor's rights and responsibilities under the Contract Documents, except as provided in Subparagraph 3.3.1.

4.2.3   The Architect will not be responsible for the Contractor's failure to perform the Work in accordance with the requirements of the Contract Documents. The Architect will not have control over or charge of and will not be responsible for acts or omissions of the Contractor, Subcontractors, or their agents or employees, or any other persons or entities performing portions of the Work.

© 1997   A I A ®
AIA DOCUMENT A201-1997
GENERAL CONDITIONS
OF THE CONTRACT FOR
CONSTRUCTION

The American Institute
of Architects
1735 New York Avenue, N.W.
Washington, D.C. 20006-5292

18

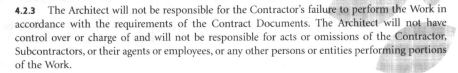

by the Owner, the Contractor shall notify the Owner in writing not less than five days or other agreed period before commencing the Work to permit the timely filing of morgages, mechanic's liens and other security interests.

8.2.3 The Contractor shall proceed expeditiously with adequate forces and shall achieve Substantial Completion within the Contract Time.

8.3 DELAYS AND EXTENSIONS OF TIME

8.3.1 If the Contractor is delayed at any time in the commencement or progress of the Work by an act or neglect of the Owner or Architect, or of an employee of either, or of a separate contractor employed by the Owner, or by changes ordered in the Work, or by labor disputes, fire, unusual delay in deliveries, unavoidable casualties or other causes beyond the Contractor's control, or by delay authorized by the Owner pending mediation and arbitration, or by other causes which the Architect determines may justify delay, then the Contract Time shall be extended by Change Order for such reasonable time as the Architect may determine.

8.3.2 Claims relating to time shall be made in accordance with applicable provisions of Paragraph 4.3.

8.3.3 This Paragraph 8.3 does not preclude recovery of damages for delay by either party under other provisions of the Contract Documents.

ARTICLE 9 PAYMENTS AND COMPLETION

9.1 CONTRACT SUM

9.1.1 The Contract Sum is stated in the Agreement and, including authorized adjustments, is the total amount payable by the Owner to the Contractor for performance of the Work under the Contract Documents.

9.2 SCHEDULE OF VALUES

9.2.1 Before the first Application for Payment, the Contractor shall submit to the Architect a schedule of values allocated to various portions of the Work, prepared in such form and supported by such data to substantiate its accuracy as the Architect may require. This schedule, unless objected to by the Architect, shall be used as a basis for reviewing the Contractor's Applications for Payment.

9.3 APPLICATIONS FOR PAYMENT

9.3.1 At least ten days before the date established for each progress payment, the Contractor shall submit to the Architect an itemized Application for Payment for operations completed in accordance with the schedule of values. Such application shall be notarized, if required, and supported by such data substantiating the Contractor's right to payment as the Owner or Architect may require, such as copies of requisitions from Subcontractors and material suppliers, and reflecting retainage if provided for in the Contract Documents.

9.3.1.1 As provided in Subparagraph 7.3.8, such applications may include requests for payment on account of changes in the Work which have been properly authorized by Construction Change Directives, or by interim determinations of the Architect, but not yet included in Change Orders.

9.3.1.2 Such applications may not include requests for payment for portions of the Work for which the Contractor does not intend to pay to a Subcontractor or material supplier, unless such Work has been performed by others whom the Contractor intends to pay.

© 1997 AIA®
AIA DOCUMENT A201-1997
GENERAL CONDITIONS
OF THE CONTRACT FOR
CONSTRUCTION

The American Institute
of Architects
1735 New York Avenue, N.W.
Washington, D.C. 20006-5292

29

9.3.2   Unless otherwise provided in the Contract Documents, payments shall be made on account of materials and equipment delivered and suitably stored at the site for subsequent incorporation in the Work. If approved in advance by the Owner, payment may similarly be made for materials and equipment suitably stored off the site at a location agreed upon in writing. Payment for materials and equipment stored on or off the site shall be conditioned upon compliance by the Contractor with procedures satisfactory to the Owner to establish the Owner's title to such materials and equipment or otherwise protect the Owner's interest, and shall include the costs of applicable insurance, storage and transportation to the site for such materials and equipment stored off the site.

9.3.3   The Contractor warrants that title to all Work covered by an Application for Payment will pass to the Owner no later than the time of payment. The Contractor further warrants that upon submittal of an Application for Payment all Work for which Certificates for Payment have been previously issued and payments received from the Owner shall, to the best of the Contractor's knowledge, information and belief, be free and clear of liens, claims, security interests or encumbrances in favor of the Contractor, Subcontractors, material suppliers, or other persons or entities making a claim by reason of having provided labor, materials and equipment relating to the Work.

9.4   CERTIFICATES FOR PAYMENT

9.4.1   The Architect will, within seven days after receipt of the Contractor's Application for Payment, either issue to the Owner a Certificate for Payment, with a copy to the Contractor, for such amount as the Architect determines is properly due, or notify the Contractor and Owner in writing of the Architect's reasons for withholding certification in whole or in part as provided in Subparagraph 9.5.1.

9.4.2   The issuance of a Certificate for Payment will constitute a representation by the Architect to the Owner, based on the Architect's evaluation of the Work and the data comprising the Application for Payment, that the Work has progressed to the point indicated and that, to the best of the Architect's knowledge, information and belief, the quality of the Work is in accordance with the Contract Documents. The foregoing representations are subject to an evaluation of the Work for conformance with the Contract Documents upon Substantial Completion, to results of subsequent tests and inspections, to correction of minor deviations from the Contract Documents prior to completion and to specific qualifications expressed by the Architect. The issuance of a Certificate for Payment will further constitute a representation that the Contractor is entitled to payment in the amount certified. However, the issuance of a Certificate for Payment will not be a representation that the Architect has (1) made exhaustive or continuous on-site inspections to check the quality or quantity of the Work, (2) reviewed construction means, methods, techniques, sequences or procedures, (3) reviewed copies of requisitions received from Subcontractors and material suppliers and other data requested by the Owner to substantiate the Contractor's right to payment, or (4) made examination to ascertain how or for what purpose the Contractor has used money previously paid on account of the Contract Sum.

9.5   DECISIONS TO WITHHOLD CERTIFICATION

9.5.1   The Architect may withhold a Certificate for Payment in whole or in part, to the extent reasonably necessary to protect the Owner, if in the Architect's opinion the representations to the Owner required by Subparagraph 9.4.2 cannot be made. If the Architect is unable to certify payment in the amount of the Application, the Architect will notify the Contractor and Owner as provided in Subparagraph 9.4.1. If the Contractor and Architect cannot agree on a revised amount, the Architect will promptly issue a Certificate for Payment for the amount for which the Architect is able to make such representations to the Owner. The Architect may also withhold a Certificate for Payment or, because of subsequently discovered evidence, may nullify the whole or a part of a Certificate for Payment previously issued, to such extent as may be necessary in the Architect's

opinion to protect the Owner from loss for which the Contractor is responsible, including loss resulting from acts and omissions described in Subparagraph 3.3.2, because of:

.1 defective Work not remedied;

.2 third party claims filed or reasonable evidence indicating probable filing of such claims unless security acceptable to the Owner is provided by the Contractor;

.3 failure of the Contractor to make payments properly to Subcontractors or for labor, materials or equipment;

.4 reasonable evidence that the Work cannot be completed for the unpaid balance of the Contract Sum;

.5 damage to the Owner or another contractor;

.6 reasonable evidence that the Work will not be completed within the Contract Time, and that the unpaid balance would not be adequate to cover actual or liquidated damages for the anticipated delay; or

.7 persistent failure to carry out the Work in accordance with the Contract Documents.

9.5.2 When the above reasons for withholding certification are removed, certification will be made for amounts previously withheld.

9.6 PROGRESS PAYMENTS

9.6.1 After the Architect has issued a Certificate for Payment, the Owner shall make payment in the manner and within the time provided in the Contract Documents, and shall so notify the Architect.

9.6.2 The Contractor shall promptly pay each Subcontractor, upon receipt of payment from the Owner, out of the amount paid to the Contractor on account of such Subcontractor's portion of the Work, the amount to which said Subcontractor is entitled, reflecting percentages actually retained from payments to the Contractor on account of such Subcontractor's portion of the Work. The Contractor shall, by appropriate agreement with each Subcontractor, require each Subcontractor to make payments to Sub-subcontractors in a similar manner.

9.6.3 The Architect will, on request, furnish to a Subcontractor, if practicable, information regarding percentages of completion or amounts applied for by the Contractor and action taken thereon by the Architect and Owner on account of portions of the Work done by such Subcontractor.

9.6.4 Neither the Owner nor Architect shall have an obligation to pay or to see to the payment of money to a Subcontractor except as may otherwise be required by law.

9.6.5 Payment to material suppliers shall be treated in a manner similar to that provided in Subparagraphs 9.6.2, 9.6.3 and 9.6.4.

9.6.6 A Certificate for Payment, a progress payment, or partial or entire use or occupancy of the Project by the Owner shall not constitute acceptance of Work not in accordance with the Contract Documents.

9.6.7 Unless the Contractor provides the Owner with a payment bond in the full penal sum of the Contract Sum, payments received by the Contractor for Work properly performed by Subcontractors and suppliers shall be held by the Contractor for those Subcontractors or suppliers who performed Work or furnished materials, or both, under contract with the Contractor for which payment was made by the Owner. Nothing contained herein shall require money to be placed in a separate account and not commingled with money of the Contractor, shall create any fiduciary liability or tort liability on the part of the Contractor for breach of trust or shall entitle any person or entity to an award of punitive damages against the Contractor for breach of the requirements of this provision.

© 1997 AIA®
AIA DOCUMENT A201-1997
GENERAL CONDITIONS
OF THE CONTRACT FOR
CONSTRUCTION

The American Institute
of Architects
1735 New York Avenue, N.W.
Washington, D.C. 20006-5292

31

9.7 FAILURE OF PAYMENT

9.7.1 If the Architect does not issue a Certificate for Payment, through no fault of the Contractor, within seven days after receipt of the Contractor's Application for Payment, or if the Owner does not pay the Contractor within seven days after the date established in the Contract Documents the amount certified by the Architect or awarded by arbitration, then the Contractor may, upon seven additional days' written notice to the Owner and Architect, stop the Work until payment of the amount owing has been received. The Contract Time shall be extended appropriately and the Contract Sum shall be increased by the amount of the Contractor's reasonable costs of shut-down, delay and start-up, plus interest as provided for in the Contract Documents.

9.8 SUBSTANTIAL COMPLETION

9.8.1 Substantial Completion is the stage in the progress of the Work when the Work or designated portion thereof is sufficiently complete in accordance with the Contract Documents so that the Owner can occupy or utilize the Work for its intended use.

9.8.2 When the Contractor considers that the Work, or a portion thereof which the Owner agrees to accept separately, is substantially complete, the Contractor shall prepare and submit to the Architect a comprehensive list of items to be completed or corrected prior to final payment. Failure to include an item on such list does not alter the responsibility of the Contractor to complete all Work in accordance with the Contract Documents.

9.8.3 Upon receipt of the Contractor's list, the Architect will make an inspection to determine whether the Work or designated portion thereof is substantially complete. If the Architect's inspection discloses any item, whether or not included on the Contractor's list, which is not sufficiently complete in accordance with the Contract Documents so that the Owner can occupy or utilize the Work or designated portion thereof for its intended use, the Contractor shall, before issuance of the Certificate of Substantial Completion, complete or correct such item upon notification by the Architect. In such case, the Contractor shall then submit a request for another inspection by the Architect to determine Substantial Completion.

9.8.4 When the Work or designated portion thereof is substantially complete, the Architect will prepare a Certificate of Substantial Completion which shall establish the date of Substantial Completion, shall establish responsibilities of the Owner and Contractor for security, maintenance, heat, utilities, damage to the Work and insurance, and shall fix the time within which the Contractor shall finish all items on the list accompanying the Certificate. Warranties required by the Contract Documents shall commence on the date of Substantial Completion of the Work or designated portion thereof unless otherwise provided in the Certificate of Substantial Completion.

9.8.5 The Certificate of Substantial Completion shall be submitted to the Owner and Contractor for their written acceptance of responsibilities assigned to them in such Certificate. Upon such acceptance and consent of surety, if any, the Owner shall make payment of retainage applying to such Work or designated portion thereof. Such payment shall be adjusted for Work that is incomplete or not in accordance with the requirements of the Contract Documents.

9.9 PARTIAL OCCUPANCY OR USE

9.9.1 The Owner may occupy or use any completed or partially completed portion of the Work at any stage when such portion is designated by separate agreement with the Contractor, provided such occupancy or use is consented to by the insurer as required under Clause 11.4.1.5 and authorized by public authorities having jurisdiction over the Work. Such partial occupancy or use may commence whether or not the portion is substantially complete, provided the Owner and Contractor have accepted in writing the responsibilities assigned to each of them for payments, retainage, if any, security, maintenance, heat, utilities, damage to the Work and insurance, and have agreed in writing concerning the period for correction of the Work and commencement of warranties required by the Contract Documents. When the Contractor considers a portion substantially complete, the Contractor shall prepare and submit a list to the Architect as provided under Subparagraph 9.8.2. Consent of the Contractor to partial occupancy or use shall not be unreasonably withheld. The stage of the progress of the Work shall be determined by written agreement between the Owner and Contractor or, if no agreement is reached, by decision of the Architect.

9.9.2 Immediately prior to such partial occupancy or use, the Owner, Contractor and Architect shall jointly inspect the area to be occupied or portion of the Work to be used in order to determine and record the condition of the Work.

9.9.3 Unless otherwise agreed upon, partial occupancy or use of a portion or portions of the Work shall not constitute acceptance of Work not complying with the requirements of the Contract Documents.

9.10 FINAL COMPLETION AND FINAL PAYMENT

9.10.1 Upon receipt of written notice that the Work is ready for final inspection and acceptance and upon receipt of a final Application for Payment, the Architect will promptly make such inspection and, when the Architect finds the Work acceptable under the Contract Documents and the Contract fully performed, the Architect will promptly issue a final Certificate for Payment stating that to the best of the Architect's knowledge, information and belief, and on the basis of the Architect's on-site visits and inspections, the Work has been completed in accordance with terms and conditions of the Contract Documents and that the entire balance found to be due the Contractor and noted in the final Certificate is due and payable. The Architect's final Certificate for Payment will constitute a further representation that conditions listed in Subparagraph 9.10.2 as precedent to the Contractor's being entitled to final payment have been fulfilled.

9.10.2 Neither final payment nor any remaining retained percentage shall become due until the Contractor submits to the Architect (1) an affidavit that payrolls, bills for materials and equipment, and other indebtedness connected with the Work for which the Owner or the Owner's property might be responsible or encumbered (less amounts withheld by Owner) have been paid or otherwise satisfied, (2) a certificate evidencing that insurance required by the Contract Documents to remain in force after final payment is currently in effect and will not be canceled or allowed to expire until at least 30 days' prior written notice has been given to the Owner, (3) a written statement that the Contractor knows of no substantial reason that the insurance will not be renewable to cover the period required by the Contract Documents, (4) consent of surety, if any, to final payment and (5), if required by the Owner, other data establishing payment or satisfaction of obligations, such as receipts, releases and waivers of liens, claims, security interests or encumbrances arising out of the Contract, to the extent and in such form as may be designated by the Owner. If a Subcontractor refuses to furnish a release or waiver required by the Owner, the Contractor may furnish a bond satisfactory to the Owner to indemnify the Owner against such lien. If such lien remains unsatisfied after payments are made, the Contractor shall refund to the Owner all money that the Owner may be compelled to pay in discharging such lien, including all costs and reasonable attorneys' fees.

9.10.3 If, after Substantial Completion of the Work, final completion thereof is materially delayed through no fault of the Contractor or by issuance of Change Orders affecting final completion, and the Architect so confirms, the Owner shall, upon application by the Contractor and certification by the Architect, and without terminating the Contract, make payment of the balance due for that portion of the Work fully completed and accepted. If the remaining balance for Work not fully completed or corrected is less than retainage stipulated in the Contract Documents, and if bonds have been furnished, the written consent of surety to payment of the balance due for that

© 1997 AIA®

AIA DOCUMENT A201-1997

GENERAL CONDITIONS OF THE CONTRACT FOR CONSTRUCTION

The American Institute of Architects
1735 New York Avenue, N.W.
Washington, D.C. 20006-5292

33

portion of the Work fully completed and accepted shall be submitted by the Contractor to the Architect prior to certification of such payment. Such payment shall be made under terms and conditions governing final payment, except that it shall not constitute a waiver of claims.

9.10.4 The making of final payment shall constitute a waiver of Claims by the Owner except those arising from:

> .1 liens, Claims, security interests or encumbrances arising out of the Contract and unsettled;
> .2 failure of the Work to comply with the requirements of the Contract Documents; or
> .3 terms of special warranties required by the Contract Documents.

9.10.5 Acceptance of final payment by the Contractor, a Subcontractor or material supplier shall constitute a waiver of claims by that payee except those previously made in writing and identified by that payee as unsettled at the time of final Application for Payment.

ARTICLE 10 PROTECTION OF PERSONS AND PROPERTY

10.1 SAFETY PRECAUTIONS AND PROGRAMS

10.1.1 The Contractor shall be responsible for initiating, maintaining and supervising all safety precautions and programs in connection with the performance of the Contract.

10.2 SAFETY OF PERSONS AND PROPERTY

10.2.1 The Contractor shall take reasonable precautions for safety of, and shall provide reasonable protection to prevent damage, injury or loss to:

> .1 employees on the Work and other persons who may be affected thereby;
> .2 the Work and materials and equipment to be incorporated therein, whether in storage on or off the site, under care, custody or control of the Contractor or the Contractor's Subcontractors or Sub-subcontractors; and
> .3 other property at the site or adjacent thereto, such as trees, shrubs, lawns, walks, pavements, roadways, structures and utilities not designated for removal, relocation or replacement in the course of construction.

10.2.2 The Contractor shall give notices and comply with applicable laws, ordinances, rules, regulations and lawful orders of public authorities bearing on safety of persons or property or their protection from damage, injury or loss.

10.2.3 The Contractor shall erect and maintain, as required by existing conditions and performance of the Contract, reasonable safeguards for safety and protection, including posting danger signs and other warnings against hazards, promulgating safety regulations and notifying owners and users of adjacent sites and utilities.

10.2.4 When use or storage of explosives or other hazardous materials or equipment or unusual methods are necessary for execution of the Work, the Contractor shall exercise utmost care and carry on such activities under supervision of properly qualified personnel.

10.2.5 The Contractor shall promptly remedy damage and loss (other than damage or loss insured under property insurance required by the Contract Documents) to property referred to in Clauses 10.2.1.2 and 10.2.1.3 caused in whole or in part by the Contractor, a Subcontractor, a Sub-subcontractor, or anyone directly or indirectly employed by any of them, or by anyone for whose acts they may be liable and for which the Contractor is responsible under Clauses 10.2.1.2 and 10.2.1.3, except damage or loss attributable to acts or omissions of the Owner or Architect or anyone directly or indirectly employed by either of them, or by anyone for whose acts either of them may be liable, and not attributable to the fault or negligence of the Contractor. The foregoing obligations of the Contractor are in addition to the Contractor's obligations under Paragraph 3.18.

10.2.6 The Contractor shall designate a responsible member of the Contractor's organization at the site whose duty shall be the prevention of accidents. This person shall be the Contractor's superintendent unless otherwise designated by the Contractor in writing to the Owner and Architect.

10.2.7 The Contractor shall not load or permit any part of the construction or site to be loaded so as to endanger its safety.

10.3   HAZARDOUS MATERIALS

10.3.1 If reasonable precautions will be inadequate to prevent foreseeable bodily injury or death to persons resulting from a material or substance, including but not limited to asbestos or polychlorinated biphenyl (PCB), encountered on the site by the Contractor, the Contractor shall, upon recognizing the condition, immediately stop Work in the affected area and report the condition to the Owner and Architect in writing.

10.3.2 The Owner shall obtain the services of a licensed laboratory to verify the presence or absence of the material or substance reported by the Contractor and, in the event such material or substance is found to be present, to verify that it has been rendered harmless. Unless otherwise required by the Contract Documents, the Owner shall furnish in writing to the Contractor and Architect the names and qualifications of persons or entities who are to perform tests verifying the presence or absence of such material or substance or who are to perform the task of removal or safe containment of such material or substance. The Contractor and the Architect will promptly reply to the Owner in writing stating whether or not either has reasonable objection to the persons or entities proposed by the Owner. If either the Contractor or Architect has an objection to a person or entity proposed by the Owner, the Owner shall propose another to whom the Contractor and the Architect have no reasonable objection. When the material or substance has been rendered harmless, Work in the affected area shall resume upon written agreement of the Owner and Contractor. The Contract Time shall be extended appropriately and the Contract Sum shall be increased in the amount of the Contractor's reasonable additional costs of shut-down, delay and start-up, which adjustments shall be accomplished as provided in Article 7.

10.3.3 To the fullest extent permitted by law, the Owner shall indemnify and hold harmless the Contractor, Subcontractors, Architect, Architect's consultants and agents and employees of any of them from and against claims, damages, losses and expenses, including but not limited to attorneys' fees, arising out of or resulting from performance of the Work in the affected area if in fact the material or substance presents the risk of bodily injury or death as described in Subparagraph 10.3.1 and has not been rendered harmless, provided that such claim, damage, loss or expense is attributable to bodily injury, sickness, disease or death, or to injury to or destruction of tangible property (other than the Work itself) and provided that such damage, loss or expense is not due to the sole negligence of a party seeking indemnity.

10.4   The Owner shall not be responsible under Paragraph 10.3 for materials and substances brought to the site by the Contractor unless such materials or substances were required by the Contract Documents.

10.5   If, without negligence on the part of the Contractor, the Contractor is held liable for the cost of remediation of a hazardous material or substance solely by reason of performing Work as required by the Contract Documents, the Owner shall indemnify the Contractor for all cost and expense thereby incurred.

10.6   EMERGENCIES

10.6.1 In an emergency affecting safety of persons or property, the Contractor shall act, at the Contractor's discretion, to prevent threatened damage, injury or loss. Additional compensation or

© 1997 AIA®
AIA DOCUMENT A201-1997
GENERAL CONDITIONS
OF THE CONTRACT FOR
CONSTRUCTION

The American Institute
of Architects
1735 New York Avenue, N.W.
Washington, D.C. 20006-5292

35

extension of time claimed by the Contractor on account of an emergency shall be determined as provided in Paragraph 4.3 and Article 7.

ARTICLE 11 INSURANCE AND BONDS

11.1 CONTRACTOR'S LIABILITY INSURANCE

11.1.1 The Contractor shall purchase from and maintain in a company or companies lawfully authorized to do business in the jurisdiction in which the Project is located such insurance as will protect the Contractor from claims set forth below which may arise out of or result from the Contractor's operations under the Contract and for which the Contractor may be legally liable, whether such operations be by the Contractor or by a Subcontractor or by anyone directly or indirectly employed by any of them, or by anyone for whose acts any of them may be liable:

   .1 claims under workers' compensation, disability benefit and other similar employee benefit acts which are applicable to the Work to be performed;

   .2 claims for damages because of bodily injury, occupational sickness or disease, or death of the Contractor's employees;

   .3 claims for damages because of bodily injury, sickness or disease, or death of any person other than the Contractor's employees;

   .4 claims for damages insured by usual personal injury liability coverage;

   .5 claims for damages, other than to the Work itself, because of injury to or destruction of tangible property, including loss of use resulting therefrom;

   .6 claims for damages because of bodily injury, death of a person or property damage arising out of ownership, maintenance or use of a motor vehicle;

   .7 claims for bodily injury or property damage arising out of completed operations; and

   .8 claims involving contractual liability insurance applicable to the Contractor's obligations under Paragraph 3.18.

11.1.2 The insurance required by Subparagraph 11.1.1 shall be written for not less than limits of liability specified in the Contract Documents or required by law, whichever coverage is greater. Coverages, whether written on an occurrence or claims-made basis, shall be maintained without interruption from date of commencement of the Work until date of final payment and termination of any coverage required to be maintained after final payment.

11.1.3 Certificates of insurance acceptable to the Owner shall be filed with the Owner prior to commencement of the Work. These certificates and the insurance policies required by this Paragraph 11.1 shall contain a provision that coverages afforded under the policies will not be canceled or allowed to expire until at least 30 days' prior written notice has been given to the Owner. If any of the foregoing insurance coverages are required to remain in force after final payment and are reasonably available, an additional certificate evidencing continuation of such coverage shall be submitted with the final Application for Payment as required by Subparagraph 9.10.2. Information concerning reduction of coverage on account of revised limits or claims paid under the General Aggregate, or both, shall be furnished by the Contractor with reasonable promptness in accordance with the Contractor's information and belief.

11.2 OWNER'S LIABILITY INSURANCE

11.2.1 The Owner shall be responsible for purchasing and maintaining the Owner's usual liability insurance.

11.3 PROJECT MANAGEMENT PROTECTIVE LIABILITY INSURANCE

11.3.1 Optionally, the Owner may require the Contractor to purchase and maintain Project Management Protective Liability insurance from the Contractor's usual sources as primary coverage for the Owner's, Contractor's and Architect's vicarious liability for construction operations under the Contract. Unless otherwise required by the Contract Documents, the Owner shall reimburse the Contractor by increasing the Contract Sum to pay the cost of purchasing and maintaining such optional insurance coverage, and the Contractor shall not be responsible for purchasing any other liability insurance on behalf of the Owner. The minimum limits of liability purchased with such coverage shall be equal to the aggregate of the limits required for Contractor's Liability Insurance under Clauses 11.1.1.2 through 11.1.1.5.

11.3.2 To the extent damages are covered by Project Management Protective Liability insurance, the Owner, Contractor and Architect waive all rights against each other for damages, except such rights as they may have to the proceeds of such insurance. The policy shall provide for such waivers of subrogation by endorsement or otherwise.

11.3.3 The Owner shall not require the Contractor to include the Owner, Architect or other persons or entities as additional insureds on the Contractor's Liability Insurance coverage under Paragraph 11.1.

11.4 PROPERTY INSURANCE

11.4.1 Unless otherwise provided, the Owner shall purchase and maintain, in a company or companies lawfully authorized to do business in the jurisdiction in which the Project is located, property insurance written on a builder's risk "all-risk" or equivalent policy form in the amount of the initial Contract Sum, plus value of subsequent Contract modifications and cost of materials supplied or installed by others, comprising total value for the entire Project at the site on a replacement cost basis without optional deductibles. Such property insurance shall be maintained, unless otherwise provided in the Contract Documents or otherwise agreed in writing by all persons and entities who are beneficiaries of such insurance, until final payment has been made as provided in Paragraph 9.10 or until no person or entity other than the Owner has an insurable interest in the property required by this Paragraph 11.4 to be covered, whichever is later. This insurance shall include interests of the Owner, the Contractor, Subcontractors and Sub-subcontractors in the Project.

11.4.1.1 Property insurance shall be on an "all-risk" or equivalent policy form and shall include, without limitation, insurance against the perils of fire (with extended coverage) and physical loss or damage including, without duplication of coverage, theft, vandalism, malicious mischief, collapse, earthquake, flood, windstorm, falsework, testing and startup, temporary buildings and debris removal including demolition occasioned by enforcement of any applicable legal requirements, and shall cover reasonable compensation for Architect's and Contractor's services and expenses required as a result of such insured loss.

11.4.1.2 If the Owner does not intend to purchase such property insurance required by the Contract and with all of the coverages in the amount described above, the Owner shall so inform the Contractor in writing prior to commencement of the Work. The Contractor may then effect insurance which will protect the interests of the Contractor, Subcontractors and Sub-subcontractors in the Work, and by appropriate Change Order the cost thereof shall be charged to the Owner. If the Contractor is damaged by the failure or neglect of the Owner to purchase or maintain insurance as described above, without so notifying the Contractor in writing, then the Owner shall bear all reasonable costs properly attributable thereto.

11.4.1.3 If the property insurance requires deductibles, the Owner shall pay costs not covered because of such deductibles.

11.4.1.4 This property insurance shall cover portions of the Work stored off the site, and also portions of the Work in transit.

11.4.1.5 Partial occupancy or use in accordance with Paragraph 9.9 shall not commence until the insurance company or companies providing property insurance have consented to such partial

© 1997 AIA®
AIA DOCUMENT A201-1997
GENERAL CONDITIONS
OF THE CONTRACT FOR
CONSTRUCTION

The American Institute
of Architects
1735 New York Avenue, N.W.
Washington, D.C. 20006-5292

37

occupancy or use by endorsement or otherwise. The Owner and the Contractor shall take reasonable steps to obtain consent of the insurance company or companies and shall, without mutual written consent, take no action with respect to partial occupancy or use that would cause cancellation, lapse or reduction of insurance.

11.4.2 Boiler and Machinery Insurance. The Owner shall purchase and maintain boiler and machinery insurance required by the Contract Documents or by law, which shall specifically cover such insured objects during installation and until final acceptance by the Owner; this insurance shall include interests of the Owner, Contractor, Subcontractors and Sub-subcontractors in the Work, and the Owner and Contractor shall be named insureds.

11.4.3 Loss of Use Insurance. The Owner, at the Owner's option, may purchase and maintain such insurance as will insure the Owner against loss of use of the Owner's property due to fire or other hazards, however caused. The Owner waives all rights of action against the Contractor for loss of use of the Owner's property, including consequential losses due to fire or other hazards however caused.

11.4.4 If the Contractor requests in writing that insurance for risks other than those described herein or other special causes of loss be included in the property insurance policy, the Owner shall, if possible, include such insurance, and the cost thereof shall be charged to the Contractor by appropriate Change Order.

11.4.5 If during the Project construction period the Owner insures properties, real or personal or both, at or adjacent to the site by property insurance under policies separate from those insuring the Project, or if after final payment property insurance is to be provided on the completed Project through a policy or policies other than those insuring the Project during the construction period, the Owner shall waive all rights in accordance with the terms of Subparagraph 11.4.7 for damages caused by fire or other causes of loss covered by this separate property insurance. All separate policies shall provide this waiver of subrogation by endorsement or otherwise.

11.4.6 Before an exposure to loss may occur, the Owner shall file with the Contractor a copy of each policy that includes insurance coverages required by this Paragraph 11.4. Each policy shall contain all generally applicable conditions, definitions, exclusions and endorsements related to this Project. Each policy shall contain a provision that the policy will not be canceled or allowed to expire, and that its limits will not be reduced, until at least 30 days' prior written notice has been given to the Contractor.

11.4.7 Waivers of Subrogation. The Owner and Contractor waive all rights against (1) each other and any of their subcontractors, sub-subcontractors, agents and employees, each of the other, and (2) the Architect, Architect's consultants, separate contractors described in Article 6, if any, and any of their subcontractors, sub-subcontractors, agents and employees, for damages caused by fire or other causes of loss to the extent covered by property insurance obtained pursuant to this Paragraph 11.4 or other property insurance applicable to the Work, except such rights as they have to proceeds of such insurance held by the Owner as fiduciary. The Owner or Contractor, as appropriate, shall require of the Architect, Architect's consultants, separate contractors described in Article 6, if any, and the subcontractors, sub-subcontractors, agents and employees of any of them, by appropriate agreements, written where legally required for validity, similar waivers each in favor of other parties enumerated herein. The policies shall provide such waivers of subrogation by endorsement or otherwise. A waiver of subrogation shall be effective as to a person or entity even though that person or entity would otherwise have a duty of indemnification, contractual or otherwise, did not pay the insurance premium directly or indirectly, and whether or not the person or entity had an insurable interest in the property damaged.

©1997 AIA®
AIA DOCUMENT A201-1997
GENERAL CONDITIONS
OF THE CONTRACT FOR
CONSTRUCTION

The American Institute
of Architects
1735 New York Avenue, N.W.
Washington, D.C. 20006-5292

38

11.4.8 A loss insured under Owner's property insurance shall be adjusted by the Owner as fiduciary and made payable to the Owner as fiduciary for the insureds, as their interests may appear, subject to requirements of any applicable mortgagee clause and of Subparagraph 11.4.10. The Contractor shall pay Subcontractors their just shares of insurance proceeds received by the Contractor, and by appropriate agreements, written where legally required for validity, shall require Subcontractors to make payments to their Sub-subcontractors in similar manner.

11.4.9 If required in writing by a party in interest, the Owner as fiduciary shall, upon occurrence of an insured loss, give bond for proper performance of the Owner's duties. The cost of required bonds shall be charged against proceeds received as fiduciary. The Owner shall deposit in a separate account proceeds so received, which the Owner shall distribute in accordance with such agreement as the parties in interest may reach, or in accordance with an arbitration award in which case the procedure shall be as provided in Paragraph 4.6. If after such loss no other special agreement is made and unless the Owner terminates the Contract for convenience, replacement of damaged property shall be performed by the Contractor after notification of a Change in the Work in accordance with Article 7.

11.4.10 The Owner as fiduciary shall have power to adjust and settle a loss with insurers unless one of the parties in interest shall object in writing within five days after occurrence of loss to the Owner's exercise of this power; if such objection is made, the dispute shall be resolved as provided in Paragraphs 4.5 and 4.6. The Owner as fiduciary shall, in the case of arbitration, make settlement with insurers in accordance with directions of the arbitrators. If distribution of insurance proceeds by arbitration is required, the arbitrators will direct such distribution.

11.5   PERFORMANCE BOND AND PAYMENT BOND

11.5.1 The Owner shall have the right to require the Contractor to furnish bonds covering faithful performance of the Contract and payment of obligations arising thereunder as stipulated in bidding requirements or specifically required in the Contract Documents on the date of execution of the Contract.

11.5.2 Upon the request of any person or entity appearing to be a potential beneficiary of bonds covering payment of obligations arising under the Contract, the Contractor shall promptly furnish a copy of the bonds or shall permit a copy to be made.

ARTICLE 12   UNCOVERING AND CORRECTION OF WORK

12.1   UNCOVERING OF WORK

12.1.1 If a portion of the Work is covered contrary to the Architect's request or to requirements specifically expressed in the Contract Documents, it must, if required in writing by the Architect, be uncovered for the Architect's examination and be replaced at the Contractor's expense without change in the Contract Time.

12.1.2 If a portion of the Work has been covered which the Architect has not specifically requested to examine prior to its being covered, the Architect may request to see such Work and it shall be uncovered by the Contractor. If such Work is in accordance with the Contract Documents, costs of uncovering and replacement shall, by appropriate Change Order, be at the Owner's expense. If such Work is not in accordance with the Contract Documents, correction shall be at the Contractor's expense unless the condition was caused by the Owner or a separate contractor in which event the Owner shall be responsible for payment of such costs.

© 1 9 9 7   A I A ®
AIA DOCUMENT A201-1997
GENERAL CONDITIONS
OF THE CONTRACT FOR
CONSTRUCTION

The American Institute
of Architects
1735 New York Avenue, N.W.
Washington, D.C. 20006-5292

39

12.2 CORRECTION OF WORK

12.2.1 BEFORE OR AFTER SUBSTANTIAL COMPLETION

12.2.1.1 The Contractor shall promptly correct Work rejected by the Architect or failing to conform to the requirements of the Contract Documents, whether discovered before or after Substantial Completion and whether or not fabricated, installed or completed. Costs of correcting such rejected Work, including additional testing and inspections and compensation for the Architect's services and expenses made necessary thereby, shall be at the Contractor's expense.

12.2.2 AFTER SUBSTANTIAL COMPLETION

12.2.2.1 In addition to the Contractor's obligations under Paragraph 3.5, if, within one year after the date of Substantial Completion of the Work or designated portion thereof or after the date for commencement of warranties established under Subparagraph 9.9.1, or by terms of an applicable special warranty required by the Contract Documents, any of the Work is found to be not in accordance with the requirements of the Contract Documents, the Contractor shall correct it promptly after receipt of written notice from the Owner to do so unless the Owner has previously given the Contractor a written acceptance of such condition. The Owner shall give such notice promptly after discovery of the condition. During the one-year period for correction of Work, if the Owner fails to notify the Contractor and give the Contractor an opportunity to make the correction, the Owner waives the rights to require correction by the Contractor and to make a claim for breach of warranty. If the Contractor fails to correct nonconforming Work within a reasonable time during that period after receipt of notice from the Owner or Architect, the Owner may correct it in accordance with Paragraph 2.4.

12.2.2.2 The one-year period for correction of Work shall be extended with respect to portions of Work first performed after Substantial Completion by the period of time between Substantial Completion and the actual performance of the Work.

12.2.2.3 The one-year period for correction of Work shall not be extended by corrective Work performed by the Contractor pursuant to this Paragraph 12.2.

12.2.3 The Contractor shall remove from the site portions of the Work which are not in accordance with the requirements of the Contract Documents and are neither corrected by the Contractor nor accepted by the Owner.

12.2.4 The Contractor shall bear the cost of correcting destroyed or damaged construction, whether completed or partially completed, of the Owner or separate contractors caused by the Contractor's correction or removal of Work which is not in accordance with the requirements of the Contract Documents.

12.2.5 Nothing contained in this Paragraph 12.2 shall be construed to establish a period of limitation with respect to other obligations which the Contractor might have under the Contract Documents. Establishment of the one-year period for correction of Work as described in Subparagraph 12.2.2 relates only to the specific obligation of the Contractor to correct the Work, and has no relationship to the time within which the obligation to comply with the Contract Documents may be sought to be enforced, nor to the time within which proceedings may be commenced to establish the Contractor's liability with respect to the Contractor's obligations other than specifically to correct the Work.

12.3 ACCEPTANCE OF NONCONFORMING WORK

12.3.1 If the Owner prefers to accept Work which is not in accordance with the requirements of the Contract Documents, the Owner may do so instead of requiring its removal and correction, in which case the Contract Sum will be reduced as appropriate and equitable. Such adjustment shall be effected whether or not final payment has been made.

ARTICLE 13 MISCELLANEOUS PROVISIONS

13.1 GOVERNING LAW

13.1.1 The Contract shall be governed by the law of the place where the Project is located.

13.2 SUCCESSORS AND ASSIGNS

13.2.1 The Owner and Contractor respectively bind themselves, their partners, successors, assigns and legal representatives to the other party hereto and to partners, successors, assigns and legal representatives of such other party in respect to covenants, agreements and obligations contained in the Contract Documents. Except as provided in Subparagraph 13.2.2, neither party to the Contract shall assign the Contract as a whole without written consent of the other. If either party attempts to make such an assignment without such consent, that party shall nevertheless remain legally responsible for all obligations under the Contract.

13.2.2 The Owner may, without consent of the Contractor, assign the Contract to an institutional lender providing construction financing for the Project. In such event, the lender shall assume the Owner's rights and obligations under the Contract Documents. The Contractor shall execute all consents reasonably required to facilitate such assignment.

13.3 WRITTEN NOTICE

13.3.1 Written notice shall be deemed to have been duly served if delivered in person to the individual or a member of the firm or entity or to an officer of the corporation for which it was intended, or if delivered at or sent by registered or certified mail to the last business address known to the party giving notice.

13.4 RIGHTS AND REMEDIES

13.4.1 Duties and obligations imposed by the Contract Documents and rights and remedies available thereunder shall be in addition to and not a limitation of duties, obligations, rights and remedies otherwise imposed or available by law.

13.4.2 No action or failure to act by the Owner, Architect or Contractor shall constitute a waiver of a right or duty afforded them under the Contract, nor shall such action or failure to act constitute approval of or acquiescence in a breach thereunder, except as may be specifically agreed in writing.

13.5 TESTS AND INSPECTIONS

13.5.1 Tests, inspections and approvals of portions of the Work required by the Contract Documents or by laws, ordinances, rules, regulations or orders of public authorities having jurisdiction shall be made at an appropriate time. Unless otherwise provided, the Contractor shall make arrangements for such tests, inspections and approvals with an independent testing laboratory or entity acceptable to the Owner, or with the appropriate public authority, and shall bear all related costs of tests, inspections and approvals. The Contractor shall give the Architect timely notice of when and where tests and inspections are to be made so that the Architect may be present for such procedures. The Owner shall bear costs of tests, inspections or approvals which do not become requirements until after bids are received or negotiations concluded.

13.5.2 If the Architect, Owner or public authorities having jurisdiction determine that portions of the Work require additional testing, inspection or approval not included under Subparagraph 13.5.1, the Architect will, upon written authorization from the Owner, instruct the Contractor to make arrangements for such additional testing, inspection or approval by an entity acceptable to the Owner, and the Contractor shall give timely notice to the Architect of when and where tests and inspections are to be made so that the Architect may be present for such procedures. Such costs, except as provided in Subparagraph 13.5.3, shall be at the Owner's expense.

© 1997 A I A ®
AIA DOCUMENT A201-1997
GENERAL CONDITIONS
OF THE CONTRACT FOR
CONSTRUCTION

The American Institute
of Architects
1735 New York Avenue, N.W.
Washington, D.C. 20006-5292

41

13.5.3 If such procedures for testing, inspection or approval under Subparagraphs 13.5.1 and 13.5.2 reveal failure of the portions of the Work to comply with requirements established by the Contract Documents, all costs made necessary by such failure including those of repeated procedures and compensation for the Architect's services and expenses shall be at the Contractor's expense.

13.5.4 Required certificates of testing, inspection or approval shall, unless otherwise required by the Contract Documents, be secured by the Contractor and promptly delivered to the Architect.

13.5.5 If the Architect is to observe tests, inspections or approvals required by the Contract Documents, the Architect will do so promptly and, where practicable, at the normal place of testing.

13.5.6 Tests or inspections conducted pursuant to the Contract Documents shall be made promptly to avoid unreasonable delay in the Work.

13.6   INTEREST
13.6.1 Payments due and unpaid under the Contract Documents shall bear interest from the date payment is due at such rate as the parties may agree upon in writing or, in the absence thereof, at the legal rate prevailing from time to time at the place where the Project is located.

13.7   COMMENCEMENT OF STATUTORY LIMITATION PERIOD
13.7.1 As between the Owner and Contractor:
    .1 **Before Substantial Completion.** As to acts or failures to act occurring prior to the relevant date of Substantial Completion, any applicable statute of limitations shall commence to run and any alleged cause of action shall be deemed to have accrued in any and all events not later than such date of Substantial Completion;
    .2 **Between Substantial Completion and Final Certificate for Payment.** As to acts or failures to act occurring subsequent to the relevant date of Substantial Completion and prior to issuance of the final Certificate for Payment, any applicable statute of limitations shall commence to run and any alleged cause of action shall be deemed to have accrued in any and all events not later than the date of issuance of the final Certificate for Payment; and
    .3 **After Final Certificate for Payment.** As to acts or failures to act occurring after the relevant date of issuance of the final Certificate for Payment, any applicable statute of limitations shall commence to run and any alleged cause of action shall be deemed to have accrued in any and all events not later than the date of any act or failure to act by the Contractor pursuant to any Warranty provided under Paragraph 3.5, the date of any correction of the Work or failure to correct the Work by the Contractor under Paragraph 12.2, or the date of actual commission of any other act or failure to perform any duty or obligation by the Contractor or Owner, whichever occurs last.

ARTICLE 14   TERMINATION OR SUSPENSION OF THE CONTRACT
14.1   TERMINATION BY THE CONTRACTOR
14.1.1 The Contractor may terminate the Contract if the Work is stopped for a period of 30 consecutive days through no act or fault of the Contractor or a Subcontractor, Sub-subcontractor or their agents or employees or any other persons or entities performing portions of the Work under direct or indirect contract with the Contractor, for any of the following reasons:
    .1 issuance of an order of a court or other public authority having jurisdiction which requires all Work to be stopped;
    .2 an act of government, such as a declaration of national emergency which requires all Work to be stopped;

.3 because the Architect has not issued a Certificate for Payment and has not notified the Contractor of the reason for withholding certification as provided in Subparagraph 9.4.1, or because the Owner has not made payment on a Certificate for Payment within the time stated in the Contract Documents; or

.4 the Owner has failed to furnish to the Contractor promptly, upon the Contractor's request, reasonable evidence as required by Subparagraph 2.2.1.

14.1.2 The Contractor may terminate the Contract if, through no act or fault of the Contractor or a Subcontractor, Sub-subcontractor or their agents or employees or any other persons or entities performing portions of the Work under direct or indirect contract with the Contractor, repeated suspensions, delays or interruptions of the entire Work by the Owner as described in Paragraph 14.3 constitute in the aggregate more than 100 percent of the total number of days scheduled for completion, or 120 days in any 365-day period, whichever is less.

14.1.3 If one of the reasons described in Subparagraph 14.1.1 or 14.1.2 exists, the Contractor may, upon seven days' written notice to the Owner and Architect, terminate the Contract and recover from the Owner payment for Work executed and for proven loss with respect to materials, equipment, tools, and construction equipment and machinery, including reasonable overhead, profit and damages.

14.1.4 If the Work is stopped for a period of 60 consecutive days through no act or fault of the Contractor or a Subcontractor or their agents or employees or any other persons performing portions of the Work under contract with the Contractor because the Owner has persistently failed to fulfill the Owner's obligations under the Contract Documents with respect to matters important to the progress of the Work, the Contractor may, upon seven additional days' written notice to the Owner and the Architect, terminate the Contract and recover from the Owner as provided in Subparagraph 14.1.3.

14.2 TERMINATION BY THE OWNER FOR CAUSE

14.2.1 The Owner may terminate the Contract if the Contractor:

.1 persistently or repeatedly refuses or fails to supply enough properly skilled workers or proper materials;

.2 fails to make payment to Subcontractors for materials or labor in accordance with the respective agreements between the Contractor and the Subcontractors;

.3 persistently disregards laws, ordinances, or rules, regulations or orders of a public authority having jurisdiction; or

.4 otherwise is guilty of substantial breach of a provision of the Contract Documents.

14.2.2 When any of the above reasons exist, the Owner, upon certification by the Architect that sufficient cause exists to justify such action, may without prejudice to any other rights or remedies of the Owner and after giving the Contractor and the Contractor's surety, if any, seven days' written notice, terminate employment of the Contractor and may, subject to any prior rights of the surety:

.1 take possession of the site and of all materials, equipment, tools, and construction equipment and machinery thereon owned by the Contractor;

.2 accept assignment of subcontracts pursuant to Paragraph 5.4; and

.3 finish the Work by whatever reasonable method the Owner may deem expedient. Upon request of the Contractor, the Owner shall furnish to the Contractor a detailed accounting of the costs incurred by the Owner in finishing the Work.

14.2.3 When the Owner terminates the Contract for one of the reasons stated in Subparagraph 14.2.1, the Contractor shall not be entitled to receive further payment until the Work is finished.

© 1997 AIA®
AIA DOCUMENT A201-1997
GENERAL CONDITIONS
OF THE CONTRACT FOR
CONSTRUCTION

The American Institute
of Architects
1735 New York Avenue, N.W.
Washington, D.C. 20006-5292

43

14.2.4 If the unpaid balance of the Contract Sum exceeds costs of finishing the Work, including compensation for the Architect's services and expenses made necessary thereby, and other damages incurred by the Owner and not expressly waived, such excess shall be paid to the Contractor. If such costs and damages exceed the unpaid balance, the Contractor shall pay the difference to the Owner. The amount to be paid to the Contractor or Owner, as the case may be, shall be certified by the Architect, upon application, and this obligation for payment shall survive termination of the Contract.

14.3 SUSPENSION BY THE OWNER FOR CONVENIENCE

14.3.1 The Owner may, without cause, order the Contractor in writing to suspend, delay or interrupt the Work in whole or in part for such period of time as the Owner may determine.

14.3.2 The Contract Sum and Contract Time shall be adjusted for increases in the cost and time caused by suspension, delay or interruption as described in Subparagraph 14.3.1. Adjustment of the Contract Sum shall include profit. No adjustment shall be made to the extent:

    .1 that performance is, was or would have been so suspended, delayed or interrupted by another cause for which the Contractor is responsible; or

    .2 that an equitable adjustment is made or denied under another provision of the Contract.

14.4 TERMINATION BY THE OWNER FOR CONVENIENCE

14.4.1 The Owner may, at any time, terminate the Contract for the Owner's convenience and without cause.

14.4.2 Upon receipt of written notice from the Owner of such termination for the Owner's convenience, the Contractor shall:

    .1 cease operations as directed by the Owner in the notice;

    .2 take actions necessary, or that the Owner may direct, for the protection and preservation of the Work; and

    .3 except for Work directed to be performed prior to the effective date of termination stated in the notice, terminate all existing subcontracts and purchase orders and enter into no further subcontracts and purchase orders.

14.4.3 In case of such termination for the Owner's convenience, the Contractor shall be entitled to receive payment for Work executed, and costs incurred by reason of such termination, along with reasonable overhead and profit on the Work not executed.

1 9 9 8 E D I T I O N

AIA DOCUMENT | A511-1998

Guide for Supplementary Conditions

The following supplements are representative of those that may be required on a typical project, but are not intended to be all-inclusive. Not all of the additions or modifications shown will be necessary for every project.

SUPPLEMENTARY CONDITIONS
SUGGESTED INTRODUCTORY PARAGRAPH

The following supplements modify the "General Conditions of the Contract for Construction," AIA Document A201-1997. Where a portion of the General Conditions is modified or deleted by these Supplementary Conditions, the unaltered portions of the General Conditions shall remain in effect.

Where A201/CMa or A271 form the General Conditions, the reference above should read as follows:

"General Conditions of the Contract for Construction," AIA Document A201/CMa, 1992 Edition

or

"General Conditions of the Contract for Furniture, Furnishings and Equipment," AIA Document A271, 1990 Edition

Where the General Conditions are those contained in AIA Document A107-1997 or AIA Document A177, the reference here should read as follows:

"Abbreviated Standard Form of Agreement Between Owner and Contractor for Construction Projects of Limited Scope," AIA Document A107-1997

or

"Abbreviated Form of Agreement Between Owner and Contractor for Furniture, Furnishings and Equipment," AIA Document A177, 1990 Edition

ARTICLE 1 GENERAL PROVISIONS
1.1 BASIC DEFINITIONS

Certain corporate clients or governmental agencies may require the use of terms such as "Project Manager," "Contracting Officer" or others which may have important and necessary connotations, and these terms should be defined here.

1.1.1 If a client requires that the bidding requirements and other documents be included in the Contract Documents, the specific documents should be enumerated in the Agreement between the Owner and Contractor and added to Subparagraph 1.1.1 of the General Conditions as Contract Documents. It may also be advisable to bring this to the attention of Bidders in the Instructions to Bidders.

©1998 AIA®
AIA DOCUMENT A511-1998
GUIDE FOR SUPPLEMENTARY
CONDITIONS

1

The American Institute
of Architects
1735 New York Avenue, N.W.
Washington, D.C. 20006-5292

Problems have been reported relating to the use of electronic versions of the Contract Documents. Incompatibilities among software and hardware can alter documents in ways that are not readily apparent. Use of electronic documents implies a significant increase in the Contractor's responsibility to manage, control and distribute this information. Architects who are required to provide documents in electronic form should consult with an attorney for appropriate indemnification language to be added to the Supplementary Conditions.

> Add the following sentence to the end of Subparagraph 1.1.1:
> The Contract Documents executed in accordance with Subparagraph 1.5.1 shall prevail in case of an inconsistency with subsequent versions made through manipulatable electronic operations involving computers.

1.1.4 THE PROJECT

If the Work to be performed by the Contractor does not constitute the total Project, the relationship of the Contractor's Work to that to be performed by other separate contractors or the Owner should be made clear in the Contract Documents. General information concerning the relationship of the Contractor's activities to construction by other separate contractors or the Owner should be specified in the General Requirements (Division 1 of the Specifications).

1.2 CORRELATION AND INTENT OF THE CONTRACT DOCUMENTS

1.2.1 A principle of the AIA General Conditions is not to establish a system of precedence among the Contract Documents, but to provide that all Documents are complementary. In the event of inconsistencies among the Documents, the Architect is to interpret them in accordance with this principle. The Architect's decisions are subject to mediation and arbitration if a decision is disputed. Establishing a fixed order of priority is not recommended, because no one document constitutes the best authority on all issues that may arise. The order shown here is suggested for consistency in the event an Owner insists on establishing a precedence. Note that this modification does not establish a precedence between Drawings and Divisions 2 through 16 of the Specifications, which together describe the Work.

> Add Clause 1.2.1.1 to Subparagraph 1.2.1:
> 1.2.1.1 In the event of conflicts or discrepancies among the Contract Documents, interpretations will be based on the following priorities:
> 1. The Agreement.
> 2. Addenda, with those of later date having precedence over those of earlier date.
> 3. The Supplementary Conditions.
> 4. The General Conditions of the Contract for Construction.
> 5. Division 1 of the Specifications.
> 6. Drawings and Divisions 2-16 of the Specifications.
> In the case of conflicts or discrepancies between Drawings and Divisions 2-16 of the Specifications or within either Document not clarified by Addendum, the Architect will determine which takes precedence in accordance with Subparagraph 4.2.11.

1.6 OWNERSHIP AND USE OF DRAWINGS, SPECIFICATIONS AND OTHER INSTRUMENTS OF SERVICE

In some instances, the Contractor may request copies of the Drawings in electronic form for use in preparing shop drawings. Use of such electronic documents may save time, but there are hazards involved for the Owner and Architect. It is not practical to suggest model language for all the situations that may arise, but some considerations may be noted.

Because certain information in the electronic documents may no longer conform to the Contract Documents in use on the site in hard copy, it is critical that the Contractor disclose, and the

Owner and Architect agree to, the use that the Contractor is to make of the electronic documents or data. The Contractor should also be required to agree not to transfer such documents or data to other formats or to other persons or entities without permission of the Owner and Architect, and to indemnify the Owner and Architect against any adverse consequences of their use. A "test run" involving a detailed sample of the electronic documents or data is also advisable to ensure software compatibility between the Architect's and Contractor's computer systems.

CAUTION: Subparagraph 4.2.4 requires that transfer of electronic documents to Subcontractors must be managed by the Contractor.

Add the following Subparagraph 1.6.2 to Paragraph 1.6:
1.6.2   Contractor's Use of Instruments of Service in Electronic Form.

1.6.2.1   The Architect may, with the concurrence of the Owner, furnish to the Contractor versions of Instruments of Service in electronic form. The Contract Documents executed or identified in accordance with Subparagraph 1.5.1 shall prevail in case of an inconsistency with subsequent versions made through manipulatable electronic means involving computers.

1.6.2.2   The Contractor shall not transfer or reuse Instruments of Service in electronic or machine readable form without the prior written consent of the Architect.

1.7   If a partnering arrangement is anticipated following execution of the Agreement, the following language may be appropriate. Involvement of representatives of the Owner, Contractor and Architect is customary, but selected Subcontractors and design consultants may also be required to participate.

Partnering arrangements should be approached with a measure of caution. The term "partnering" should be avoided because it implies a contractual relationship. The term "partnering" has no defined meaning, and could not have one without adding to the contractual obligations of the parties. By the language given below, the parties simply agree to meet in furtherance of their common interests. Such meetings may be used to obtain interpretations of the Contract Documents, and to resolve disputes short of mediation or arbitration. It should be recognized, however, that actions taken at such meetings may modify the rights and obligations of the parties under the Contract, even though the agreement to meet does not do so.

If the participants intend to retain the use of a facilitator to further this arrangement, language should be added to address who shall bear the cost of the facilitator.

Add the following Paragraph 1.7 to Article 1:
1.7   Representatives of the Owner, Contractor and Architect shall meet periodically at mutually agreed-upon intervals for the purpose of establishing procedures to facilitate cooperation, communication and timely responses among the participants. By participating in this arrangement, the parties do not intend to create additional contractual obligations or modify the legal relationships which may otherwise exist.

ARTICLE 2 OWNER
2.2   INFORMATION AND SERVICES REQUIRED OF THE OWNER
2.2.2   When the Project is subject, after award of the Contract, to a prolonged review or approval process by governmental or other agencies, it is desirable, if possible, to describe this process and to state whether the Contractor is expected to play any role in the process and the effect this process may be expected to have on the commencement of the Work and the progress schedule.

2.2.3   It may be necessary in some instances to describe more fully the surveys which the

© 1998 AIA®
AIA DOCUMENT A511-1998
GUIDE FOR SUPPLEMENTARY
CONDITIONS

The American Institute
of Architects
1735 New York Avenue, N.W.
Washington, D.C. 20006-5292

3

354

Owner is furnishing (i.e., metes and bounds only or topographical). This should be done as a supplement to this subparagraph.

When the Owner identified in the Agreement between Owner and Contractor is not the owner of record of the Project site or the legal description of the property is not set forth in the Contract Documents, this subparagraph should be modified in the Supplementary Conditions to advise the Contractor where the legal description is available.

2.2.5   If the Contractor is to be furnished only a specific number of sets of Drawings and Project Manuals without charge, this should be stated here, with the basis on which the Contractor will be charged for additional sets.

Use of this supplement requires the section in the General Requirements (Division 1 of the Specifications) that contains the list of items to be identified.

> Delete Subparagraph 2.2.5 and substitute the following:
> 2.2.5   The Contractor will be furnished, free of charge, _____ copies of Drawings and Project Manuals. Additional sets will be furnished at the cost of reproduction, postage and handling.

2.2.6   In some states, Uniform Building Code language is interpreted to mean that the Contractor may not employ or be responsible for special inspectors.

> Add Subparagraph 2.2.6 to Paragraph 2.2:
> 2.2.6   The Owner will procure and bear costs of structural tests and special inspections as required by the applicable building code.

ARTICLE 3   CONTRACTOR
3.2   REVIEW OF CONTRACT DOCUMENTS AND FIELD CONDITIONS BY CONTRACTOR

In B141-1997, Subparagraph 2.6.1.5 provides for the Architect's review of the Contractor's requests for information. In addition, Subparagraph 2.8.2.2 of B141-1997 provides for a change in the Architect's services for responses to Contractor's requests for information where such information is already available to the Contractor. The following model language may be used to provide consistency between A201 and B141 provisions for the Architect's review of Contractor's requests for information. Note that it may be desirable for the Architect to develop a standard request for information form for the Contractor's use on the Project.

> Add the following Subparagraph 3.2.4 to Paragraph 3.2:
> 3.2.4   The Owner shall be entitled to deduct from the Contract Sum amounts paid to the Architect for the Architect to evaluate and respond to the Contractor's requests for information, where such information was available to the Contractor from a careful study and comparison of the Contract Documents, field conditions, other Owner-provided information, Contractor-prepared coordination drawings, or prior Project correspondence or documentation.

3.4   LABOR AND MATERIALS

3.4.2   The following language may be used in situations where it is intended that substitutions should not necessarily require a Change Order. This subparagraph establishes the criteria for submission and evaluation of such substitutions. Such language should be included in the General Requirements (Division 1 of the Specifications) as well as the Supplementary Conditions. With regard to Clause 3.4.2.3, note that when multiple contracts are employed, substitutions may expose the Owner to claims from other separate contractors.

> Delete Subparagraph 3.4.2 and substitute the following:
> 3.4.2 After the Contract has been executed, the Owner and Architect will consider a formal request for the substitution of products in place of those specified only under the conditions set forth in the

4

General Requirements (Division 1 of the Specifications). By making requests for substitutions, the Contractor:

.1 represents that the Contractor has personally investigated the proposed substitute product and determined that it is equal or superior in all respects to that specified;

.2 represents that the Contractor will provide the same warranty for the substitution that the Contractor would for that specified;

.3 certifies that the cost data presented is complete and includes all related costs under this Contract except the Architect's redesign costs, and waives all claims for additional costs related to the substitution which subsequently become apparent; and

.4 will coordinate the installation of the accepted substitute, making such changes as may be required for the Work to be complete in all respects.

3.4.4   Substitutions proposed by the Contractor must be evaluated by the Architect and, if accepted, may require revision of the Drawings and Specifications. The resulting demands on the Architect's time and other resources may entitle the Architect to an adjustment in compensation, as is the case under Subparagraph 2.8.2.6 of AIA Document B141-1997. The following language allows the Owner to pass this expense on to the Contractor. This language should only be used on Projects where the Owner is fully prepared to deal with disputes that may arise from enforcement of this provision—for example, in situations where the Architect evaluates and then rejects the Contractor's proposed substitution. The Owner and Architect should also be prepared to deal with proposed substitutions that benefit the Owner.

Add the following Subparagraph 3.4.4 to Paragraph 3.4:
3.4.4   The Owner shall be entitled to deduct from the Contract Sum amounts paid to the Architect to evaluate the Contractor's proposed substitutions and to make agreed-upon changes in the Drawings and Specifications made necessary by the Owner's acceptance of such substitutions.

3.5   WARRANTY
3.5.1   Note that the terms of the warranty under Subparagraph 3.5.1 are separate and distinct from the Contractor's obligation to correct the Work, as required under Paragraph 12.2. Special warranties in the technical sections of the Specifications may also limit or expand obligations under this warranty. It is strongly suggested that Subparagraph 3.5.1 only be modified with legal advice.

3.6   TAXES
3.6.1   Certain non-profit organizations may be wholly or partially tax-exempt. Since the degree of tax exemption varies from jurisdiction to jurisdiction, the Owner should provide the exact language for statements concerning tax exemption to be included in the Supplementary Conditions.

3.7   PERMITS, FEES AND NOTICES
3.7.1   Where separate Contracts are used, list the permits and governmental fees, licenses and inspections each Contractor is required to obtain and pay for to avoid duplication or error.
In certain circumstances the Owner may elect to obtain the building permit, certain permits may not be required or the Owner may elect to pay for other fees, and this subparagraph should be appropriately modified. Attention should be given to Subparagraph 2.2.2 which relates to this issue. In any case, deviation from the provisions of the General Conditions should be specifically noted. The model language below can be expanded upon by adding other items, such as environmental impact fees.

© 1998   AIA®
AIA DOCUMENT A511-1998
GUIDE FOR SUPPLEMENTARY CONDITIONS

5

The American Institute
of Architects
1735 New York Avenue, N.W.
Washington, D.C. 20006-5292

Add the following two sentences to Subparagraph 3.7.1 :
The Owner shall pay fees for public or private water, gas, electrical, and other utility extensions at the site. The Contractor shall secure and arrange for all necessary utility connections.

3.8   ALLOWANCES

3.8.1   Allowances should be specified in the General Requirements (Division 1 of the Specifications) with appropriate references in the particular sections of the Specifications. If allowances are to be expended by Subcontractors rather than directly by the Contractor (for example, an allowance for the purchase of special light fixtures), the information in the General Requirements (Division 1 of the Specifications) should clarify that the Subcontractor's overhead, profit, handling and other costs are included in the Contract Sum and that the allowance covers only the net cost to the Subcontractor.

3.8.2.2   If installation costs are to be included in the allowances, they should be excluded from the Contractor's costs under Clause 3.8.2.2. Similar exclusions should be used if allowances include costs of testing or inspection. Quantity allowances or contingency allowances require further revision of Subparagraph 3.8.2.

3.8.2.2   Delete the semicolon at the end of Clause 3.8.2.2 and add the following:
, except that if installation is included as part of an allowance in Divisions 1-16 of the Specifications, the installation and labor cost for greater or lesser quantities of Work shall be determined in accordance with Subparagraph 7.3.6;

Renovation projects often require implementation of contractual techniques to manage unknown conditions. Quantity allowances may be established for such conditions and coupled with unit pricing mechanisms that will be triggered in the event that greater or lesser quantities of Work than those anticipated by the quantity allowance become necessary. If the potential range of variation is large, the Owner may wish to include overhead and profit in the quantity allowance, but not in the unit price. Since the quantity allowance is an assumed amount of Work in the Contract Sum and the unit price is the amount proposed by the Contractor to perform a greater or lesser increment of Work, the fair overhead and profit percentage for greater quantities is usually different from the percentage for lesser quantities of Work, which would result in a credit to the Contract Sum. If such conditions exist on a Project, Clause 3.8.2.2 may be modified accordingly.

3.9   SUPERINTENDENT

3.9.2   During construction, coordination of mechanical and electrical Work may require additional attention to avoid conflicts. Specific technical requirements for coordination submittals are usually stated in the General Requirements (Division 1 of the Specifications). The language shown here deals with mechanical and electrical Work only, but may serve as a model if similar services are required with respect to other systems.

Add the following Subparagraph 3.9.2 to 3.9:
3.9.2   The Contractor shall employ a superintendent or an assistant to the superintendent who will perform as a coordinator for mechanical and electrical Work. The coordinator shall be knowledgeable in mechanical and electrical systems and capable of reading, interpreting and coordinating Drawings, Specifications, and shop drawings pertaining to such systems. The coordinator shall assist the Subcontractors in arranging space conditions to eliminate interference between the mechanical and electrical systems and other Work and shall supervise the preparation of coordination drawings documenting the spatial arrangements for such systems within restricted spaces. The coordinator shall assist in planning and expediting the proper sequence of delivery of mechanical and electrical equipment to the site.

© 1998   A I A ®
AIA DOCUMENT A511-1998
GUIDE FOR SUPPLEMENTARY
CONDITIONS

The American Institute
of Architects
1735 New York Avenue, N.W.
Washington, D.C. 20006-5292

6

3.10 CONTRACTOR'S CONSTRUCTION SCHEDULES

3.10.1 A detailed description of the Contractor's construction and submittal schedules (CPM, bar graph or other), the process by which they are to be prepared and updated, and the extent of information required should be specified in the General Requirements (Division 1 of the Specifications).

3.10.1.1 FAST-TRACK SCHEDULE REPORTING

"Fast-track" is the term applied to a process in which certain portions of the Architect's design services overlap with the Contractor's performance of the Work. This process can save time if the activities of the Owner, Contractor and Architect are closely coordinated. Coordination is vital, however, because changes may be expected in the Drawings and Specifications as construction proceeds. The costs of materials, equipment and labor may be changing as well. Note that use of the model language below may require appropriate modifications to the Owner-Architect Agreement.

> Add the following Clause 3.10.1.1 to Subparagraph 3.10.1:
>
> 3.10.1.1 The Owner may authorize construction activities to commence prior to completion of the Drawings and Specifications. If the Drawings and Specifications require further development at the time the initial construction schedule is prepared, the Contractor shall 1) allow time in the schedule for further development of the Drawings and Specifications by the Architect, including time for review by the Owner and Contractor and for the Contractor's coordination of Subcontractors' Work, and 2) furnish to the Owner in a timely manner information regarding anticipated market conditions and construction cost; availability of labor, materials and equipment; and proposed methods, sequences and time schedules for construction of the Work.

3.11 DOCUMENTS AND SAMPLES AT THE SITE

3.11.1 The documents required here constitute "record documents" and their function is limited to showing actual changes made in the Work during construction.

Specific detailed requirements for record drawings, especially for mechanical and electrical portions of the Work, should be specified in the General Requirements (Division 1 of the Specifications), or the appropriate section of the Specifications. The term "as-built drawings" is ambiguous and should not be used because it makes the obligations of the Contractor and the Architect under Subparagraph 3.11.1 uncertain.

3.12 SHOP DRAWINGS, PRODUCT DATA AND SAMPLES

3.12.5 Detailed procedures for handling Shop Drawings, Product Data and Samples (e.g., use of reproducibles, number of copies, identification and labeling, revision and resubmission procedures) should be specified in the General Requirements (Division 1 of the Specifications).

3.12.11 Review of multiple resubmittals can be a serious drain on the Architect's time and other resources. If the Architect is entitled to an adjustment in compensation for such services under the Owner-Architect agreement (for example, under Subparagraph 2.8.1.1 of AIA Document B141-1997), language such as that shown below may be appropriate.

> Add Subparagraph 3.12.11 to Paragraph 3.12:
>
> 3.12.11 The Architect's review of Contractor's submittals will be limited to examination of an initial submittal and _____ (____) resubmittals. The Architect's review of additional submittals will be made only with the consent of the Owner after notification by the Architect. The Owner shall be entitled to deduct from the Contract Sum amounts paid to the Architect for evaluation of such additional resubmittals.

©1998 AIA®
AIA DOCUMENT A511-1998
GUIDE FOR SUPPLEMENTARY CONDITIONS

7

The American Institute
of Architects
1735 New York Avenue, N.W.
Washington, D.C. 20006-5292

3.13   USE OF SITE
3.13.1 Detailed requirements may need to be specified in the General Requirements (Division 1 of the Specifications) if an existing building will remain occupied or require access by the public.

3.14.   CUTTING AND PATCHING
3.14.1 Special requirements for Work involving renovation, remodeling, or historic restoration or other detailed requirements should be specified in Divisions 1-16 of the Specifications.

3.15   CLEANING UP
3.15.1 Detailed requirements for the cleaning should be specified in the General Requirements (Division 1 of the Specifications).

3.18   INDEMNIFICATION
In some jurisdictions statutory requirements may modify this indemnification clause or void it completely. The Owner should seek the advice of legal counsel for modifications to this paragraph.

If model language was added in Paragraph 1.1 or 1.6 to address Drawings and Specifications in electronic form, the Owner would be well-advised to seek legal counsel to draft indemnification language for inclusion in Paragraph 3.18 that would be enforceable in the applicable jurisdiction.

ARTICLE 4 ADMINISTRATION OF THE CONTRACT
4.1   ARCHITECT
If the Architect's construction administration duties vary from those identified in A201-1997, use the following model language to identify the variations.

> 4.1.2   The Architect's duties, responsibilities and limitations of authority are modified as follows: (List or attach as an exhibit.)

4.2   ARCHITECT'S ADMINISTRATION OF THE CONTRACT
Some clients, especially public authorities, may not engage the Architect for contract administration services or only for specified portions of them. In this situation, the entire Paragraph 4.2 should be reviewed carefully and correlated with the provisions of the Agreement between Owner and Architect. The Architect should be especially alert to the possible delegation of the Architect's duties or authority to someone else, and should specify who will assume each function normally assigned to the Architect under this paragraph. Other provisions of the General Conditions may have to be modified as well.

4.2.1 AIA Document B141-1997 addresses instances when the Architect makes site visits as a result of Contractor actions. The following language may be added to the Supplementary Conditions for consistency between Subparagraph 4.2.2 of A201-1997 and Paragraph 2.8.1 of B141-1997.

> Add Clause 4.2.2.1 to Subparagraph 4.2.1:
> 4.2.2.1 The Contractor shall reimburse the Owner for compensation paid to the Architect for additional site visits made necessary by the fault, neglect or request of the Contractor.

4.2.10 This modification advises the Contractor in advance that a Project Representative will be employed for the Project. In addition, a copy of AIA Document B352, which enumerates the duties, responsibilities and limitations of authority of the Project Representative, should be bound into the Project Manual or furnished separately to the Contractor.

> Delete Subparagraph 4.2.10 and substitute the following:

8

4.2.10 A Project Representative will be employed at the site by the Architect. The Project Representative's duties, responsibilities and limitations of authority are as set forth in AIA Document B352, Duties, Responsibilities and Limitations of Authority of the Project Representative, a copy of which is bound in this Project Manual.

4.3.7 CLAIMS FOR ADDITIONAL TIME

On projects where time is especially critical, or where delays are especially likely to occur, the Owner may require added protection in this area. In the language suggested below, Clause 4.3.7.3 strengthens the documentation requirements for Claims for additional time, and Clause 4.3.7.4 requires the Contractor to demonstrate that the delay was on the critical path. It is advisable to further describe the scheduling, documentation and submittal timing required of the Contractor in Division 1 of the Specifications.

Add the following Clauses 4.3.7.3 and 4.3.7.4 to Subparagraph 4.3.7:

4.3.7.3 Claims for increase in the Contract Time shall set forth in detail the circumstances that form the basis for the Claim, the date upon which each cause of delay began to affect the progress of the Work, the date upon which each cause of delay ceased to affect the progress of the Work and the number of days' increase in the Contract Time claimed as a consequence of each such cause of delay. The Contractor shall provide such supporting documentation as the Owner may require including, where appropriate, a revised construction schedule indicating all the activities affected by the circumstances forming the basis of the Claim.

4.3.7.4 The Contractor shall not be entitled to a separate increase in the Contract Time for each one of the number of causes of delay which may have concurrent or interrelated effects on the progress of the Work, or for concurrent delays due to the fault of the Contractor.

4.3.10 CLAIMS FOR CONSEQUENTIAL DAMAGES

Under this subparagraph, the Owner and Contractor waive consequential damages arising out of the Contract for Construction. Generally, a rule of law known as the economic loss doctrine would bar independent tort claims relating to the Contract. In some states, however, the economic loss doctrine has been weakened or discarded; in that situation the Architect (against whom the Contractor does not waive consequential damages) would be exposed to tort claims by the Contractor for such damages. Where the law of such a state applies, the following language (and, of course, compliance with the stated conditions) is recommended.

4.3.10 Add the following sentence to Subparagraph 4.3.10:

If, before expiration of 30 days from the date of execution for this Agreement, the Owner obtains by separate agreement and furnishes to the Contractor a similar mutual waiver of all claims from the Architect against the Contractor for consequential damages which the Architect may incur as a result of any act or omission of the Owner or Contractor, then the waiver of consequential damages by the Owner and Contractor contained in this Subparagraph 4.3.10 shall be applicable to claims by the Contractor against the Architect.

4.6 ARBITRATION

The General Conditions do not require an arbitration to be held in any particular jurisdiction. If it is desired to require that the demand for arbitration be filed with a specific office of the American Arbitration Association and that the arbitration be held in a particular place, unless otherwise mutually agreed, this requirement should be stated in the Supplementary Conditions. These provisions should be reviewed by the Owner's legal counsel, in view of the variance of the rules with respect to such requirements from one jurisdiction to another.

©1998 AIA®
AIA DOCUMENT A511-1998
GUIDE FOR SUPPLEMENTARY CONDITIONS

9

The American Institute
of Architects
1735 New York Avenue, N.W.
Washington, D.C. 20006-5292

4.6.1 On some projects, the parties may wish to place a dollar limit on Claims subject to arbitration. The rationale for doing this is to make the procedural safeguards of the legal system available for Claims exceeding that specified amount. Possible drawbacks are the costs and delays involved in litigation.

Delete the period at the end of the first sentence of Subparagraph 4.6.1 and add:
, provided such Claim involves an amount less than or equal to _____ Dollars ($_____).

ARTICLE 5 SUBCONTRACTORS

5.2 AWARD OF SUBCONTRACTS AND OTHER CONTRACTS FOR PORTIONS OF THE WORK

5.2.1 If the principal Subcontractors are to be identified and selected prior to execution or award of the Contract, this should be set forth in the bidding requirements (e.g., AIA Document A701-1997, Instructions to Bidders). If this procedure is followed, it will be necessary to modify Paragraph 5.2 to conform to the stipulations in the bidding requirements. This should be done by a supplement to this paragraph.

If the Owner wishes to take sub-bids on certain parts of the Work or to require the Contractor to employ certain Subcontractors or material suppliers chosen by the Owner, this should be explained in detail in the Instructions to Bidders.

5.2.1.1 If the Owner wishes to review proposed manufacturers or fabricators, Clause 5.2.1.1 should be included in the Supplementary Conditions. It is recommended that not more than 60 days be allowed; shorter times may be practicable on smaller projects.

5.2.1.1 Not later than _____ days after the date of commencement of the Work, the Contractor shall furnish in writing to the Owner through the Architect the names of persons or entities proposed as manufacturers, fabricators or material suppliers for the products, equipment and systems identified in the General Requirements (Division 1 of the Specifications) and, where applicable, the name of the installing Subcontractor.

ARTICLE 6 CONSTRUCTION BY OWNER OR BY SEPARATE CONTRACTORS

6.1 OWNER'S RIGHT TO PERFORM CONSTRUCTION AND TO AWARD SEPARATE CONTRACTS

6.1.3 If separate contracts are to be awarded, or if the Owner's forces are to perform construction or operations related to the Project, Subparagraph 6.1.3 of the General Conditions requires that the Owner coordinate this construction with the Work of the Contractor. The details of this coordination should be set forth in the General Requirements (Division 1 of the Specifications), including the enumeration of those portions of the Work to be provided under this Article, and identification of separate contractors, when known.

ARTICLE 7 CHANGES IN THE WORK

The modification of Paragraph 7.1 applies only to contracts where the basis of payment is a stipulated sum, such as AIA Document A101-1997. They do not apply to contracts where the basis of payment is the cost of the Work plus a fee, such as AIA Document A111-1997.

7.1.4 Overhead and profit may be stated separately or combined but, in either case, should distinguish between:

©1998 AIA®
AIA DOCUMENT A511-1998
GUIDE FOR SUPPLEMENTARY
CONDITIONS

The American Institute
of Architects
1735 New York Avenue, N.W.
Washington, D.C. 20006-5292

10

a) The amounts to be paid to the Contractor for Work performed by the Contractor with that Contractor's own forces and for materials purchased directly by the Contractor (not through a Subcontractor).

b) The amounts to be paid to the Contractor and Subcontractor for Work performed by the Subcontractor with that Subcontractor's own forces or purchased directly by that Subcontractor (not through a Sub-subcontractor).

c) The amounts to be paid to the Contractor, Subcontractor and Sub-subcontractor for Work performed by the Sub-subcontractor with that Sub-subcontractor's own forces and for material and labor purchased by that Sub-subcontractor.

CAUTION: On some projects it may be desirable to add more specific information concerning items to be considered as part of "cost" as opposed to "overhead." Items that might be defined as one or the other may include costs for preparing Shop Drawings, reserves for future service liability, engineering and estimating costs, added costs for bonds and insurance, and travel and transportation expenses.

Add the following Subparagraph 7.1.4 to Paragraph 7.1:

7.1.4   The combined overhead and profit included in the total cost to the Owner of a change in the Work shall be based on the following schedule:

.1   For the Contractor, for Work performed by the Contractor's own forces, _____ percent of the cost.

.2   For the Contractor, for Work performed by the Contractor's Subcontractors, _____ percent of the amount due the Subcontractors.

.3   For each Subcontractor involved, for Work performed by that Subcontractor's own forces, _____ percent of the cost.

.4   For each Subcontractor involved, for Work performed by the Subcontractor's Sub-subcontractors, _____ percent of the amount due the Sub-subcontractor.

.5   Cost to which overhead and profit is to be applied shall be determined in accordance with Subparagraph 7.3.6.

.6   In order to facilitate checking of quotations for extras or credits, all proposals, except those so minor that their propriety can be seen by inspection, shall be accompanied by a complete itemization of costs including labor, materials and Subcontracts. Labor and materials shall be itemized in the manner prescribed above. Where major cost items are Subcontracts, they shall be itemized also. In no case will a change involving over $_____ be approved without such itemization.

7.3.6   If the amounts to be paid for overhead and profit on changes are to be established in the Contract Documents rather than being negotiated at the time of the changes, the figures to be used should be stated in the Supplementary Conditions.

ARTICLE 8 TIME
8.1   DEFINITIONS
8.1.4   This modification is necessary if there is a requirement or preference to measure time related to the Contract in actual working days rather than calendar days.

© 1998   A I A ®
AIA DOCUMENT A511-1998
GUIDE FOR SUPPLEMENTARY
CONDITIONS

11

The American Institute
of Architects
1735 New York Avenue, N.W.
Washington, D.C. 20006-5292

Delete Subparagraph 8.1.4 and substitute the following:
8.1.4 The term "day" as used in the Contract Documents shall mean working day, excluding weekends and legal holidays.

Occasionally an Owner will want no Work to be performed on certain days when Work might normally be carried on (i.e., special religious holidays); it would be appropriate to list these in a supplement to this subparagraph.

ARTICLE 9 PAYMENTS AND COMPLETION
9.2 SCHEDULE OF VALUES
9.2.1 Requirements concerning the format and data required for the schedule of values should be stated in the General Requirements (Division 1 of the Specifications), rather than by inserting language here to modify the General Conditions.

A frequent requirement is that the schedule must be prepared in such a manner that each major item of Work and each subcontracted item of Work is shown as a single line item on AIA Document G703, Certificate for Payment, Continuation Sheet.

9.3 APPLICATIONS FOR PAYMENT
9.3.1 Detailed requirements concerning the format (and notarization, if required) of the Contractor's Application for Payment should be specified in the General Requirements (Division 1 of the Specifications) rather than by inserting language here to modify the General Conditions.

A frequent requirement is the use of AIA Document G702, Application and Certificate for Payment, and G703, Continuation Sheet. Public authorities often have their own forms which must be used. There are unauthorized facsimiles of AIA documents G702 and G703 that the Architect can reject if the following language is used.

9.3.1   Add the following sentence to Subparagraph 9.3.1:
The form of Application for Payment, duly notarized, shall be a current authorized edition of AIA Document G702, Application and Certificate for Payment, supported by a current authorized edition of AIA Document G703, Continuation Sheet.

The Owner may wish to consider the reduction of retained sums. Various methods for this procedure are set out in OPTIONS A and B which follow. OPTION C is used for constant retainage only. When reduction in retainage is provided for in the Supplementary Conditions, the Architect should recommend that the Agreement between the Owner and Contractor include (in Article 5 of AIA Document A101-1997) this supplement by reference rather than adding the language to the Agreement. This will avoid typographical or textual errors that could occur in the transcribing. Since reduction in retainage affects Subcontractors and certain material suppliers as well as the Contractor, the procedure recommended here for including this provision in the Supplementary Conditions is considered preferable to describing the reduction arrangement in the Owner-Contractor Agreement. The terms and conditions of the Agreement are not generally made available to other interested persons, but the Supplementary Conditions should be.

OPTION A
Option A provides for progress payments in full to the Contractor after the Work is 50% complete. This method can have the disadvantage of applying retainage unequally to the Subcontracts, requiring full retainage on Work performed during the early stages of construction while the amount of retainage withheld on Work in the later stage of construction may be reduced or perhaps even eliminated. The net effect of this method is a sliding reduction to 50% of the basic retainage at the time of Substantial Completion. This supplement should be coordinated with Subparagraph 9.8.5.

©1998 AIA®
AIA DOCUMENT A511-1998
GUIDE FOR SUPPLEMENTARY
CONDITIONS

The American Institute
of Architects
1735 New York Avenue, N.W.
Washington, D.C. 20006-5292

12

Add the following Clause 9.3.1.3 to Subparagraph 9.3.1:

9.3.1.3   Until the Work is 50 percent complete, the Owner shall pay _____ percent of the amount due the Contractor on account of progress payments. At the time the Work is 50 percent complete and thereafter, the Architect will authorize remaining partial payments to be paid in full.

OPTION B

Option B provides for line item retainage. This method applies retainage and any reduction thereof equally to all phases of the Work. Thus, early finishing Subcontractors (e.g. foundations, structural steel) can have their retained funds reduced when they have satisfactorily performed 50% of their Subcontracts without waiting for the entire Project to be 50% complete. This supplement should be coordinated with Subparagraph 9.8.5 because that subparagraph requires release of retainage at Substantial Completion.

Add the following Clause 9.3.1.3 to Subparagraph 9.3.1:

9.3.1.3   Until final payment, the Owner shall pay _____ percent of the amount due the Contractor on account of progress payments. For each Work category shown to be 50 percent or more complete in the Application for Payment, the Architect will, without reduction of previous retainage, certify any remaining progress payments for each Work category to be paid in full.

OPTION C

Option C is used if payment to the Contractor will be made with a constant percentage retained until the Date of Substantial Completion. The percentage called for here and that shown in Article 5 of AIA Document A101-1997, the Agreement between the Owner and Contractor, must be identical.

Add the following Clause 9.3.1.3 to Subparagraph 9.3.1:

9.3.1.3   Until Substantial Completion, the Owner shall pay _____ percent of the amount due the Contractor on account of progress payments.

9.3.2   If it is not intended that stored materials and equipment, either on or off the site, will be paid for until incorporated in the Work, this subparagraph needs to be modified appropriately. This should also be reflected in the provisions of Article 5 of AIA Document A101-1997, the Agreement between the Owner and Contractor, which must likewise be modified to omit reference to stored materials. In addition, modifications should also be made to Clause 11.4.1.4.

9.4   CERTIFICATES FOR PAYMENT
9.4.2   If the Agreement between the Owner and Contractor is other than on a stipulated-sum basis (such as cost-plus fee where payments are made based on invoices or vouchers submitted to the Architect), this subparagraph should be qualified to limit the extent and meaning of the Architect's Certificate for Payment.

9.6   PROGRESS PAYMENTS
Placing retained funds in an escrow account that earns interest provides a method of compensating the Contractor for money earned but not made available for the Contractor's use. Several government entities have enacted legislation requiring escrow accounts for retainage on public work, but it can be equally appropriate for private projects. Before using this supplement, the Owner's legal counsel must review it for conformance with local laws. An escrow account can be used with all the various methods of retainage recommended above.

Add the following Subparagraphs 9.6.8 through 9.6.14 to Paragraph 9.6:

9.6.8   Upon commencement of the Work, an escrow account shall be established in a financial institution chosen by the Contractor and approved by the Owner.

© 1998   AIA®
AIA DOCUMENT A511-1998
GUIDE FOR SUPPLEMENTARY
CONDITIONS

13

The American Institute
of Architects
1735 New York Avenue, N.W.
Washington, D.C. 20006-5292

9.6.9   The escrow agreement shall provide that the financial institution will act as escrow agent, will pay interest on funds deposited in such account in accordance with the provisions of the escrow agreement and will disburse funds from the account upon the direction of the Owner as set forth below. Compensation to the escrow agent for establishing and maintaining the escrow account shall be paid from interest accrued in the escrow account.

9.6.10   As each progress payment is made, the retainage with respect to that payment shall be deposited by the Owner in the escrow account.

9.6.11   The interest earned on funds in the account shall accrue for the benefit of the Contractor until the date of Substantial Completion. Interest earned after such date shall accrue for the benefit of the Owner. Cost of compensation to the escrow agent paid out of interest earned shall be borne by the Contractor.

9.6.12   When the Contractor has fulfilled all of the requirements of the Contract providing for reduction of retained funds, the escrow agent shall release to the Contractor one-half of the accrued funds but none of the interest thereon. When the Work has been fully completed in a satisfactory manner and the Architect has issued a final Certificate for Payment, the escrow agent shall pay to the Contractor the full amount of funds remaining in the account, including net balance of the interest paid to the account, but less any interest that may have accrued for the benefit of the Owner, which shall be paid to the Owner.

9.6.13   If, after Substantial Completion of the Work, final completion thereof is materially delayed through no fault of the Contractor, the escrow agent shall make payment to the Contractor as provided in Subparagraph 9.10.3.

9.6.14   Sums owed to the Owner by the Contractor may be deducted from payments otherwise due the Contractor pursuant to Article 9.

9.8   SUBSTANTIAL COMPLETION
9.8.1   If designated portions of the Work are to be accepted separately by the Owner, include appropriate information in the General Requirements (Division 1 of the Specifications).

9.8.3.1   Multiple reinspections can be a serious drain on the Architect's time and other resources. If the Architect is entitled to an adjustment in compensation for such services under the Owner-Architect agreement (for example, under Subparagraph 2.8.1.3 of AIA Document B141-1997), the following language may be appropriate.

Add the following Clause 9.8.3.1 to Subparagraph 9.8.3:
9.8.3.1   Except with the consent of the Owner, the Architect will perform no more than _____ (_____) inspections to determine whether the Work or a designated portion thereof has attained Substantial Completion in accordance with the Contract Documents. The Owner shall be entitled to deduct from the Contract Sum amounts paid to the Architect for any additional inspections.

9.8.5   Subparagraph 9.8.5 of A201-1997 requires release of all retainage on completed Work at Substantial Completion. If it is intended instead that a partial release be made, the following modification may be used to indicate the percentage. It is important to coordinate with Clause 9.3.1.3.

9.8.5   Delete the second sentence and substitute the following:

Upon such acceptance and consent of surety, if any, the Owner shall make payment sufficient to increase the total payments to _____ percent (%) of the Contract Sum, less such amounts as the Architect shall determine for incomplete Work and unsettled claims.

9.10.1.1 Multiple reinspections can be a serious drain on the Architect's time and other resources. If the Architect is entitled to an adjustment in compensation for such services under the Owner-Architect agreement (for example, under Subparagraph 2.8.1.4 of AIA Document B141-1997), the following language may be appropriate.

Add the following Clause 9.10.1.1 to Subparagraph 9.10.1:
9.10.1.1 Except with the consent of the Owner, the Architect will perform no more than _____ (_____) inspections to determine whether the Work or a designated portion thereof has attained Final Completion in accordance with the Contract Documents. The Owner shall be entitled to deduct from the Contract Sum amounts paid to the Architect for any additional inspections.

9.11 LIQUIDATED DAMAGES

CAUTION: Such a provision would normally appear in the Agreement; for example, space is provided in the Agreement between the Owner and Contractor (AIA Document A101-1997) under Article 3, relating to time, for insertion of appropriate terms and conditions related to liquidated damages. However, it is important for Subcontractors and others to be aware of such a provision, and it is not unusual for this requirement to be set out in the Supplementary Conditions.

The language shown here is a suggested guide. It should not be included as Supplementary Conditions without review by the Owner's attorney and concurrence of the Owner. Repetition should be avoided. If the provision is written in the Supplementary Conditions, a cross-reference should appear in Article 3 of the Agreement between the Owner and Contractor.

Care must be taken to avoid even the appearance that a provision is used to extract a penalty rather than for liquidated damages. Liquidated damages are enforceable if the amount inserted into Subparagraph 9.11.1 is a reasonable measure of the anticipated harm. An advantage of liquidated damages is the elimination of the cost entailed to prove the actual damages.

On the other hand, if the amount inserted into Subparagraph 9.11.1 is grossly disproportionate to the anticipated harm or if there is no anticipated harm, the provision may be deemed a penalty and unenforceable by a court. Penalties in contracts are not generally enforceable because of the public policy that the use of a court to inflict punishment is a function of society to discourage criminal and other antisocial behavior and should not inure to the benefit of an individual using a private agreement. The few exceptions to this policy are typically made by statute and grant authority to public entities, such as cities and municipalities, to enforce forfeiture of posted bonds. It is not a good idea to use the term "penalty" in Supplementary Conditions. The model language below assumes substantial completion of the entire Work.

It is necessary to ensure that liquidated damages are not speculative and do not conflict with the mutual waiver of consequential damages in Subparagraph 4.3.10.

Add the following Paragraph 9.11 to Article 9:
9.11   The Contractor and the Contractor's surety, if any, shall be liable for and shall pay the Owner the sums hereinafter stipulated as liquidated damages for each calendar day of delay after the date established for Substantial Completion in the Contract Documents until the Work is substantially complete:

_____ Dollars ($_____)

©1998 AIA®
AIA DOCUMENT A511-1998
GUIDE FOR SUPPLEMENTARY CONDITIONS

15

The American Institute
of Architects
1735 New York Avenue, N.W.
Washington, D.C. 20006-5292

9.12   BONUS

This is an example of a bonus provision which may be counterbalanced with a liquidated damages provision such as that shown in Paragraph 9.11. Often such a provision is erroneously referred to as a "penalty bonus" provision. To overcome the public policy objection against penalties in contracts, some believe that a bonus counterpoint will cause a court to look more favorably on a penalty. The liquidated damages provision becomes a penalty when an arbitrarily high amount is inserted into the provision to add pressure on the Contractor to complete the Work within the Contract Time. There is little or no legal precedent to support this proposition of linking a bonus with a penalty.

It is not a recommended practice to employ such a clause without specific advice from local legal counsel.

Bonus provisions should be used only when the Owner will obtain a specific benefit if the Contractor completes the construction prior to the time set for Substantial Completion. On occasion the Owner may not desire early completion because of the timing requirements of other commitments, such as mortgage closings or the commencement of tenant leases. The model language below assumes substantial completion of the entire Work.

> Add the following Paragraph 9.12 to Article 9:
> 9.12   Bonus
>
> 9.12.1   The Owner shall pay as a bonus to the Contractor a sum of _____ Dollars ($_____) for each calendar day preceding the date established for Substantial Completion in the Contract Documents that the Work is determined to be substantially complete by the Architect.

ARTICLE 10 PROTECTION OF PERSONS AND PROPERTY
10.2   SAFETY OF PERSONS AND PROPERTY
10.2.4   It is usually desirable that the Owner and Contractor inform each other of known potential hazards on the site. The Owner and Contractor may be held liable to third parties who are harmed by them, and may therefore wish to take precautions against unauthorized access.

> 10.2.4.1   When use or storage of explosives, or other hazardous materials, substances or equipment, or unusual methods are necessary for execution of the Work, the Contractor shall give the Owner reasonable advance notice.

The Contract Documents may require the Contractor to handle materials that under certain circumstances may be designated as hazardous.

> 10.2.4.1   If the Contract Documents require the Contractor to handle materials or substances that under certain circumstances may be designated as hazardous, the Contractor shall handle such materials in an appropriate manner.

10.3   For renovations, remodeling, or work on an existing site, disclosure language may be needed that requires the Owner to notify the Contractor of possible hazardous materials, such as lead or fiberglass, on the site. In addition, supplementary language may be added requiring the Contractor to comply with all applicable statutes in working with such materials. Coordinate with Article 11 regarding insurance for special hazards or pollutants.

Materials that the Contractor alleges to be hazardous should be handled in an appropriate manner in accordance with the model language provided for Subparagraph 10.2.4.

ARTICLE 11 INSURANCE AND BONDS

©1998 AIA®
AIA DOCUMENT A511-1998
GUIDE FOR SUPPLEMENTARY
CONDITIONS

The American Institute
of Architects
1735 New York Avenue, N.W.
Washington, D.C. 20006-5292

16

CAUTION: THE ARCHITECT IS NOT QUALIFIED AS AN INSURANCE COUNSELOR, AND ARCHITECTURAL PROFESSIONAL LIABILITY INSURANCE MAY NOT COVER PROVIDING INSURANCE ADVICE. THE ARCHITECT SHOULD NOT MAKE RECOMMENDATIONS ABOUT INSURANCE OR APPROVE INSURANCE CERTIFICATES OR POLICIES. It is in the best interests of all parties that insurance matters be placed in the hands of the Owner's insurance counselor. See the AIA Architect's Handbook and the AIA Construction Bonds and Insurance Guide for additional information regarding the insurance protection required under Article 11.

The Owner's insurance counselor must review the Contractor's submittals regarding insurance to determine that the required coverages are in place.

11.1 CONTRACTOR'S LIABILITY INSURANCE

11.1.1.1 In some states, some business entities may not be required by statute to carry worker's compensation insurance. Such exempted employers, however, can be required by the Contract Documents to maintain voluntary compensation coverage. The Owner's insurance advisor should determine whether or not this coverage should be a contract requirement. In most states, an exempted employer, by maintaining such voluntary coverage, is entitled to indemnity from normal tort liability and is not subject to other tort liability to employees for job-related injuries.

In addition to each state having applicable workers' compensation laws, federal and foreign laws may apply to the Contractor's or Subcontractor's employees. Where the Work includes construction involving the following categories, specific coverage may be required: maritime work, longshoremen, harbor work, work at or outside U.S. boundaries, and benefits required by labor union contracts.

11.1.1.1 Delete the semicolon at the end of Clause 11.1.1.1 and add:
, including private entities performing Work at the site and exempt from the coverage on account of number of employees or occupation, which entities shall maintain voluntary compensation coverage at the same limits specified for mandatory coverage for the duration of the Project;

11.1.1.2 This requires Employers' Liability Coverage, which is normally afforded as separate coverage under the workers' compensation policy, but evidence of such coverage should be shown on the certificate of insurance. If the modification to Clause 11.1.1.1 shown above is used, this modification to Clause 11.11.2 must be used as well.

11.1.1.2 Delete the semicolon at the end of Clause 11.1.1.2 and add:
or persons or entities excluded by statute from the requirements of Clause 11.1.1.1 but required by the Contract Documents to provide the insurance required by that clause;

11.1.1.4 There is a difference between the bodily injury coverage required in Clause 11.1.1.3 and the personal injury coverage required by this clause. Bodily injury is, as its name implies, physical harm to a person, including death, while personal injury includes libel, slander, false arrest and similar wrongs. Both bodily and personal injury coverages are required; hence the need for a careful review of the original insurance certificates by the Owner's insurance counselor.

11.1.1.6 Business Auto Liability Insurance is normally issued as a separate policy. It is generally advisable to have this policy and the Commercial General Liability policy written by the same insurance company to avoid disputes as to which insurer is responsible for a particular loss.

11.1.1.7 Products and Completed Operations insurance is specified to cover claims arising out of or resulting from the Contractor's operations when the injury or damage occurs after the

17

Contractor's Work at the site has been completed, the Project has been put to its intended use and the Contractor is no longer at the site.

11.1.1.8   In some jurisdictions statutory requirements may modify the indemnification clause of Paragraph 3.18 or void it completely. The Owner should seek the advice of legal counsel for modifications to Paragraph 3.18 or this Clause 11.1.1.8.

11.1.1.9   Some projects or jurisdictions may require special types of coverages. The Owner should seek the advice of insurance counsel for the nature of coverage required. The coverages listed below are common on construction projects:

1. Premises-Operations
2. Independent Contractors' Protective
3. Products-Completed Operations
4. Personal Injury Liability
5. Contractual Liability
6. Personal and Advertising Injury
7. Owned, Non-Owned and Hired Motor Vehicles
8. Excess or Umbrella Liability

11.1.2   The following supplements represent sample modifications. It is recommended that the Architect use AIA Document G612, Owner's Instructions Regarding the Construction Contract, Insurance and Bonds, and Bidding Procedures, requesting the Owner to furnish to the Architect such information as well as forms of coverage required when this information needs to be included in the Contract Documents. See the AIA Construction Bonds and Insurance Guide for a more detailed discussion of the coverages and amounts of insurance involved. While location, size and potential exposure have bearing on the limits of coverage for each project, it must be remembered that serious injury or loss of life may result in the same amount of damages no matter what the size, cost or location of the project. THE OWNER, NOT THE ARCHITECT, MUST ESTABLISH THE AMOUNTS AND TIME LIMITS OF INSURANCE REQUIRED BY THIS ARTICLE 11.

Explanatory material about Workers' Compensation is provided in Clauses 11.1.1.1 and 11.1.1.2 above.

A period of time should be stated for Products and Completed Operations insurance, commencing with issuance of the final Certificate for Payment, during which this insurance will be kept in force. The procedure for ascertaining continuation of this coverage is set out in Subparagraph 9.10.2.

The Commercial General Liability (CGL) policy combines several coverage aggregates into a single General Aggregate; this is the maximum amount that will be paid under the policy. The General Aggregate may be modified to apply to an individual project, and this endorsement should be called for as shown in the suggested language. Note that the "per project" limit of liability called for in Subclause 11.1.2.2.2 requires an endorsement amending the standard CGL policy. In some circumstances, this may be difficult to obtain.

If Umbrella or Excess Liability insurance coverage is required over the primary insurance, insert the coverage limits. Commercial General Liability and Automobile Liability limits may be attained by individual policies or by a combination of primary policies and Umbrella or Excess Liability policies.

©1998 AIA®
AIA DOCUMENT A511-1998
GUIDE FOR SUPPLEMENTARY
CONDITIONS

The American Institute
of Architects
1735 New York Avenue, N.W.
Washington, D.C. 20006-5292

18

Add the following Clauses 11.1.2.1 through 11.1.2.4 to Subparagraph 11.1.2:

11.1.2.1 The limits for Worker's Compensation and Employers' Liability insurance shall meet statutory limits mandated by State and Federal Laws. If (1) limits in excess of those required by statute are to be provided or (2) the employer is not statutorily bound to obtain such insurance coverage or (3) additional coverages are required, additional coverages and limits for such insurance shall be as follows:

11.1.2.2 The limits for Commercial General Liability insurance including coverage for Premises-Operations, Independent Contractors' Protective, Products-Completed Operations, Contractual Liability, Personal Injury and Broad Form Property Damage (including coverage for Explosion, Collapse and Underground hazards) shall be as follows:

$_____ Each Occurrence
$_____ General Aggregate
$_____ Personal and Advertising Injury
$_____ Products-Completed Operations Aggregate

.1 The policy shall be endorsed to have the General Aggregate apply to this Project only.

.2 The Contractual Liability insurance shall include coverage sufficient to meet the obligations in AIA Document A201-1997 under Paragraph 3.18.

.3 Products and Completed Operations insurance shall be maintained for a minimum period of at least   (   ) year(s) after either 90 days following Substantial Completion or final payment, whichever is earlier.

11.1.2.3 Automobile Liability insurance (owned, non-owned and hired vehicles) for bodily injury and property damage:
$_____ Each Accident

11.1.2.4 Umbrella or Excess Liability coverage:

11.1.3 If a Commercial General Liability form is used for this insurance, ACORD form 25-S is written specifically to list required coverages under those policies.

11.1.3   Add the following sentence to Subparagraph 11.1.3:
If this insurance is written on a Commercial General Liability policy form, the certificates shall be ACORD form 25-S, completed and supplemented in accordance with AIA Document G715, Instruction Sheet and Supplemental Attachment for ACORD Certificate of Insurance 25-S.

11.3   PROJECT MANAGEMENT PROTECTIVE LIABILITY INSURANCE
If the Contractor will be required to carry Project Management Protective Liability insurance, this paragraph must be modified since, as written, this insurance is optional. The Owner should first confirm that this coverage is available in the jurisdiction of the Project.

11.3.1   Delete Subparagraph 11.3.1 and substitute the following:
11.3.1 The Contractor shall purchase and maintain Project Management Protective Liability insurance, from the Contractor's usual sources, as primary coverage for the Owner's, Contractor's and Architect's vicarious liability for construction operations under the Contract. The Contractor shall not be responsible for purchasing any other liability insurance on behalf of the Owner. The minimum limits of liability for such coverage shall be equal to the aggregate limits required for Contractor's liability insurance under Clauses 11.1.1.2 through 11.1.1.5.

© 1998 AIA®
AIA DOCUMENT A511-1998
GUIDE FOR SUPPLEMENTARY
CONDITIONS

19   The American Institute
of Architects
1735 New York Avenue, N.W.
Washington, D.C. 20006-5292

11.4   PROPERTY INSURANCE

During preparation of the Supplementary Conditions the Owner's insurance counselor should carefully compare the terms of Paragraph 11.4 with the actual policy form to be purchased to make certain all requirements for property insurance are, or can be, met, and certainly before any major changes in these requirements are proposed. If modifications are desirable, all other subparagraphs and clauses of Paragraph 11.4, in addition to the one being changed, should be reviewed and conflicting requirements reconciled. Every effort should be made to obtain the broadest protection possible.

This paragraph, as it appears in the General Conditions, requires the Owner to provide the property insurance on an "all-risk" or equivalent policy form. "All-risk" is the term commonly used to refer to a policy that includes coverage for earthquake, flood and windstorm, testing and start-up may be included in the policy or by separate endorsement. The actual policy form is now called open perils.

The supplements suggested in this Guide are in two parts: OPTION A, modifications to Paragraph 11.4 as written; and OPTION B, modifications to Paragraph 11.4 when the Contractor is required to furnish this coverage.

When considering the use of Option B, it is important to consider that, by the terms of Subparagraph 9.3.3 of the General Conditions, Work completed and paid for is the property of the Owner and insurance proceeds for such property damaged or destroyed by a covered loss are rightfully the Owner's. Special consideration should be given to property insurance requirements when the Project involves additions or alterations to existing buildings.

Other exclusions under the property insurance for which waivers may be equally desirable are those that relate to claims based upon the Architect's professional acts or omissions or to claims based upon the Contractor's faulty workmanship or materials. Such exclusions are typically standard in most property insurance policies.

The amount for property insurance coverage is established in this paragraph as being equal to the Contract Sum, plus the value of subsequent modifications and cost of materials supplied or installed by others, on a replacement cost basis. There may be occasions when an Owner is advised to carry less insurance, in which instance the Contractor must be so advised in the Supplementary Conditions so that the Contractor can include the cost of any additional necessary coverage in the Contract Sum.

As previously noted, A201-1997 requires insurance for physical loss or damage on an "all-risk" or equivalent policy form. Since these policies vary in their exclusions, it is important for the Owner's insurance counselor to carefully review the policy that may be used to be certain it provides the desired protection. If any of the excluded risks are to be included in the coverage by endorsement, they should be specifically noted here so that the Contractor and Subcontractors will be aware of the extent of the coverage.

An example of when removal of an excluded risk might be necessary is in regard to theft coverage. Most "all-risk" policy forms include coverage for theft of materials and equipment (excluding the Contractor's own equipment) stored on the site or in transit. However, some policies may restrict the theft protection to property that is an integral part of a building or structure. In this case, the Owner normally can obtain expanded coverage to provide the necessary protection for materials and equipment stored on the site but not yet incorporated in the Work. If the Owner does not obtain this coverage, then the Contractor should be required to obtain an installation floater to make certain the necessary insurance is in effect. If the Project is

©1998 AIA®
AIA DOCUMENT A511-1998
GUIDE FOR SUPPLEMENTARY
CONDITIONS

The American Institute
of Architects
1735 New York Avenue, N.W.
Washington, D.C. 20006-5292

20

located in a high crime area, it may not be possible, or financially feasible, to acquire theft insurance.

Certain policy forms prohibit any waiver of the insured's rights. When this is the case, it is necessary to provide a further endorsement to acknowledge the contractual provision for waiver of subrogation (11.4.7) before any loss occurs.

OPTION A
(When the Owner carries property insurance as required by AIA Document A201-1997)

11.4.1.3 Most property or fire insurance policies are written with a deductible. This clause describes how deductibles are to be handled. It is necessary to show amount of deductible per occurrence unless no deductible is, or will be, established in the policy, in which case the added sentence would read, "This property insurance shall be written with no deductibles." There is no need to identify the voluntary deductibles described in this Clause 11.4.1.3 since they will be paid by the Owner in the event of an insured occurrence.

> 11.4.1.3 Add the following sentence to Clause 11.4.1.3:
> This property insurance is written with a deductible of $ _____ per occurrence with a deductible aggregate of $ _____ .

11.4.1.4 If the Owner does not intend to secure coverage for off-site storage or materials in transit, the Contractor must be advised so that this coverage can be obtained. See also Subparagraph 9.3.2.

> Delete Clause 11.4.1.4 and substitute the following:
> 11.4.1.4 The Contractor shall at the Contractor's own expense provide insurance coverage for materials stored off the site after written approval of the Owner at the value established in the approval, and also for portions of the Work in transit until such materials are permanently attached to the Work.

The property insurance is to cover the entire Work, which is defined under Subparagraph 1.1.3 as including "all other labor, materials, equipment and services provided or to be provided by the Contractor to fulfill the Contractor's obligations." The following provision clarifies the Contractor's responsibility to provide insurance coverage for the Contractor's machinery, tools and equipment that remain the property of the Contractor upon completion of the Project.

> Add the following Clause 11.4.1.6 to Subparagraph 11.4.1:
> 11.4.1.6 The insurance required by Paragraph 11.4 is not intended to cover machinery, tools or equipment owned or rented by the Contractor that are utilized in the performance of the Work but not incorporated into the permanent improvements. The Contractor shall, at the Contractor's own expense, provide insurance coverage for owned or rented machinery, tools or equipment, which shall be subject to the provisions of Subparagraph 11.4.7.

OPTION B
(When the Contractor is required to carry the property insurance, the following modifications to Paragraph 11.4 should be included.)

Because deletion of Paragraph 11.4 in its entirety and substitution of a simplified single paragraph requiring the Contractor to carry this insurance (a frequent method used to effect this

©1998 AIA®
AIA DOCUMENT A511-1998
GUIDE FOR SUPPLEMENTARY
CONDITIONS

21

The American Institute
of Architects
1735 New York Avenue, N.W.
Washington, D.C. 20006-5292

change) results in omission of important statements, it is recommended that Paragraph 11.4 be modified only where appropriate, with the balance of the wording left intact. The following discussion of the subparagraphs under Paragraph 11.4 is offered as a guide for use under those circumstances.

Separate contractors other than the designated Contractor should be reminded of the necessity for Inland Marine coverage in the event of the need to insure their materials and equipment in transit. If a Project includes two or more buildings, each under separate contract, and the Owner does not intend to provide a single policy covering all buildings, each separate contractor should be required to carry the property insurance covering that contractor's Work on the Project. Possible areas of duplication of coverage should be assigned to one specific separate contractor.

The Owner's insurance counselor should carefully evaluate the possible advantages or disadvantages if separate insurance is being provided by more than one contractor under the above circumstances.

Under certain circumstances, it is possible that the Contractor may be unable to obtain "all-risk" insurance, or would propose providing named-perils property insurance . In this situation, the Owner's insurance counselor would be well-advised to have the Contractor consider supplementing the named-perils property insurance policy with a Difference In Conditions contract (D.I.C.) to make the named-perils policy coverage consistent with the "all-risk" requirement.

If the Owner does not intend to carry the property insurance, the Supplementary Conditions should designate which Contractor shall purchase this coverage.

> 11.4.1 Modify the first sentence of Subparagraph 11.4.1 as follows: Delete "Unless otherwise provided, the Owner" and substitute "The Contractor." Add the following sentences:

> If the Owner is damaged by the failure of the Contractor to purchase and maintain such insurance without so notifying the Owner in writing, then the Contractor shall bear all reasonable costs attributable thereto.

> 11.4.1.2 Delete Clause 11.4.1.2.

Since the Owner pays the deductible when responsible for property insurance, it may be advisable to modify Clause 11.4.1.3 when the Contractor is responsible for purchasing and maintaining the property insurance.

> 11.4.1.3 Modify Clause 11.4.1.3 by substituting "Contractor" for "Owner."

11.4.4 Since it relates to the Contractor's own insurance requirements, this subparagraph can be omitted.

> 11.4.4 Delete Subparagraph 11.4.4.

> 11.4.6 Modify Subparagraph 11.4.6 by making the following substitutions: (1) in the first sentence, substitute "Contractor" for "Owner" and "Owner" for "Contractor," and (2) substitute "Owner" for "Contractor" at the end of the last sentence.

> 11.4.7 Modify Subparagraph 11.4.7 by substituting "Contractor" for "Owner" at the end of the first sentence.

©1998 AIA®
AIA DOCUMENT A511-1998
GUIDE FOR SUPPLEMENTARY
CONDITIONS

The American Institute
of Architects
1735 New York Avenue, N.W.
Washington, D.C. 20006-5292

22

11.4.8   Modify Subparagraph 11.4.8 by substituting "Contractor" for "Owner" ; except that at the first reference to "Owner" in the first sentence, the word "this" should be substituted for "Owner's."

11.4.9   Modify Subparagraph 11.4.9 by substituting "Contractor" for "Owner" each time the latter word appears except in the last sentence.

11.4.10   Modify Subparagraph 11.4.10 by substituting "Contractor" for "Owner" each time the latter word appears.

11.5   PERFORMANCE BOND AND PAYMENT BOND
The requirements for a Performance Bond and Payment Bond should be noted in the Instructions to Bidders. (See AIA Document A701-1997, Instructions to Bidders.)

The amount of the bonds should be determined by the Owner's legal counsel.

Delete Subparagraph 11.5.1 and substitute the following:
11.5.1   The Contractor shall furnish bonds covering faithful performance of the Contract and payment of obligations arising thereunder. Bonds may be obtained through the Contractor's usual source and the cost thereof shall be included in the Contract Sum. The amount of each bond shall be equal to _____ percent of the Contract Sum.

11.5.1.1 The Contractor shall deliver the required bonds to the Owner not later than three days following the date the Agreement is entered into, or if the Work is to be commenced prior thereto in response to a letter of intent, the Contractor shall, prior to the commencement of the Work, submit evidence satisfactory to the Owner that such bonds will be furnished.

11.5.1.2 The Contractor shall require the attorney-in-fact who executes the required bonds on behalf of the surety to affix thereto a certified and current copy of the power of attorney.

ARTICLE 12 UNCOVERING AND CORRECTION OF WORK
12.2   CORRECTION OF WORK

12.2.2 Note that the time limit of one year within which the Contractor is obliged to correct the Work may be modified by special warranties required by the Contract Documents. This one-year time limit should not be construed as a limitation of the Contractor's warranty under Subparagraph 3.5.1.

Paragraph 2.7.2 of AIA Document B141-1997 provides for a meeting to be held with the Owner, Owner's Designated Representative and the Architect prior to the expiration of one year from the date of Substantial Completion to review facility operations and performance and to make appropriate recommendations. It may be desirable to require the Contractor to attend this meeting, as the recommendations from this meeting may form the basis for the written notice required by Clause 12.2.2.1 of A201-1997 of Work that is not in accordance with the Contract Documents.

Add the following Clause 12.2.2.4 to Subparagraph 12.2.2:
12.2.2.4 Upon request by the Owner and prior to the expiration of one year from the date of Substantial Completion, the Architect will conduct and the Contractor shall attend a meeting with the Owner to review the facility operations and performance.

ARTICLE 13 MISCELLANEOUS PROVISIONS
13.5   TESTS AND INSPECTIONS
13.5.1 If the Owner is to employ the testing agency directly and pay for any inspections, tests or approvals other than those conducted by public authorities, this should be stated in the General Requirements (Division 1 of the Specifications) or elsewhere in the Specifications rather than by inserting language here to modify the General Conditions.

© 1998 AIA®
AIA DOCUMENT A511-1998
GUIDE FOR SUPPLEMENTARY CONDITIONS

23

The American Institute
of Architects
1735 New York Avenue, N.W.
Washington, D.C. 20006-5292

13.6 INTEREST
13.6.1 Usury laws and requirements under the Federal Truth in Lending Act, similar consumer credit laws at the Owner's and Contractor's principal places of business, the location of the Project and elsewhere may affect the validity of this provision. Legal advice should be obtained with respect to deletion, modification, or other requirements such as written disclosures or waivers.

13.8 The Architect must be alert to provisions of local non-discrimination and affirmative action statutes in force at the Project location. If a supplementary condition is required, it should be added as Paragraph 13.8.

ARTICLE 14 TERMINATION OR SUSPENSION OF THE CONTRACT
14.4 Many Owners reserve to themselves the right to terminate the Contract for convenience—that is, without cause. Termination for convenience is provided for in A201-1997, but not in all AIA General Conditions.

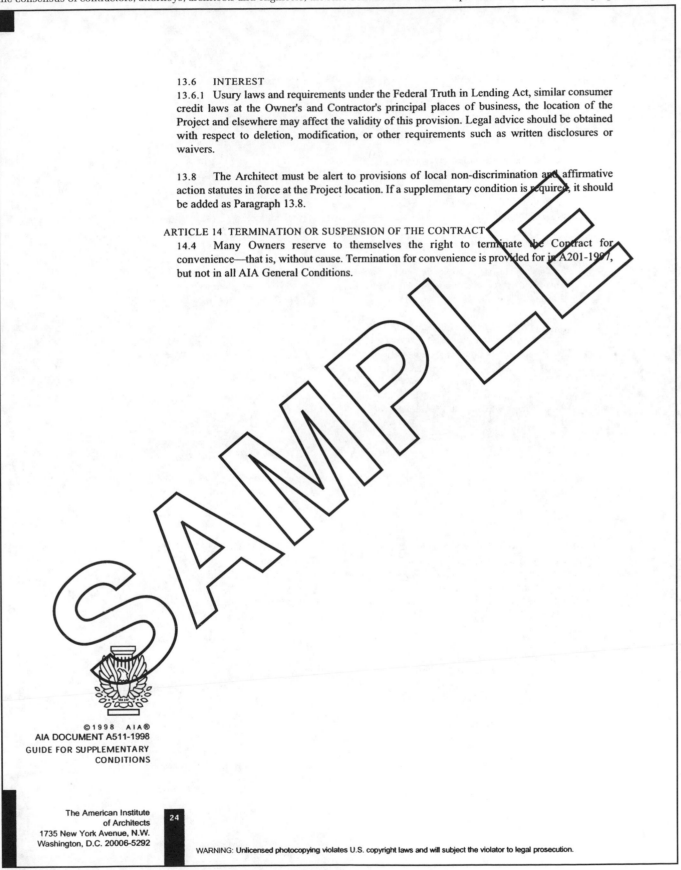

The American Institute
of Architects
1735 New York Avenue, N.W.
Washington, D.C. 20006-5292

24

Appendix F

Sample EJCDC Documents*

STANDARD
GENERAL CONDITIONS
OF THE
CONSTRUCTION CONTRACT

Prepared by

ENGINEERS JOINT CONTRACT DOCUMENTS COMMITTEE

and

Issued and Published Jointly By

PROFESSIONAL ENGINEERS IN PRIVATE PRACTICE
a practice division of the
NATIONAL SOCIETY OF PROFESSIONAL ENGINEERS

AMERICAN COUNCIL OF ENGINEERING COMPANIES

AMERICAN SOCIETY OF CIVIL ENGINEERS

This document has been approved and endorsed by

The Associated General Contractors of America

Knowledge for Creating
and Sustaining
the Built Environment

Construction Specifications Institute

Copyright ©2002

National Society of Professional Engineers
1420 King Street, Alexandria, VA 22314

American Council of Engineering Companies
1015 15th Street, N.W., Washington, DC 20005

American Society of Civil Engineers
1801 Alexander Bell Drive, Reston, VA 20191-4400

These General Conditions have been prepared for use with the Suggested Forms of Agreement Between Owner and Contractor Nos. C-520 or C-525 (2002 Editions). Their provisions are interrelated and a change in one may necessitate a change in the other. Comments concerning their usage are contained in the EJCDC Construction Documents, General and Instructions (No. C-001) (2002 Edition). For guidance in the preparation of Supplementary Conditions, see Guide to the Preparation of Supplementary Conditions (No. C-800) (2002 Edition).

TABLE OF CONTENTS

GENERAL CONDITIONS

ARTICLE 1 - DEFINITIONS AND TERMINOLOGY

1.01   *Defined Terms*

A. Wherever used in the Bidding Requirements or Contract Documents and printed with initial capital letters, the terms listed below will have the meanings indicated which are applicable to both the singular and plural thereof. In addition to terms specifically defined, terms with initial capital letters in the Contract Documents include references to identified articles and paragraphs, and the titles of other documents or forms.

1. *Addenda*--Written or graphic instruments issued prior to the opening of Bids which clarify, correct, or change the Bidding Requirements or the proposed Contract Documents.

2. *Agreement*--The written instrument which is evidence of the agreement between Owner and Contractor covering the Work.

3. *Application for Payment*--The form acceptable to Engineer which is to be used by Contractor during the course of the Work in requesting progress or final payments and which is to be accompanied by such supporting documentation as is required by the Contract Documents.

4. *Asbestos*--Any material that contains more than one percent asbestos and is friable or is releasing asbestos fibers into the air above current action levels established by the United States Occupational Safety and Health Administration.

5. *Bid*--The offer or proposal of a Bidder submitted on the prescribed form setting forth the prices for the Work to be performed.

6. *Bidder*--The individual or entity who submits a Bid directly to Owner.

7. *Bidding Documents*--The Bidding Requirements and the proposed Contract Documents (including all Addenda).

8. *Bidding Requirements*--The Advertisement or Invitation to Bid, Instructions to Bidders, bid security of acceptable form, if any, and the Bid Form with any supplements.

9. *Change Order*--A document recommended by Engineer which is signed by Contractor and Owner and authorizes an addition, deletion, or revision in the Work or an adjustment in the Contract Price or the Contract Times, issued on or after the Effective Date of the Agreement.

10. *Claim*--A demand or assertion by Owner or Contractor seeking an adjustment of Contract Price or Contract Times, or both, or other relief with respect to the terms of the Contract. A demand for money or services by a third party is not a Claim.

11. *Contract*--The entire and integrated written agreement between the Owner and Contractor concerning the Work. The Contract supersedes prior negotiations, representations, or agreements, whether written or oral.

12. *Contract Documents*--Those items so designated in the Agreement. Only printed or hard copies of the items listed in the Agreement are Contract Documents. Approved Shop Drawings, other Contractor's submittals, and the reports and drawings of subsurface and physical conditions are not Contract Documents.

13. *Contract Price*--The moneys payable by Owner to Contractor for completion of the Work in accordance with the Contract Documents as stated in the Agreement (subject to the provisions of Paragraph 11.03 in the case of Unit Price Work).

14. *Contract Times*--The number of days or the dates stated in the Agreement to: (i) achieve Milestones, if any; (ii) achieve Substantial Completion; and (iii) complete the Work so that it is ready for final payment as evidenced by Engineer's written recommendation of final payment.

15. *Contractor*--The individual or entity with whom Owner has entered into the Agreement.

16. *Cost of the Work*--See Paragraph 11.01.A for definition.

17. *Drawings*--That part of the Contract Documents prepared or approved by Engineer which graphically shows the scope, extent, and character of the Work to be performed by Contractor. Shop Drawings and other Contractor submittals are not Drawings as so defined.

18. *Effective Date of the Agreement*--The date indicated in the Agreement on which it becomes effective, but if no such date is indicated, it means the date on which the Agreement is signed and delivered by the last of the two parties to sign and deliver.

19. *Engineer*--The individual or entity named as such in the Agreement.

20. *Field Order*--A written order issued by Engineer which requires minor changes in the Work but which does not involve a change in the Contract Price or the Contract Times.

21. *General Requirements*--Sections of Division 1 of the Specifications. The General Requirements pertain to all sections of the Specifications.

22. *Hazardous Environmental Condition*--The presence at the Site of Asbestos, PCBs, Petroleum, Hazardous Waste, or Radioactive Material in such quantities or circumstances that may present a substantial danger to persons or property exposed thereto in connection with the Work.

23. *Hazardous Waste*--The term Hazardous Waste shall have the meaning provided in Section 1004 of the Solid Waste Disposal Act (42 USC Section 6903) as amended from time to time.

24. *Laws and Regulations; Laws or Regulations*--Any and all applicable laws, rules, regulations, ordinances, codes, and orders of any and all governmental bodies, agencies, authorities, and courts having jurisdiction.

25. *Liens*--Charges, security interests, or encumbrances upon Project funds, real property, or personal property.

26. *Milestone*--A principal event specified in the Contract Documents relating to an intermediate completion date or time prior to Substantial Completion of all the Work.

27. *Notice of Award*--The written notice by Owner to the Successful Bidder stating that upon timely compliance by the Successful Bidder with the conditions precedent listed therein, Owner will sign and deliver the Agreement.

28. *Notice to Proceed*--A written notice given by Owner to Contractor fixing the date on which the Contract Times will commence to run and on which Contractor shall start to perform the Work under the Contract Documents.

29. *Owner*--The individual or entity with whom Contractor has entered into the Agreement and for whom the Work is to be performed.

30. *PCBs*--Polychlorinated biphenyls.

31. *Petroleum*--Petroleum, including crude oil or any fraction thereof which is liquid at standard conditions of temperature and pressure (60 degrees Fahrenheit and 14.7 pounds per square inch absolute), such as oil, petroleum, fuel oil, oil sludge, oil refuse, gasoline, kerosene, and oil mixed with other non-Hazardous Waste and crude oils.

32. *Progress Schedule*--A schedule, prepared and maintained by Contractor, describing the sequence and duration of the activities comprising the Contractor's plan to accomplish the Work within the Contract Times.

33. *Project*--The total construction of which the Work to be performed under the Contract Documents may be the whole, or a part.

34. *Project Manual*--The bound documentary information prepared for bidding and constructing the Work. A listing of the contents of the Project Manual, which may be bound in one or more volumes, is contained in the table(s) of contents.

35. *Radioactive Material*--Source, special nuclear, or byproduct material as defined by the Atomic Energy Act of 1954 (42 USC Section 2011 et seq.) as amended from time to time.

36. *Related Entity* -- An officer, director, partner, employee, agent, consultant, or subcontractor.

37. *Resident Project Representative*--The authorized representative of Engineer who may be assigned to the Site or any part thereof.

38. *Samples*--Physical examples of materials, equipment, or workmanship that are representative of some portion of the Work and which establish the standards by which such portion of the Work will be judged.

39. *Schedule of Submittals*--A schedule, prepared and maintained by Contractor, of required submittals and the time requirements to support scheduled performance of related construction activities.

40. *Schedule of Values*--A schedule, prepared and maintained by Contractor, allocating portions of the Contract Price to various portions of the Work and used as the basis for reviewing Contractor's Applications for Payment.

41. *Shop Drawings*--All drawings, diagrams, illustrations, schedules, and other data or information which are specifically prepared or assembled by or for Contractor and submitted by Contractor to illustrate some portion of the Work.

42. *Site*--Lands or areas indicated in the Contract Documents as being furnished by Owner upon which the Work is to be performed, including rights-of-way and easements for access thereto, and such other lands furnished by Owner which are designated for the use of Contractor.

43. *Specifications*--That part of the Contract Documents consisting of written requirements for materials, equipment, systems, standards and workmanship as applied to the Work, and certain

administrative requirements and procedural matters applicable thereto.

44. *Subcontractor*--An individual or entity having a direct contract with Contractor or with any other Subcontractor for the performance of a part of the Work at the Site.

45. *Substantial Completion*--The time at which the Work (or a specified part thereof) has progressed to the point where, in the opinion of Engineer, the Work (or a specified part thereof) is sufficiently complete, in accordance with the Contract Documents, so that the Work (or a specified part thereof) can be utilized for the purposes for which it is intended. The terms "substantially complete" and "substantially completed" as applied to all or part of the Work refer to Substantial Completion thereof.

46. *Successful Bidder*--The Bidder submitting a responsive Bid to whom Owner makes an award.

47. *Supplementary Conditions*--That part of the Contract Documents which amends or supplements these General Conditions.

48. *Supplier*--A manufacturer, fabricator, supplier, distributor, materialman, or vendor having a direct contract with Contractor or with any Subcontractor to furnish materials or equipment to be incorporated in the Work by Contractor or any Subcontractor.

49. *Underground Facilities*--All underground pipelines, conduits, ducts, cables, wires, manholes, vaults, tanks, tunnels, or other such facilities or attachments, and any encasements containing such facilities, including those that convey electricity, gases, steam, liquid petroleum products, telephone or other communications, cable television, water, wastewater, storm water, other liquids or chemicals, or traffic or other control systems.

50. *Unit Price Work*--Work to be paid for on the basis of unit prices.

51. *Work*--The entire construction or the various separately identifiable parts thereof required to be provided under the Contract Documents. Work includes and is the result of performing or providing all labor, services, and documentation necessary to produce such construction, and furnishing, installing, and incorporating all materials and equipment into such construction, all as required by the Contract Documents.

52. *Work Change Directive*--A written statement to Contractor issued on or after the Effective Date of the Agreement and signed by Owner and recommended by Engineer ordering an addition, deletion, or revision in the Work, or responding to differing or unforeseen subsurface or physical conditions under which the Work is to be performed or to emergencies. A Work Change Directive will not change the Contract Price or the Contract Times

but is evidence that the parties expect that the change ordered or documented by a Work Change Directive will be incorporated in a subsequently issued Change Order following negotiations by the parties as to its effect, if any, on the Contract Price or Contract Times.

1.02 *Terminology*

A. The following words or terms are not defined but, when used in the Bidding Requirements or Contract Documents, have the following meaning.

B. *Intent of Certain Terms or Adjectives*

1. The Contract Documents include the terms "as allowed," "as approved," "as ordered", "as directed" or terms of like effect or import to authorize an exercise of professional judgment by Engineer. In addition, the adjectives "reasonable," "suitable," "acceptable," "proper," "satisfactory," or adjectives of like effect or import are used to describe an action or determination of Engineer as to the Work. It is intended that such exercise of professional judgment, action or determination will be solely to evaluate, in general, the Work for compliance with the requirements of and information in the Contract Documents and conformance with the design concept of the completed Project as a functioning whole as shown or indicated in the Contract Documents (unless there is a specific statement indicating otherwise). The use of any such term or adjective is not intended to and shall not be effective to assign to Engineer any duty or authority to supervise or direct the performance of the Work or any duty or authority to undertake responsibility contrary to the provisions of Paragraph 9.09 or any other provision of the Contract Documents.

C. *Day*

1. The word "day" means a calendar day of 24 hours measured from midnight to the next midnight.

D. *Defective*

1. The word "defective," when modifying the word "Work," refers to Work that is unsatisfactory, faulty, or deficient in that it:

a. does not conform to the Contract Documents, or

b. does not meet the requirements of any applicable inspection, reference standard, test, or approval referred to in the Contract Documents, or

c. has been damaged prior to Engineer's - recommendation of final payment (unless responsibility for the protection thereof has been assumed by Owner at Substantial Completion in accordance with Paragraph 14.04 or 14.05).

E. Furnish, Install, Perform, Provide

1. The word "furnish," when used in connection with services, materials, or equipment, shall mean to supply and deliver said services, materials, or equipment to the Site (or some other specified location) ready for use or installation and in usable or operable condition.

2. The word "install," when used in connection with services, materials, or equipment, shall mean to put into use or place in final position said services, materials, or equipment complete and ready for intended use.

3. The words "perform" or "provide," when used in connection with services, materials, or equipment, shall mean to furnish and install said services, materials, or equipment complete and ready for intended use.

4. When "furnish," "install," "perform," or "provide" is not used in connection with services, materials, or equipment in a context clearly requiring an obligation of Contractor, "provide" is implied.

F. Unless stated otherwise in the Contract Documents, words or phrases which have a well-known technical or construction industry or trade meaning are used in the Contract Documents in accordance with such recognized meaning.

ARTICLE 2 - PRELIMINARY MATTERS

2.01   *Delivery of Bonds and Evidence of Insurance*

A. When Contractor delivers the executed counterparts of the Agreement to Owner, Contractor shall also deliver to Owner such bonds as Contractor may be required to furnish.

B. *Evidence of Insurance:* Before any Work at the Site is started, Contractor and Owner shall each deliver to the other, with copies to each additional insured identified in the Supplementary Conditions, certificates of insurance (and other evidence of insurance which either of them or any additional insured may reasonably request) which Contractor and Owner respectively are required to purchase and maintain in accordance with Article 5.

2.02   *Copies of Documents*

A. Owner shall furnish to Contractor up to ten printed or hard copies of the Drawings and Project Manual. Additional copies will be furnished upon request at the cost of reproduction.

2.03   *Commencement of Contract Times; Notice to Proceed*

A. The Contract Times will commence to run on the thirtieth day after the Effective Date of the Agreement

or, if a Notice to Proceed is given, on the day indicated in the Notice to Proceed. A Notice to Proceed may be given at any time within 30 days after the Effective Date of the Agreement. In no event will the Contract Times commence to run later than the sixtieth day after the day of Bid opening or the thirtieth day after the Effective Date of the Agreement, whichever date is earlier.

2.04   *Starting the Work*

A. Contractor shall start to perform the Work on the date when the Contract Times commence to run. No Work shall be done at the Site prior to the date on which the Contract Times commence to run.

2.05   *Before Starting Construction*

A. *Preliminary Schedules:* Within 10 days after the Effective Date of the Agreement (unless otherwise specified in the General Requirements), Contractor shall submit to Engineer for timely review:

1. a preliminary Progress Schedule; indicating the times (numbers of days or dates) for starting and completing the various stages of the Work, including any Milestones specified in the Contract Documents;

2. a preliminary Schedule of Submittals; and

3. a preliminary Schedule of Values for all of the Work which includes quantities and prices of items which when added together equal the Contract Price and subdivides the Work into component parts in sufficient detail to serve as the basis for progress payments during performance of the Work. Such prices will include an appropriate amount of overhead and profit applicable to each item of Work.

2.06   *Preconstruction Conference*

A. Before any Work at the Site is started, a conference attended by Owner, Contractor, Engineer, and others as appropriate will be held to establish a working understanding among the parties as to the Work and to discuss the schedules referred to in Paragraph 2.05.A, procedures for handling Shop Drawings and other submittals, processing Applications for Payment, and maintaining required records.

2.07   *Initial Acceptance of Schedules*

A. At least 10 days before submission of the first Application for Payment a conference attended by Contractor, Engineer, and others as appropriate will be held to review for acceptability to Engineer as provided below the schedules submitted in accordance with Paragraph 2.05.A. Contractor shall have an additional 10 days to make corrections and adjustments and to complete and resubmit the schedules. No progress payment shall be made to Contractor until acceptable schedules are submitted to Engineer.

1. The Progress Schedule will be acceptable to Engineer if it provides an orderly progression of the Work to completion within the Contract Times. Such acceptance will not impose on Engineer responsibility for the Progress Schedule, for sequencing, scheduling, or progress of the Work nor interfere with or relieve Contractor from Contractor's full responsibility therefor.

2. Contractor's Schedule of Submittals will be acceptable to Engineer if it provides a workable arrangement for reviewing and processing the required submittals.

3. Contractor's Schedule of Values will be acceptable to Engineer as to form and substance if it provides a reasonable allocation of the Contract Price to component parts of the Work.

ARTICLE 3 - CONTRACT DOCUMENTS: INTENT, AMENDING, REUSE

3.01   *Intent*

A. The Contract Documents are complementary; what is required by one is as binding as if required by all.

B. It is the intent of the Contract Documents to describe a functionally complete Project (or part thereof) to be constructed in accordance with the Contract Documents. Any labor, documentation, services, materials, or equipment that may reasonably be inferred from the Contract Documents or from prevailing custom or trade usage as being required to produce the intended result will be provided whether or not specifically called for at no additional cost to Owner.

C. Clarifications and interpretations of the Contract Documents shall be issued by Engineer as provided in Article 9.

3.02   *Reference Standards*

A. Standards, Specifications, Codes, Laws, and Regulations

1. Reference to standards, specifications, manuals, or codes of any technical society, organization, or association, or to Laws or Regulations, whether such reference be specific or by implication, shall mean the standard, specification, manual, code, or Laws or Regulations in effect at the time of opening of Bids (or on the Effective Date of the Agreement if there were no Bids), except as may be otherwise specifically stated in the Contract Documents.

2. No provision of any such standard, specification, manual or code, or any instruction of a

Supplier shall be effective to change the duties or responsibilities of Owner, Contractor, or Engineer, or any of their subcontractors, consultants, agents, or employees from those set forth in the Contract Documents. No such provision or instruction shall be effective to assign to Owner, or Engineer, or any of, their Related Entities, any duty or authority to supervise or direct the performance of the Work or any duty or authority to undertake responsibility inconsistent with the provisions of the Contract Documents.

3.03   *Reporting and Resolving Discrepancies*

A. Reporting Discrepancies

1. *Contractor's Review of Contract Documents Before Starting Work*: Before undertaking each part of the Work, Contractor shall carefully study and compare the Contract Documents and check and verify pertinent figures therein and all applicable field measurements. Contractor shall promptly report in writing to Engineer any conflict, error, ambiguity, or discrepancy which Contractor may discover and shall obtain a written interpretation or clarification from Engineer before proceeding with any Work affected thereby.

2. *Contractor's Review of Contract Documents During Performance of Work*: If, during the performance of the Work, Contractor discovers any conflict, error, ambiguity, or discrepancy within the Contract Documents or between the Contract Documents and any provision of any Law or Regulation applicable to the performance of the Work or of any standard, specification, manual or code, or of any instruction of any Supplier, Contractor shall promptly report it to Engineer in writing. Contractor shall not proceed with the Work affected thereby (except in an emergency as required by Paragraph 6.16.A) until an amendment or supplement to the Contract Documents has been issued by one of the methods indicated in Paragraph 3.04.

3. Contractor shall not be liable to Owner or Engineer for failure to report any conflict, error, ambiguity, or discrepancy in the Contract Documents unless Contractor knew or reasonably should have known thereof.

B. Resolving Discrepancies

1. Except as may be otherwise specifically stated in the Contract Documents, the provisions of the Contract Documents shall take precedence in resolving any conflict, error, ambiguity, or discrepancy between the provisions of the Contract Documents and:

a. the provisions of any standard, specification, manual, code, or instruction (whether or not specifically incorporated by reference in the Contract Documents); or

b. the provisions of any Laws or Regulations applicable to the performance of the Work (unless such an interpretation of the provisions of the Contract Documents would result in violation of such Law or Regulation).

3.04   *Amending and Supplementing Contract Documents*

A. The Contract Documents may be amended to provide for additions, deletions, and revisions in the Work or to modify the terms and conditions thereof by either a Change Order or a Work Change Directive.

B. The requirements of the Contract Documents may be supplemented, and minor variations and deviations in the Work may be authorized, by one or more of the following ways:

1. A Field Order;

2. Engineer's approval of a Shop Drawing or Sample; (Subject to the provisions of Paragraph 6.17.D.3); or

3. Engineer's written interpretation or clarification.

3.05   *Reuse of Documents*

A. Contractor and any Subcontractor or Supplier or other individual or entity performing or furnishing all of the Work under a direct or indirect contract with Contractor, shall not:

1. have or acquire any title to or ownership rights in any of the Drawings, Specifications, or other documents (or copies of any thereof) prepared by or bearing the seal of Engineer or Engineer's consultants, including electronic media editions; or

2. reuse any of such Drawings, Specifications, other documents, or copies thereof on extensions of the Project or any other project without written consent of Owner and Engineer and specific written verification or adaption by Engineer.

B. The prohibition of this Paragraph 3.05 will survive final payment or termination of the Contract. Nothing herein shall preclude Contractor from retaining copies of the Contract Documents for record purposes.

3.06   *Electronic Data*

A. Copies of data furnished by Owner or Engineer to Contractor or Contractor to Owner or Engineer that may be relied upon are limited to the printed copies (also known as hard copies). Files in electronic media format of text, data, graphics, or other types are furnished only for the convenience of the receiving party. Any conclusion or information obtained or derived from such electronic files will be at the user's sole risk. If there is a discrepancy between the electronic files and the hard copies, the hard copies govern.

B. Because data stored in electronic media format can deteriorate or be modified inadvertently or otherwise without authorization of the data's creator, the party receiving electronic files agrees that it will perform acceptance tests or procedures within 60 days, after which the receiving party shall be deemed to have accepted the data thus transferred. Any errors detected within the 60-day acceptance period will be corrected by the transferring party..

C. When transferring documents in electronic media format, the transferring party makes no representations as to long term compatibility, usability, or readability of documents resulting from the use of software application packages, operating systems, or computer hardware differing from those used by the data's creator.

ARTICLE 4 - AVAILABILITY OF LANDS; SUBSURFACE AND PHYSICAL CONDITIONS; HAZARDOUS ENVIRONMENTAL CONDITIONS; REFERENCE POINTS

4.01   *Availability of Lands*

A. Owner shall furnish the Site. Owner shall notify Contractor of any encumbrances or restrictions not of general application but specifically related to use of the Site with which Contractor must comply in performing the Work. Owner will obtain in a timely manner and pay for easements for permanent structures or permanent changes in existing facilities. If Contractor and Owner are unable to agree on entitlement to or on the amount or extent, if any, of any adjustment in the Contract Price or Contract Times, or both, as a result of any delay in Owner's furnishing the Site or a part thereof, Contractor may make a Claim therefor as provided in Paragraph 10.05.

B. Upon reasonable written request, Owner shall furnish Contractor with a current statement of record legal title and legal description of the lands upon which the Work is to be performed and Owner's interest therein as necessary for giving notice of or filing a mechanic's or construction lien against such lands in accordance with applicable Laws and Regulations.

C. Contractor shall provide for all additional lands and access thereto that may be required for temporary construction facilities or storage of materials and equipment.

4.02 *Subsurface and Physical Conditions*

A. *Reports and Drawings:* The Supplementary Conditions identify:

1. those reports of explorations and tests of subsurface conditions at or contiguous to the Site that Engineer has used in preparing the Contract Documents; and

2. those drawings of physical conditions in or relating to existing surface or subsurface structures at or contiguous to the Site (except Underground Facilities) that Engineer has used in preparing the Contract Documents.

B. *Limited Reliance by Contractor on Technical Data Authorized:* Contractor may rely upon the general accuracy of the "technical data" contained in such reports and drawings, but such reports and drawings are not Contract Documents. Such "technical data" is identified in the Supplementary Conditions. Except for such reliance on such "technical data," Contractor may not rely upon or make any claim against Owner or Engineer, or any of their Related Entities with respect to:

1. the completeness of such reports and drawings for Contractor's purposes, including, but not limited to, any aspects of the means, methods, techniques, sequences, and procedures of construction to be employed by Contractor, and safety precautions and programs incident thereto; or

2. other data, interpretations, opinions, and information contained in such reports or shown or indicated in such drawings; or

3. any Contractor interpretation of or conclusion drawn from any "technical data" or any such other data, interpretations, opinions, or information.

4.03 *Differing Subsurface or Physical Conditions*

A. *Notice:* If Contractor believes that any subsurface or physical condition at or contiguous to the Site that is uncovered or revealed either:

1. is of such a nature as to establish that any "technical data" on which Contractor is entitled to rely as provided in Paragraph 4.02 is materially inaccurate; or

2. is of such a nature as to require a change in the Contract Documents; or

3. differs materially from that shown or indicated in the Contract Documents; or

4. is of an unusual nature, and differs materially from conditions ordinarily encountered and generally recognized as inherent in work of the character provided for in the Contract Documents;

then Contractor shall, promptly after becoming aware thereof and before further disturbing the subsurface or physical conditions or performing any Work in connection therewith (except in an emergency as required by Paragraph 6.16.A), notify Owner and Engineer in writing about such condition. Contractor shall not further disturb such condition or perform any Work in connection therewith (except as aforesaid) until receipt of written order to do so.

B. *Engineer's Review:* After receipt of written notice as required by Paragraph 4.03.A, Engineer will promptly review the pertinent condition, determine the necessity of Owner's obtaining additional exploration or tests with respect thereto, and advise Owner in writing (with a copy to Contractor) of Engineer's findings and conclusions.

C. *Possible Price and Times Adjustments*

1. The Contract Price or the Contract Times, or both, will be equitably adjusted to the extent that the existence of such differing subsurface or physical condition causes an increase or decrease in Contractor's cost of, or time required for, performance of the Work; subject, however, to the following:

a. such condition must meet any one or more of the categories described in Paragraph 4.03.A; and

b. with respect to Work that is paid for on a Unit Price Basis, any adjustment in Contract Price will be subject to the provisions of Paragraphs 9.07 and 11.03.

2. Contractor shall not be entitled to any adjustment in the Contract Price or Contract Times if:

a. Contractor knew of the existence of such conditions at the time Contractor made a final commitment to Owner with respect to Contract Price and Contract Times by the submission of a Bid or becoming bound under a negotiated contract; or

b. the existence of such condition could reasonably have been discovered or revealed as a result of any examination, investigation, exploration, test, or study of the Site and contiguous areas required by the Bidding Requirements or Contract Documents to be conducted by or for Contractor prior to Contractor's making such final commitment; or

c. Contractor failed to give the written notice as required by Paragraph 4.03.A.

3. If Owner and Contractor are unable to agree on entitlement to or on the amount or extent, if any, of any adjustment in the Contract Price or Contract Times, or both, a Claim may be made therefor as provided in Paragraph 10.05. However, Owner and Engineer, and any of their Related Entities shall not be liable to Contractor for any claims, costs, losses, or damages (including but not limited to all fees and charges of engineers, architects, attorneys, and other professionals and all court or arbitration or other dispute resolution costs) sustained by Contractor on or in connection with any other project or anticipated project.

4.04 *Underground Facilities*

A. *Shown or Indicated:* The information and data shown or indicated in the Contract Documents with respect to existing Underground Facilities at or contiguous to the Site is based on information and data furnished to Owner or Engineer by the owners of such Underground Facilities, including Owner, or by others. Unless it is otherwise expressly provided in the Supplementary Conditions:

1. Owner and Engineer shall not be responsible for the accuracy or completeness of any such information or data; and

2. the cost of all of the following will be included in the Contract Price, and Contractor shall have full responsibility for:

a. reviewing and checking all such information and data,

b. locating all Underground Facilities shown or indicated in the Contract Documents,

c. coordination of the Work with the owners of such Underground Facilities, including Owner, during construction, and

d. the safety and protection of all such Underground Facilities and repairing any damage thereto resulting from the Work.

B. *Not Shown or Indicated*

1. If an Underground Facility is uncovered or revealed at or contiguous to the Site which was not shown or indicated, or not shown or indicated with reasonable accuracy in the Contract Documents, Contractor shall, promptly after becoming aware thereof and before further disturbing conditions affected thereby or performing any Work in connection therewith (except in an emergency as required by Paragraph 6.16.A), identify the owner of such Underground Facility and give written notice to that owner and to Owner and Engineer. Engineer will promptly review the Underground Facility and determine the extent, if any, to which a change is required in the Contract Documents to reflect and document the consequences of the existence or location of the Underground Facility. During such time, Contractor shall be responsible for the safety and protection of such Underground Facility.

2. If Engineer concludes that a change in the Contract Documents is required, a Work Change Directive or a Change Order will be issued to reflect and document such consequences. An equitable adjustment shall be made in the Contract Price or Contract Times, or both, to the extent that they are attributable to the existence or location of any Underground Facility that was not shown or indicated or not shown or indicated with reasonable accuracy in the Contract Documents and that Contractor did not know of and could not reasonably have been expected to be aware of or to have anticipated. If Owner and Contractor are unable to agree on entitlement to or on the amount or extent, if any, of any such adjustment in Contract Price or Contract Times, Owner or Contractor may make a Claim therefor as provided in Paragraph 10.05.

4.05 *Reference Points*

A. Owner shall provide engineering surveys to establish reference points for construction which in Engineer's judgment are necessary to enable Contractor to proceed with the Work. Contractor shall be responsible for laying out the Work, shall protect and preserve the established reference points and property monuments, and shall make no changes or relocations without the prior written approval of Owner. Contractor shall report to Engineer whenever any reference point or property monument is lost or destroyed or requires relocation because of necessary changes in grades or locations, and shall be responsible for the accurate replacement or relocation of such reference points or property monuments by professionally qualified personnel.

4.06 *Hazardous Environmental Condition at Site*

A. *Reports and Drawings:* Reference is made to the Supplementary Conditions for the identification of those reports and drawings relating to a Hazardous Environmental Condition identified at the Site, if any, that have been utilized by the Engineer in the preparation of the Contract Documents.

B. *Limited Reliance by Contractor on Technical Data Authorized:* Contractor may rely upon the general accuracy of the "technical data" contained in such reports and drawings, but such reports and drawings are not Contract Documents. Such "technical data" is identified in the Supplementary Conditions. Except for such reliance on such "technical data," Contractor may not rely upon or make any claim against Owner or Engineer, or any of their Related Entities with respect to:

1. the completeness of such reports and drawings for Contractor's purposes, including, but not limited to, any aspects of the means, methods, techniques, sequences and procedures of construction to be employed by Contractor and safety precautions and programs incident thereto; or

2. other data, interpretations, opinions and information contained in such reports or shown or indicated in such drawings; or

3. any Contractor interpretation of or conclusion drawn from any "technical data" or any such other data, interpretations, opinions or information.

C. Contractor shall not be responsible for any Hazardous Environmental Condition uncovered or revealed at the Site which was not shown or indicated in Drawings or Specifications or identified in the Contract Documents to be within the scope of the Work. Contractor shall be responsible for a Hazardous Environmental Condition created with any materials brought to the Site by Contractor, Subcontractors, Suppliers, or anyone else for whom Contractor is responsible.

D. If Contractor encounters a Hazardous Environmental Condition or if Contractor or anyone for whom Contractor is responsible creates a Hazardous Environmental Condition, Contractor shall immediately: (i) secure or otherwise isolate such condition; (ii) stop all Work in connection with such condition and in any area affected thereby (except in an emergency as required by Paragraph 6.16.A); and (iii) notify Owner and Engineer (and promptly thereafter confirm such notice in writing). Owner shall promptly consult with Engineer concerning the necessity for Owner to retain a qualified expert to evaluate such condition or take corrective action, if any.

E. Contractor shall not be required to resume Work in connection with such condition or in any affected area until after Owner has obtained any required permits related thereto and delivered to Contractor written notice: (i) specifying that such condition and any affected area is or has been rendered safe for the resumption of Work; or (ii) specifying any special conditions under which such Work may be resumed safely. If Owner and Contractor cannot agree as to entitlement to or on the amount or extent, if any, of any adjustment in Contract Price or Contract Times, or both, as a result of such Work stoppage or such special conditions under which Work is agreed to be resumed by Contractor, either party may make a Claim therefor as provided in Paragraph 10.05.

F. If after receipt of such written notice Contractor does not agree to resume such Work based on a reasonable belief it is unsafe, or does not agree to resume such Work under such special conditions, then Owner may order the portion of the Work that is in the area affected by such condition to be deleted from the Work. If Owner and Contractor cannot agree as to

entitlement to or on the amount or extent, if any, of an adjustment in Contract Price or Contract Times as a result of deleting such portion of the Work, then either party may make a Claim therefor as provided in Paragraph 10.05. Owner may have such deleted portion of the Work performed by Owner's own forces or others in accordance with Article 7.

G. To the fullest extent permitted by Laws and Regulations, Owner shall indemnify and hold harmless Contractor, Subcontractors, and Engineer, and the officers, directors, partners, employees, agents, consultants, and subcontractors of each and any of them from and against all claims, costs, losses, and damages (including but not limited to all fees and charges of engineers, architects, attorneys, and other professionals and all court or arbitration or other dispute resolution costs) arising out of or relating to a Hazardous Environmental Condition, provided that such Hazardous Environmental Condition: (i) was not shown or indicated in the Drawings or Specifications or identified in the Contract Documents to be included within the scope of the Work, and (ii) was not created by Contractor or by anyone for whom Contractor is responsible. Nothing in this Paragraph 4.06.G shall obligate Owner to indemnify any individual or entity from and against the consequences of that individual's or entity's own negligence.

H. To the fullest extent permitted by Laws and Regulations, Contractor shall indemnify and hold harmless Owner and Engineer, and the officers, directors, partners, employees, agents, consultants, and subcontractors of each and any of them from and against all claims, costs, losses, and damages (including but not limited to all fees and charges of engineers, architects, attorneys, and other professionals and all court or arbitration or other dispute resolution costs) arising out of or relating to a Hazardous Environmental Condition created by Contractor or by anyone for whom Contractor is responsible. Nothing in this Paragraph 4.06.H shall obligate Contractor to indemnify any individual or entity from and against the consequences of that individual's or entity's own negligence.

I. The provisions of Paragraphs 4.02, 4.03, and 4.04 do not apply to a Hazardous Environmental Condition uncovered or revealed at the Site.

ARTICLE 5 - BONDS AND INSURANCE

5.01   *Performance, Payment, and Other Bonds*

A. Contractor shall furnish performance and payment bonds, each in an amount at least equal to the Contract Price as security for the faithful performance and payment of all of Contractor's obligations under the Contract Documents. These bonds shall remain in effect until one year after the date when final payment becomes due or until completion of the correction period specified in Paragraph 13.07, whichever is later, except as provided otherwise by Laws or Regulations or by the Contract Documents. Contractor shall also furnish such other bonds as are required by the Contract Documents.

B. All bonds shall be in the form prescribed by the Contract Documents except as provided otherwise by Laws or Regulations, and shall be executed by such sureties as are named in the current list of "Companies Holding Certificates of Authority as Acceptable Sureties on Federal Bonds and as Acceptable Reinsuring Companies" as published in Circular 570 (amended) by the Financial Management Service, Surety Bond Branch, U.S. Department of the Treasury. All bonds signed by an agent must be accompanied by a certified copy of the agent's authority to act.

C. If the surety on any bond furnished by Contractor is declared bankrupt or becomes insolvent or its right to do business is terminated in any state where any part of the Project is located or it ceases to meet the requirements of Paragraph 5.01.B, Contractor shall promptly notify Owner and Engineer and shall, within 20 days after the event giving rise to such notification, provide another bond and surety, both of which shall comply with the requirements of Paragraphs 5.01.B and 5.02.

5.02   *Licensed Sureties and Insurers*

A. All bonds and insurance required by the Contract Documents to be purchased and maintained by Owner or Contractor shall be obtained from surety or insurance companies that are duly licensed or authorized in the jurisdiction in which the Project is located to issue bonds or insurance policies for the limits and coverages so required. Such surety and insurance companies shall also meet such additional requirements and qualifications as may be provided in the Supplementary Conditions.

5.03   *Certificates of Insurance*

A. Contractor shall deliver to Owner, with copies to each additional insured identified in the Supplementary Conditions, certificates of insurance (and other evidence of insurance requested by Owner or any other additional insured) which Contractor is required to purchase and maintain.

B. Owner shall deliver to Contractor, with copies to each additional insured identified in the Supplementary Conditions, certificates of insurance (and other evidence of insurance requested by Contractor or any other additional insured) which Owner is required to purchase and maintain.

5.04   *Contractor's Liability Insurance*

A. Contractor shall purchase and maintain such liability and other insurance as is appropriate for the Work being performed and as will provide protection from claims set forth below which may arise out of or result from Contractor's performance of the Work and Contractor's other obligations under the Contract Documents, whether it is to be performed by Contractor, any Subcontractor or Supplier, or by anyone directly or indirectly employed by any of them to perform any of the Work, or by anyone for whose acts any of them may be liable:

1. claims under workers' compensation, disability benefits, and other similar employee benefit acts;

2. claims for damages because of bodily injury, occupational sickness or disease, or death of Contractor's employees;

3. claims for damages because of bodily injury, sickness or disease, or death of any person other than Contractor's employees;

4. claims for damages insured by reasonably available personal injury liability coverage which are sustained:

a. by any person as a result of an offense directly or indirectly related to the employment of such person by Contractor, or

b. by any other person for any other reason;

5. claims for damages, other than to the Work itself, because of injury to or destruction of tangible property wherever located, including loss of use resulting therefrom; and

6. claims for damages because of bodily injury or death of any person or property damage arising out of the ownership, maintenance or use of any motor vehicle.

B. The policies of insurance required by this Paragraph 5.04 shall:

1. with respect to insurance required by Paragraphs 5.04.A.3 through 5.04.A.6 inclusive, include as additional insured (subject to any customary exclusion regarding professional liability) Owner and Engineer, and any other individuals or entities identified in the Supplementary Conditions, all of whom shall be listed as additional insureds, and include coverage for the respective officers, directors, partners, employees, agents, consultants and subcontractors of each and any of all such additional insureds, and the insurance afforded to these additional insureds shall provide primary coverage for all claims covered thereby;

2. include at least the specific coverages and be written for not less than the limits of liability provided in the Supplementary Conditions or required by Laws or Regulations, whichever is greater;

3. include completed operations insurance;

4. include contractual liability insurance covering Contractor's indemnity obligations under Paragraphs 6.11 and 6.20;

5. contain a provision or endorsement that the coverage afforded will not be canceled, materially changed or renewal refused until at least 30 days prior written notice has been given to Owner and Contractor and to each other additional insured identified in the Supplementary Conditions to whom a certificate of insurance has been issued (and the certificates of insurance furnished by the Contractor pursuant to Paragraph 5.03 will so provide);

6. remain in effect at least until final payment and at all times thereafter when Contractor may be correcting, removing, or replacing defective Work in accordance with Paragraph 13.07; and

7. with respect to completed operations insurance, and any insurance coverage written on a claims-made basis, remain in effect for at least two years after final payment.

a. Contractor shall furnish Owner and each other additional insured identified in the Supplementary Conditions, to whom a certificate of insurance has been issued, evidence satisfactory to Owner and any such additional insured of continuation of such insurance at final payment and one year thereafter.

5.05   *Owner's Liability Insurance*

A. In addition to the insurance required to be provided by Contractor under Paragraph 5.04, Owner, at Owner's option, may purchase and maintain at Owner's expense Owner's own liability insurance as will protect Owner against claims which may arise from operations under the Contract Documents.

5.06   *Property Insurance*

A. Unless otherwise provided in the Supplementary Conditions, Owner shall purchase and maintain property insurance upon the Work at the Site in the amount of the full replacement cost thereof (subject to such deductible amounts as may be provided in the Supplementary Conditions or required by Laws and Regulations). This insurance shall:

1. include the interests of Owner, Contractor, Subcontractors, and Engineer, and any other individuals or entities identified in the Supplementary Conditions, and the officers, directors, partners, employees, agents, consultants and subcontractors of each and any of them, each of whom is deemed to have an insurable interest and shall be listed as an insured or additional insured;

2. be written on a Builder's Risk "all-risk" or open peril or special causes of loss policy form that shall at least include insurance for physical loss or damage to the Work, temporary buildings, false work, and materials and equipment in transit, and shall insure against at least the following perils or causes of loss: fire, lightning, extended coverage, theft, vandalism and malicious mischief, earthquake, collapse, debris removal, demolition occasioned by enforcement of Laws and Regulations, water damage, (other than caused by flood) and such other perils or causes of loss as may be specifically required by the Supplementary Conditions;

3. include expenses incurred in the repair or replacement of any insured property (including but not limited to fees and charges of engineers and architects);

4. cover materials and equipment stored at the Site or at another location that was agreed to in writing by Owner prior to being incorporated in the Work, provided that such materials and equipment have been included in an Application for Payment recommended by Engineer;

5. allow for partial utilization of the Work by Owner;

6. include testing and startup; and

7. be maintained in effect until final payment is made unless otherwise agreed to in writing by Owner, Contractor, and Engineer with 30 days written notice to each other additional insured to whom a certificate of insurance has been issued.

B. Owner shall purchase and maintain such boiler and machinery insurance or additional property insurance as may be required by the Supplementary Conditions or Laws and Regulations which will include the interests of Owner, Contractor, Subcontractors, and Engineer, and any other individuals or entities identified in the Supplementary Conditions, and the officers, directors, partners, employees, agents, consultants and subcontractors of each and any of them, each of whom is deemed to have an insurable interest and shall be listed as an insured or additional insured.

C. All the policies of insurance (and the certificates or other evidence thereof) required to be purchased and maintained in accordance with Paragraph 5.06 will contain a provision or endorsement that the coverage afforded will not be canceled or materially changed or renewal refused until at least 30 days prior written notice has been given to Owner and Contractor and to each other additional insured to whom a certificate of insurance has been issued and will contain waiver provisions in accordance with Paragraph 5.07.

D. Owner shall not be responsible for purchasing and maintaining any property insurance specified in this Paragraph 5.06 to protect the interests of Contractor, Subcontractors, or others in the Work to the extent of any deductible amounts that are identified in the Supplementary Conditions. The risk of loss within such identified deductible amount will be borne by Contractor, Subcontractors, or others suffering any such loss, and if any of them wishes property insurance coverage within the limits of such amounts, each may purchase and maintain it at the purchaser's own expense.

E. If Contractor requests in writing that other special insurance be included in the property insurance policies provided under Paragraph 5.06, Owner shall, if possible, include such insurance, and the cost thereof will be charged to Contractor by appropriate Change Order. Prior to commencement of the Work at the Site, Owner shall in writing advise Contractor whether or not such other insurance has been procured by Owner.

5.07 *Waiver of Rights*

A. Owner and Contractor intend that all policies purchased in accordance with Paragraph 5.06 will protect Owner, Contractor, Subcontractors, and Engineer, and all other individuals or entities identified in the Supplementary Conditions to be listed as insureds or additional insureds (and the officers, directors, partners, employees, agents, consultants and subcontractors of each and any of them) in such policies and will provide primary coverage for all losses and damages caused by the perils or causes of loss covered thereby. All such policies shall contain provisions to the effect that in the event of payment of any loss or damage the insurers will have no rights of recovery against any of the insureds or additional insureds thereunder. Owner and Contractor waive all rights against each other and their respective officers, directors, partners, employees, agents, consultants and subcontractors of each and any of them for all losses and damages caused by, arising out of or resulting from any of the perils or causes of loss covered by such policies and any other property insurance applicable to the Work; and, in addition, waive all such rights against Subcontractors, and Engineer, and all other individuals or entities identified in the Supplementary Conditions to be listed as insured or additional insured (and the officers, directors, partners, employees, agents, consultants and subcontractors of each and any of them) under such policies for losses and damages so caused. None of the above waivers shall extend to the rights that any party making such waiver may have to the proceeds of insurance held by Owner as trustee or otherwise payable under any policy so issued.

B. Owner waives all rights against Contractor, Subcontractors, and Engineer, and the officers, directors, partners, employees, agents, consultants and subcontractors of each and any of them for:

1. loss due to business interruption, loss of use, or other consequential loss extending beyond direct physical loss or damage to Owner's property or the Work caused by, arising out of, or resulting from fire or other perils whether or not insured by Owner; and

2. loss or damage to the completed Project or part thereof caused by, arising out of, or resulting from fire or other insured peril or cause of loss covered by any property insurance maintained on the completed Project or part thereof by Owner during partial utilization pursuant to Paragraph 14.05, after Substantial Completion pursuant to Paragraph 14.04, or after final payment pursuant to Paragraph 14.07.

C. Any insurance policy maintained by Owner covering any loss, damage or consequential loss referred to in Paragraph 5.07.B shall contain provisions to the effect that in the event of payment of any such loss, damage, or consequential loss, the insurers will have no rights of recovery against Contractor, Subcontractors, or Engineer, and the officers, directors, partners, employees, agents, consultants and subcontractors of each and any of them.

5.08 *Receipt and Application of Insurance Proceeds*

A. Any insured loss under the policies of insurance required by Paragraph 5.06 will be adjusted with Owner and made payable to Owner as fiduciary for the insureds, as their interests may appear, subject to the requirements of any applicable mortgage clause and of Paragraph 5.08.B. Owner shall deposit in a separate account any money so received and shall distribute it in accordance with such agreement as the parties in interest may reach. If no other special agreement is reached, the damaged Work shall be repaired or replaced, the moneys so received applied on account thereof, and the Work and the cost thereof covered by an appropriate Change Order .

B. Owner as fiduciary shall have power to adjust and settle any loss with the insurers unless one of the parties in interest shall object in writing within 15 days after the occurrence of loss to Owner's exercise of this power. If such objection be made, Owner as fiduciary shall make settlement with the insurers in accordance with such agreement as the parties in interest may reach. If no such agreement among the parties in interest is reached, Owner as fiduciary shall adjust and settle the loss with the insurers and, if required in writing by any party in interest, Owner as fiduciary shall give bond for the proper performance of such duties.

5.09 *Acceptance of Bonds and Insurance; Option to Replace*

A. If either Owner or Contractor has any objection to the coverage afforded by or other provisions of the bonds or insurance required to be purchased and maintained by the other party in accordance with Article 5 on the basis of non-conformance with the Contract

Documents, the objecting party shall so notify the other party in writing within 10 days after receipt of the certificates (or other evidence requested) required by Paragraph 2.01.B. Owner and Contractor shall each provide to the other such additional information in respect of insurance provided as the other may reasonably request. If either party does not purchase or maintain all of the bonds and insurance required of such party by the Contract Documents, such party shall notify the other party in writing of such failure to purchase prior to the start of the Work, or of such failure to maintain prior to any change in the required coverage. Without prejudice to any other right or remedy, the other party may elect to obtain equivalent bonds or insurance to protect such other party's interests at the expense of the party who was required to provide such coverage, and a Change Order shall be issued to adjust the Contract Price accordingly.

5.10 *Partial Utilization, Acknowledgment of Property Insurer*

A. If Owner finds it necessary to occupy or use a portion or portions of the Work prior to Substantial Completion of all the Work as provided in Paragraph 14.05, no such use or occupancy shall commence before the insurers providing the property insurance pursuant to Paragraph 5.06 have acknowledged notice thereof and in writing effected any changes in coverage necessitated thereby. The insurers providing the property insurance shall consent by endorsement on the policy or policies, but the property insurance shall not be canceled or permitted to lapse on account of any such partial use or occupancy.

ARTICLE 6 - CONTRACTOR'S RESPONSIBILITIES

6.01 *Supervision and Superintendence*

A. Contractor shall supervise, inspect, and direct the Work competently and efficiently, devoting such attention thereto and applying such skills and expertise as may be necessary to perform the Work in accordance with the Contract Documents. Contractor shall be solely responsible for the means, methods, techniques, sequences, and procedures of construction. Contractor shall not be responsible for the negligence of Owner or Engineer in the design or specification of a specific means, method, technique, sequence, or procedure of construction which is shown or indicated in and expressly required by the Contract Documents.

B. At all times during the progress of the Work, Contractor shall assign a competent resident superintendent who shall not be replaced without written notice to Owner and Engineer except under extraordinary circumstances. The superintendent will be Contractor's representative at the Site and shall have authority to act on behalf of Contractor. All communications given to or received from the superintendent shall be binding on Contractor.

6.02 *Labor; Working Hours*

A. Contractor shall provide competent, suitably qualified personnel to survey and lay out the Work and perform construction as required by the Contract Documents. Contractor shall at all times maintain good discipline and order at the Site.

B. Except as otherwise required for the safety or protection of persons or the Work or property at the Site or adjacent thereto, and except as otherwise stated in the Contract Documents, all Work at the Site shall be performed during regular working hours. Contractor will not permit the performance of Work on a Saturday, Sunday, or any legal holiday without Owner's written consent (which will not be unreasonably withheld) given after prior written notice to Engineer.

6.03 *Services, Materials, and Equipment*

A. Unless otherwise specified in the Contract Documents, Contractor shall provide and assume full responsibility for all services, materials, equipment, labor, transportation, construction equipment and machinery, tools, appliances, fuel, power, light, heat, telephone, water, sanitary facilities, temporary facilities, and all other facilities and incidentals necessary for the performance, testing, start-up, and completion of the Work.

B. All materials and equipment incorporated into the Work shall be as specified or, if not specified, shall be of good quality and new, except as otherwise provided in the Contract Documents. All special warranties and guarantees required by the Specifications shall expressly run to the benefit of Owner. If required by Engineer, Contractor shall furnish satisfactory evidence (including reports of required tests) as to the source, kind, and quality of materials and equipment.

C. All materials and equipment shall be stored, applied, installed, connected, erected, protected, used, cleaned, and conditioned in accordance with instructions of the applicable Supplier, except as otherwise may be provided in the Contract Documents.

6.04 *Progress Schedule*

A. Contractor shall adhere to the Progress Schedule established in accordance with Paragraph 2.07 as it may be adjusted from time to time as provided below.

1. Contractor shall submit to Engineer for acceptance (to the extent indicated in Paragraph 2.07) proposed adjustments in the Progress Schedule that will not result in changing the Contract Times. Such adjustments will comply with any provisions of the General Requirements applicable thereto.

2. Proposed adjustments in the Progress Schedule that will change the Contract Times shall be submitted in accordance with the requirements of Article 12. Adjustments in Contract Times may only be made by a Change Order.

6.05   *Substitutes and "Or-Equals"*

A. Whenever an item of material or equipment is specified or described in the Contract Documents by using the name of a proprietary item or the name of a particular Supplier, the specification or description is intended to establish the type, function, appearance, and quality required. Unless the specification or description contains or is followed by words reading that no like, equivalent, or "or-equal" item or no substitution is permitted, other items of material or equipment or material or equipment of other Suppliers may be submitted to Engineer for review under the circumstances described below.

1. *"Or-Equal" Items:* If in Engineer's sole discretion an item of material or equipment proposed by Contractor is functionally equal to that named and sufficiently similar so that no change in related Work will be required, it may be considered by Engineer as an "or-equal" item, in which case review and approval of the proposed item may, in Engineer's sole discretion, be accomplished without compliance with some or all of the requirements for approval of proposed substitute items. For the purposes of this Paragraph 6.05.A.1, a proposed item of material or equipment will be considered functionally equal to an item so named if:

a. in the exercise of reasonable judgment Engineer determines that:

1) it is at least equal in materials of construction, quality, durability, appearance, strength, and design characteristics;

2) it will reliably perform at least equally well the function and achieve the results imposed by the design concept of the completed Project as a functioning whole,

3) it has a proven record of performance and availability of responsive service; and

b. Contractor certifies that, if approved and incorporated into the Work:

1) there will be no increase in cost to the Owner or increase in Contract Times, and

2) it will conform substantially to the detailed requirements of the item named in the Contract Documents.

2. Substitute Items

a. If in Engineer's sole discretion an item of material or equipment proposed by Contractor does not qualify as an "or-equal" item under Paragraph 6.05.A.1, it will be considered a proposed substitute item.

b. Contractor shall submit sufficient information as provided below to allow Engineer to determine that the item of material or equipment proposed is essentially equivalent to that named and an acceptable substitute therefor. Requests for review of proposed substitute items of material or equipment will not be accepted by Engineer from anyone other than Contractor.

c. The requirements for review by Engineer will be as set forth in Paragraph 6.05.A.2.d, as supplemented in the General Requirements and as Engineer may decide is appropriate under the circumstances.

d. Contractor shall make written application to Engineer for review of a proposed substitute item of material or equipment that Contractor seeks to furnish or use. The application:

1) shall certify that the proposed substitute item will:

a) perform adequately the functions and achieve the results called for by the general design,

b) be similar in substance to that specified, and

c) be suited to the same use as that specified;

2) will state:

a) the extent, if any, to which the use of the proposed substitute item will prejudice Contractor's achievement of Substantial Completion on time;

b) whether or not use of the proposed substitute item in the Work will require a change in any of the Contract Documents (or in the provisions of any other direct contract with Owner for other work on the Project) to adapt the design to the proposed substitute item; and

c) whether or not incorporation or use of the proposed substitute item in connection with the Work is subject to payment of any license fee or royalty;

3) will identify:

a) all variations of the proposed substitute item from that specified , and

b) available engineering, sales, maintenance, repair, and replacement services;

4) and shall contain an itemized estimate of all costs or credits that will result directly or indirectly from use of such substitute item, including costs of redesign and claims of other contractors affected by any resulting change,

B. *Substitute Construction Methods or Procedures:* If a specific means, method, technique, sequence, or procedure of construction is expressly required by the Contract Documents, Contractor may furnish or utilize a substitute means, method, technique, sequence, or procedure of construction approved by Engineer. Contractor shall submit sufficient information to allow Engineer, in Engineer's sole discretion, to determine that the substitute proposed is equivalent to that expressly called for by the Contract Documents. The requirements for review by Engineer will be similar to those provided in Paragraph 6.05.A.2.

C. *Engineer's Evaluation:* Engineer will be allowed a reasonable time within which to evaluate each proposal or submittal made pursuant to Paragraphs 6.05.A and 6.05.B. Engineer may require Contractor to furnish additional data about the proposed substitute item. Engineer will be the sole judge of acceptability. No "or equal" or substitute will be ordered, installed or utilized until Engineer's review is complete, which will be evidenced by either a Change Order for a substitute or an approved Shop Drawing for an "or equal." Engineer will advise Contractor in writing of any negative determination.

D. *Special Guarantee:* Owner may require Contractor to furnish at Contractor's expense a special performance guarantee or other surety with respect to any substitute.

E. *Engineer's Cost Reimbursement:* Engineer will record Engineer's costs in evaluating a substitute proposed or submitted by Contractor pursuant to Paragraphs 6.05.A.2 and 6.05.B Whether or not Engineer approves a substitute item so proposed or submitted by Contractor, Contractor shall reimburse Owner for the charges of Engineer for evaluating each such proposed substitute. Contractor shall also reimburse Owner for the charges of Engineer for making changes in the Contract

Documents (or in the provisions of any other direct contract with Owner) resulting from the acceptance of each proposed substitute.

F. *Contractor's Expense:* Contractor shall provide all data in support of any proposed substitute or "or-equal" at Contractor's expense.

6.06 *Concerning Subcontractors, Suppliers, and Others*

A. Contractor shall not employ any Subcontractor, Supplier, or other individual or entity (including those acceptable to Owner as indicated in Paragraph 6.06.B), whether initially or as a replacement, against whom Owner may have reasonable objection. Contractor shall not be required to employ any Subcontractor, Supplier, or other individual or entity to furnish or perform any of the Work against whom Contractor has reasonable objection.

B. If the Supplementary Conditions require the identity of certain Subcontractors, Suppliers, or other individuals or entities to be submitted to Owner in advance for acceptance by Owner by a specified date prior to the Effective Date of the Agreement, and if Contractor has submitted a list thereof in accordance with the Supplementary Conditions, Owner's acceptance (either in writing or by failing to make written objection thereto by the date indicated for acceptance or objection in the Bidding Documents or the Contract Documents) of any such Subcontractor, Supplier, or other individual or entity so identified may be revoked on the basis of reasonable objection after due investigation. Contractor shall submit an acceptable replacement for the rejected Subcontractor, Supplier, or other individual or entity, and the Contract Price will be adjusted by the difference in the cost occasioned by such replacement, and an appropriate Change Order will be issued . No acceptance by Owner of any such Subcontractor, Supplier, or other individual or entity, whether initially or as a replacement, shall constitute a waiver of any right of Owner or Engineer to reject defective Work.

C. Contractor shall be fully responsible to Owner and Engineer for all acts and omissions of the Subcontractors, Suppliers, and other individuals or entities performing or furnishing any of the Work just as Contractor is responsible for Contractor's own acts and omissions. Nothing in the Contract Documents:

1. shall create for the benefit of any such Subcontractor, Supplier, or other individual or entity any contractual relationship between Owner or Engineer and any such Subcontractor, Supplier or other individual or entity, nor

2. shall anything in the Contract Documents create any obligation on the part of Owner or Engineer to pay or to see to the payment of any moneys due any such Subcontractor, Supplier, or other individual

or entity except as may otherwise be required by Laws and Regulations.

D. Contractor shall be solely responsible for scheduling and coordinating the Work of Subcontractors, Suppliers, and other individuals or entities performing or furnishing any of the Work under a direct or indirect contract with Contractor.

E. Contractor shall require all Subcontractors, Suppliers, and such other individuals or entities performing or furnishing any of the Work to communicate with Engineer through Contractor.

F. The divisions and sections of the Specifications and the identifications of any Drawings shall not control Contractor in dividing the Work among Subcontractors or Suppliers or delineating the Work to be performed by any specific trade.

G. All Work performed for Contractor by a Subcontractor or Supplier will be pursuant to an appropriate agreement between Contractor and the Subcontractor or Supplier which specifically binds the Subcontractor or Supplier to the applicable terms and conditions of the Contract Documents for the benefit of Owner and Engineer. Whenever any such agreement is with a Subcontractor or Supplier who is listed as an additional insured on the property insurance provided in Paragraph 5.06, the agreement between the Contractor and the Subcontractor or Supplier will contain provisions whereby the Subcontractor or Supplier waives all rights against Owner, Contractor, and Engineer, and all other individuals or entities identified in the Supplementary Conditions to be listed as insureds or additional insureds (and the officers, directors, partners, employees, agents, consultants and subcontractors of each and any of them) for all losses and damages caused by, arising out of, relating to, or resulting from any of the perils or causes of loss covered by such policies and any other property insurance applicable to the Work. If the insurers on any such policies require separate waiver forms to be signed by any Subcontractor or Supplier, Contractor will obtain the same.

6.07   *Patent Fees and Royalties*

A. Contractor shall pay all license fees and royalties and assume all costs incident to the use in the performance of the Work or the incorporation in the Work of any invention, design, process, product, or device which is the subject of patent rights or copyrights held by others. If a particular invention, design, process, product, or device is specified in the Contract Documents for use in the performance of the Work and if to the actual knowledge of Owner or Engineer its use is subject to patent rights or copyrights calling for the payment of any license fee or royalty to others, the existence of such rights shall be disclosed by Owner in the Contract Documents.

B. To the fullest extent permitted by Laws and Regulations, Contractor shall indemnify and hold harmless Owner and Engineer, and the officers, directors, partners, employees, agents, consultants and subcontractors of each and any of them from and against all claims, costs, losses, and damages (including but not limited to all fees and charges of engineers, architects, attorneys, and other professionals and all court or arbitration or other dispute resolution costs) arising out of or relating to any infringement of patent rights or copyrights incident to the use in the performance of the Work or resulting from the incorporation in the Work of any invention, design, process, product, or device not specified in the Contract Documents.

6.08   *Permits*

A. Unless otherwise provided in the Supplementary Conditions, Contractor shall obtain and pay for all construction permits and licenses. Owner shall assist Contractor, when necessary, in obtaining such permits and licenses. Contractor shall pay all governmental charges and inspection fees necessary for the prosecution of the Work which are applicable at the time of opening of Bids, or, if there are no Bids, on the Effective Date of the Agreement. Owner shall pay all charges of utility owners for connections for providing permanent service to the Work.

6.09   *Laws and Regulations*

A. Contractor shall give all notices required by and shall comply with all Laws and Regulations applicable to the performance of the Work. Except where otherwise expressly required by applicable Laws and Regulations, neither Owner nor Engineer shall be responsible for monitoring Contractor's compliance with any Laws or Regulations.

B. If Contractor performs any Work knowing or having reason to know that it is contrary to Laws or Regulations, Contractor shall bear all claims, costs, losses, and damages (including but not limited to all fees and charges of engineers, architects, attorneys, and other professionals and all court or arbitration or other dispute resolution costs) arising out of or relating to such Work. However, it shall not be Contractor's primary responsibility to make certain that the Specifications and Drawings are in accordance with Laws and Regulations, but this shall not relieve Contractor of Contractor's obligations under Paragraph 3.03.

C. Changes in Laws or Regulations not known at the time of opening of Bids (or, on the Effective Date of the Agreement if there were no Bids) having an effect on the cost or time of performance of the Work shall be the subject of an adjustment in Contract Price or Contract Times. If Owner and Contractor are unable to agree on entitlement to or on the amount or extent, if any, of any such adjustment, a Claim may be made therefor as provided in Paragraph 10.05.

6.10 *Taxes*

A. Contractor shall pay all sales, consumer, use, and other similar taxes required to be paid by Contractor in accordance with the Laws and Regulations of the place of the Project which are applicable during the performance of the Work.

6.11 *Use of Site and Other Areas*

A. Limitation on Use of Site and Other Areas

1. Contractor shall confine construction equipment, the storage of materials and equipment, and the operations of workers to the Site and other areas permitted by Laws and Regulations, and shall not unreasonably encumber the Site and other areas with construction equipment or other materials or equipment. Contractor shall assume full responsibility for any damage to any such land or area, or to the owner or occupant thereof, or of any adjacent land or areas resulting from the performance of the Work.

2. Should any claim be made by any such owner or occupant because of the performance of the Work, Contractor shall promptly settle with such other party by negotiation or otherwise resolve the claim by arbitration or other dispute resolution proceeding or at law.

3. To the fullest extent permitted by Laws and Regulations, Contractor shall indemnify and hold harmless Owner and Engineer, and the officers, directors, partners, employees, agents, consultants and subcontractors of each and any of them from and against all claims, costs, losses, and damages (including but not limited to all fees and charges of engineers, architects, attorneys, and other professionals and all court or arbitration or other dispute resolution costs) arising out of or relating to any claim or action, legal or equitable, brought by any such owner or occupant against Owner, Engineer or any other party indemnified hereunder to the extent caused by or based upon Contractor's performance of the Work.

B. *Removal of Debris During Performance of the Work:* During the progress of the Work Contractor shall keep the Site and other areas free from accumulations of waste materials, rubbish, and other debris. Removal and disposal of such waste materials, rubbish, and other debris shall conform to applicable Laws and Regulations.

C. *Cleaning:* Prior to Substantial Completion of the Work Contractor shall clean the Site and the Work and make it ready for utilization by Owner. At the completion of the Work Contractor shall remove from the Site all tools, appliances, construction equipment and machinery, and surplus materials and shall restore to original condition all property not designated for alteration by the Contract Documents.

D. *Loading Structures:* Contractor shall not load nor permit any part of any structure to be loaded in any manner that will endanger the structure, nor shall Contractor subject any part of the Work or adjacent property to stresses or pressures that will endanger it.

6.12 *Record Documents*

A. Contractor shall maintain in a safe place at the Site one record copy of all Drawings, Specifications, Addenda, Change Orders, Work Change Directives, Field Orders, and written interpretations and clarifications in good order and annotated to show changes made during construction. These record documents together with all approved Samples and a counterpart of all approved Shop Drawings will be available to Engineer for reference. Upon completion of the Work, these record documents, Samples, and Shop Drawings will be delivered to Engineer for Owner.

6.13 *Safety and Protection*

A. Contractor shall be solely responsible for initiating, maintaining and supervising all safety precautions and programs in connection with the Work. Contractor shall take all necessary precautions for the safety of, and shall provide the necessary protection to prevent damage, injury or loss to:

1. all persons on the Site or who may be affected by the Work;

2. all the Work and materials and equipment to be incorporated therein, whether in storage on or off the Site; and

3. other property at the Site or adjacent thereto, including trees, shrubs, lawns, walks, pavements, roadways, structures, utilities, and Underground Facilities not designated for removal, relocation, or replacement in the course of construction.

B. Contractor shall comply with all applicable Laws and Regulations relating to the safety of persons or property, or to the protection of persons or property from damage, injury, or loss; and shall erect and maintain all necessary safeguards for such safety and protection. Contractor shall notify owners of adjacent property and of Underground Facilities and other utility owners when prosecution of the Work may affect them, and shall cooperate with them in the protection, removal, relocation, and replacement of their property.

C. All damage, injury, or loss to any property referred to in Paragraph 6.13.A.2 or 6.13.A.3 caused, directly or indirectly, in whole or in part, by Contractor, any Subcontractor, Supplier, or any other individual or entity directly or indirectly employed by any of them to perform any of the Work, or anyone for whose acts any of them may be liable, shall be remedied by Contractor (except damage or loss attributable to the fault of Drawings or Specifications or to the acts or omissions of Owner or Engineer or , or anyone employed by any of them, or anyone for whose acts any of them may be liable, and not attributable, directly or indirectly, in whole or in part, to the fault or negligence of Contractor or any Subcontractor, Supplier, or other individual or entity directly or indirectly employed by any of them).

D. Contractor's duties and responsibilities for safety and for protection of the Work shall continue until such time as all the Work is completed and Engineer has issued a notice to Owner and Contractor in accordance with Paragraph 14.07.B that the Work is acceptable (except as otherwise expressly provided in connection with Substantial Completion).

6.14   *Safety Representative*

A. Contractor shall designate a qualified and experienced safety representative at the Site whose duties and responsibilities shall be the prevention of accidents and the maintaining and supervising of safety precautions and programs.

6.15   *Hazard Communication Programs*

A. Contractor shall be responsible for coordinating any exchange of material safety data sheets or other hazard communication information required to be made available to or exchanged between or among employers at the Site in accordance with Laws or Regulations.

6.16   *Emergencies*

A. In emergencies affecting the safety or protection of persons or the Work or property at the Site or adjacent thereto, Contractor is obligated to act to prevent threatened damage, injury or loss. Contractor shall give Engineer prompt written notice if Contractor believes that any significant changes in the Work or variations from the Contract Documents have been caused thereby or are required as a result thereof. If Engineer determines that a change in the Contract Documents is required because of the action taken by Contractor in response to such an emergency, a Work Change Directive or Change Order will be issued.

6.17   *Shop Drawings and Samples*

A. Contractor shall submit Shop Drawings and Samples to Engineer for review and approval in accordance with the acceptable Schedule of Submittals (as required by Paragraph 2.07). Each submittal will be identified as Engineer may require.

1. Shop Drawings

a. Submit number of copies specified in the General Requirements.

b. Data shown on the Shop Drawings will be complete with respect to quantities, dimensions, specified performance and design criteria, materials, and similar data to show Engineer the services, materials, and equipment Contractor proposes to provide and to enable Engineer to review the information for the limited purposes required by Paragraph 6.17.D.

2. *Samples:* Contractor shall also submit Samples to Engineer for review and approval in accordance with the acceptable schedule of Shop Drawings and Sample submittals.

a. Submit number of Samples specified in the Specifications.

b. Clearly identify each Sample as to material, Supplier, pertinent data such as catalog numbers, the use for which intended and other data as Engineer may require to enable Engineer to review the submittal for the limited purposes required by Paragraph 6.17.D.

B. Where a Shop Drawing or Sample is required by the Contract Documents or the Schedule of Submittals , any related Work performed prior to Engineer's review and approval of the pertinent submittal will be at the sole expense and responsibility of Contractor.

C. Submittal Procedures

1. Before submitting each Shop Drawing or Sample, Contractor shall have determined and verified:

a. all field measurements, quantities, dimensions, specified performance and design criteria, installation requirements, materials, catalog numbers, and similar information with respect thereto;

b. the suitability of all materials with respect to intended use, fabrication, shipping, handling, storage, assembly, and installation pertaining to the performance of the Work;

c. all information relative to Contractor's responsibilities for means, methods, techniques, sequences, and procedures of construction, and safety precautions and programs incident thereto; and

d. shall also have reviewed and coordinated each Shop Drawing or Sample with other Shop Drawings and Samples and with the requirements of the Work and the Contract Documents.

2. Each submittal shall bear a stamp or specific written certification that Contractor has satisfied Contractor's obligations under the Contract Documents

with respect to Contractor's review and approval of that submittal.

3. With each submittal, Contractor shall give Engineer specific written notice of any variations, that the Shop Drawing or Sample may have from the requirements of the Contract Documents. This notice shall be both a written communication separate from the Shop Drawing's or Sample Submittal; and, in addition, by a specific notation made on each Shop Drawing or Sample submitted to Engineer for review and approval of each such variation.

D. *Engineer's Review*

1. Engineer will provide timely review of Shop Drawings and Samples in accordance with the Schedule of Submittals acceptable to Engineer. Engineer's review and approval will be only to determine if the items covered by the submittals will, after installation or incorporation in the Work, conform to the information given in the Contract Documents and be compatible with the design concept of the completed Project as a functioning whole as indicated by the Contract Documents.

2. Engineer's review and approval will not extend to means, methods, techniques, sequences, or procedures of construction (except where a particular means, method, technique, sequence, or procedure of construction is specifically and expressly called for by the Contract Documents) or to safety precautions or programs incident thereto. The review and approval of a separate item as such will not indicate approval of the assembly in which the item functions.

3. Engineer's review and approval shall not relieve Contractor from responsibility for any variation from the requirements of the Contract Documents unless Contractor has complied with the requirements of Paragraph 6.17.C.3 and Engineer has given written approval of each such variation by specific written notation thereof incorporated in or accompanying the Shop Drawing or Sample. Engineer's review and approval shall not relieve Contractor from responsibility for complying with the requirements of Paragraph 6.17.C.1.

E. *Resubmittal Procedures*

1. Contractor shall make corrections required by Engineer and shall return the required number of corrected copies of Shop Drawings and submit, as required, new Samples for review and approval. Contractor shall direct specific attention in writing to revisions other than the corrections called for by Engineer on previous submittals.

6.18 *Continuing the Work*

A. Contractor shall carry on the Work and adhere to the Progress Schedule during all disputes or disagreements with Owner. No Work shall be delayed or postponed pending resolution of any disputes or disagreements, except as permitted by Paragraph 15.04 or as Owner and Contractor may otherwise agree in writing.

6.19 *Contractor's General Warranty and Guarantee*

A. Contractor warrants and guarantees to Owner that all Work will be in accordance with the Contract Documents and will not be defective. Engineer and its Related Entities shall be entitled to rely on representation of Contractor's warranty and guarantee.

B. Contractor's warranty and guarantee hereunder excludes defects or damage caused by:

1. abuse, modification, or improper maintenance or operation by persons other than Contractor, Subcontractors, Suppliers, or any other individual or entity for whom Contractor is responsible; or

2. normal wear and tear under normal usage.

C. Contractor's obligation to perform and complete the Work in accordance with the Contract Documents shall be absolute. None of the following will constitute an acceptance of Work that is not in accordance with the Contract Documents or a release of Contractor's obligation to perform the Work in accordance with the Contract Documents:

1. observations by Engineer;

2. recommendation by Engineer or payment by Owner of any progress or final payment;

3. the issuance of a certificate of Substantial Completion by Engineer or any payment related thereto by Owner;

4. use or occupancy of the Work or any part thereof by Owner;

5. any review and approval of a Shop Drawing or Sample submittal or the issuance of a notice of acceptability by Engineer;

6. any inspection, test, or approval by others; or

7. any correction of defective Work by Owner.

6.20 *Indemnification*

A. To the fullest extent permitted by Laws and Regulations, Contractor shall indemnify and hold harmless Owner and Engineer, and the officers, directors, partners, employees, agents, consultants and subcontractors of each and any of them from and against all claims, costs, losses, and damages (including but not limited to all fees and charges of engineers, architects, attorneys, and other professionals and all court or arbitration or other dispute resolution costs) arising out of or relating to the performance of the Work, provided that any such claim, cost, loss, or damage is attributable to bodily injury, sickness, disease, or death, or to injury to or destruction of tangible property (other than the Work itself), including the loss of use resulting therefrom but only to the extent caused by any negligent act or omission of Contractor, any Subcontractor, any Supplier, or any individual or entity directly or indirectly employed by any of them to perform any of the Work or anyone for whose acts any of them may be liable .

B. In any and all claims against Owner or Engineer or any of their respective consultants, agents, officers, directors, partners, or employees by any employee (or the survivor or personal representative of such employee) of Contractor, any Subcontractor, any Supplier, or any individual or entity directly or indirectly employed by any of them to perform any of the Work, or anyone for whose acts any of them may be liable, the indemnification obligation under Paragraph 6.20.A shall not be limited in any way by any limitation on the amount or type of damages, compensation, or benefits payable by or for Contractor or any such Subcontractor, Supplier, or other individual or entity under workers' compensation acts, disability benefit acts, or other employee benefit acts.

C. The indemnification obligations of Contractor under Paragraph 6.20.A shall not extend to the liability of Engineer and Engineer's officers, directors, partners, employees, agents, consultants and subcontractors arising out of:

1. the preparation or approval of or the failure to prepare or approve, maps, Drawings, opinions, reports, surveys, Change Orders, designs, or Specifications; or

2. giving directions or instructions, or failing to give them, if that is the primary cause of the injury or damage.

6.21   *Delegation of Professional Design Services*

A. Contractor will not be required to provide professional design services unless such services are specifically required by the Contract Documents for a portion of the Work or unless such services are required to carry out Contractor's responsibilities for construction means, methods, techniques, sequences and procedures. Contractor shall not be required to provide professional services in violation of applicable law.

B. If professional design services or certifications by a design professional related to systems, materials or equipment are specifically required of Contractor by the Contract Documents, Owner and Engineer will specify all performance and design criteria that such services must satisfy. Contractor shall cause such services or certifications to be provided by a properly licensed professional, whose signature and seal

shall appear on all drawings, calculations, specifications, certifications, Shop Drawings and other submittals prepared by such professional. Shop Drawings and other submittals related to the Work designed or certified by such professional, if prepared by others, shall bear such professional's written approval when submitted to Engineer.

C. Owner and Engineer shall be entitled to rely upon the adequacy, accuracy and completeness of the services, certifications or approvals performed by such design professionals, provided Owner and Engineer have specified to Contractor all performance and design criteria that such services must satisfy.

D. Pursuant to this Paragraph 6.21, Engineer's review and approval of design calculations and design drawings will be only for the limited purpose of checking for conformance with performance and design criteria given and the design concept expressed in the Contract Documents. Engineer's review and approval of Shop Drawings and other submittals (except design calculations and design drawings) will be only for the purpose stated in Paragraph 6.17.D.1.

E. Contractor shall not be responsible for the adequacy of the performance or design criteria required by the Contract Documents.

ARTICLE 7 - OTHER WORK AT THE SITE

7.01   *Related Work at Site*

A. Owner may perform other work related to the Project at the Site with Owner's employees, or via other direct contracts therefor, or have other work performed by utility owners. If such other work is not noted in the Contract Documents, then:

1. written notice thereof will be given to Contractor prior to starting any such other work; and

2. if Owner and Contractor are unable to agree on entitlement to or on the amount or extent, if any, of any adjustment in the Contract Price or Contract Times that should be allowed as a result of such other work, a Claim may be made therefor as provided in Paragraph 10.05.

B. Contractor shall afford each other contractor who is a party to such a direct contract, each utility owner and Owner, if Owner is performing other work with Owner's employees, proper and safe access to the Site, a reasonable opportunity for the introduction and storage of materials and equipment and the execution of such other work, and shall properly coordinate the Work with theirs. Contractor shall do all cutting, fitting, and patching of the Work that may be required to properly connect or otherwise make its several parts come together and properly integrate with such other work. Contractor shall not endanger any work of others by cutting, excavating, or otherwise altering their work and will only cut or alter their work with the written consent of Engineer and the others whose work will be affected. The duties and responsibilities of Contractor under this Paragraph are for the benefit of such utility owners and other contractors to the extent that there are comparable provisions for the benefit of Contractor in said direct contracts between Owner and such utility owners and other contractors.

C. If the proper execution or results of any part of Contractor's Work depends upon work performed by others under this Article 7, Contractor shall inspect such other work and promptly report to Engineer in writing any delays, defects, or deficiencies in such other work that render it unavailable or unsuitable for the proper execution and results of Contractor's Work. Contractor's failure to so report will constitute an acceptance of such other work as fit and proper for integration with Contractor's Work except for latent defects and deficiencies in such other work.

7.02   *Coordination*

A. If Owner intends to contract with others for the performance of other work on the Project at the Site, the following will be set forth in Supplementary Conditions:

1. the individual or entity who will have authority and responsibility for coordination of the activities among the various contractors will be identified;

2. the specific matters to be covered by such authority and responsibility will be itemized; and

3. the extent of such authority and responsibilities will be provided.

B. Unless otherwise provided in the Supplementary Conditions, Owner shall have sole authority and responsibility for such coordination.

7.03   *Legal Relationships*

A. Paragraphs 7.01.A and 7.02 are not applicable for utilities not under the control of Owner.

B. Each other direct contract of Owner under Paragraph 7.01.A shall provide that the other contractor is liable to Owner and Contractor for the reasonable direct delay and disruption costs incurred by Contractor as a result of the other contractor's actions or inactions.

C. Contractor shall be liable to Owner and any other contractor for the reasonable direct delay and disruption costs incurred by such other contractor as a result of Contractor's action or inactions.

ARTICLE 8 - OWNER'S RESPONSIBILITIES

8.01   *Communications to Contractor*

A. Except as otherwise provided in these General Conditions, Owner shall issue all communications to Contractor through Engineer.

8.02   *Replacement of Engineer*

A. In case of termination of the employment of Engineer, Owner shall appoint an engineer to whom Contractor makes no reasonable objection, whose status under the Contract Documents shall be that of the former Engineer.

8.03   *Furnish Data*

A. Owner shall promptly furnish the data required of Owner under the Contract Documents.

8.04   *Pay When Due*

A. Owner shall make payments to Contractor when they are due as provided in Paragraphs 14.02.C and 14.07.C.

8.05   *Lands and Easements; Reports and Tests*

A. Owner's duties in respect of providing lands and easements and providing engineering surveys to establish reference points are set forth in Paragraphs 4.01 and 4.05. Paragraph 4.02 refers to Owner's identifying and making available to Contractor copies of reports of explorations and tests of subsurface conditions and drawings of physical conditions in or relating to existing surface or subsurface structures at or contiguous to the Site that have been utilized by Engineer in preparing the Contract Documents.

8.06   *Insurance*

A. Owner's responsibilities, if any, in respect to purchasing and maintaining liability and property insurance are set forth in Article 5.

8.07   *Change Orders*

A. Owner is obligated to execute Change Orders as indicated in Paragraph 10.03.

8.08   *Inspections, Tests, and Approvals*

A. Owner's responsibility in respect to certain inspections, tests, and approvals is set forth in Paragraph 13.03.B.

8.09 *Limitations on Owner's Responsibilities*

A. The Owner shall not supervise, direct, or have control or authority over, nor be responsible for, Contractor's means, methods, techniques, sequences, or procedures of construction, or the safety precautions and programs incident thereto, or for any failure of Contractor to comply with Laws and Regulations applicable to the performance of the Work. Owner will not be responsible for Contractor's failure to perform the Work in accordance with the Contract Documents.

8.10 *Undisclosed Hazardous Environmental Condition*

A. Owner's responsibility in respect to an undisclosed Hazardous Environmental Condition is set forth in Paragraph 4.06.

8.11 *Evidence of Financial Arrangements*

A. If and to the extent Owner has agreed to furnish Contractor reasonable evidence that financial arrangements have been made to satisfy Owner's obligations under the Contract Documents, Owner's responsibility in respect thereof will be as set forth in the Supplementary Conditions.

ARTICLE 9 - ENGINEER'S STATUS DURING CONSTRUCTION

9.01 *Owner's Representative*

A. Engineer will be Owner's representative during the construction period. The duties and responsibilities and the limitations of authority of Engineer as Owner's representative during construction are set forth in the Contract Documents and will not be changed without written consent of Owner and Engineer.

9.02 *Visits to Site*

A. Engineer will make visits to the Site at intervals appropriate to the various stages of construction as Engineer deems necessary in order to observe as an experienced and qualified design professional the progress that has been made and the quality of the various aspects of Contractor's executed Work. Based on information obtained during such visits and observations, Engineer, for the benefit of Owner, will determine, in general, if the Work is proceeding in accordance with the Contract Documents. Engineer will not be required to make exhaustive or continuous inspections on the Site to check the quality or quantity of the Work. Engineer's efforts will be directed toward providing for Owner a greater degree of confidence that the completed Work will conform generally to the Contract Documents. On the basis of such visits and observations, Engineer will keep Owner informed of the progress of the Work and will endeavor to guard Owner against defective Work.

B. Engineer's visits and observations are subject to all the limitations on Engineer's authority and responsibility set forth in Paragraph 9.09. Particularly, but without limitation, during or as a result of Engineer's visits or observations of Contractor's Work Engineer will not supervise, direct, control, or have authority over or be responsible for Contractor's means, methods, techniques, sequences, or procedures of construction, or the safety precautions and programs incident thereto, or for any failure of Contractor to comply with Laws and Regulations applicable to the performance of the Work.

9.03 *Project Representative*

A. If Owner and Engineer agree, Engineer will furnish a Resident Project Representative to assist Engineer in providing more extensive observation of the Work. The authority and responsibilities of any such Resident Project Representative and assistants will be as provided in the Supplementary Conditions, and limitations on the responsibilities thereof will be as provided in Paragraph 9.09. If Owner designates another representative or agent to represent Owner at the Site who is not Engineer's consultant, agent or employee, the responsibilities and authority and limitations thereon of such other individual or entity will be as provided in the Supplementary Conditions.

9.04 *Authorized Variations in Work*

A. Engineer may authorize minor variations in the Work from the requirements of the Contract Documents which do not involve an adjustment in the Contract Price or the Contract Times and are compatible with the design concept of the completed Project as a functioning whole as indicated by the Contract Documents. These may be accomplished by a Field Order and will be binding on Owner and also on Contractor, who shall perform the Work involved promptly. If Owner or Contractor believes that a Field Order justifies an adjustment in the Contract Price or Contract Times, or both, and the parties are unable to agree on entitlement to or on the amount or extent, if any, of any such adjustment, a Claim may be made therefor as provided in Paragraph 10.05.

9.05 *Rejecting Defective Work*

A. Engineer will have authority to reject Work which Engineer believes to be defective, or that Engineer believes will not produce a completed Project that conforms to the Contract Documents or that will prejudice the integrity of the design concept of the completed Project as a functioning whole as indicated by the Contract Documents. Engineer will also have authority to require special inspection or testing of the Work as provided in Paragraph 13.04, whether or not the Work is fabricated, installed, or completed.

9.06 *Shop Drawings, Change Orders and Payments*

A. In connection with Engineer's authority, and limitations thereof, as to Shop Drawings and Samples, see Paragraph 6.17.

B. In connection with Engineer's authority, and limitations thereof, as to design calculations and design drawings submitted in response to a delegation of professional design services, if any, see Paragraph 6.21.

C. In connection with Engineer's authority as to Change Orders, see Articles 10, 11, and 12.

D. In connection with Engineer's authority as to Applications for Payment, see Article 14.

9.07 *Determinations for Unit Price Work*

A. Engineer will determine the actual quantities and classifications of Unit Price Work performed by Contractor. Engineer will review with Contractor the Engineer's preliminary determinations on such matters before rendering a written decision thereon (by recommendation of an Application for Payment or otherwise). Engineer's written decision thereon will be final and binding (except as modified by Engineer to reflect changed factual conditions or more accurate data) upon Owner and Contractor, subject to the provisions of Paragraph 10.05.

9.08 *Decisions on Requirements of Contract Documents and Acceptability of Work*

A. Engineer will be the initial interpreter of the requirements of the Contract Documents and judge of the acceptability of the Work thereunder. All matters in question and other matters between Owner and Contractor arising prior to the date final payment is due relating to the acceptability of the Work, and the interpretation of the requirements of the Contract Documents pertaining to the performance of the Work, will be referred initially to Engineer in writing within 30 days of the event giving rise to the question.

B. Engineer will, with reasonable promptness, render a written decision on the issue referred. If Owner or Contractor believe that any such decision entitles them to an adjustment in the Contract Price or Contract Times or both, a Claim may be made under Paragraph 10.05. The date of Engineer's decision shall be the date of the event giving rise to the issues referenced for the purposes of Paragraph 10.05.B.

C. Engineer's written decision on the issue referred will be final and binding on Owner and Contractor, subject to the provisions of Paragraph 10.05.

D. When functioning as interpreter and judge under this Paragraph 9.08, Engineer will not show partiality to Owner or Contractor and will not be liable in connection with any interpretation or decision rendered in good faith in such capacity.

9.09 *Limitations on Engineer's Authority and Responsibilities*

A. Neither Engineer's authority or responsibility under this Article 9 or under any other provision of the Contract Documents nor any decision made by Engineer in good faith either to exercise or not exercise such authority or responsibility or the undertaking, exercise, or performance of any authority or responsibility by Engineer shall create, impose, or give rise to any duty in contract, tort, or otherwise owed by Engineer to Contractor, any Subcontractor, any Supplier, any other individual or entity, or to any surety for or employee or agent of any of them.

B. Engineer will not supervise, direct, control, or have authority over or be responsible for Contractor's means, methods, techniques, sequences, or procedures of construction, or the safety precautions and programs incident thereto, or for any failure of Contractor to comply with Laws and Regulations applicable to the performance of the Work. Engineer will not be responsible for Contractor's failure to perform the Work in accordance with the Contract Documents.

C. Engineer will not be responsible for the acts or omissions of Contractor or of any Subcontractor, any Supplier, or of any other individual or entity performing any of the Work.

D. Engineer's review of the final Application for Payment and accompanying documentation and all maintenance and operating instructions, schedules, guarantees, bonds, certificates of inspection, tests and approvals, and other documentation required to be delivered by Paragraph 14.07.A will only be to determine generally that their content complies with the requirements of, and in the case of certificates of inspections, tests, and approvals that the results certified indicate compliance with the Contract Documents.

E. The limitations upon authority and responsibility set forth in this Paragraph 9.09 shall also apply to, the Resident Project Representative, if any, and assistants, if any.

ARTICLE 10 - CHANGES IN THE WORK; CLAIMS

10.01 *Authorized Changes in the Work*

A. Without invalidating the Contract and without notice to any surety, Owner may, at any time or from time to time, order additions, deletions, or revisions in the Work by a Change Order, or a Work Change Directive. Upon receipt of any such document, Contractor shall promptly proceed with the Work involved which will be performed under the applicable conditions of the Contract Documents (except as otherwise specifically provided).

B. If Owner and Contractor are unable to agree on entitlement to, or on the amount or extent, if any, of an adjustment in the Contract Price or Contract Times, or both, that should be allowed as a result of a Work Change Directive, a Claim may be made therefor as provided in Paragraph 10.05.

10.02   Unauthorized Changes in the Work

A. Contractor shall not be entitled to an increase in the Contract Price or an extension of the Contract Times with respect to any work performed that is not required by the Contract Documents as amended, modified, or supplemented as provided in Paragraph 3.04, except in the case of an emergency as provided in Paragraph 6.16 or in the case of uncovering Work as provided in Paragraph 13.04.B.

10.03   Execution of Change Orders

A. Owner and Contractor shall execute appropriate Change Orders recommended by Engineer covering:

1. changes in the Work which are: (i) ordered by Owner pursuant to Paragraph 10.01.A, (ii) required because of acceptance of defective Work under Paragraph 13.08.A or Owner's correction of defective Work under Paragraph 13.09, or (iii) agreed to by the parties;

2. changes in the Contract Price or Contract Times which are agreed to by the parties, including any undisputed sum or amount of time for Work actually performed in accordance with a Work Change Directive; and

3. changes in the Contract Price or Contract Times which embody the substance of any written decision rendered by Engineer pursuant to Paragraph 10.05; provided that, in lieu of executing any such Change Order, an appeal may be taken from any such decision in accordance with the provisions of the Contract Documents and applicable Laws and Regulations, but during any such appeal, Contractor shall carry on the Work and adhere to the Progress Schedule as provided in Paragraph 6.18.A.

10.04   Notification to Surety

A. If notice of any change affecting the general scope of the Work or the provisions of the Contract Documents (including, but not limited to, Contract Price or Contract Times) is required by the provisions of any bond to be given to a surety, the giving of any such notice will be Contractor's responsibility. The amount of each applicable bond will be adjusted to reflect the effect of any such change.

10.05   Claims

A. *Engineer's Decision Required*: All Claims, except those waived pursuant to Paragraph 14.09, shall be referred to the Engineer for decision. A decision by Engineer shall be required as a condition precedent to any exercise by Owner or Contractor of any rights or remedies either may otherwise have under the Contract Documents or by Laws and Regulations in respect of such Claims.

B. *Notice:* Written notice stating the general nature of each Claim, shall be delivered by the claimant to Engineer and the other party to the Contract promptly (but in no event later than 30 days) after the start of the event giving rise thereto. The responsibility to substantiate a Claim shall rest with the party making the Claim. Notice of the amount or extent of the Claim, with supporting data shall be delivered to the Engineer and the other party to the Contract within 60 days after the start of such event (unless Engineer allows additional time for claimant to submit additional or more accurate data in support of such Claim). A Claim for an adjustment in Contract Price shall be prepared in accordance with the provisions of Paragraph 12.01.B. A Claim for an adjustment in Contract Time shall be prepared in accordance with the provisions of Paragraph 12.02.B. Each Claim shall be accompanied by claimant's written statement that the adjustment claimed is the entire adjustment to which the claimant believes it is entitled as a result of said event. The opposing party shall submit any response to Engineer and the claimant within 30 days after receipt of the claimant's last submittal (unless Engineer allows additional time).

C. *Engineer's Action*: Engineer will review each Claim and, within 30 days after receipt of the last submittal of the claimant or the last submittal of the opposing party, if any, take one of the following actions in writing:

1. deny the Claim in whole or in part,

2. approve the Claim, or

3. notify the parties that the Engineer is unable to resolve the Claim if, in the Engineer's sole discretion, it would be inappropriate for the Engineer to do so. For purposes of further resolution of the Claim, such notice shall be deemed a denial.

D. In the event that Engineer does not take action on a Claim within said 30 days, the Claim shall be deemed denied.

E. Engineer's written action under Paragraph 10.05.C or denial pursuant to Paragraphs 10.05.C.3 or 10.05.D will be final and binding upon Owner and Contractor, unless Owner or Contractor invoke the dispute resolution procedure set forth in Article 16 within 30 days of such action or denial.

F. No Claim for an adjustment in Contract Price or Contract Times will be valid if not submitted in accordance with this Paragraph 10.05.

ARTICLE 11 - COST OF THE WORK; ALLOWANCES; UNIT PRICE WORK

11.01 *Cost of the Work*

A. *Costs Included:* The term Cost of the Work means the sum of all costs, except those excluded in Paragraph 11.01.B, necessarily incurred and paid by Contractor in the proper performance of the Work. When the value of any Work covered by a Change Order or when a Claim for an adjustment in Contract Price is determined on the basis of Cost of the Work, the costs to be reimbursed to Contractor will be only those additional or incremental costs required because of the change in the Work or because of the event giving rise to the Claim. Except as otherwise may be agreed to in writing by Owner, such costs shall be in amounts no higher than those prevailing in the locality of the Project, shall include only the following items, and shall not include any of the costs itemized in Paragraph 11.01.B.

1. Payroll costs for employees in the direct employ of Contractor in the performance of the Work under schedules of job classifications agreed upon by Owner and Contractor. Such employees shall include, without limitation, superintendents, foremen, and other personnel employed full time at the Site. Payroll costs for employees not employed full time on the Work shall be apportioned on the basis of their time spent on the Work. Payroll costs shall include, but not be limited to, salaries and wages plus the cost of fringe benefits, which shall include social security contributions, unemployment, excise, and payroll taxes, workers' compensation, health and retirement benefits, bonuses, sick leave, vacation and holiday pay applicable thereto. The expenses of performing Work outside of regular working hours, on Saturday, Sunday, or legal holidays, shall be included in the above to the extent authorized by Owner.

2. Cost of all materials and equipment furnished and incorporated in the Work, including costs of transportation and storage thereof, and Suppliers' field services required in connection therewith. All cash discounts shall accrue to Contractor unless Owner deposits funds with Contractor with which to make payments, in which case the cash discounts shall accrue to Owner. All trade discounts, rebates and refunds and returns from sale of surplus materials and equipment shall accrue to Owner, and Contractor shall make provisions so that they may be obtained.

3. Payments made by Contractor to Subcontractors for Work performed by Subcontractors. If required by Owner, Contractor shall obtain competitive bids from subcontractors acceptable to Owner and Contractor and shall deliver such bids to Owner, who will then determine, with the advice of Engineer, which bids, if any, will be acceptable. If any subcontract provides that the Subcontractor is to be paid on the basis of Cost of the Work plus a fee, the Subcontractor's Cost of the Work and fee shall be determined in the same manner as Contractor's Cost of the Work and fee as provided in this Paragraph 11.01.

4. Costs of special consultants (including but not limited to Engineers, architects, testing laboratories, surveyors, attorneys, and accountants) employed for services specifically related to the Work.

5. Supplemental costs including the following:

a. The proportion of necessary transportation, travel, and subsistence expenses of Contractor's employees incurred in discharge of duties connected with the Work.

b. Cost, including transportation and maintenance, of all materials, supplies, equipment, machinery, appliances, office, and temporary facilities at the Site, and hand tools not owned by the workers, which are consumed in the performance of the Work, and cost, less market value, of such items used but not consumed which remain the property of Contractor.

c. Rentals of all construction equipment and machinery, and the parts thereof whether rented from Contractor or others in accordance with rental agreements approved by Owner with the advice of Engineer, and the costs of transportation, loading, unloading, assembly, dismantling, and removal thereof. All such costs shall be in accordance with the terms of said rental agreements. The rental of any such equipment, machinery, or parts shall cease when the use thereof is no longer necessary for the Work.

d. Sales, consumer, use, and other similar taxes related to the Work, and for which Contractor is liable, imposed by Laws and Regulations.

e. Deposits lost for causes other than negligence of Contractor, any Subcontractor, or anyone directly or indirectly employed by any of them or for whose acts any of them may be liable, and royalty payments and fees for permits and licenses.

f. Losses and damages (and related expenses) caused by damage to the Work, not compensated by insurance or otherwise, sustained by Contractor in connection with the performance of the Work (except losses and damages within the deductible amounts of property insurance established in accordance with Paragraph

5.06.D), provided such losses and damages have resulted from causes other than the negligence of Contractor, any Subcontractor, or anyone directly or indirectly employed by any of them or for whose acts any of them may be liable. Such losses shall include settlements made with the written consent and approval of Owner. No such losses, damages, and expenses shall be included in the Cost of the Work for the purpose of determining Contractor's fee.

g. The cost of utilities, fuel, and sanitary facilities at the Site.

h. Minor expenses such as telegrams, long distance telephone calls, telephone service at the Site, expresses, and similar petty cash items in connection with the Work.

i. The costs of premiums for all bonds and insurance Contractor is required by the Contract Documents to purchase and maintain.

B. *Costs Excluded:* The term Cost of the Work shall not include any of the following items:

1. Payroll costs and other compensation of Contractor's officers, executives, principals (of partnerships and sole proprietorships), general managers, safety managers, engineers, architects, estimators, attorneys, auditors, accountants, purchasing and contracting agents, expediters, timekeepers, clerks, and other personnel employed by Contractor, whether at the Site or in Contractor's principal or branch office for general administration of the Work and not specifically included in the agreed upon schedule of job classifications referred to in Paragraph 11.01.A.1 or specifically covered by Paragraph 11.01.A.4, all of which are to be considered administrative costs covered by the Contractor's fee.

2. Expenses of Contractor's principal and branch offices other than Contractor's office at the Site.

3. Any part of Contractor's capital expenses, including interest on Contractor's capital employed for the Work and charges against Contractor for delinquent payments.

4. Costs due to the negligence of Contractor, any Subcontractor, or anyone directly or indirectly employed by any of them or for whose acts any of them may be liable, including but not limited to, the correction of defective Work, disposal of materials or equipment wrongly supplied, and making good any damage to property.

5. Other overhead or general expense costs of any kind and the costs of any item not specifically and expressly included in Paragraphs 11.01.A and 11.01.B.

C. *Contractor's Fee:* When all the Work is performed on the basis of cost-plus, Contractor's fee shall be determined as set forth in the Agreement. When the value of any Work covered by a Change Order or when a Claim for an adjustment in Contract Price is determined on the basis of Cost of the Work, Contractor's fee shall be determined as set forth in Paragraph 12.01.C.

D. *Documentation:* Whenever the Cost of the Work for any purpose is to be determined pursuant to Paragraphs 11.01.A and 11.01.B, Contractor will establish and maintain records thereof in accordance with generally accepted accounting practices and submit in a form acceptable to Engineer an itemized cost breakdown together with supporting data.

11.02 *Allowances*

A. It is understood that Contractor has included in the Contract Price all allowances so named in the Contract Documents and shall cause the Work so covered to be performed for such sums and by such persons or entities as may be acceptable to Owner and Engineer.

B. *Cash Allowances*

1. Contractor agrees that:

a. the cash allowances include the cost to Contractor (less any applicable trade discounts) of materials and equipment required by the allowances to be delivered at the Site, and all applicable taxes; and

b. Contractor's costs for unloading and handling on the Site, labor, installation , overhead, profit, and other expenses contemplated for the cash allowances have been included in the Contract Price and not in the allowances, and no demand for additional payment on account of any of the foregoing will be valid.

C. Contingency Allowance

1. Contractor agrees that a contingency allowance, if any, is for the sole use of Owner to cover unanticipated costs.

D. Prior to final payment, an appropriate Change Order will be issued as recommended by Engineer to reflect actual amounts due Contractor on account of Work covered by allowances, and the Contract Price shall be correspondingly adjusted.

11.03 *Unit Price Work*

A. Where the Contract Documents provide that all or part of the Work is to be Unit Price Work, initially the Contract Price will be deemed to include for all Unit Price Work an amount equal to the sum of the unit price for each separately identified item of Unit Price Work times the estimated quantity of each item as indicated in the Agreement.

B. The estimated quantities of items of Unit Price Work are not guaranteed and are solely for the purpose of comparison of Bids and determining an initial Contract Price. Determinations of the actual quantities and classifications of Unit Price Work performed by Contractor will be made by Engineer subject to the provisions of Paragraph 9.07.

C. Each unit price will be deemed to include an amount considered by Contractor to be adequate to cover Contractor's overhead and profit for each separately identified item.

D. Owner or Contractor may make a Claim for an adjustment in the Contract Price in accordance with Paragraph 10.05 if:

1. the quantity of any item of Unit Price Work performed by Contractor differs materially and significantly from the estimated quantity of such item indicated in the Agreement; and

2. there is no corresponding adjustment with respect any other item of Work; and

3. Contractor believes that Contractor is entitled to an increase in Contract Price as a result of having incurred additional expense or Owner believes that Owner is entitled to a decrease in Contract Price and the parties are unable to agree as to the amount of any such increase or decrease.

ARTICLE 12 - CHANGE OF CONTRACT PRICE; CHANGE OF CONTRACT TIMES

12.01   *Change of Contract Price*

A. The Contract Price may only be changed by a Change Order. Any Claim for an adjustment in the Contract Price shall be based on written notice submitted by the party making the Claim to the Engineer and the other party to the Contract in accordance with the provisions of Paragraph 10.05.

B. The value of any Work covered by a Change Order or of any Claim for an adjustment in the Contract Price will be determined as follows:

1. where the Work involved is covered by unit prices contained in the Contract Documents, by application of such unit prices to the quantities of the items involved (subject to the provisions of Paragraph 11.03); or

2. where the Work involved is not covered by unit prices contained in the Contract Documents, by a mutually agreed lump sum (which may include an allowance for overhead and profit not necessarily in accordance with Paragraph 12.01.C.2); or

3. where the Work involved is not covered by unit prices contained in the Contract Documents and agreement to a lump sum is not reached under Paragraph 12.01.B.2, on the basis of the Cost of the Work (determined as provided in Paragraph 11.01) plus a Contractor's fee for overhead and profit (determined as provided in Paragraph 12.01.C).

C. *Contractor's Fee:* The Contractor's fee for overhead and profit shall be determined as follows:

1. a mutually acceptable fixed fee; or

2. if a fixed fee is not agreed upon, then a fee based on the following percentages of the various portions of the Cost of the Work:

a. for costs incurred under Paragraphs 11.01.A.1 and 11.01.A.2 the Contractor's fee shall be 15 percent;

b. for costs incurred under Paragraph 11.01.A.3, the Contractor's fee shall be five percent;

c. where one or more tiers of subcontracts are on the basis of Cost of the Work plus a fee and no fixed fee is agreed upon, the intent of Paragraph 12.01.C.2.a is that the Subcontractor who actually performs the Work, at whatever tier, will be paid a fee of 15 percent of the costs incurred by such Subcontractor under Paragraphs 11.01.A.1 and 11.01.A.2 and that any higher tier Subcontractor and Contractor will each be paid a fee of five percent of the amount paid to the next lower tier Subcontractor;

d. no fee shall be payable on the basis of costs itemized under Paragraphs 11.01.A.4, 11.01.A.5, and 11.01.B;

e. the amount of credit to be allowed by Contractor to Owner for any change which results in a net decrease in cost will be the amount of the actual net decrease in cost plus a deduction in Contractor's fee by an amount equal to five percent of such net decrease; and

f. when both additions and credits are involved in any one change, the adjustment in Contractor's fee shall be computed on the basis of the net change in accordance with Paragraphs 12.01.C.2.a through 12.01.C.2.e, inclusive.

12.02 Change of Contract Times

A. The Contract Times may only be changed by a Change Order. Any Claim for an adjustment in the Contract Times shall be based on written notice submitted by the party making the Claim to the Engineer and the other party to the Contract in accordance with the provisions of Paragraph 10.05.

B. Any adjustment of the Contract Times covered by a Change Order or any Claim for an adjustment in the Contract Times will be determined in accordance with the provisions of this Article 12.

12.03 Delays

A. Where Contractor is prevented from completing any part of the Work within the Contract Times due to delay beyond the control of Contractor, the Contract Times will be extended in an amount equal to the time lost due to such delay if a Claim is made therefor as provided in Paragraph 12.02.A. Delays beyond the control of Contractor shall include, but not be limited to, acts or neglect by Owner, acts or neglect of utility owners or other contractors performing other work as contemplated by Article 7, fires, floods, epidemics, abnormal weather conditions, or acts of God.

B. If Owner, Engineer, or other contractors or utility owners performing other work for Owner as contemplated by Article 7, or anyone for whom Owner is responsible, delays, disrupts, or interferes with the performance or progress of the Work, then Contractor shall be entitled to an equitable adjustment in the Contract Price or the Contract Times , or both. Contractor's entitlement to an adjustment of the Contract Times is conditioned on such adjustment being essential to Contractor's ability to complete the Work within the Contract Times.

C If Contractor is delayed in the performance or progress of the Work by fire, flood, epidemic, abnormal weather conditions, acts of God, acts or failures to act of utility owners not under the control of Owner, or other causes not the fault of and beyond control of Owner and Contractor, then Contractor shall be entitled to an equitable adjustment in Contract Times, if such adjustment is essential to Contractor's ability to complete the Work within the Contract Times. Such an adjustment shall be Contractor's sole and exclusive remedy for the delays described in this Paragraph 12.03.C.

D. Owner, Engineer and the Related Entities of each of them shall not be liable to Contractor for any claims, costs, losses, or damages (including but not limited to all fees and charges of Engineers, architects, attorneys, and other professionals and all court or arbitration or other dispute resolution costs) sustained by Contractor on or in connection with any other project or anticipated project.

E. Contractor shall not be entitled to an adjustment in Contract Price or Contract Times for delays within the control of Contractor. Delays attributable to and within the control of a Subcontractor or Supplier shall be deemed to be delays within the control of Contractor.

ARTICLE 13 - TESTS AND INSPECTIONS; CORRECTION, REMOVAL OR ACCEPTANCE OF DEFECTIVE WORK

13.01 Notice of Defects

A. Prompt notice of all defective Work of which Owner or Engineer has actual knowledge will be given to Contractor. All defective Work may be rejected, corrected, or accepted as provided in this Article 13.

13.02 Access to Work

A. Owner, Engineer, their consultants and other representatives and personnel of Owner, independent testing laboratories, and governmental agencies with jurisdictional interests will have access to the Site and the Work at reasonable times for their observation, inspecting, and testing. Contractor shall provide them proper and safe conditions for such access and advise them of Contractor's Site safety procedures and programs so that they may comply therewith as applicable.

13.03 Tests and Inspections

A. Contractor shall give Engineer timely notice of readiness of the Work for all required inspections, tests, or approvals and shall cooperate with inspection and testing personnel to facilitate required inspections or tests.

B. Owner shall employ and pay for the services of an independent testing laboratory to perform all inspections, tests, or approvals required by the Contract Documents except:

1. for inspections, tests, or approvals covered by Paragraphs 13.03.C and 13.03.D below;

2. that costs incurred in connection with tests or inspections conducted pursuant to Paragraph 13.04.B shall be paid as provided in said Paragraph 13.04.C; and

3. as otherwise specifically provided in the Contract Documents.

C. If Laws or Regulations of any public body having jurisdiction require any Work (or part thereof) specifically to be inspected, tested, or approved by an employee or other representative of such public body, Contractor shall assume full responsibility for arranging and obtaining such inspections, tests, or approvals, pay all costs in connection therewith, and furnish Engineer the required certificates of inspection or approval.

D. Contractor shall be responsible for arranging and obtaining and shall pay all costs in connection with any inspections, tests, or approvals required for Owner's and Engineer's acceptance of materials or equipment to be incorporated in the Work; or acceptance of materials, mix designs, or equipment submitted for approval prior to Contractor's purchase thereof for incorporation in the Work. Such inspections, tests, or approvals shall be performed by organizations acceptable to Owner and Engineer.

E. If any Work (or the work of others) that is to be inspected, tested, or approved is covered by Contractor without written concurrence of Engineer, it must, if requested by Engineer, be uncovered for observation.

F. Uncovering Work as provided in Paragraph 13.03.E shall be at Contractor's expense unless Contractor has given Engineer timely notice of Contractor's intention to cover the same and Engineer has not acted with reasonable promptness in response to such notice.

13.04   *Uncovering Work*

A. If any Work is covered contrary to the written request of Engineer, it must, if requested by Engineer, be uncovered for Engineer's observation and replaced at Contractor's expense.

B. If Engineer considers it necessary or advisable that covered Work be observed by Engineer or inspected or tested by others, Contractor, at Engineer's request, shall uncover, expose, or otherwise make available for observation, inspection, or testing as Engineer may require, that portion of the Work in question, furnishing all necessary labor, material, and equipment.

C. If it is found that the uncovered Work is defective, Contractor shall pay all claims, costs, losses, and damages (including but not limited to all fees and charges of engineers, architects, attorneys, and other professionals and all court or arbitration or other dispute resolution costs) arising out of or relating to such uncovering, exposure, observation, inspection, and testing, and of satisfactory replacement or reconstruction (including but not limited to all costs of repair or replacement of work of others); and Owner shall be entitled to an appropriate decrease in the Contract Price. If the parties are unable to agree as to the amount thereof, Owner may make a Claim therefor as provided in Paragraph 10.05.

D. If, the uncovered Work is not found to be defective, Contractor shall be allowed an increase in the Contract Price or an extension of the Contract Times, or both, directly attributable to such uncovering, exposure, observation, inspection, testing, replacement, and reconstruction. If the parties are unable to agree as to the amount or extent thereof, Contractor may make a Claim therefor as provided in Paragraph 10.05.

13.05   *Owner May Stop the Work*

A. If the Work is defective, or Contractor fails to supply sufficient skilled workers or suitable materials or equipment, or fails to perform the Work in such a way that the completed Work will conform to the Contract Documents, Owner may order Contractor to stop the Work, or any portion thereof, until the cause for such order has been eliminated; however, this right of Owner to stop the Work shall not give rise to any duty on the part of Owner to exercise this right for the benefit of Contractor, any Subcontractor, any Supplier, any other individual or entity, or any surety for, or employee or agent of any of them.

13.06   *Correction or Removal of Defective Work*

A. Promptly after receipt of notice, Contractor shall correct all defective Work, whether or not fabricated, installed, or completed, or, if the Work has been rejected by Engineer, remove it from the Project and replace it with Work that is not defective. Contractor shall pay all claims, costs, losses, and damages (including but not limited to all fees and charges of engineers, architects, attorneys, and other professionals and all court or arbitration or other dispute resolution costs) arising out of or relating to such correction or removal (including but not limited to all costs of repair or replacement of work of others).

B. When correcting defective Work under the terms of this Paragraph 13.06 or Paragraph 13.07, Contractor shall take no action that would void or otherwise impair Owner's special warranty and guarantee, if any, on said Work.

13.07   *Correction Period*

A. If within one year after the date of Substantial Completion (or such longer period of time as may be prescribed by the terms of any applicable special guarantee required by the Contract Documents) or by any specific provision of the Contract Documents, any Work is found to be defective, or if the repair of any damages to the land or areas made available for Contractor's use by Owner or permitted by Laws and Regulations as contemplated in Paragraph 6.11.A is found to be defective, Contractor shall promptly, without cost to Owner and in accordance with Owner's written instructions:

1. repair such defective land or areas; or

2. correct such defective Work; or

3. if the defective Work has been rejected by Owner, remove it from the Project and replace it with Work that is not defective, and

4. satisfactorily correct or repair or remove and replace any damage to other Work, to the work of others or other land or areas resulting therefrom.

B. If Contractor does not promptly comply with the terms of Owner's written instructions, or in an emergency where delay would cause serious risk of loss or damage, Owner may have the defective Work corrected or repaired or may have the rejected Work removed and replaced. All claims, costs, losses, and damages (including but not limited to all fees and charges of engineers, architects, attorneys, and other professionals and all court or arbitration or other dispute resolution costs) arising out of or relating to such correction or repair or such removal and replacement (including but not limited to all costs of repair or replacement of work of others) will be paid by Contractor.

C. In special circumstances where a particular item of equipment is placed in continuous service before Substantial Completion of all the Work, the correction period for that item may start to run from an earlier date if so provided in the Specifications .

D. Where defective Work (and damage to other Work resulting therefrom) has been corrected or removed and replaced under this Paragraph 13.07, the correction period hereunder with respect to such Work will be extended for an additional period of one year after such correction or removal and replacement has been satisfactorily completed.

E. Contractor's obligations under this Paragraph 13.07 are in addition to any other obligation or warranty. The provisions of this Paragraph 13.07 shall not be construed as a substitute for or a waiver of the provisions of any applicable statute of limitation or repose.

13.08 *Acceptance of Defective Work*

A. If, instead of requiring correction or removal and replacement of defective Work, Owner (and, prior to Engineer's recommendation of final payment, Engineer) prefers to accept it, Owner may do so. Contractor shall pay all claims, costs, losses, and damages (including but not limited to all fees and charges of engineers, architects, attorneys, and other professionals and all court or arbitration or other dispute resolution costs) attributable to Owner's evaluation of and determination to accept such defective Work (such costs to be approved by Engineer as to reasonableness) and the diminished value of the Work to the extent not otherwise paid by Contractor pursuant to this sentence. If any such acceptance occurs prior to Engineer's recommendation of final payment, a Change Order will be issued incorporating the necessary revisions in the Contract Documents with respect to the Work, and Owner shall be entitled to an appropriate decrease in the Contract Price, reflecting the diminished value of Work so accepted. If the parties are unable to agree as to the amount thereof, Owner may make a Claim therefor as provided in Paragraph 10.05. If the acceptance occurs after such recommendation, an appropriate amount will be paid by Contractor to Owner.

13.09 *Owner May Correct Defective Work*

A. If Contractor fails within a reasonable time after written notice from Engineer to correct defective Work or to remove and replace rejected Work as required by Engineer in accordance with Paragraph 13.06.A, or if Contractor fails to perform the Work in accordance with the Contract Documents, or if Contractor fails to comply with any other provision of the Contract Documents, Owner may, after seven days written notice to Contractor, correct or remedy any such deficiency.

B. In exercising the rights and remedies under this Paragraph 13.09, Owner shall proceed expeditiously. In connection with such corrective or remedial action, Owner may exclude Contractor from all or part of the Site, take possession of all or part of the Work and suspend Contractor's services related thereto, take possession of Contractor's tools, appliances, construction equipment and machinery at the Site, and incorporate in the Work all materials and equipment stored at the Site or for which Owner has paid Contractor but which are stored elsewhere. Contractor shall allow Owner, Owner's representatives, agents and employees, Owner's other contractors, and Engineer and Engineer's consultants access to the Site to enable Owner to exercise the rights and remedies under this Paragraph.

C. All claims, costs, losses, and damages (including but not limited to all fees and charges of engineers, architects, attorneys, and other professionals and all court or arbitration or other dispute resolution costs) incurred or sustained by Owner in exercising the rights and remedies under this Paragraph 13.09 will be charged against Contractor, and a Change Order will be issued incorporating the necessary revisions in the Contract Documents with respect to the Work; and Owner shall be entitled to an appropriate decrease in the Contract Price. If the parties are unable to agree as to the amount of the adjustment, Owner may make a Claim therefor as provided in Paragraph 10.05. Such claims, costs, losses and damages will include but not be limited to all costs of repair, or replacement of work of others destroyed or damaged by correction, removal, or replacement of Contractor's defective Work.

D. Contractor shall not be allowed an extension of the Contract Times because of any delay in the performance of the Work attributable to the exercise by Owner of Owner's rights and remedies under this Paragraph 13.09.

ARTICLE 14 - PAYMENTS TO CONTRACTOR AND COMPLETION

14.01 Schedule of Values

A. The Schedule of Values established as provided in Paragraph 2.07.A will serve as the basis for progress payments and will be incorporated into a form of Application for Payment acceptable to Engineer. Progress payments on account of Unit Price Work will be based on the number of units completed.

14.02 Progress Payments

A. Applications for Payments

1. At least 20 days before the date established in the Agreement for each progress payment (but not more often than once a month), Contractor shall submit to Engineer for review an Application for Payment filled out and signed by Contractor covering the Work completed as of the date of the Application and accompanied by such supporting documentation as is required by the Contract Documents. If payment is requested on the basis of materials and equipment not incorporated in the Work but delivered and suitably stored at the Site or at another location agreed to in writing, the Application for Payment shall also be accompanied by a bill of sale, invoice, or other documentation warranting that Owner has received the materials and equipment free and clear of all Liens and evidence that the materials and equipment are covered by appropriate property insurance or other arrangements to protect Owner's interest therein, all of which must be satisfactory to Owner.

2. Beginning with the second Application for Payment, each Application shall include an affidavit of Contractor stating that all previous progress payments received on account of the Work have been applied on account to discharge Contractor's legitimate obligations associated with prior Applications for Payment.

3. The amount of retainage with respect to progress payments will be as stipulated in the Agreement.

B. Review of Applications

1. Engineer will, within 10 days after receipt of each Application for Payment, either indicate in writing a recommendation of payment and present the Application to Owner or return the Application to Contractor indicating in writing Engineer's reasons for refusing to recommend payment. In the latter case, Contractor may make the necessary corrections and resubmit the Application.

2. Engineer's recommendation of any payment requested in an Application for Payment will constitute a representation by Engineer to Owner, based on Engineer's observations on the Site of the executed Work as an experienced and qualified design professional and on Engineer's review of the Application for Payment and the accompanying data and schedules, that to the best of Engineer's knowledge, information and belief:

a. the Work has progressed to the point indicated;

b. the quality of the Work is generally in accordance with the Contract Documents (subject to an evaluation of the Work as a functioning whole prior to or upon Substantial Completion, to the results of any subsequent tests called for in the Contract Documents, to a final determination of quantities and classifications for Unit Price Work under Paragraph 9.07 and to any other qualifications stated in the recommendation); and

c. the conditions precedent to Contractor's being entitled to such payment appear to have been fulfilled in so far as it is Engineer's responsibility to observe the Work.

3. By recommending any such payment Engineer will not thereby be deemed to have represented that:

a. inspections made to check the quality or the quantity of the Work as it has been performed have been exhaustive, extended to every aspect of the Work in progress, or involved detailed inspections of the Work beyond the responsibilities specifically assigned to Engineer in the Contract Documents; or

b. that there may not be other matters or issues between the parties that might entitle Contractor to be paid additionally by Owner or entitle Owner to withhold payment to Contractor.

4. Neither Engineer's review of Contractor's Work for the purposes of recommending payments nor Engineer's recommendation of any payment, including final payment, will impose responsibility on Engineer:

a. to supervise, direct, or control the Work, or

b. for the means, methods, techniques, sequences, or procedures of construction, or the safety precautions and programs incident thereto, or

c. for Contractor's failure to comply with Laws and Regulations applicable to Contractor's performance of the Work, or

d. to make any examination to ascertain how or for what purposes Contractor has used the moneys paid on account of the Contract Price, or

e. to determine that title to any of the Work, materials, or equipment has passed to Owner free and clear of any Liens.

5. Engineer may refuse to recommend the whole or any part of any payment if, in Engineer's opinion, it would be incorrect to make the representations to Owner stated in Paragraph 14.02.B.2. Engineer may also refuse to recommend any such payment or, because of subsequently discovered evidence or the results of subsequent inspections or tests, revise or revoke any such payment recommendation previously made, to such extent as may be necessary in Engineer's opinion to protect Owner from loss because:

a. the Work is defective, or completed Work has been damaged, requiring correction or replacement;

b. the Contract Price has been reduced by Change Orders;

c. Owner has been required to correct defective Work or complete Work in accordance with Paragraph 13.09; or

d. Engineer has actual knowledge of the occurrence of any of the events enumerated in Paragraph 15.02.A.

C. *Payment Becomes Due*

1. Ten days after presentation of the Application for Payment to Owner with Engineer's recommendation, the amount recommended will (subject to the provisions of Paragraph 14.02.D) become due, and when due will be paid by Owner to Contractor.

D. *Reduction in Payment*

1. Owner may refuse to make payment of the full amount recommended by Engineer because:

a. claims have been made against Owner on account of Contractor's performance or furnishing of the Work;

b. Liens have been filed in connection with the Work, except where Contractor has delivered a specific bond satisfactory to Owner to secure the satisfaction and discharge of such Liens;

c. there are other items entitling Owner to a set-off against the amount recommended; or

d. Owner has actual knowledge of the occurrence of any of the events enumerated in Paragraphs 14.02.B.5.a through 14.02.B.5.c or Paragraph 15.02.A.

2. If Owner refuses to make payment of the full amount recommended by Engineer, Owner will give Contractor immediate written notice (with a copy to Engineer) stating the reasons for such action and promptly pay Contractor any amount remaining after deduction of the amount so withheld. Owner shall promptly pay Contractor the amount so withheld, or any adjustment thereto agreed to by Owner and Contractor, when Contractor corrects to Owner's satisfaction the reasons for such action.

3. If it is subsequently determined that Owner's refusal of payment was not justified, the amount wrongfully withheld shall be treated as an amount due as determined by Paragraph 14.02.C.1.

14.03 *Contractor's Warranty of Title*

A. Contractor warrants and guarantees that title to all Work, materials, and equipment covered by any Application for Payment, whether incorporated in the Project or not, will pass to Owner no later than the time of payment free and clear of all Liens.

14.04 *Substantial Completion*

A. When Contractor considers the entire Work ready for its intended use Contractor shall notify Owner and Engineer in writing that the entire Work is substantially complete (except for items specifically listed by Contractor as incomplete) and request that Engineer issue a certificate of Substantial Completion.

B. Promptly after Contractor's notification, , Owner, Contractor, and Engineer shall make an inspection of the Work to determine the status of completion. If Engineer does not consider the Work substantially complete, Engineer will notify Contractor in writing giving the reasons therefor.

C. If Engineer considers the Work substantially complete, Engineer will deliver to Owner a tentative certificate of Substantial Completion which shall fix the date of Substantial Completion. There shall be attached to the certificate a tentative list of items to be completed or corrected before final payment. Owner shall have seven days after receipt of the tentative certificate during which to make written objection to Engineer as to any provisions of the certificate or attached list. If, after considering such objections, Engineer concludes that the Work is not substantially complete, Engineer will within 14 days after submission of the tentative certificate to Owner notify Contractor in writing, stating the reasons therefor. If, after consideration of Owner's objections, Engineer considers the Work substantially complete, Engineer will within said 14 days execute and deliver to Owner and Contractor a definitive certificate of Substantial Completion (with a revised tentative list of items to be completed or corrected) reflecting such changes from the tentative certificate as Engineer believes justified after consideration of any objections from Owner.

D. At the time of delivery of the tentative certificate of Substantial Completion, Engineer will deliver to Owner and Contractor a written recommendation as to division of responsibilities pending final payment between Owner and Contractor with respect to security, operation, safety, and protection of the Work, maintenance, heat, utilities, insurance, and warranties and guarantees. Unless Owner and Contractor agree otherwise in writing and so inform Engineer in writing prior to Engineer's issuing the definitive certificate of Substantial Completion, Engineer's aforesaid recommendation will be binding on Owner and Contractor until final payment.

E. Owner shall have the right to exclude Contractor from the Site after the date of Substantial Completion subject to allowing Contractor reasonable access to complete or correct items on the tentative list.

14.05   *Partial Utilization*

A. Prior to Substantial Completion of all the Work, Owner may use or occupy any substantially completed part of the Work which has specifically been identified in the Contract Documents, or which Owner, Engineer, and Contractor agree constitutes a separately functioning and usable part of the Work that can be used by Owner for its intended purpose without significant interference with Contractor's performance of the remainder of the Work, subject to the following conditions.

1. Owner at any time may request Contractor in writing to permit Owner to use or occupy any such part of the Work which Owner believes to be ready for its intended use and substantially complete. If and when Contractor agrees that such part of the Work is substantially complete, Contractor will certify to Owner and Engineer that such part of the Work is substantially complete and request Engineer to issue a certificate of Substantial Completion for that part of the Work.

2. Contractor at any time may notify Owner and Engineer in writing that Contractor considers any such part of the Work ready for its intended use and substantially complete and request Engineer to issue a certificate of Substantial Completion for that part of the Work.

3. Within a reasonable time after either such request, Owner, Contractor, and Engineer shall make an inspection of that part of the Work to determine its status of completion. If Engineer does not consider that part of the Work to be substantially complete, Engineer will notify Owner and Contractor in writing giving the reasons therefor. If Engineer considers that part of the Work to be substantially complete, the provisions of Paragraph 14.04 will apply with respect to certification of Substantial Completion of that part of the Work and the division of responsibility in respect thereof and access thereto.

4. No use or occupancy or separate operation of part of the Work may occur prior to compliance with the requirements of Paragraph 5.10 regarding property insurance.

14.06   *Final Inspection*

A. Upon written notice from Contractor that the entire Work or an agreed portion thereof is complete, Engineer will promptly make a final inspection with Owner and Contractor and will notify Contractor in writing of all particulars in which this inspection reveals that the Work is incomplete or defective. Contractor shall immediately take such measures as are necessary to complete such Work or remedy such deficiencies.

14.07   *Final Payment*

A. Application for Payment

1. After Contractor has, in the opinion of Engineer, satisfactorily completed all corrections identified during the final inspection and has delivered, in accordance with the Contract Documents, all maintenance and operating instructions, schedules, guarantees, bonds, certificates or other evidence of insurance certificates of inspection, marked-up record documents (as provided in Paragraph 6.12), and other documents, Contractor may make application for final payment following the procedure for progress payments.

2. The final Application for Payment shall be accompanied (except as previously delivered) by:

a. all documentation called for in the Contract Documents, including but not limited to the evidence of insurance required by Paragraph 5.04.B.7;

b. consent of the surety, if any, to final payment;

c. a list of all Claims against Owner that Contractor believes are unsettled; and

d. complete and legally effective releases or waivers (satisfactory to Owner) of all Lien rights arising out of or Liens filed in connection with the Work.

3. In lieu of the releases or waivers of Liens specified in Paragraph 14.07.A.2 and as approved by Owner, Contractor may furnish receipts or releases in full and an affidavit of Contractor that: (i) the releases and receipts include all labor, services, material, and equipment for which a Lien could be filed; and (ii) all payrolls, material and equipment bills, and other indebtedness connected with the Work for which Owner or Owner's property might in any way be responsible have been paid or otherwise satisfied. If any Subcontractor or Supplier fails to furnish such a release or receipt in full, Contractor may furnish a bond or other collateral

satisfactory to Owner to indemnify Owner against any Lien.

B. *Engineer's Review of Application and Acceptance*

1. If, on the basis of Engineer's observation of the Work during construction and final inspection, and Engineer's review of the final Application for Payment and accompanying documentation as required by the Contract Documents, Engineer is satisfied that the Work has been completed and Contractor's other obligations under the Contract Documents have been fulfilled, Engineer will, within ten days after receipt of the final Application for Payment, indicate in writing Engineer's recommendation of payment and present the Application for Payment to Owner for payment. At the same time Engineer will also give written notice to Owner and Contractor that the Work is acceptable subject to the provisions of Paragraph 14.09. Otherwise, Engineer will return the Application for Payment to Contractor, indicating in writing the reasons for refusing to recommend final payment, in which case Contractor shall make the necessary corrections and resubmit the Application for Payment.

C. Payment Becomes Due

1. Thirty days after the presentation to Owner of the Application for Payment and accompanying documentation, the amount recommended by Engineer, less any sum Owner is entitled to set off against Engineer's recommendation, including but not limited to liquidated damages, will become due and , will be paid by Owner to Contractor.

14.08   *Final Completion Delayed*

A. If, through no fault of Contractor, final completion of the Work is significantly delayed, and if Engineer so confirms, Owner shall, upon receipt of Contractor's final Application for Payment (for Work fully completed and accepted) and recommendation of Engineer, and without terminating the Contract, make payment of the balance due for that portion of the Work fully completed and accepted. If the remaining balance to be held by Owner for Work not fully completed or corrected is less than the retainage stipulated in the Agreement, and if bonds have been furnished as required in Paragraph 5.01, the written consent of the surety to the payment of the balance due for that portion of the Work fully completed and accepted shall be submitted by Contractor to Engineer with the Application for such payment. Such payment shall be made under the terms and conditions governing final payment, except that it shall not constitute a waiver of Claims.

14.09   *Waiver of Claims*

A. The making and acceptance of final payment will constitute:

1. a waiver of all Claims by Owner against Contractor, except Claims arising from unsettled Liens, from defective Work appearing after final inspection pursuant to Paragraph 14.06, from failure to comply with the Contract Documents or the terms of any special guarantees specified therein, or from Contractor's continuing obligations under the Contract Documents; and

2. a waiver of all Claims by Contractor against Owner other than those previously made in accordance with the requirements herein and expressly acknowledged by Owner in writing as still unsettled.

ARTICLE 15 - SUSPENSION OF WORK AND TERMINATION

15.01   *Owner May Suspend Work*

A. At any time and without cause, Owner may suspend the Work or any portion thereof for a period of not more than 90 consecutive days by notice in writing to Contractor and Engineer which will fix the date on which Work will be resumed. Contractor shall resume the Work on the date so fixed. Contractor shall be granted an adjustment in the Contract Price or an extension of the Contract Times, or both, directly attributable to any such suspension if Contractor makes a Claim therefor as provided in Paragraph 10.05.

15.02   *Owner May Terminate for Cause*

A. The occurrence of any one or more of the following events will justify termination for cause:

1. Contractor's persistent failure to perform the Work in accordance with the Contract Documents (including, but not limited to, failure to supply sufficient skilled workers or suitable materials or equipment or failure to adhere to the Progress Schedule established under Paragraph 2.07 as adjusted from time to time pursuant to Paragraph 6.04);

2. Contractor's disregard of Laws or Regulations of any public body having jurisdiction;

3. Contractor's disregard of the authority of Engineer; or

4. Contractor's violation in any substantial way of any provisions of the Contract Documents.

B. If one or more of the events identified in Paragraph 15.02.A occur, Owner may, after giving Contractor (and surety) seven days written notice of its intent to terminate the services of Contractor:

1. exclude Contractor from the Site, and take possession of the Work and of all Contractor's tools, appliances, construction equipment, and machinery at the Site, and use the same to the full extent they could be used by Contractor (without liability to Contractor for trespass or conversion),

2. incorporate in the Work all materials and equipment stored at the Site or for which Owner has paid Contractor but which are stored elsewhere, and

3. complete the Work as Owner may deem expedient.

C. If Owner proceeds as provided in Paragraph 15.02.B, Contractor shall not be entitled to receive any further payment until the Work is completed. If the unpaid balance of the Contract Price exceeds all claims, costs, losses, and damages (including but not limited to all fees and charges of engineers, architects, attorneys, and other professionals and all court or arbitration or other dispute resolution costs) sustained by Owner arising out of or relating to completing the Work, such excess will be paid to Contractor. If such claims, costs, losses, and damages exceed such unpaid balance, Contractor shall pay the difference to Owner. Such claims, costs, losses, and damages incurred by Owner will be reviewed by Engineer as to their reasonableness and, when so approved by Engineer, incorporated in a Change Order. When exercising any rights or remedies under this Paragraph Owner shall not be required to obtain the lowest price for the Work performed.

D. Notwithstanding Paragraphs 15.02.B and 15.02.C, Contractor's services will not be terminated if Contractor begins within seven days of receipt of notice of intent to terminate to correct its failure to perform and proceeds diligently to cure such failure within no more than 30 days of receipt of said notice.

E. Where Contractor's services have been so terminated by Owner, the termination will not affect any rights or remedies of Owner against Contractor then existing or which may thereafter accrue. Any retention or payment of moneys due Contractor by Owner will not release Contractor from liability.

F. If and to the extent that Contractor has provided a performance bond under the provisions of Paragraph 5.01.A, the termination procedures of that bond shall supersede the provisions of Paragraphs 15.02.B, and 15.02.C.

15.03   *Owner May Terminate For Convenience*

A. Upon seven days written notice to Contractor and Engineer, Owner may, without cause and without prejudice to any other right or remedy of Owner, terminate the Contract. In such case, Contractor shall be paid for (without duplication of any items):

1. completed and acceptable Work executed in accordance with the Contract Documents prior to the effective date of termination, including fair and reasonable sums for overhead and profit on such Work;

2. expenses sustained prior to the effective date of termination in performing services and furnishing labor, materials, or equipment as required by the Contract Documents in connection with uncompleted Work, plus fair and reasonable sums for overhead and profit on such expenses;

3. all claims, costs, losses, and damages (including but not limited to all fees and charges of engineers, architects, attorneys, and other professionals and all court or arbitration or other dispute resolution costs) incurred in settlement of terminated contracts with Subcontractors, Suppliers, and others; and

4. reasonable expenses directly attributable to termination.

B. Contractor shall not be paid on account of loss of anticipated profits or revenue or other economic loss arising out of or resulting from such termination.

15.04   *Contractor May Stop Work or Terminate*

A. If, through no act or fault of Contractor, (i) the Work is suspended for more than 90 consecutive days by Owner or under an order of court or other public authority, or (ii) Engineer fails to act on any Application for Payment within 30 days after it is submitted, or (iii) Owner fails for 30 days to pay Contractor any sum finally determined to be due, then Contractor may, upon seven days written notice to Owner and Engineer, and provided Owner or Engineer do not remedy such suspension or failure within that time, terminate the Contract and recover from Owner payment on the same terms as provided in Paragraph 15.03.

B. In lieu of terminating the Contract and without prejudice to any other right or remedy, if Engineer has failed to act on an Application for Payment within 30 days after it is submitted, or Owner has failed for 30 days to pay Contractor any sum finally determined to be due, Contractor may, seven days after written notice to Owner and Engineer, stop the Work until payment is made of all such amounts due Contractor, including interest thereon. The provisions of this Paragraph 15.04 are not intended to preclude Contractor from making a Claim under Paragraph 10.05 for an adjustment in Contract Price or Contract Times or otherwise for expenses or damage directly attributable to Contractor's stopping the Work as permitted by this Paragraph.

ARTICLE 16 - DISPUTE RESOLUTION

16.01 *Methods and Procedures*

A. Either Owner or Contractor may request mediation of any Claim submitted to Engineer for a decision under Paragraph 10.05 before such decision becomes final and binding. The mediation will be governed by the Construction Industry Mediation Rules of the American Arbitration Association in effect as of the Effective Date of the Agreement. The request for mediation shall be submitted in writing to the American Arbitration Association and the other party to the Contract. Timely submission of the request shall stay the effect of Paragraph 10.05.E.

B. Owner and Contractor shall participate in the mediation process in good faith. The process shall be concluded within 60 days of filing of the request. The date of termination of the mediation shall be determined by application of the mediation rules referenced above.

C. If the Claim is not resolved by mediation, Engineer's action under Paragraph 10.05.C or a denial pursuant to Paragraphs 10.05.C.3 or 10.05.D shall become final and binding 30 days after termination of the mediation unless, within that time period, Owner or Contractor:

1. elects in writing to invoke any dispute resolution process provided for in the Supplementary Conditions, or

2. agrees with the other party to submit the Claim to another dispute resolution process, or

3. gives written notice to the other party of their intent to submit the Claim to a court of competent jurisdiction.

ARTICLE 17 - MISCELLANEOUS

17.01 *Giving Notice*

A. Whenever any provision of the Contract Documents requires the giving of written notice, it will be deemed to have been validly given if:

1. delivered in person to the individual or to a member of the firm or to an officer of the corporation for whom it is intended, or

2. delivered at or sent by registered or certified mail, postage prepaid, to the last business address known to the giver of the notice.

17.02 *Computation of Times*

A. When any period of time is referred to in the Contract Documents by days, it will be computed to exclude the first and include the last day of such period. If the last day of any such period falls on a Saturday or Sunday or on a day made a legal holiday by the law of the applicable jurisdiction, such day will be omitted from the computation.

17.03 *Cumulative Remedies*

A. The duties and obligations imposed by these General Conditions and the rights and remedies available hereunder to the parties hereto are in addition to, and are not to be construed in any way as a limitation of, any rights and remedies available to any or all of them which are otherwise imposed or available by Laws or Regulations, by special warranty or guarantee, or by other provisions of the Contract Documents. The provisions of this Paragraph will be as effective as if repeated specifically in the Contract Documents in connection with each particular duty, obligation, right, and remedy to which they apply.

17.04 *Survival of Obligations*

A. All representations, indemnifications, warranties, and guarantees made in, required by, or given in accordance with the Contract Documents, as well as all continuing obligations indicated in the Contract Documents, will survive final payment, completion, and acceptance of the Work or termination or completion of the Contract or termination of the services of Contractor.

17.05 *Controlling Law*

A. This Contract is to be governed by the law of the state in which the Project is located.

17.06 *Headings*

A. Article and paragraph headings are inserted for convenience only and do not constitute parts of these General Conditions.

INDEX